Time Series with PyTorch

Modern Deep Learning Toolkit for Real-World Forecasting Challenges

Graeme Davidson

Lei Ma

<packt>

Time Series with PyTorch

Portfolio Director: Sunith Shetty
Relationship Leads: Tushar Gupta, Vaideeshwari Roshan
Project Manager: Shashank Desai
Content Engineer: Ayushi Bulani
Technical Editor: Seemanjay Ameriya
Indexer: Hemangini Bari
Production Designer: Shankar Kalbhor
Growth Lead: Abhishek Kaushik

First published: May 2026

Production reference: 2060326

Published by Packt Publishing Ltd.
Grosvenor House
11 St Paul's Square
Birmingham
B3 1RB, UK.

ISBN 978-1-80512-818-2

`www.packtpub.com`

To Claire and Sebastian, every day with you both is a blessing that lightens every struggle and gives it purpose.

– Graeme Davidson

To Han and Yoyo, for this beautiful life together.

– Lei Ma

Acknowledgments

During my (Graeme Davidson) undergraduate degree in the late 2000s, I encountered a BBC documentary which discussed demand forecasting within Tesco. The scale at which they were working was incredible to me; it still is. Unfortunately, the documentary only superficially covered how statistics were applied to prediction, but I was fascinated by how we apply statistics over time. Why don't many process, or research experiments account for the context and passage of time? Clearly, the answer was that time series forecasting is difficult. Of course, data management plays a huge role in that. Around the same period, every day, the news was full of time series as financial markets crashed globally, in part because people who paid a lot of money to analyze market data were not paying attention to the details of their data, along with several other factors. But overall, this is what Michael Burry capitalized on in 2007.

Several years have passed since then, and I would say that time series forecasting is somewhat easier to implement than it was. It does help that one can now spin up an autoregressive foundation model to churn out code, but as to how well they do this, I think they still leave a lot to be desired. And it is increasingly obvious that *vibe coding* undermines learning and understanding. That isn't to say you shouldn't use them, you should! But rather that you should make sure you understand what you are doing. Beforehand, I used to search the internet trawling for people who knew what they were talking about: Abhishek Thakur, Rob Hyndman, Stephan Kolassa, and Konrad Banachewicz, etc. In particular, around the time I started in demand forecasting, Abhishek Thakur hosted a series of YouTube videos on time series forecasting by Konrad Banachewicz. A few years later, Packt contacted me on LinkedIn to ask if I'd write a book on time series forecasting. I certainly wasn't as skilled then, so I suggested they contact Konrad. For some reason they then came back to me and said that Konrad would like to work on the book but needed a co-author. I wasn't going to turn down the opportunity to work with a Kaggle grandmaster. Unfortunately, as time went on, Konrad became too busy with other projects and work to continue. However, during my online searches, I came across Lei Ma's website; his approach to deep learning and modeling in general appealed to me. He was precise and concise, providing intuition about theory that so many blogs fail to convey; of course, he is a physicist by trade. Luckily, I was able to persuade him to take on the project.

Lei and I have very different writing styles, in part due to our backgrounds (I have a cognitive neuroscience background). While he provides the intuition of a physicist and tends to draw out how deep learning systems make sense of data, I tend to focus on systems and experimentation. I believe that this gives you the best of both perspectives so that you can confidently design and test your own forecasting systems, with the understanding you need to make good decisions.

I hope you enjoy the book!

Graeme Davidson

Contributors

About the authors

Graeme Davidson is a Lead Data Scientist at Retail Express, where he redesigned the company's demand forecasting framework in line with contemporary statistical learning practices. His background spans cognitive neuroscience, researching implicit reward processing and human decision-making, through advertising analytics to research-focused demand forecasting. He is an active contributor to several data science Slack and Discord communities, an occasional competitor in forecasting competitions, and was approached by Packt in late 2022 to write the book he wished had existed when he first fell down an ARIMA rabbit hole chasing answers about how supermarkets actually forecast demand, and how a quantitative researcher models financial markets.

I would like to thank the team at Packt who worked on this book: Ayushi Bulani, Shashank Desai, and Seemanjay Ameriya. A special thank you to Lei Ma for taking on this project as co-author, and to Joe Xiao, whose technical review genuinely strengthened our work. Thank you to my brother Andrew and my best friend Ben Goodhart for their encouragement throughout, and to my 'sweat the death out' group — you know who you are. I owe a great deal to the managers I have had over the years: Gareth Jenkins for his patience, Andy Mahood for changing my perspective, and Simon Duckworth, who was my manager when this project began in 2023 and actively promoted my development throughout. I would also like to thank Tim Kirkham for developing my critical reasoning skills during my PhD — I am still, and always will be, grateful. Finally, thank you to my family for accepting that this book consumes our valuable time together. Claire and Sebastian, none of this happens without you.

Lei Ma is a physicist-turned data scientist specializing in time series forecasting. He is a theorist but has tackled real-world forecasting challenges across a variety of industries like housing, logistics, ecommerce, and manufacturing. Lei has led and delivered numerous forecasting projects where he combines deep expertise in building advanced time series models with a strategic approach to delivering holistic business insights. Lei creates time series forecasting tutorials online and joined the venture when Graeme approached him to collaborate on this book.

I would like to thank Graeme Davidson for this exciting and fun collaboration, as well as the Packt team—Ayushi Bulani, Shashank Desai, Seemanjay Ameriya, Tushar Gupta, and everyone who worked on this book—for all their guidance and support. The reviewers, Jack Lee and Joe Xiao, have shaped the chapters with their expertise, and I would like to express my sincere thanks to them. I also want to express my gratitude to the open-source community for driving tech forward and providing the foundational knowledge for this book. Finally, thank you to my wife, Han Lu, for being the driving force that pushed me past every doubt, and to our daughter, Yoyo, for the peaceful nights and joyful days that made this journey enjoyable.

About the reviewers

YiZi (Joe) Xiao, Ph.D., is a Senior Principal AI/ML Scientist at Optum (UnitedHealth Group), where he leads teams developing machine learning and LLM-driven solutions in healthcare. He has over a decade of experience in AI and Data Science across industry and academia, including roles at Target, General Mills, Starkey Hearing Technologies, and Medtronic. Dr. Xiao holds a Ph.D. in Biomedical Engineering from the University of Minnesota, an M.S. in Computer Science from the University of Illinois Urbana-Champaign, and a B.S. in Bioengineering from UC Berkeley. He has authored numerous publications and patents and serves as a reviewer for leading journals in biomedical engineering.

Sriram is an AI architect, consultant, and technical author with two decades of industry experience. He specializes in quantum AI, advanced machine learning, and secure model deployment. He is a co-author of Time Series Forecasting Using Generative AI and serves as a reviewer for technical books, bringing rigorous, research-driven evaluation to every manuscript. As a mentor and visiting faculty member, he teaches Python and AI to diverse learners and contributes actively to the academic community through publications, peer review, and ongoing research in quantum reservoir computing. He is also an invited speaker at industry events.

Table of Contents

Chapter 3: Evaluating Time-Series Models 53

Chapter 6: Optimization 143

Chapter 7: Conformal Prediction 189

Chapter 10: Other Neural Structures 269

Chapter 11: Transfer Learning and Global Modelling 295

Chapter 12: Synthetic Time Series Data 329

Chapter 13: Diffusion Models 363

Chapter 14: Time Series Classification 387

Chapter 16: Embeddings for Time Series 469

Chapter 17: Supervised and Unsupervised Anomaly Detection 479

Preface

Temporally embedded data is fundamental to our world; it tells us about what has been, why it has been and what it may become. You see it everywhere, from the melodies of your favorite musician/composer and the neural activity in your brain to the energy powering your computer and the critical supply chain supplying your food. Often, people only see them when something bad is happening, e.g., a financial crash, average global temperature increases, COVID-19 cases, and hospital beds. On the opposite side of dualism, we see life expectancy increasing, poverty decreasing, literacy increasing, and an abundance of food supply. More information is available to us than ever before to explain our lives and environments, and with that data, we can understand and improve the world around us.

Modeling this data is a fascinating and challenging area of data science and statistics, and I think it is fair to say that many people find it so. Temporal dependencies, non-stationarity, missing observations, modeling residual data, and a variety of data-generating processes all conspire to make forecasting difficult for novices. This is not helped by the fact that books like Hamilton's *Time Series Analysis* (which is excellent) are impenetrable without developing a strong mathematical and statistical knowledge. However, not everyone needs to be at the level of a quantitative analyst for Renaissance Technologies. Many people, such as Rob Hyndman, have worked hard to develop and improve knowledge around time series forecasting.

Neural networks and Deep learning (depending on how you define it) entered the scene around the late 1980s, when **Feed-Forward Networks (FFNs)** started to be used for sequential modeling, followed by LSTMs in 1997. And there they sat, useful but not making much headway in forecasting until 2017, when Transformers were created by a team at Google, followed by Oreshkin et al.'s N-BEATS, N-HiTS and a huge range of new architectures and techniques for training that allow us to apply cross-learning and transfer-learning on untrained and trained models.

Currently, many of these networks/algorithms can be, or are, written in PyTorch, which gives us a flexible and expressive framework for building, training, and evaluating forecasting models, from simple FFNs to foundation models trained across thousands of series. Of course, the kind people at NeuralForecast and Darts have done a lot of the work for us, so you only need to start from scratch occasionally.

It has never been easier to experiment with neural networks, but how do these architectures work, how do we decide between them, and when are they useful? In time series forecasting there is no free lunch; this book aims to help you understand that, and to guide your intuition by explaining how these architectures build functions for temporal data, so that you can be confident in your choices.

This book is a practical guide. We wrote it for data scientists, machine learning engineers, and researchers who already have some experience with Python and statistics but want to build a working understanding of modern deep learning approaches to time series. Our goal is not to catalog every published architecture, but to give you the conceptual foundations, hands-on code, and evaluation discipline you need to design, build, and critically assess forecasting systems in practice.

The book is organized in three broad sections. The first third (*Chapters 1 to 7*) establishes foundations: what makes time series data challenging, how to evaluate models honestly, the mechanics of PyTorch, and how to build, optimize, and calibrate neural networks with conformal prediction intervals. The middle section (*Chapters 8 to 13*) covers major architectural families — recurrent networks, transformers, **Convolutional Neural Networks** (**CNNs**), graph neural networks, and generative models, including **Variational Autoencoders** (**VAEs**) and diffusion — alongside transfer learning and global modeling strategies. The final third (*Chapters 14 to 18*) moves beyond point forecasting into classification, clustering, embeddings, anomaly detection, and self-supervised learning, reflecting the breadth of tasks practitioners often encounter.

Throughout, we emphasize understanding over black-box application. We want you to know why an architecture works the way it does, not just how to call its API. Where possible, we build models from first principles in PyTorch before showing how libraries like Nixtla's NeuralForecast and statsforecast can accelerate your experimentation. We hope you find it useful.

Who this book is for

This book is written for data scientists, machine learning engineers, and statisticians who have a working knowledge of Python and are comfortable with core machine learning concepts and statistics, but may have limited experience applying deep learning to time-series data. We rely on pandas, NumPy, Matplotlib, Seaborn and PyTorch; knowledge of these is helpful but not strictly required (especially in the age of generative foundation models). We cover PyTorch fundamentals in *Chapter 4*. While prior exposure to classical time series methods (ARIMA, exponential smoothing) will help contextualise approaches with neural architectures, it is not a prerequisite. As the book progresses in complexity, we tend to use more libraries to avoid presenting too much code, with Neuralforecast and Aeon among the main ones.

What this book covers

Chapter 1, Time Series for Everyone, introduces time series and the shift from statistical modeling to applying deep learning to language and then to time series.

Chapter 2, The Challenge of Time Series, explains the basics of temporal data structure and modeling on residuals. It covers stationarity, missing data, seasonality, trends, and exploratory data analysis.

Chapter 3, Evaluating Time-Series Models, focuses on evaluation methodology including the bias–variance trade-off, backtesting strategies, error metrics, and the principles of honest model assessment.

Chapter 4, PyTorch Fundamentals, introduces you to tensors, computational graphs, automatic differentiation, and core PyTorch abstractions that will build your understanding.

Chapter 5, Simple Neural Architecture, we build simple neural networks from first principles — perceptrons, activation functions, and feed-forward networks. We also train a basic forecasting model.

Chapter 6, Optimization, explores the *art* of optimizing network performance via learning rate schedules, regularization, activation functions, and hyperparameter tuning.

Chapter 7, Conformal Prediction, introduces uncertainty quantification for forecasts, covering prediction intervals and conformal prediction methods that provide distribution-free coverage guarantees.

Chapter 8, Recurrent Neural Networks, takes a dynamical-systems view of sequential data and builds RNN, LSTM, and GRU architectures for forecasting, progressing from manual implementations to practical training pipelines.

Chapter 9, Transformers, explains attention mechanisms and transformer architecture, going on to implement transformer-based forecasting models and benchmark them on data.

Chapter 10, Other Neural Structures, surveys additional families of neural architectures, including graph neural networks, Kolmogorov–Arnold Networks, and foundation models.

Chapter 11, Transfer Learning and Global Modeling, covers how to train models across multiple time series simultaneously, make use of pre-trained representations, and build global forecasting models.

Chapter 12, Synthetic Time Series Data, explains data-generating processes and shows how to create realistic synthetic time series using variational autoencoders, with practical implementations of TimeVAE.

Chapter 13, Diffusion Models, introduces diffusion as a generative framework for time series, covering the mathematical foundations of forward and reverse diffusion processes and their application to probabilistic forecasting.

Chapter 14, Time Series Classification, covers distance-based, feature-based, shapelet, dictionary, convolution, and deep learning approaches to classifying temporal data, using the aeon library throughout.

Chapter 15, Time Series Clustering, extends classification concepts into unsupervised territory, covering featurisation, clustering algorithms, deep-learning-based approaches, and cluster evaluation for temporal data.

Chapter 16, Embeddings for Time Series, explores how to represent time series in learned vector spaces, from dynamical-system perspectives to neural embedding techniques and their applications across downstream tasks.

Chapter 17, Supervised and Unsupervised Anomaly Detection, covers rule-based, statistical, and machine-learning approaches to detecting anomalies in time series, including adaptive thresholds, isolation forests, autoencoders, and Matrix Profiling.

Chapter 18, Self-Supervised Learning for Time Series, introduces self-supervised learning paradigms for temporal data, focusing on contrastive predictive coding and representation learning techniques that reduce dependence on labeled data.

To get the most out of this book

To follow the code examples in this book, you will need a working Python 3.10+ environment with PyTorch installed. We recommend using a virtual environment manager such as Poetry, uv or conda. A CUDA-capable GPU is beneficial for the later chapters but is not strictly required — most examples can be run on CPU, though training times will be longer.

The book assumes you are comfortable reading and writing Python, including working with pandas DataFrames and NumPy arrays. Some familiarity with basic statistics (means, standard deviations, hypothesis testing) and machine learning concepts (overfitting, cross-validation, loss functions) will help, though we revisit these where relevant. You do not need prior experience with PyTorch or deep learning — Chapter 4 covers the fundamentals.

Key libraries used throughout include PyTorch, Nixtla's statsforecast, NeuralForecast, and MLForecast, scikit-learn, aeon, stumpy, and matplotlib. Installation instructions and version requirements are provided at the start of each chapter.

Download the example code files

The code bundle for the book is hosted on GitHub at `https://github.com/PacktPublishing/Time-Series-with-PyTorch`. We also have other code bundles from our rich catalog of books and videos available at `https://github.com/PacktPublishing`. Check them out!

Download the color images

We also provide a PDF file that has color images of the screenshots and diagrams used in this book. You can download it here: `https://packt.link/gbp/9781805128182`.

Conventions used

There are a number of text conventions used throughout this book.

`CodeInText`: Indicates code words in text, package names, folder names, filenames, file extensions, pathnames, and variable names. For example: "We instantiate a `TimeSeriesDataSet` object and pass it to our `DataLoader`."

A block of code is set as follows:

```
plot_component(
    decomposition.resid, 'Residual',
    custom_palette[3], plot_type='scatter'
)
```

When we wish to draw your attention to a particular part of a code block, the relevant lines or items are set in bold.

Any command-line input or output is written as follows:

```
pip install neuralforecast
```

Bold: Indicates a new term, an important word, or words that you see on screen. "A critical aspect of this approach involved providing algorithms with direct or **transformed** time series data relative to known labels."

Warnings or important notes appear like this.

Tips and tricks appear like this.

Get in touch

Feedback from our readers is always welcome.

General feedback: If you have questions about any aspect of this book, email us at `customercare@packt.com` and mention the book's title in the subject of your message.

Errata: Although we have taken every care to ensure the accuracy of our content, mistakes do happen. If you have found a mistake in this book, we would be grateful if you would report this to us. Please visit `http://www.packt.com/submit-errata`, click Submit Errata, and fill in the form.

Piracy: If you come across any illegal copies of our works in any form on the internet, we would be grateful if you would provide us with the location address or website name. Please contact us at `copyright@packt.com` with a link to the material.

If you are interested in becoming an author: If there is a topic that you have expertise in and you are interested in either writing or contributing to a book, please visit `http://authors.packt.com/`.

Share your thoughts

Once you've read *Time Series with PyTorch*, we'd love to hear your thoughts! Scan the QR code below to go straight to the Amazon review page for this book and share your feedback.

https://packt.link/r/1805128183

Your review is important to us and the tech community and will help us make sure we're delivering excellent quality content.

1

Time Series for Everyone

"It's difficult to make predictions, especially about the future."

Chances are you have heard this line before (attributed to Niels Bohr or Yogi Berra, depending on whom you trust). Attempting to predict the future has been a human activity for a very long time. Numerous cultures around the world, across time, have attempted to foretell the future based on the distribution of sacrificial animal entrails. In the Bible, the Book of Daniel discusses magicians, enchanters, astrologers, and diviners employed at the court. The Oracle of Delphi provided her predictions while intoxicated by vapors.

The fortunes of forecasters varied throughout history (Emperor Constantine banned soothsayers, mathematicians, and forecasters), but in our day and age, forecasting is a crucial activity. Forecasts and estimations are not always accurate, however; here are some of the most famous examples of people really missing the mark:

- In 1962, Decca Records executives rejected the Beatles: "We don't like their sound, and guitar music is on the way out."
- In 1977, DEC founder Ken Olson announced: "There is no reason anyone would want a computer in their home."
- In 1996, J.K. Rowling got a rejection letter for her first *Harry Potter* novel: "Children just aren't interested in witches and wizards anymore."
- In 1998, Paul Krugman announced: "The growth of the Internet will slow drastically, as the flaw in 'Metcalfe's law' becomes apparent: most people have nothing to say to each other! By 2005, it will become clear that the Internet's impact on the economy has been no greater than the fax machines."
- In 2007, Steve Ballmer (CEO of Microsoft) had this to say about the iPhone: "[Apple's iPhone] is the most expensive phone in the world, and it doesn't appeal to business customers because it doesn't have a keyboard, which makes it not a very good email machine."

This list has not been provided to call out the authors of those failed predictions but to stress an important point: it is quite easy to be very certain and very wrong—and there is quite a lot at stake. "How much?" you may ask. Eric Wilson from the Institute of Business Forecasting demonstrated in a study that a 15% forecast accuracy improvement will deliver a 3% or higher improvement in pre-tax profit. For a company with $10 billion turnover, a 1% forecasting accuracy improvement is worth $300 million. Estimating the impact across Fortune 500 companies is left as an exercise to the reader.

With good forecasting models, we can make predictions that improve our world! Apart from business benefits, forecasting results can also help companies and institutes optimize their resources and reduce waste. Forecasting is also integrated into our daily lives; for example, weather forecasting helps us plan our trips and daily lives.

Initially, we'll focus the chapters of this book on developing an appreciation of time-series data and forecasting predictions using this data. As you progress through the chapters, our simple goals are for you to understand how we forecast using neural architectures such as **feedforward neural networks** (**FFNs**), **recurrent neural networks** (**RNNs**), and NBEATSx and how these models may be built in PyTorch, along with how to evaluate and target your modeling. Of course, we cannot build every model and pipeline in a way that will work for every use case, but by the end of the book, we hope you understand enough about neural networks, PyTorch, and time-series modeling to experiment on your own.

In this chapter, we'll take you through a brief history of time-series forecasting, which can be roughly divided into three periods: early origins, classical, and modern. You might wonder why we're looking at this history—isn't it just old news? Not at all; as you'll discover, many older modeling approaches have stood the test of time, being shaped by many fields of study. As we trace the evolution from early astronomical predictions to today's **machine learning** (**ML**) techniques, you'll gain a richer appreciation of the foundations underpinning the methods we use. More specifically, we will cover the following topics:

- Why time series?
- The early period
- The classical era
- The promise of deep learning

> **Your purchase includes a free PDF copy + exclusive extras**
>
> Your purchase includes a DRM-free PDF copy of this book, 7-day trial to the Packt+ library (no credit card required), and additional exclusive extras. See the *Free benefits with your book* section in the *Preface* to unlock them instantly and maximize your learning.

Why time series?

Humans have always had to contend with uncertainty in the world: in our earliest days, survival depended on rough predictions based on our experience to plan ahead, be it stocking wood for the winter or storing enough food for dry seasons. But even a small attempt at predicting the weather for the next few weeks can tell you that this is not a trivial task.

Before we move on, let's quickly recap what a time series could look like:

- The chemoelectrical signals of your brain as you read this book, collected every microsecond
- The amount of electricity a power plant produces in each hour
- Sales of coffee and tea at your local cafe every day of the year

All of these are forms of data collected over time, and in the simplest terms, this is all that time series are. Details follow in the coming pages.

Before we can attempt to forecast a probable future event, we must first capture measurements of the past, which is essentially what a time series is: logged data over time. The earliest attempts at recording data over time come from the Domesday Book (1086 AD) and the Chinese imperial archives, which were used to capture information on assets and crop yields in order to tax members of a population.

Since these early examples, there has been an explosion of time-series usages, particularly in the last few years, as data logging and storage have become cheaper, easier, and more accurate (sometimes). Time series are everywhere in our world, from trivial daily activities to finding new planets far, far away. So often, temporal information is entangled with other temporal information. For example, weather data is used to predict energy consumption, the food we eat, and the clothes we wear. At the level of governments and global markets, economic activity influences investment and monetary value, which interacts with education and health. Organizations, big and small, that are in tune with time-series data can create new strategies and improve efficiency and profitability, which in turn affects product prices and inflation.

Regardless of what the time-series data is, we need to make good use of it. Commonly, we see data analysts and scientists having to report simple month-on-month and year-on-year metrics to describe the data. While descriptives are useful, the true value of time series is drawn from modeling the data to uncover the underlying patterns in time-series data, in an attempt to get close to the **data-generating process** (**DGP**). In doing so, we can predict and compare probable futures and scenarios. Forecasting well is crucial; when forecasting is not well thought out, it can go spectacularly badly. Let's look at some examples of this:

- **2007–2008 financial crisis**: This crisis cost people all over the world their livelihoods, but why? Because the brilliant minds in many of the banks were buying up securities without understanding what was in the packages. Of course, people such as Michael Burry famously did predict this impending disaster, but that is because they looked at the subprime mortgage data.
- **2010s Arab Spring**: While not the only cause, the price of wheat/grain was a significant factor in the destabilization of several countries in the Arab world, leading to job losses, increasing poverty in Middle East countries due to speculation in the commodities market. These pressures created unrest, toppling governments. Forecasting the price well could have helped the international institutes mitigate some of the disruptions.

- **2020s COVID-19 pandemic**: This pandemic is an incredible example of poor modeling of infection dynamics and healthcare demand due to a lack of sophistication and, critically, a lack of information.
- **2021 GME short squeeze**: GME is the ticker symbol on the **New York Stock Exchange** (**NYSE**) for GameStop Corp. Several funds were bankrupted when retail investors such as Keith Gill learned and shared information about institutional investors shorting GME to approximately 140% of the available shares, leading to a gamma squeeze (infinite money glitch for those with diamond hands). Accurate forecasting of short interest dynamics and market sentiment could have warned institutional investors of the unsustainable positions they were building, leading to better risk management and potentially avoiding such catastrophic losses.
- **Online shopping**: This has become part of our daily lives. Large e-commerce platforms usually stocks up items to ensure availability and fast delivery. Behind the scenes, time-series forecasting models for supply and demand are key. Forecasting it wrong would lead to serious issues. Target piled up too many pandemic related items and had to mark down massive amount of such unwanted items after the Covid period to make space for more up to date demands. It damaged the operating margin rate of Target and Target stock prices dropped.

We've seen why time-series forecasting is crucial, with wide-ranging examples. To better appreciate today's sophisticated forecasting techniques, it helps to understand where the field began. Let's briefly trace the origins of time-series analysis.

The early period

Time-series analysis is a statistical method that has been used for centuries to identify patterns in data that changes over time. The earliest known use of time-series analysis dates back to the 17th century, when Sir Isaac Newton used it to predict the movement of planets. However, time-series analysis did not start developing as a field for another two centuries. In the early 19th century, Carl Friedrich Gauss developed the concept of least squares regression, which is a statistical technique used to model relationships between variables. This technique was later applied to time-series data, leading to the development of early time-series models. In the mid-1800s, mathematicians such as Adolphe Quetelet and Francis Galton developed statistical techniques to analyze time-series data.

George Airy, Britain's seventh Astronomer Royal, used a statistical approach known as the harmonic method to predict tidal patterns—a rather important topic for a naval power like Britain. The core idea was to represent cyclic patterns with a Fourier series, which is an idea still used in forecasting today. In the early 20th century, another Brit—Sir Ronald Fisher—made significant contributions to the development of time-series analysis. Fisher developed methods for analyzing the autocorrelation function, which measures the degree of correlation between observations at different time points. These methods provided a powerful tool for understanding the behavior of time-series data and were widely used in the development of early time-series models.

The early pioneers in the field of time-series analysis include mathematicians such as Norbert Wiener, who is the name behind the famous Wiener process; Andrey Kolmogorov, who developed further theories about stochastic processes; and Udny Yule, known for his contribution to the autoregressive model. These early researchers developed mathematical techniques for analyzing time-series data and made significant contributions to the understanding of the underlying processes that generate such data.

Dan Carlin famously said, "If you could read a history book from 500 years from now, our entire century would be smashed into a paragraph"—and this is not unlike what we have done in this section with our summary of our epic scientific past. In the mid-20th century, the advent of computers led to the development of more advanced statistical techniques, such as autoregression and moving averages, which we will explore a bit in the next section.

The classical era

As important as the lineage discussed in the previous section was, the classical era of time series started with George Yule, who introduced the world to autoregression. The Yule-Walker equation demonstrated that autoregression could forecast sunspots, ushering in a new era of time-series forecasting.

The next major step was the Box-Jenkins method (named, unsurprisingly, after its authors): it provided a framework for modeling and forecasting time-series data based on a combination of autoregressive and moving average components. The Box-Jenkins method became widely used in industry and academia and remains an important tool in time-series modeling today; here, we list major variants of the approach that are very much used to this day:

- **Autoregressive integrated moving average (ARIMA) model:** ARIMA is a method that combines autoregression, differencing, and moving averages to model the underlying structure and trends of a time series. The following are some variations:
    - **Seasonal ARIMA (SARIMA):** Extends ARIMA to include seasonal components (cyclic patterns, i.e., tides).
    - **Autoregressive integrated moving average with exogenous variables (ARIMAX):** Includes exogenous variables that can help improve the accuracy of the forecasts.
    - **Seasonal autoregressive integrated moving average with exogenous variables (SARIMAX):** This is an extension of the ARIMAX model that includes seasonal components and exogenous variables.
    - **Generalized autoregressive conditional heteroscedasticity (GARCH):** Models the error terms by modeling their variance in an autoregressive fashion. GARCH brings in the capability of modeling time-series data with time-varying noise.
- **Seasonal and Trend decomposition using Loess (STL):** A decomposition method that separates time series into the underlying components of trend, seasonality, and residuals/error.
- **Exponential smoothing:** Covers a family of forecasting methods that assign exponentially decreasing weights to older observations. Common examples are simple exponential smoothing, Holt's linear trend, and the Holt-Winters seasonal method. These techniques are particularly effective for forecasting data with trends and seasonal patterns.
- **Vector autoregression (VAR):** Models interdependencies between multiple time series, allowing for more complex and dynamic relationships.
- **Bayesian structural time series (BSTS):** This is a probabilistic Bayesian time-series approach that can handle complex trends, seasonality, and other time-varying effects. Unlike deterministic methods, it quantifies uncertainty in its forecasts, providing probability distributions with prior knowledge, rather than point estimates.

Another important technique developed during the classical era of time-series modeling is state-space modeling. State-space models are a flexible class of models that can be used to capture complex patterns in time-series data. These models are particularly useful when modeling non-linear and non-stationary time-series data, which are difficult to model using traditional linear models such as ARIMA. A massive advantage of state-space models is that other statistical approaches (ARIMA, linear regression, and exponential smoothing) can be seen as special cases of this more general approach. Kalman filtering is a very successful recursive estimation method that can be used to perform predictive inference in a state-space framework: filtering, prediction, and smoothing are all conducted in closely related manners. The original algorithm was developed out of necessity: in the 1950s, computers had negligible (for 21st-century standards) amounts of memory, so minimizing the memory footprint was crucial. With computational power becoming cheaper over subsequent decades, the motivation for efficiency decreased—but the advent of applications requiring low latency led to a resurgence of sorts for variations around the original idea of Rudolf Kalman.

Throughout the 1970s, building a time-series model was pretty much synonymous with fitting an appropriate order for an ARIMA model. In the next decade, an interesting problem arose in the financial time-series context: Engle (later with Bollerslev) observed volatility clustering—the phenomenon of financial time series alternating between periods of high and low variation. Attempts to capture this with ARIMA models failed, for (with hindsight) obvious reasons: equations of an ARIMA model specify a time-varying conditional expectation, but the focus needed to be on conditional variance. GARCH models are a class of models used to model time-series data that exhibit volatility clustering, or the tendency of volatility to persist for short periods of time. GARCH models are a natural extension of the ARIMA model, which is widely used in time-series analysis.

GARCH models have become particularly useful in finance because financial time-series data often exhibits volatility clustering. For example, in financial markets, periods of high volatility tend to be followed by periods of high volatility, and periods of low volatility tend to be followed by periods of low volatility. GARCH models allow analysts to model and forecast this volatility clustering, which is important for risk management and portfolio optimization. GARCH models have been widely used in finance for a variety of applications, including volatility forecasting, option pricing, and risk management.

> We think it is worth noting at this point that simpler methods are not necessarily inferior, and in many cases can be the optimal approach: statistical methods can generate excellent forecasts in areas from financial markets to energy supply requirement; even when they are not used directly, they often are utilized as base forecasts, as a means to compare the performance of more complex models. While we cover them briefly in this book, we highly recommend reading Rob Hyndman and George Athanasopoulos' excellent book *Forecasting: Principles and Practice* (`https://otexts.com/fpp3/`) in order to get a good background on autoregressive methods. Despite the continued usefulness of statistical methods, several changes have led to the rise and, arguably, current dominance of ML methods.

While classical methods remain indispensable, the increasing complexity of data and advances in computation have pushed the field toward ML. This shift set the stage for the rise of **deep learning** (**DL**) in time-series forecasting.

The promise of deep learning

In recent years, DL techniques have revolutionized the field of time-series analysis. DL utilizes artificial neural networks and data to achieve predefined tasks, for example, forecasting the weather for next week. The first DL-based time-series models came from the field of **natural language processing** (**NLP**). NLP models natural languages by classification, interpretation, and even generation (e.g., the well-known **large language models** (**LLMs**)). The prevalence of neural networks such as RNN (and later LSTM) exploded in NLP from 2015 onward: the ability to model long sequences proved to be a game-changer.

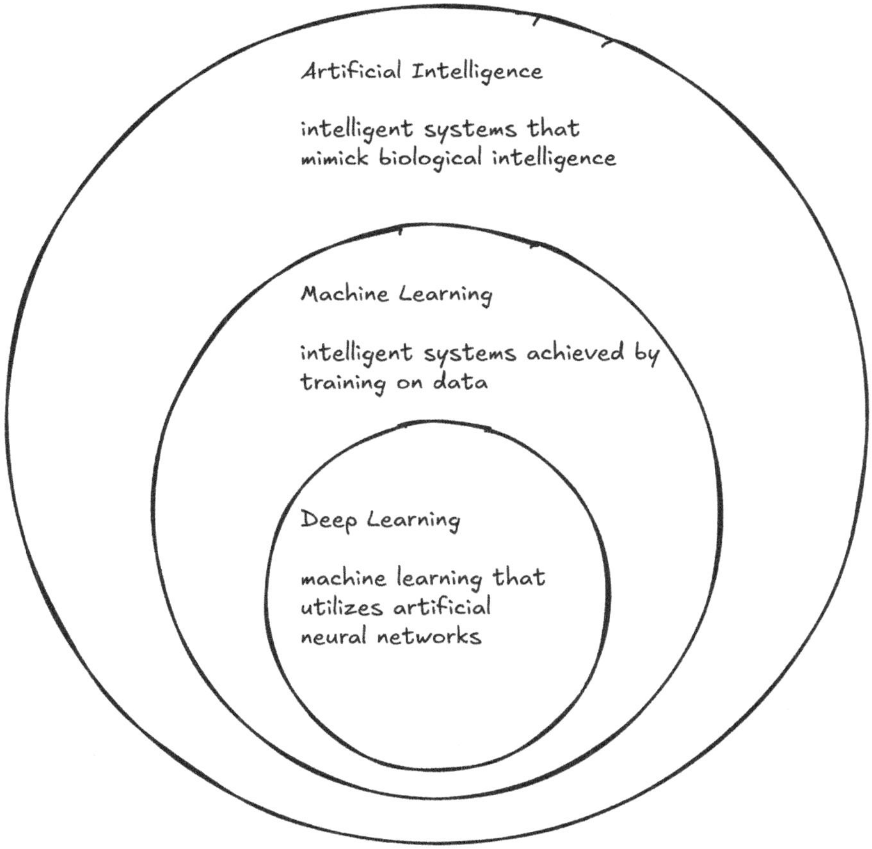

Figure 1.1: Relation between artificial intelligence (AI), ML, and DL

Let's take a closer look at this evolution.

From language to time

The developments in NLP have made progress for time series possible by providing the inspiration for new DL techniques. In particular, the success of RNNs in modeling sequences of words in NLP tasks led to the development of similar techniques for modeling time-series data. RNNs are a powerful tool for modeling sequential data because they can capture long-term dependencies between observations. This makes them particularly useful for modeling time-series data, which often exhibits complex temporal patterns that are difficult to capture using traditional statistical methods. It did not take long for researchers and practitioners to observe that the same methods could be applied to time-series problems. The one major difference from NLP was the arrow of time—it is perfectly fine to check the word at the end of a sentence you are translating from English to Polish, but not to peek forward in a time series you are forecasting.

Another important development in NLP that has influenced DL for time series is the use of attention mechanisms. Attention mechanisms allow neural networks to focus on specific parts of the input sequence, which is useful for modeling time-series data that exhibits complex temporal patterns. Attention mechanisms have been used in a wide range of DL models for time-series data, including LSTM networks and transformer networks. The developments in DL for time series have the potential to transform a wide range of industries, including finance, healthcare, and manufacturing. By providing more accurate predictions and a better understanding of the underlying processes that generate time-series data, DL can help companies make better decisions and optimize their operations. As the field of DL continues to evolve, new techniques and applications for time-series analysis will undoubtedly emerge, paving the way for even more exciting developments in the years to come.

In addition, DL models can be more flexible in their input representation, and can handle a wide range of data types, including sequential, categorical, and numerical data, which can be useful in time-series forecasting tasks that involve multiple variables.

However, DL models can also be more complex and computationally intensive than traditional ML algorithms, and they may require more data to train effectively. Additionally, DL models may be more difficult to interpret than traditional ML algorithms, which can make it challenging to understand how they are making predictions.

Forecasting and performance

Statistical methods dominated the early years of forecasting, which is reflected in the domination of data science competitions such as the M Competitions, named after Spyros Makridakis. There have been six competitions up to the time of writing, which are named M (published in 1982), M2 (published in 1993), M3 (published in 2000), M4 (published in 2018), M5 (published in 2022), and M6 (published in 2024). These competitions were mostly focused on product demand over time, a dataset type we will look at later. "Pure" statistical approaches were winning the M competition up until M3 in 2000. By contrast, until recently, ML, that is, decision trees, and DL approaches did not always perform well. Neural networks in particular performed poorly in M3, and a subsequent competition, NN3, in 2006, conducted to encourage neural network approaches, failed to produce a method that outperformed the original statistical winner of M3. It wasn't until 2017's web traffic competition, launched on Kaggle, that we started to see neural networks outperform statistical models.

By the time 2018's M4 competition was completed, there was a marked change in computational power and, consequently, hybrid models ruled the day, specifically a variant of RNN named ES-RNN, where **ES** stands for **exponential smoothing**. In second place was a boosted decision tree method called XGBoost.

In the follow-up M5 competition, ML methods, specifically tree-based approaches, won the accuracy competition with a LightGBM (gradient-boosted decision tree) approach coming top, followed by approaches utilizing DL methods such as N-BEATS. Tree-based methods continue to do well in tabular data competitions, for example, data that is organized into a table, in which rows represent observations, such as sales in a day, and columns represent variables/features, such as price, bank holidays, and sporting events.

ML methods such as LightGBM currently dominate forecasting competition leaderboards and are becoming more common in industry. Plenty has been written about them and the methods required to generate good performances. We have listed some good references at the end of the chapter. Neural architectures are starting to emerge as strong options for forecasting, as the last decade shows us. You only need to look at the change in language bots to appreciate the progress toward OpenAI's ChatGPT-3. For example, in a blog post (`https://medium.com/deep-writing/harry-potter-written-by-artificial-intelligence-8a9431803da6`), the author trained an LSTM model on the first four Harry Potter books to generate a short excerpt.

Taken together, the competitions and parallel advances in NLP illustrate the dramatic shift from classical statistics to ML and DL.

Summary

This concludes our crash course on time-series history—quite possibly the one part of this book with no formulas, graphs, or tables. Let's briefly recap our journey; we began by exploring the early origins of time-series analysis, from ancient attempts at predicting crop yields to the first recorded instances of data logging in the Domesday Book and Chinese imperial archives. We then moved on to the classical era, examining the development of fundamental techniques that still form the backbone of many forecasting activities today.

We've traced the evolution from simple descriptive statistics to more sophisticated modeling techniques such as ARIMA and its variants. We discussed the emergence of state-space models and the Kalman filter, showcasing the field's adaptability to difficult-to-model data. We moved on to talk about the development of GARCH models to address volatility clustering in financial time series. Finally, we introduced ML and DL techniques, marking the transition to the modern era of time-series analysis.

You've seen how time-series forecasting has been shaped by interdisciplinary contributions, from astronomy to economics. We've also observed how advancements in computational power have driven innovations in the field. In the following chapters of this book, we will discuss the fundamentals of time-series data and modeling, PyTorch and DL fundamentals, followed by time-series forecasting and classification models using PyTorch. You will not only read about how DL is applied to time-series data but also understand how we have taken the leap from the classical methods to DL-based methods. At this point in our journey, we are ready to discuss time-series structures and how we analyze them. In the next chapter, we will start our journey through time series proper beginning with the building blocks we need.

Get this book's PDF version and more

Scan the QR code (or go to `packtpub.com/unlock`). Search for this book by name, confirm the edition, and then follow the steps on the page.

UNLOCK NOW

Note: Keep your invoice handy. Purchases made directly from Packt don't require an invoice.

2

The Challenge of Time Series

"Without data, you're just another person with an opinion." —W. Edwards Deming

In a world full of data, it is very easy for us to churn out models with the latest algorithm, or best publicized (don't use Prophet), without giving much thought to the data that goes in. Just get the data in the right format and you are ready to go, with `.fit` and `.predict`! This is an easy trap to fall into when you are under pressure to provide sales team members with predictions, but ultimately leads to poor models and overconfidence in the data and forecasts built upon them. To make useful predictions, we must have a strong understanding of our data. We need to understand how data is formed and structured, and recognize the inherent strengths and weaknesses of the information we are using to build our models. This highlights the need for **exploratory data analysis** (**EDA**). EDA may seem like a laborious task, but in reality, it's a time-saving and efficient task that can improve our data and the development of robust forecasts.

To conduct effective EDA, we must first recognize the structures that underlie time series data. Different structures have different implications for analysis and forecasting. In this chapter, we briefly explore these structures and their impact on modeling. You will learn about the following:

- What is time series data?
- Key components of time series
- Variations in time series formats
- Handling irregularities in data
- Preparing time series for modeling
- The importance of transformations

This chapter is intended as a brief introduction or refresher on the basics of time series. You may be familiar with many of the concepts here and wish to skip some of the content. The end of the chapter covers preparing data for forecasting, and recommends reviewing this section.

What is time-series data?

In essence, a time series is observations (measurements) of data (i.e., sales, temperature, etc.) collected in a sequential order with a temporal structure; these can be regular or irregular. Typically, we deal with regular time series, meaning the intervals between measurements are the same (i.e., hourly, daily, weekly, monthly, quarterly, etc.). While regular time series may contain missing data, we expect the intervals between measurements to remain constant throughout most of the series. You'll often see series as charts of data over time in newspapers or on news programs, especially when the economy is not performing so well, or unemployment figures are high. What's strange about these charts is that they often show only small sections of a time series, rather than the whole thing. Because of that, it's hard to spot important patterns. For example, financial crashes actually occur fairly frequently—some are smaller, while others are larger—but if we only look at a short period, we might miss this regularity. Our world is full of an incredible variety of data (e.g., Kaggle, Monash Time Series Repository, Yahoo Finance). Simply seeing and labeling the data gives us a good starting point for understanding data, which moves us beyond sensationalist headlines. Now is a good time to start up your computer and follow along. Feel free to use your own data!

In the following table, we have used data from the `Gapminder` dataset, which presents a dataframe of average life expectancy for each year from 1952 to 2007 by continent.

Continent	Year	Life expectancy
Africa	1952	38.8330
Africa	1957	40.5925
Africa	1962	42.6305
Africa	1967	44.6985
Africa	1972	47.0315

Table 2.1: The Gapminder dataset

The data that underlies this time series is simply a grouping of life expectancy values for all countries in each continent by year, and we have then collected the median values of each year.

We then visualize this using Seaborn to create a line plot that displays an encouraging increase in the average life expectancy of humans from each year from 1952 to 2007. This is a very simple and clean set of time-series data.

Refer to color images on the Graphic bundle or the ebook.

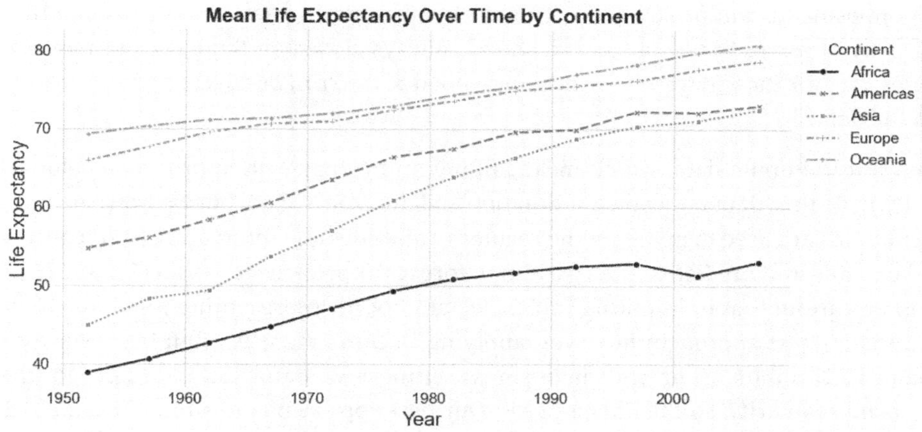

Figure 2.1: Time series life expectancy by continent over time

Missing data

Typically, you will deal with regular time series like this, where we have data collected at regular intervals. By contrast, irregular time series are built with data that, while recorded sequentially, varies in the intervals of collection. For example, natural disasters are infrequent and can vary from days apart to years. Regular data can degrade into irregular data if we have a lot of missing data. This is not unusual in retail settings, where we often see low-frequency items/products, such as the sales of Bugatti in the UK (169 in 2019). Let's look at a practical example to illustrate this:

```
df = pd.read_csv(Path.cwd().parent / "data" / "bugatti_sales.csv")
print(df.head(n=5))
```

The output for this code is as follows:

```
   year month  sales
0  2016   Mar      1
1  2016   Apr      2
2  2016   Sep      2
3  2016   Nov      1
4  2016   Dec      1
```

We can see from the table that several months are missing (no data) between particular dates.

Types of missing data

Missing data can have a big impact on our analysis and the performance of our models, so it's something we need to deal with. How we handle it depends on the type of missingness, which usually falls into two categories:

- **Missing at random** (**MAR**): This occurs when the likelihood of data being missing depends on other data that can be observed. For example, Bugatti sales might appear "noisy" because they're so rare, not because the data is missing at random.

- **Not missing at random** (**NMAR**): This means the missing data is tied to something we can't directly observe. For instance, if a shop is always closed on Sundays and national holidays, sales data for those days will be missing—not due to error, but because we didn't record them in the data.

Often, we want to impute data, which means filling in the gaps with appropriate values. There are several ways to do this, depending on the situation and the type of data, but we won't go into the details here. Imputations are used to maintain the regularity of values and intervals between observations, to aid model training and predictions. As we aim to forecast a pattern we "expect" to see (i.e., demand), we need to be careful that in imputing values, we are not misrepresenting relationships within our data. There is no best approach; however, applying Occam's razor is often best. For example, our Bugatti data is a common occurrence in business settings; we simply have no sales for this product, and there is no regularity, so the data is MAR. Our best approach is to simply resample the data to monthly (lowest temporal resolution), and zero-fill the missing data. Let's do that now:

```
# Convert year value to a string and then combine with month
df['date'] = pd.to_datetime(df['year'].astype(str) + '-' + df['month'],
format='%Y-%b')

# Set new month variable to index and resample to get missing dates, filled
with 0
df = df.set_index('date').resample('ME').sum().fillna(0)

# Drop used columns
df = df.drop(['year', 'month'], axis=1)

# Reset index
df = df.reset_index()
print(df.head(n=5))
```

The output is the following:

```
        date  sales
0 2016-03-31      1
1 2016-04-30      2
2 2016-05-31      0
3 2016-06-30      0
4 2016-07-31      0
```

You can see in the resampled data that we have added several dates that now have `0` sales. We could also resample to a higher resolution, such as quarterly, depending on business requirements. This would smooth out the sales over time, but could be less informative. You should check the frequency of your datetime columns when resampling.

Charting our Bugatti data enables us to visualize the full range of sales over time, including zeros. Another option would be to aggregate upward to the next level (i.e., quarterly or yearly frequency instead); this would smooth out the data, which may or may not be more informative, depending on your business needs.

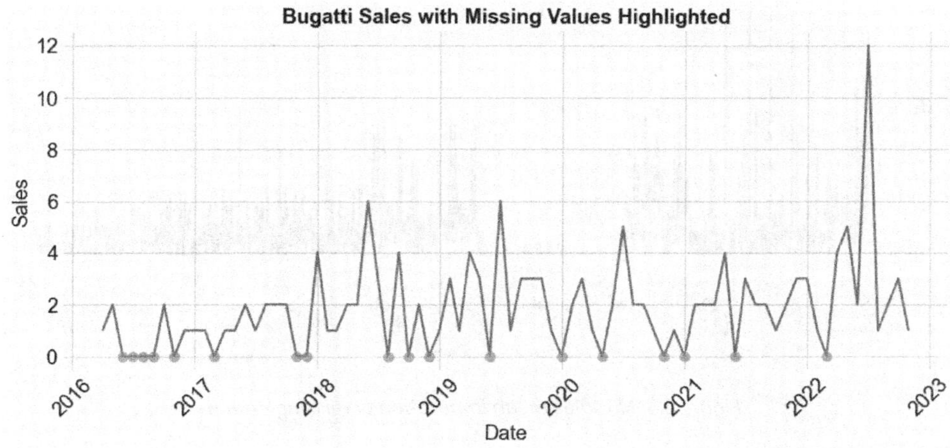

Figure 2.2: Bugatti sales time series with missing values highlighted

This approach would also be valid for NMAR data. For instance, if a shop is regularly closed on Sundays, we might consider treating those days consistently by either excluding them or leaving them as they are. However, it could be more appropriate to work only with business days (Monday to Friday) and avoid adding zeros for non-operating days. Sometimes, though, data is simply missing—either MAR or NMAR. A good example of this can be seen in the following plot of some data from the M5 Kaggle competition dataset (generously provided by Walmart). In this dataset, some time series contain flat spots—periods of no recorded activity. This is a fairly common phenomenon in real-world time series data:

```
series_df = panel_df.loc[panel_df.item_id == 'FOODS_3_424'].copy()

# Let's make this 2013 onwards
series_df = series_df[series_df['date'].dt.year >= 2013].copy()

fig, ax = plt.subplots(figsize=(12, 6))
sns.lineplot(data=series_df, x="date", y="sold", color="black")
ax.set_xlabel('Date')
ax.set_ylabel('Sales')
ax.set_title('M5 Sales - FOODS_3_424 (2013 onwards)')
plt.rcParams['font.size'] = 16
plt.xticks(rotation=45)
plt.tight_layout()
plt.show()
```

Here's what the plot looks like:

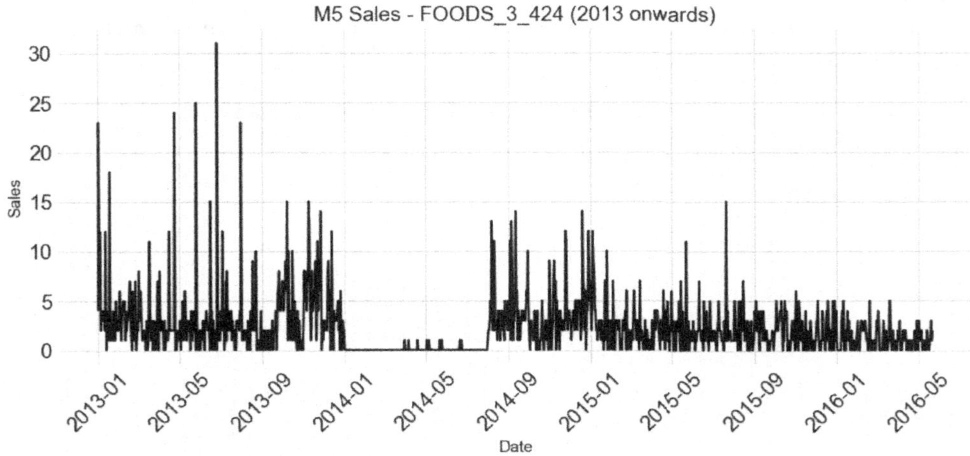

Figure 2.3: M5 sales mean imputation of missing period

This is MAR data, which has been filled with zeros, due to the low frequency of sales. But we also see a systematic drop-off from 2014-01 to 2014-09, with some occasional sales. This could be due to a lack of stock, reporting issues, and so on; we don't actually know, but it is NMAR and would likely affect modeling. So, how can we address this?

> Sometimes, the best thing to do is to just cut older data from our time series data when we see dramatic changes in the temporal pattern, such as sudden **drop-offs** of **uplifts**. While this reduces our data, if there is enough data left to model, it is often the best thing to do.

Simple imputations

A simple approach is mean imputation: we take the mean within a range of periods (i.e., 30 time points) and fill the values. Determining which values/gaps to apply imputation to falls under the purview of **time series anomaly detection** (**TSAD**), which we discuss in *Chapters 17* and *18*. You also cannot go wrong by looking at Eamonn Keogh's work on the subject of missingness and anomalies: `https://www.cs.ucr.edu/~eamonn/`.

For our data, we employ a direct approach to TSAD.

We identify the gap and create a Boolean mask, which is a series of `0`s and `1`s (`False` and `True`) for inclusion in the dates we are interested in:

```
# Define the gap period
gap_start = '2014-01-01'
gap_end = '2014-08-01'

# Boolean to identify gap
```

```
gap_mask = (series_df['date'] >= gap_start) & (series_df['date'] < gap_end)

# create dataframe to hold imputed data
df_imputed = series_df.copy()
df_imputed['mean_imputed'] = series_df['sold'].replace(0, df_imputed['sold'].
mean())
```

Next, we create a function for plotting our imputations of the gap:

```
def plot_imputed_sales(df, y_variable, title_suffix=""):
    fig, ax = plt.subplots(figsize=(12, 6))
    sns.lineplot(data=df, x="date", y=y_variable, color="black")
    ax.set_xlabel('Date')
    ax.set_ylabel('Sales')
    ax.set_title(f'M5 Sales - FOODS_3_424 (2013 onwards) - {title_suffix}')
    plt.rcParams['font.size'] = 18
    plt.xticks(rotation=45)
    # Highlight the imputed region
    gap_start = '2014-01-01'
    gap_end = '2014-08-01'
    ax.axvspan(gap_start, gap_end, color='#b512b8', alpha=0.3)
    plt.tight_layout()
    plt.show()
```

Now, we run the code for mean imputation:

```
plot_imputed_sales(df_imputed, "mean_imputed", "Mean Imputation")
```

This is the output:

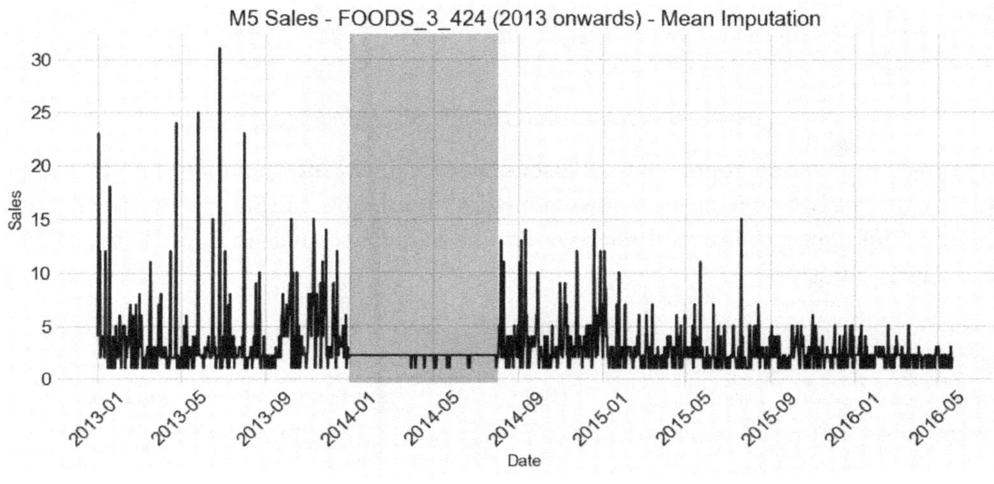

Figure 2.4: M5 sales mean imputation of missing period

This isn't an ideal approach—although it lifts the data above zero, it fails to capture any of the underlying patterns within the missing sections. A more common method is forward filling, which uses the last known value before the gap to fill missing entries. Back filling works in the opposite direction, filling gaps with the next known value.

In pandas, we can use `ffill` (forward fill) or `bfill` (back fill) if the missing values are represented as NaN. However, using a Boolean mask gives us more control over how and where we apply the fill.

In the following code, we've implemented forward filling by using the last value before the drop-off. The result is somewhat similar to mean imputation with a constant value of 2—but once again, it lacks variation and may not reflect true trends in the data:

```
df_imputed['ffill_imputed'] = df_imputed['sold'].copy()
df_imputed.loc[gap_mask, 'ffill_imputed'] = df_imputed.loc[
    ~gap_mask, 'sold'].iloc[-1]
plot_imputed_sales(df_imputed, "ffill_imputed", "Forward Fill Imputation")
```

We get the following plot on running the code:

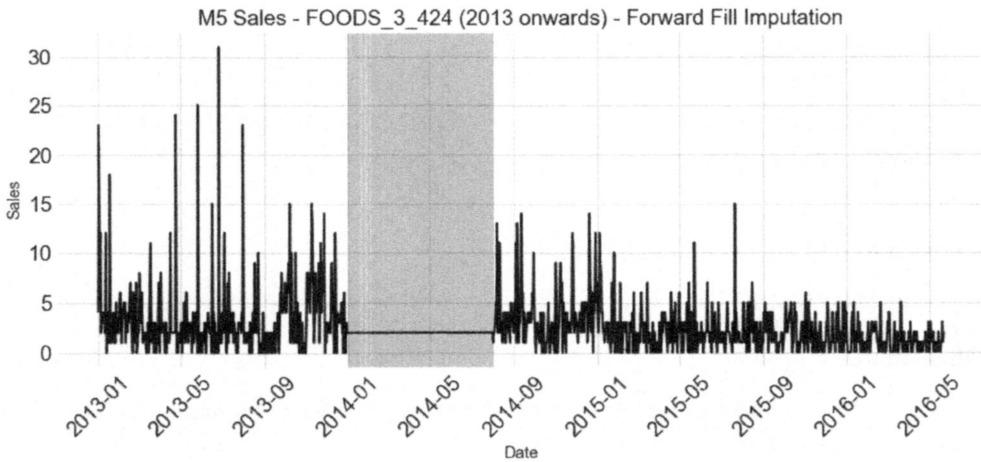

Figure 2.5: M5 sales forward-fill imputation of missing period

The last simple imputation approach we'll discuss here is **linear interpolation**. Linear interpolation draws a straight line between the last known non-zero value before the flat spot and the first non-zero value after it. This approach is useful when we assume a gradual change in values between the period before and after our gap:

```
df_imputed['interp_imputed'] = df_imputed['sold'].copy()
df_imputed.loc[gap_mask, 'interp_imputed'] = np.nan
df_imputed['interp_imputed'] = df_imputed['interp_imputed'].interpolate()
plot_imputed_sales(df_imputed, "interp_imputed", "Linear Interpolation")
```

The following is the output for the code:

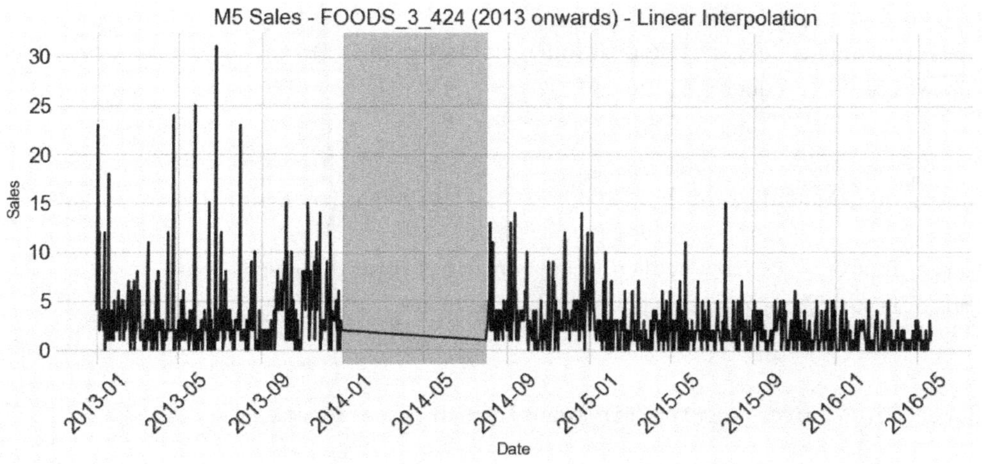

Figure 2.6: M5 sales linear interpolation imputation of missing period

Again, the imputation is simple, adding a linear downward projection, but this is not a great representation of what the missing data might look like. Despite this, we might find these simple approaches to be good enough for initial exploration and modeling.

MICE imputation

There are more complex methods we could use, such as **multiple imputation by chained equations** (**MICE**). MICE uses an iterative process to fill in missing values, considering the relationships between multiple variables in our dataset. It begins by filling in missing values with initial estimates, then iteratively refines these estimates using an algorithm such as the random forest algorithm. For each variable with missing data, it predicts values based on all other variables, cycling through this process multiple times. This allows MICE to capture dependencies in the data, potentially reflecting seasonal trends or day-of-week effects in our sales data. We can use various models for prediction (i.e., linear regression, logistic regression, or more complex models). Here, we've chosen to use a **random forest** (**RF**) with scikit-learn's `IterativeImputer`. We chose this because RF allows us to capture non-linear relationships without making strong assumptions about data distribution:

```
from sklearn.experimental import enable_iterative_imputer
from sklearn.impute import IterativeImputer
from sklearn.ensemble import RandomForestRegressor

# Add additional features to data to help RF
df_imputed['year'] = df_imputed['date'].dt.year
df_imputed['month'] = df_imputed['date'].dt.month
df_imputed['day'] = df_imputed['date'].dt.day
df_imputed['dayofweek'] = df_imputed['date'].dt.dayofweek
df_imputed['is_weekend'] = df_imputed['dayofweek'].isin([5, 6]).astype(int)

# Instantiate imputer
```

```
mice_imputer = IterativeImputer(
    estimator=RandomForestRegressor(n_estimators=300),
    max_iter=12, random_state=3962)

# Prepare data for imputation
columns_for_imputation = ['sold', 'year', 'month', 'day',
                          'dayofweek', 'is_weekend']
df_mice_impute = df_imputed[columns_for_imputation].copy()
df_mice_impute.loc[gap_mask, 'sold'] = np.nan

# Run imputation
imputed_data = mice_imputer.fit_transform(df_mice_impute)

# Update df_imputed dataframe with MICE imputed values
df_imputed['mice_imputed'] = df_imputed['sold'].copy()
df_imputed.loc[gap_mask, 'mice_imputed'] = imputed_data[
    gap_mask, 0].astype(df_imputed['sold'].dtype)

# Plot
plot_imputed_sales(df_imputed, "mice_imputed", "Iterative Imputation")
```

The code renders the following plot:

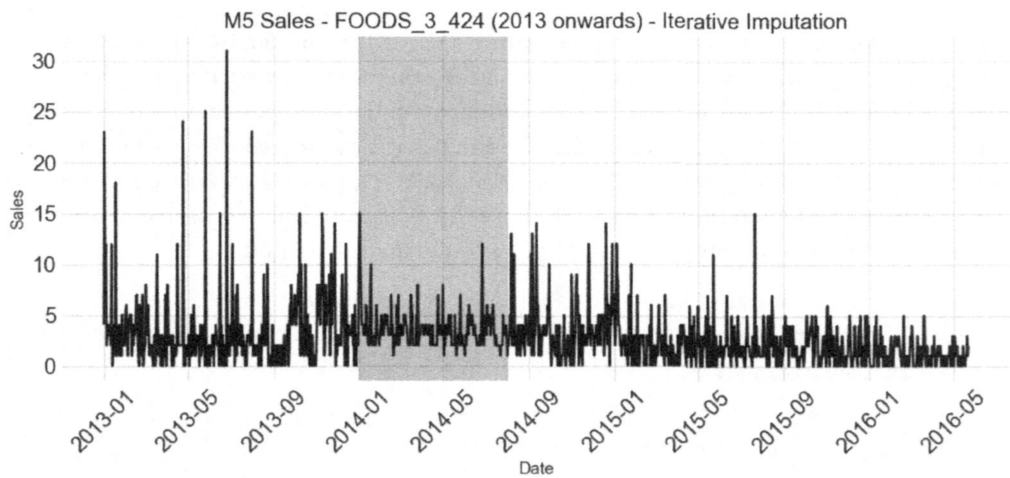

Figure 2.7: M5 sales MICE imputation with RF

This is not perfect as there are no zero values, which occur naturally throughout the series, but it looks like a better fit than the previous methods. Generally, MICE works best with data that is **missing completely at random** (**MCAR**) and MAR. It is not unusual for it to struggle with NMAR data, unless our additional variables can proxy for the missing mechanisms are included.

This is the tip of the iceberg for imputation with MICE. For a full discussion, we recommend reading Stef van Buuren's book on the subject: `https://stefvanbuuren.name/fimd/`. The `fancyimpute` library provides additional options such as matrix imputation.

Matrix imputation utilizes relationships between multiple related time series to fill missing data. It assumes the existence of a low-rank matrix that can represent the information in the larger matrix. Methods such as matrix factorization, truncated **support vector decomposition** (**SVD**), iterative SVD, soft impute, and centroid decomposition decompose the original matrix into smaller matrices, capturing patterns to reconstruct missing values. While this is an effective technique, it does rely on selecting appropriate hyperparameters for the underlying models, allowing them to capture complex patterns such as seasonality.

Theoretical foundations of time series analysis

Having addressed the practical aspects of handling missing data, we now turn to the underlying theoretical principles governing time series analysis. These principles explain why we can make meaningful inferences and predictions from discrete observations of continuous processes, such as retail sales over time.

When we collect time series data, we're essentially capturing discrete observations of a continuous stochastic process that generates our data. The Kolmogorov extension theorem provides the underlying reason of why we can infer from the former about the latter. It states that *any consistent collection of finite-dimensional probability distributions can be extended to a unique probability measure on an infinite-dimensional space*. By assuming that observed data is a sample of a larger stochastic process, we can use the Kolmogorov extension theorem to infer the probability distribution of the process at unobserved points in time.

Put simply, our motive for collecting information over time as a "time-series" data object is that it allows us to capture the probable temporal nature of what we are measuring (some stochastic process). We can, therefore, break down time-series data into features that make up a stochastic process (i.e., trends, seasonality, and cyclical patterns), which helps us both understand and mimic (model) the process underlying the generation of that data. While capturing the exact data generating process is generally unattainable, investigating common temporal dynamics and then building a model on that research helps us get predictive models closer to the **Data Generating Process** (**DGP**), resulting in useful predictions (sometimes).

The proper mathematical treatment of this theory is beyond this book's scope. Mathematically inclined readers are advised to explore terms such as the *Kolmogorov extension theorem* further in specialized literature.

Patterns in time

The first step in understanding the patterns and variations of time-series data (or stochastic processes if you will) is conducting EDA: checking data ranges, looking for missing data, calculating basic statistics, and so on.

Although we haven't done it here, it is also helpful to look at distributions of numeric variables, such as sales, price, and so on:

```
# Check date-range
print(f"Date Range: {df_imputed['date'].min()} to {df_imputed['date'].max()}")
print(f"Number of Observations: {len(df_imputed)}")

# Check for any gaps in date sequence
df_imputed['date_diff'] = df_imputed['date'].diff()
gaps = df_imputed[df_imputed['date_diff'] > pd.Timedelta(days=1)]
if not gaps.empty:
    print(f"Number of gaps: {len(gaps)}")

# Check for missing values in time series
missing = df_imputed.isnull().sum()
if missing.any():
    print(missing[missing > 0])

# Basic descriptive statistics of numerical columns
print(df_imputed.describe())
```

Understanding the patterns and variations of time-series data goes beyond these simple checks. In time-series EDA, we need to break down our series into their constituent components, referred to as decomposition. These are the following:

- **Trend:** A long-term increase or decrease in the data
- **Seasonality:** Regular patterns of ups and downs related to seasonal factors
- **Cycles:** Patterns that occur but with less regularity than seasonal patterns

Understanding these patterns is crucial because it helps in identifying the underlying data-generating process. This facilitates our interpretation and intuition of past data, and, in turn, informs our forecasting methods and models.

Decomposition

Conducting decomposition as part of our EDA presents us with several advantages:

- **Improved understanding:** It isolates the trend and seasonal components, providing clearer insights into the underlying structure
- **Simplified modeling:** The decomposed elements can be modeled separately, often leading to more accurate forecasts
- **Enhanced forecasting:** By capturing the essence of each component, we can make more informed predictions about future behavior
- **Anomaly detection:** Isolating the random noise (residuals) makes it easier to identify significant deviations from a more general pattern

Decomposition techniques, such as classical decomposition, X-11, and **seasonal and trend decomposition using Loess (STL)**, can be employed to perform time series decomposition. While each method has its strengths, STL is often used due to its flexibility. It uses locally weighted scatterplot smoothing (Loess) to estimate the trend and seasonal components of the data. What remains after extracting these components is the residual, representing the remaining variation in the data.

It's important to note that these decomposition methods assume that the time series is primarily influenced by its own past values and regular patterns. In reality, we often have strong exogenous variables (such as price, advertising, competitor actions, and economic indicators) which can significantly impact the time series, such as sales or stock prices. In such cases, traditional decomposition methods may not fully capture or explain these external influences. Analysts may need to consider additional analytical techniques to identify and understand the effects of these exogenous factors on the time series patterns.

We are going to use a mathematical formula next. While this can be intimidating, we will, as throughout the book, break down the formula. There are two common forms of decomposition: **additive** and **multiplicative**. The components of the formulas are similar; it is really just the operators that change.

Additive decomposition

We'll first take a look at the additive form alongside the graphs by decomposing `df_imputed['mice_imputed']` with the `statsmodels` library:

```
# Set date to index
df_imputed = df_imputed.set_index('date')

# sliced data to aid visualisation of seasonal patterns (e.g., last year of
data)
end_date = df_imputed.index[-1]
start_date = end_date - pd.DateOffset(years=1)
df_sliced = df_imputed.loc[start_date:end_date]

# from statsmodels.tsa.seasonal import seasonal_decompose
decomposition = sm.tsa.seasonal_decompose(df_sliced['mice_imputed'],period =28,
model='additive')

# Plotting function
def plot_component(data, title, color, plot_type='line'):
    plt.figure(figsize=(12, 4))
    if plot_type == 'line':
        plt.plot(data, color=color, linewidth=2)
    elif plot_type == 'scatter':
        plt.scatter(data.index, data, color=color, s=5)
    plt.title(title)
```

```
    plt.tight_layout()
    plt.show()

# Plot observed time series
plot_component(decomposition.observed, 'Observed', custom_palette[0])
```

The following is our output:

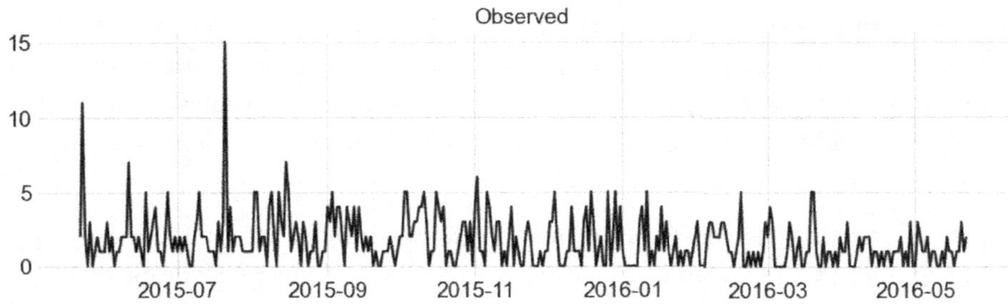

Figure 2.8: The observed time series

The code here instructs `statsmodels` to decompose using moving averages to estimate the trend and seasonal components. This is a naive approach as it assumes a strict periodicity does not adapt well to irregular seasonal patterns. This is also unlikely to effectively address boundary effects, noise, or non-stationary behaviors in many time series; using models to decompose a series can help address these issues. But we'll keep with this for demonstration purposes.

The formula for an additive decomposition is as follows:

$$X_t = T_t + S_t + C_t + I_t$$

This breaks down into the following components:

- T_t: Represents a time series trend, the long-term progression/direction (increase or decrease) of the series over the length of the dataset. When graphing this, we have time on the *X* axis and the target/dependent variable mapped to the *Y* axis:

  ```
  plot_component(decomposition.trend, 'Trend', custom_palette[1])
  ```

  The output is as follows:

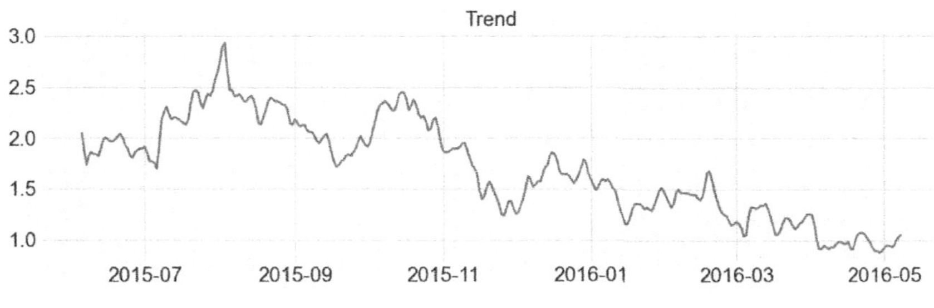

Figure 2.9: Additive decomposition trend component

- S_t: The seasonal component with regular and predictable patterns. It is typically observed when there are distinct repeated patterns observed with regular intervals: hourly, weekly, monthly, annually, and so on. Prominent examples include daily power consumption patterns or annual sales of ice cream:

```
plot_component(decomposition.seasonal, 'Seasonal', custom_palette[2])
```

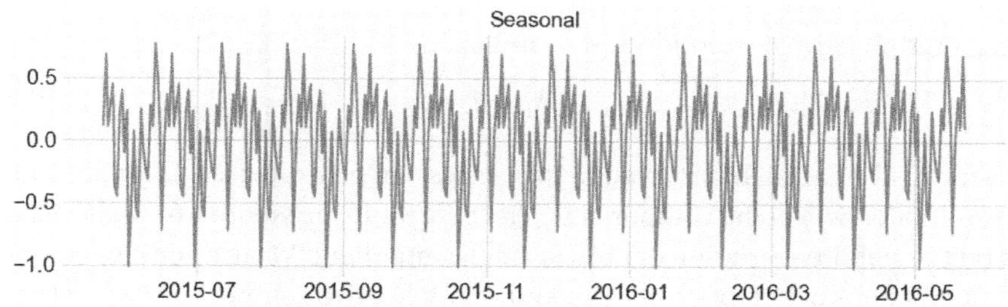

Figure 2.10: Additive decomposition seasonal component

- C_t (optional): Cyclical component that represents repetitive fluctuations that do not occur at fixed intervals. Cyclic patterns are often seen in business cycles and market demand. If we wanted to see this, we would need to use a more sophisticated approach, such as a Hodrick-Prescott model.
- I_t: The irregular component, which we also refer to as residuals. Residuals of a decomposition are the fluctuations in the time series that are left after removing trend and seasonal/cyclical variations:

```
plot_component(
    decomposition.resid, 'Residual',
    custom_palette[3], plot_type='scatter'
)
```

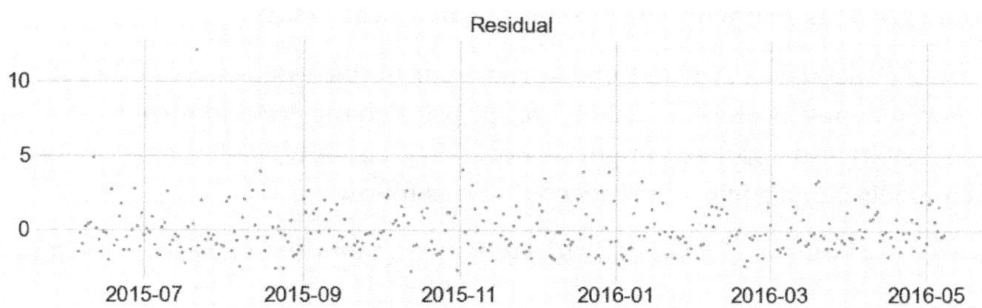

Figure 2.11: Additive decomposition residuals

If you prefer, it is possible to show all the components at once, with the following code. However, this does make it more difficult to see the patterns being pulled out of the data:

```
figure = decomposition.plot()
plt.tight_layout()
plt.show()
```

Next up, we look at multiplicative decomposition.

Multiplicative decomposition

As our data-generating processes can vary in nature, series decompositions may have different combinations of trends and seasonality. Depending on the nature of the trends and seasonality, a time series can be modeled as an additive or multiplicative time series.

This is the multiplicative decomposition formula:

$$X_t = T_t * S_t * I_t$$

What this means is that each observation in a series can be expressed as either a sum (additive) or a product (multiplicative) of the components. The case for using additive or multiplicative composition depends on the stability/variation of the seasonal component. We use additive composition when the seasonal fluctuations are stable/constant, and multiplicative composition when the seasonal fluctuations are dependent on the strength of the trend.

To think of this another way, let's think through two examples:

- **Additive seasonality**: Consider a staple product such as milk in a stable market. The difference between peak and trough sales remains relatively constant over time, regardless of any overall trend. If summer sales are typically 1,000 units and winter sales are 800 units, this 200-unit difference might remain consistent year after year, even if overall sales are growing or declining. This constant amplitude of seasonal fluctuations is indicative of additive seasonality.
- **Multiplicative seasonality**: Now, let's imagine an ice cream shop in a growing tourist destination. As the town becomes more popular over the years, not only do overall sales increase, but the difference between peak summer sales and winter sales grows larger each year. In the first year, summer sales might be 1,000 units and winter sales 200 units (a difference of 800 units). By the fifth year, summer sales could reach 5,000 units and winter sales 1,000 units (a difference of 4,000 units). This increasing amplitude of seasonal fluctuations as the overall trend grows is characteristic of multiplicative seasonality.

In more complex scenarios, we may encounter a mix of these patterns or other factors that influence our data. For instance, promotional activities or price changes could introduce additional variations. In such cases, a hybrid approach might be necessary, combining elements of both additive and multiplicative models. Such a hybrid model could be as follows:

$$Observed\ Value = (Trend\ \times\ Seasonality) + Promotional\ Effect\ +\ Random\ Noise$$

This formulation allows for multiplicative interaction between trend and seasonality, while promotional effects and random noise are treated as additive components.

In the following code let's look at an example of multiplicative decomposition

```
# Multiplicative decomposition

# We need to shift the values in the series due to 0s causing inf values
```

```
df_sliced['mice_imputed_shifted'] = df_sliced['mice_imputed'] + 1

mult_decomposition = sm.tsa.seasonal_decompose(
    df_sliced['mice_imputed_shifted'], period=28, model='multiplicative')

# Plot components
plot_component(mult_decomposition.observed,
               'Observed (Multiplicative)', custom_palette[0])
plot_component(mult_decomposition.trend,
               'Trend (Multiplicative)', custom_palette[1])
plot_component(mult_decomposition.seasonal,
               'Seasonal (Multiplicative)', custom_palette[2])
plot_component(mult_decomposition.resid,
               'Residual (Multiplicative)', custom_palette[3],
               plot_type='scatter')
```

The multiplicative approach doesn't actually do much for our data, so we haven't plotted it here.

Alternative decompositions

While additive and multiplicative decompositions are helpful for many time series, more sophisticated methods can offer deeper insights and handle complex data characteristics. These advanced techniques are particularly valuable when dealing with intricate seasonal patterns, multiple trends, or non-stationary variance. Some common approaches are as follows:

- **STL**: STL/ **Multiple Seasonal-Trend decomposition using LOESS** (**MSTL**) uses LOESS smoothers to estimate trend and season(s), making it versatile for various patterns and somewhat robust to outliers. Works well with clear seasonality.
- **Hodrick-Prescott** (**HP**) **filter**: We use these filters for separating trend and cyclical components, but not seasonality. It uses a λ penalisation to control smoothness, which helps us identify long-term trends.
- **Wavelet decomposition**: This is useful when we have multiple seasonal patterns or non-stationary variance. Wavelet transforms split signals with low or high-pass filters, capturing changing periodicities and bursts.
- **Trigonometric, Box-Cox transform, AutoRegressive Moving Average** (**ARMA**) **errors, trend, and seasonal components** (**TBATS**): This handles complex seasonality, including multiple seasonal periods, making it good for intricate time series data. It does this by using state-space forecasting decomposition with Box-Cox, damped trend, ARMA errors, and Fourier seasonalities.

STL and TBATS are perhaps more popular, but it is worth your time investigating these methods, as they provide a greater understanding of your data, revealing patterns and components that simpler methods might miss.

In the following code, we use a Hodrick-Prescott filter to build trends and cyclic patterns:

```
from statsmodels.tsa.filters.hp_filter import hpfilter

# Apply HP Filter
cycle, trend = hpfilter(df_sliced['mice_imputed'], lamb=1600)

# Plot
plt.figure(figsize=(CFG.img_dim1, CFG.img_dim2))
plt.subplot(211)
plt.plot(df_sliced['mice_imputed'], label='Original', color = custom_
palette[0])
plt.plot(trend, label='Trend')
plt.legend()
plt.title('Hodrick-Prescott Filter: Trend')

plt.subplot(212)
plt.plot(cycle, label='Cycle',color = custom_palette[1])
plt.legend()
plt.title('Hodrick-Prescott Filter: Cycle')
plt.tight_layout()
plt.show()
```

Here is our output:

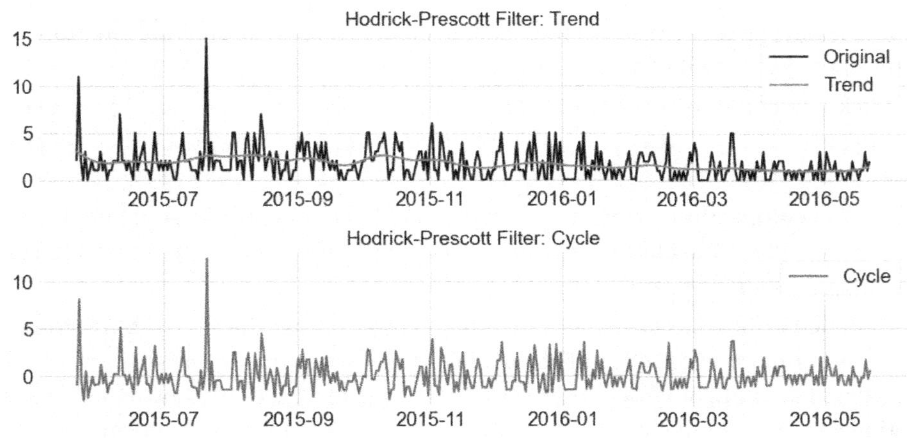

Figure 2.12: Hodrick-Prescott filter decomposition into trend and cycle

Let's look at some examples of decomposition now!

Examples of decomposition

Although we prefer to use "realistic" datasets, such as the M5 data, which is fairly typical in retail, its complexity doesn't showcase multiplicative decomposition so well. Every field of knowledge has particular datasets for demonstrating key concepts—clustering has Iris, differential equations uses Canadian lynx data, and forecasting statistics often uses the Johnson & Johnson quarterly earnings dataset. This dataset (1960–1980) demonstrates differences in decomposition:

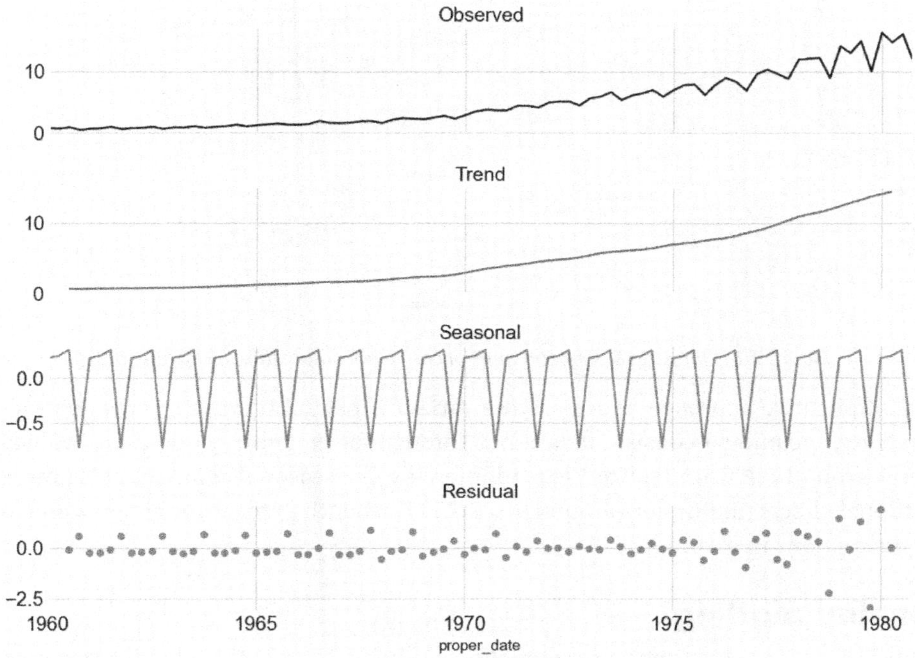

Figure 2.13: Johnson & Johnson additive decomposition (quarterly data)

Plotting our additive decomposition shows **heteroskedasticity** (variance grows over time). The residuals early in the series (1960–1965) are small and clustered around zero, but toward the end, we see much larger residuals, including extreme values around `-2.5`. This growing variance indicates that our additive decomposition failed to capture the relationship between trend and seasonality properly. Let's try a multiplicative decomposition.

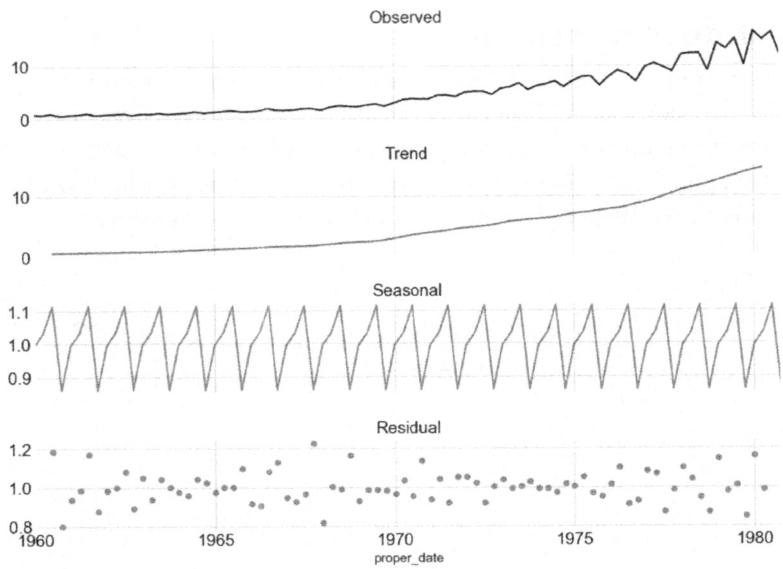

Figure 2.14: Johnson & Johnson multiplicative decomposition (quarterly data)

Using a multiplicative approach addresses the variance instability, with the residuals showing **homoskedasticity**, meaning a consistent variance throughout the entire series, with values scattered fairly evenly around `1.0` from 1960 to 1980. Additionally, the seasonal component is now interpreted as proportional effects (multipliers around `0.9` to `1.1`) rather than fixed additions, which reflects the seasonal fluctuations in earnings.

Seasonality plotting

Seasonality is a key factor in time series modeling, as the patterns act as an indication of the relationships in data over time. One way we can help ourselves understand seasonality is to break down the data into higher temporal resolutions and visualize them. Let's go back to our M5 retail data. This time, we will use `FOODS_3_714`, as it is a cleaner time series with less noise. We'll use this to demonstrate how box plots can be used to visualize different seasonal patterns:

First, we'll build a day of the week boxplot, with `showfliers=False`, which removes outliers. These can squash graphs, making it harder to see patterns. Note that we could use violin, hex, and beeswarm plots instead; these are useful to see more about the distributions with the same data:

```
plt.figure(figsize=(CFG.img_dim1, CFG.img_dim2))
sns.boxplot(
    x='dayofweek', y='sold',
    data=series_df, hue='dayofweek',
    palette=custom_palette, showfliers=False, legend=False
)
plt.xlabel('Day of Week', fontsize=18)
plt.ylabel('Sales', fontsize=18)
```

```
plt.xticks(rotation=45)
plt.tight_layout()
plt.show()
```

We get the following plot:

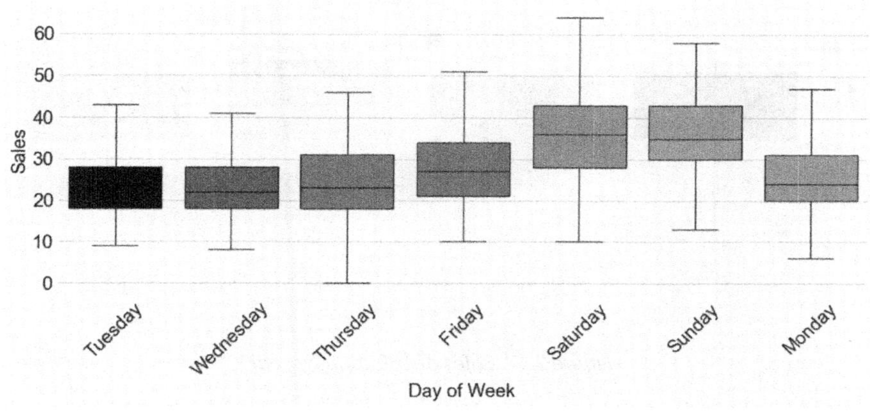

Figure 2.15: Sales distribution by day of week

Our boxplot shows how sales vary across days of the week, building up toward the weekend. We can quickly identify which days tend to have higher or lower sales, and how much variability (whisker sizes) exists within each day. We can do the same with `month`:

```
sns.boxplot(
    x='month', y='sold',
    data=series_df, hue='month',
    palette=custom_palette, showfliers=False, legend=False
)
```

Here is the output:

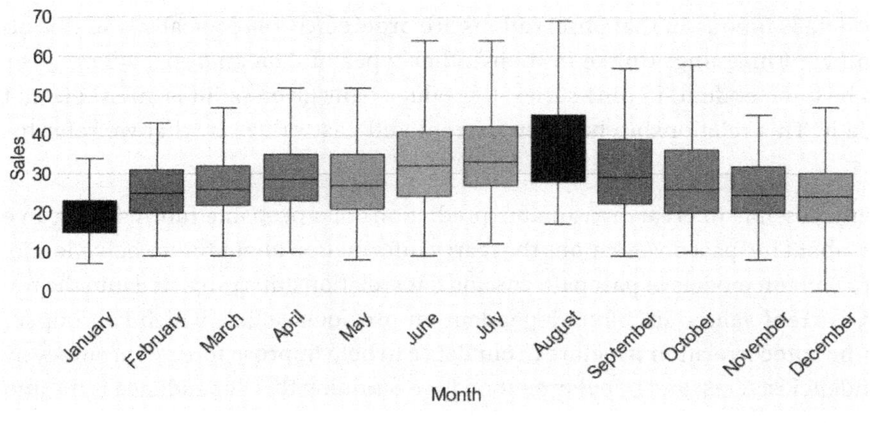

Figure 2.16: Sales distribution by month

This shows how sales demand varies by month. Here, it is likely related to the temperature (i.e., with June to August being summer). Less obvious here than in the previous data is a trend:

Figure 2.17: Sales distribution by year

We see little difference across the years, which indicates a fairly stable stochastic process. Boxplots can reveal more than the seasonality composition, potentially highlighting distributions across different temporal resolutions (i.e., days, weeks, months, etc.). This can be incredibly useful when building features and deciding on algorithms for modeling. They expand our earlier analysis by providing a different perspective on seasonality, and are particularly useful for investigating multiple seasonal patterns.

Beyond decomposition patterns

While decomposition provides valuable insights for understanding trend, seasonality, and residual components, there is another pattern that is fundamental to our understanding of time-series data: temporal dependence.

Dependence in time series

Time series data is unique in that observations are ordered chronologically, and this ordering often carries significant meaning. Unlike in many other types of data analysis, where observations are assumed to be independent, in time series, the value at any given point is often related to the values that precede it. This relationship between current and past values is what we refer to as **temporal dependence**.

It is only truly possible to create reasonable predictions of a probable future if we can extract some information about the past observations that carry information about future behavior; in other words, forecasts are built on models of past patterns and data relationships projected into the future. In many time series, current values are often dependent on previous values, which can impact forecasting. Identifying the autocorrelation structure in our data can help improve forecast accuracy by considering these dependencies. A fast and popular method for examining this dependence is the **autocorrelation** and **partial-autocorrelation** functions, which depend on what we refer to as **lags**. Before we formalize these, it might help to define lags. Lags are simply the values of a series (i.e., sales, energy units, etc.)

shifted. In our current `series_df`, we could say, let's look at lags of `sold` (sale units) from seven days ago (previous week). This is more easily understood by looking at the lags directly:

```
series_df['lag_1'] = series_df['sold'].shift(1)  # Lag of 1 day
series_df['lag_2'] = series_df['sold'].shift(2)  # Lag of 2 days
series_df['lag_3'] = series_df['sold'].shift(3)  # Lag of 3 day
series_df['lag_4'] = series_df['sold'].shift(4)  # Lag of 4 days
series_df['lag_5'] = series_df['sold'].shift(5)  # Lag of 5 day
series_df['lag_6'] = series_df['sold'].shift(6)  # Lag of 6 days
series_df['lag_7'] = series_df['sold'].shift(7)  # Lag of 7 day i.e. week

print(series_df[[
    'date','sold',
    'lag_1','lag_2','lag_3', 'lag_4','lag_5','lag_6','lag_7'
]].head(n=10))
```

```
        date  sold  lag_1  lag_2  lag_3  lag_4  lag_5  lag_6  lag_7
0 2013-01-01    17    NaN    NaN    NaN    NaN    NaN    NaN    NaN
1 2013-01-02    16   17.0    NaN    NaN    NaN    NaN    NaN    NaN
2 2013-01-03    24   16.0   17.0    NaN    NaN    NaN    NaN    NaN
3 2013-01-04    26   24.0   16.0   17.0    NaN    NaN    NaN    NaN
4 2013-01-05    21   26.0   24.0   16.0   17.0    NaN    NaN    NaN
5 2013-01-06    21   21.0   26.0   24.0   16.0   17.0    NaN    NaN
6 2013-01-07    16   21.0   21.0   26.0   24.0   16.0   17.0    NaN
7 2013-01-08    19   16.0   21.0   21.0   26.0   24.0   16.0   17.0
```

You can see we are simply moving the column across and down by the shifted value.

> Lags are extremely useful features for forecast models. But as you can see in the preceding table, adding lags creates NA (`NaN`) values; this is a problem in most models. There are ways of dealing with missing values, as we have discussed, but doing so can build relationships in your data that do not exist. The best approach is to cut the start of your dataset off. Of course, this shortens the data you have to work with. This is one reason why we don't tend to build models with, for example, 365 lags for 2 years' worth of data. That is not to say you can't, but you should be cautious in doing so.
>
> Another factor is computational cost. Building models with a large number of lag features can be computationally expensive, mainly because we need to build lags recursively as we make predictions. It can be better to create features that capture considerably large lags rather than cut your data short.

We can look at individual correlations of lags simply by building the features, as we did earlier for `series_df`, and plotting them in a scatter plot, as shown after the following code:

```
fig, axs = plt.subplots(2, 2, figsize=(12, 12))
fig.tight_layout(pad=4.0)

lags = [1, 3, 5, 7]
colors = [custom_palette[i] for i in range(4)]  # Use first 4 colors from
custom_palette

for i, (lag, ax, color) in enumerate(zip(lags, axs.ravel(), colors)):
    ax.scatter(series_df['sold'], series_df[f'lag_{lag}'],
               alpha=0.5, color=color)
    ax.set_xlabel('Current Sales')
    ax.set_ylabel(f'Sales {lag} day{"s" if lag > 1 else ""} ago')
    ax.grid(True)

plt.show()
```

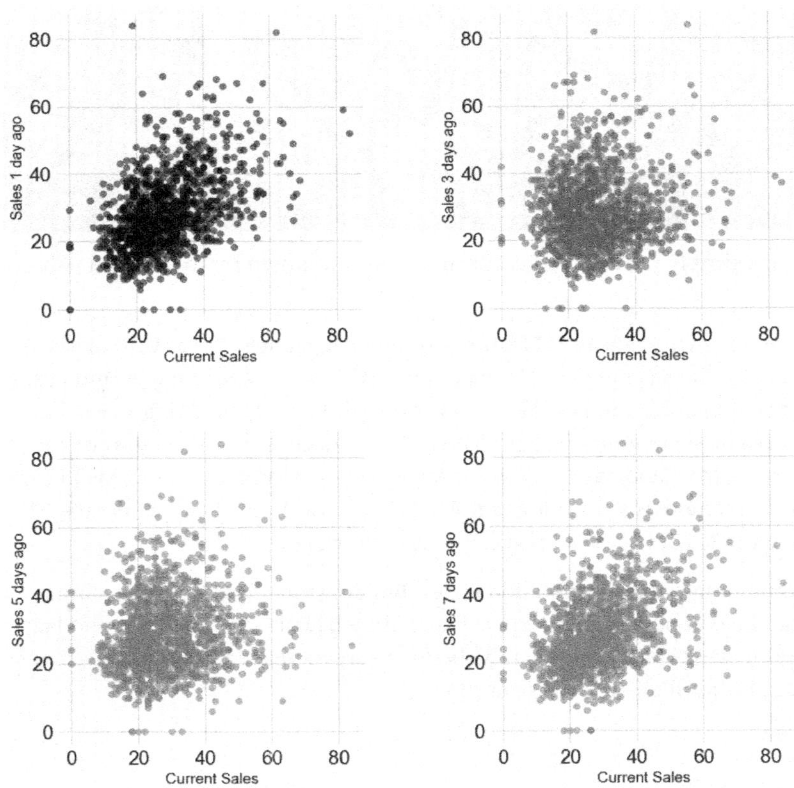

Figure 2.18: Scatter plots of sales against lagged features

The relationship between the lagged features and sales (`series_df['sold']`) is largely positive, especially for lags 1 and 7. However, for lags 3 and 5, the relationship becomes less apparent, as the data points are more widely spread. Now that we understand what lags are and how they can correlate with the original series, let's formally define autocorrelation:

- **Mean function of a time series:** Represents the expected value of the time series at time *t*:

$$\mu_t = E[X_t]$$

- **Autocovariance function:** Measures how the series covaries with itself at different time points:

$$\gamma(s,t) = Cov(X_s, X_t) = E[X_s X_t] - E[X_s]E[X_t]$$

- **Autocorrelation formula:** Normalizes autocovariance, providing a measure of linear dependence between observations at different time lags:

$$\rho(u, t+u) = Cor(X_u, X_{t+u}) = \frac{Cov(X_t, X_{t+u})}{Var(X_t)Var(X_{t+u})}$$

Revealing the patterns of autocorrelation helps us with the following:

- **Feature engineering**, particularly when we're creating new input variables that represent past values of a time series, known as lagged features
- Determining optimal **window sizes** for window approaches to data inputs (i.e., moving averages and tensor structures)
- Guiding **preprocessing** steps and data transformations (i.e., difference or detrend)
- Selecting appropriate model **architectures and hyperparameters** (i.e., number of layers in a **recursive neural network** (**RNN**))
- **Analyzing model residuals** to ensure all temporal structure has been captured

Autocorrelation function

The two most widely used approaches for examining temporal dependence are the **autocorrelation function** (**ACF**) and **partial autocorrelation function** (**PACF**). These functions provide a quantitative measure of the relationship between an observation and its lagged values.

The ACF measures the correlation between a time series and its lagged versions. Continuing with the same M5 sales data, let's take a look at the autocorrelations with the `plot_acf` function from `statsmodels`. We use 40 lags for the ACF, which, in this case, translates to 40 days due to the frequency of the data:

```
from statsmodels.graphics.tsaplots import plot_acf
fig = plot_acf(series_df['sold'], lags=40, color=custom_palette[2])
plt.xlabel('Lag/Shift', fontsize=18)
plt.ylabel('Correlation Coefficient', fontsize=18)
plt.tight_layout()
```

Here's our output:

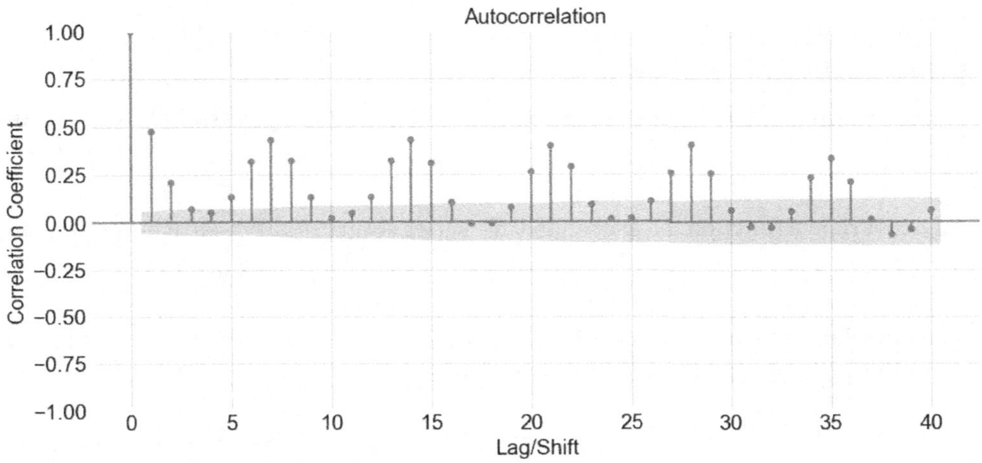

Figure 2.19: ACF of sales

In the ACF plot in *Figure 2.19*, the *x* axis represents the number of lags (i.e., time periods), which are 40 in this case, between the original values in the time series and lagged versions of itself. On the *y* axis, we have the strength of the correlation between the two series, with a strong positive correlation being `1` and a strong negative being `-1`. The blue vertical bars represent the strength of the correlation for each lag away from the unlagged data.

Correlations that extend beyond the blue shaded area are considered statistically significant. This area represents a confidence interval, typically 95%. You can adjust this and pull out a table of values instead if you prefer.

From a production standpoint, it's better to do these calculations directly, but you can pull them from `statsmodels` like so:

```
from statsmodels.tsa.stattools import acf, pacf
autocorrelations = acf(series_df['sold'], nlags=40, fft=True)
autocorr_df = pd.DataFrame({'Lag': range(len(autocorrelations)),
                            'Autocorrelation': autocorrelations})
print(autocorr_df.head(10))  # Show first 10 lags
```

In our data, you'll notice a spike at lag 0, which is always `1` and represents a perfect correlation with itself. We also see a repeating wave pattern every seven lags, suggesting a weekly pattern, although the correlation generally decreases as we look further back in time.

Now let's look at the PACF.

Partial autocorrelation

Partial autocorrelation measures the correlation between a time series and its lags after controlling for the effects of intermediate lags. It is defined as follows:

$$\phi(u) = Cor(X_t, X_{t+u} | X_{t+1}, \dots, X_{t+u-1})$$

Let's look at the code for generating a PACF plot:

```
from statsmodels.graphics.tsaplots import plot_pacf
plot_pacf(series_df['sold'], lags=40, color=custom_palette[2])
plt.xlabel('Lag/Shift', fontsize=18)
plt.ylabel('Correlation Coefficient', fontsize=18)
plt.tight_layout()
```

Here's what the PACF plot looks like:

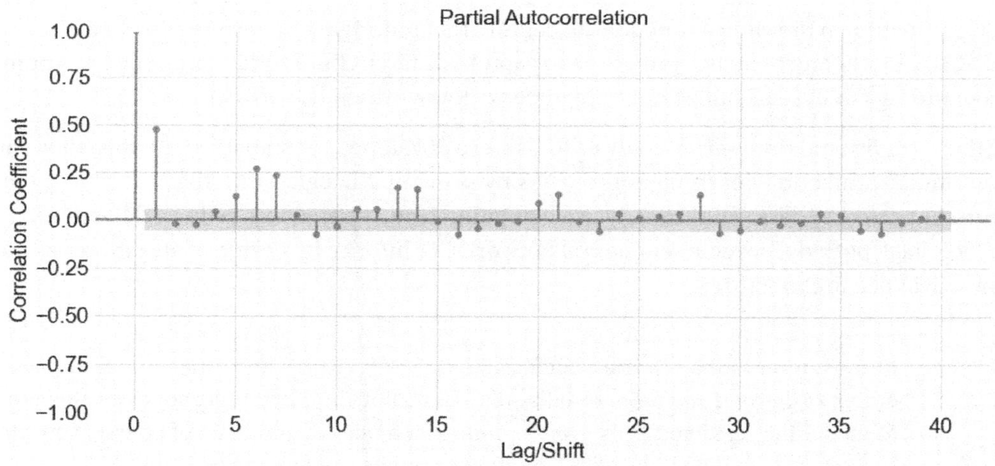

Figure 2.20: PACF of sales

PACF builds on autocorrelation by removing the influence of any correlations at shorter lags, which helps us identify the appropriate lags to use when modeling. Essentially, PACF shows us the correlation among residual values after removing information that is "explained" by earlier lagged values, which makes the lags in a PACF plot *partial* and not *complete,* as we remove already found variations before we find the next correlation.

Our plot reveals a spike at lag 1, confirming the strong relationship with the previous day, and smaller spikes at lags 7 and 14, hinting at weekly influences.

Any significant spikes at lag k in a PACF plot suggest that a lagged term should be used. We tend to use this for selecting **AutoRegressive** (**AR**) terms because we don't want to keep too many correlated features in a dataset; we need to retain only the relevant features. This helps reduce computational processing and issues with multicollinearity. PACF is particularly useful for identifying the order of an AR model.

For example, if the PACF plot shows significant spikes at lags 1 and 2, and then decays quickly, it indicates that an AR(2) model might be appropriate for the time series data. In contrast, a gradual decay in the PACF plot may suggest that a **moving average** (**MA**) process is more suitable.

An intuitive way to think about the difference is that ACF at lag k measures a linear dependence between X_t and X_{t+k}, while PACF captures the dependence between those values, correcting for all the intermediate effects.

It's worth noting that the interpretation of ACF and PACF plots assumes the time series is stationary. For non-stationary series, differencing or other transformations may be necessary before analyzing these plots.

Applications in deep learning

In deep learning, we can use information about decomposition and lags to guide architectures and feature engineering. Creating lagged values often helps models learn relationships between the current and past observations. Seasonality and trend components can also be configured for some models, allowing us to inform the choice functions to capture the underlying patterns in the data. For example, the N-BEATS architecture can be configured to capture different time series components. Another use is to take the lags to deseasonalize our data once we know which lags to use.

While these traditional time series analysis tools were developed for statistical models, they can complement modern machine learning approaches by providing insights into the underlying temporal structure of temporal data. Additionally, we can guide feature selection, help in determining appropriate look-back periods for recurrent neural networks, or inform the design of attention mechanisms in more advanced architectures.

> In most forecasting libraries, such as Darts, Nixtla, Fable, and ModelTime, we have the option to detrend and deseasonalize our data before building more sophisticated models to model the residuals. This is often done in a modeling pipeline; of course, you can do this separately, but you need to add the decomposed elements back to a forecast before conducting predictions.
>
> We can't always use this; deseasonalization, in particular, is not always advised or possible if one has strong exogenous variables in the past/future covariates.

Other features

We often have many exogenous variables available for modeling, and this can help us explore the relationships between these variables and our target variable. A simple approach that can be helpful is an **ordinary least squares** (**OLS**) regression. More critically, OLS assumes independent errors, but time series data exhibit autocorrelation where residuals are correlated across time. This violates a key assumption and can lead to overly optimistic significance tests, causing us to overestimate the strength of relationships.

A summary from OLS shows us which features have significant relationships with sales, and the R-squared value indicates how much variance in the target variable (`sold`) these features explain. But it's crucial to interpret these results cautiously. The presence of autocorrelation in time series data can lead to biased standard errors and potentially misleading significance tests. This simple model doesn't account for lagged effects or complex seasonal patterns that might be present in the data. Despite these limitations, this exploration can guide feature selection and highlight potential relationships that more sophisticated time series models could exploit:

```
                            OLS Regression Results
==============================================================================
Dep. Variable:                   sold   R-squared:                       0.476
Model:                            OLS   Adj. R-squared:                  0.465
Method:                 Least Squares   F-statistic:                     40.53
Date:                Wed, 27 Aug 2025   Prob (F-statistic):          3.57e-148
Time:                        11:27:42   Log-Likelihood:                -4332.3
No. Observations:                1231   AIC:                             8721.
Df Residuals:                    1203   BIC:                             8864.
Df Model:                          27
Covariance Type:            nonrobust
==============================================================================
```

According to the output, the OLS explains about 47.6% of the variance in sales (`R-squared: 0.476`), which is moderate.

Post-hoc comparisons show the `lag_1` variable (previous day's sales) to be highly significant (p<0.001) with a positive coefficient, indicating strong day-to-day sales correlation. The `year` variable is also significant (p<0.05), suggesting a slight upward trend over time. Interestingly, `sell_price` doesn't show a significant relationship with sales, which might be counterintuitive. The `week_of_year` variable is marginally significant (p<0.1), hinting at possible seasonal effects. The large condition number (`7.48e+06`) indicates severe multicollinearity, particularly from the lagged variables, which are naturally correlated with each other (yesterday's sales predict today's, which predict tomorrow's). This makes it difficult to isolate individual lag effects and can lead to unstable coefficient estimates.

While OLS provides initial insights despite its limitations, we can supplement this analysis with other exploratory approaches:

- **Plot features alongside the target variable:** Visualizing relationships can often reveal patterns that are not immediately obvious from statistical metrics alone.
- **Correlational analysis:** Spearman's is a good choice if we expect non-linear relationships.
- **Feature importance with a tree-based model:** Random forests provide feature importance scores (SHAP values) based on how much each feature contributes to decreasing node impurity across trees in the model. But these results should be interpreted cautiously, as feature importance doesn't directly translate to predictive value when features are removed from the model.
- **Regularized regression:** Techniques such as Ridge regression (L2 regularization) or Lasso (L1 regularization) can help with feature selection.
- **Granger causality:** This test can help determine whether one time series is useful in forecasting another.

- **Information criteria**: Both **Akaike information criterion** (**AIC**) and **Bayesian information criterion** (**BIC**) can be used to compare models with different feature sets.

The problem with all of these approaches is that they fall short in one way or another. The best approach is to build experiments that use models with and without different permutations of the features in your data, and to do so for many time series that are similar. We'll discuss this in the chapter on time series features.

Stationarity

Stationarity in time series occurs when statistical properties don't depend on the time of observation. A stationary time series has an unconditional joint probability distribution that doesn't change when shifted in time. This means parameters such as mean and variance remain constant over time. The starting point for this is often to look at examples of stationary and non-stationary series, starting with Gaussian white noise.

In the following code and figure, we plot examples of stationary and non-stationary series, starting with the simplest non-trivial stationary series, which is Gaussian white noise:

```
series = pd.DataFrame(
    data = np.random.normal(0, 1, 10000),
    columns = ['noise']
)
series.plot()
print()
```

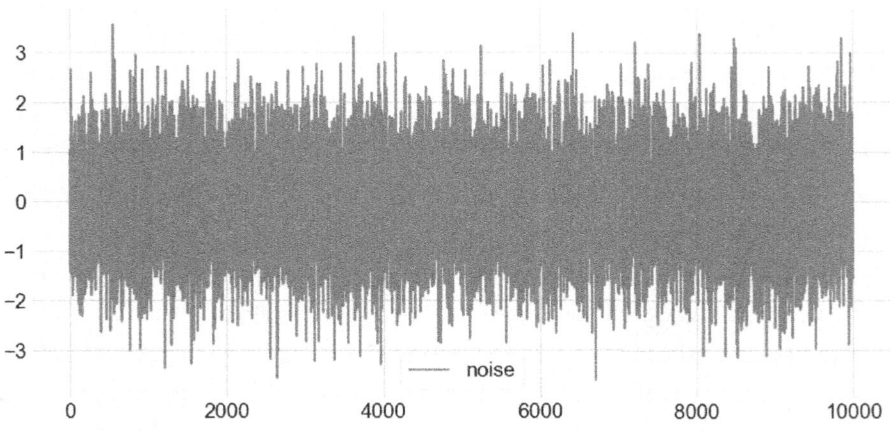

Figure 2.21: Gaussian white noise time series

This does not seem very interesting, does it? Let's build an autocorrelation plot:

```
plot_acf(series['noise'], lags=25, color=custom_palette[2])
plt.xlabel('Lag/Shift', fontsize=18)
plt.ylabel('Correlation Coefficient', fontsize=18)
plt.tight_layout()
```

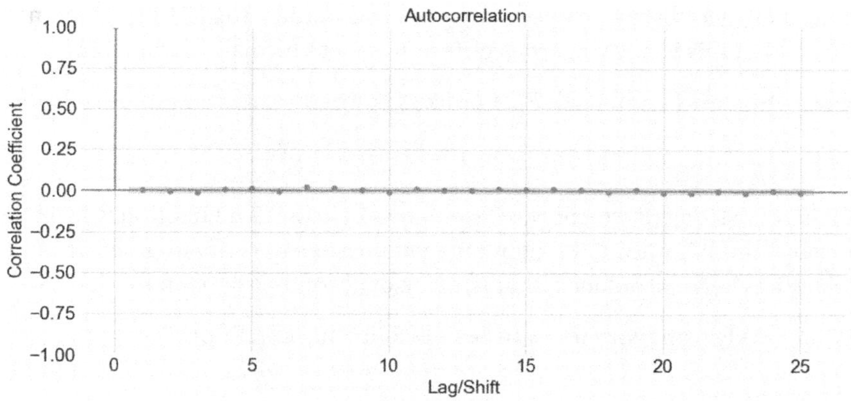

Figure 2.22: Autocorrelation function of white noise

You can see that there is no significant autocorrelation in the series, which is perfect. Why is it perfect? Because it has no underlying pattern. White noise is a perfect example of a stationary series because its statistical properties (mean, variance, and autocorrelation) remain constant over time.

The air passengers dataset is a good toy set for demonstrating non-stationarity to stationarity, due to its strong trend. Examining its distribution with a histogram shows that the data is not normally distributed because of its strong seasonal and trend components.

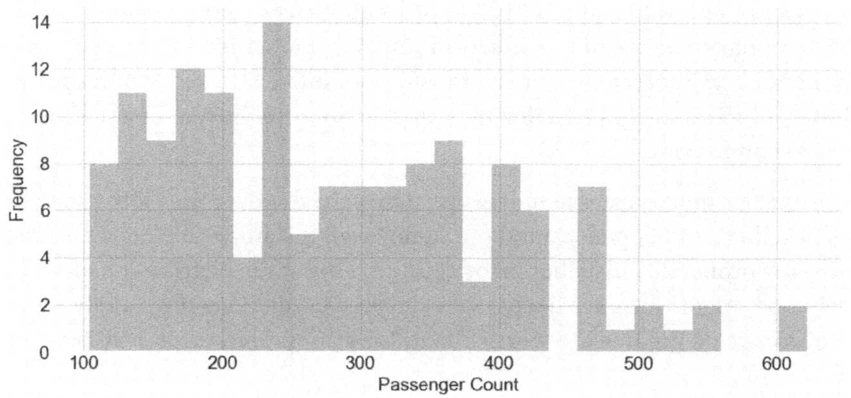

Figure 2.23: Distribution of airline passenger counts

Passengers' mean and variation split check

Splitting the series into two contiguous parts and computing summary statistics such as the mean, variance, and autocorrelation can also be helpful. If the statistics are quite different, then a series is unlikely to be stationary:

```
mean:
chunk1: 182.90 vs chunk2: 377.69
variance:
chunk1: 2244.09 vs chunk2: 7367.96
```

Here, the values between the two datasets are clearly different, suggesting strong non-stationarity (which it is). However, this is not a rigorous criterion, so let's formally define things.

Given a stochastic process $\{X_t\}$ and cdf F_X, a process is strictly stationary if the following applies:

$$F_X(X_{t_1}, \ldots, X_{t_n}) = F_X(X_{t_1+\tau}, \ldots, X_{t_n+\tau})$$

The definition is very strong (if we know a **Cumulative Distribution Function** (**CDF**) of a distribution, we can infer everything); however, it is generally not possible to verify in practice. For that reason, a less strict variant has been introduced, that is, weak or second-order stationarity.

A process $\{X_t\}$ is weakly stationary if it satisfies the following conditions:

- $\mu_t = \mu_{t+\tau}$
- $\gamma(s,t) = \gamma(s-t)$
- $Var(X_t) < \infty$

A weakly stationary process has constant mean, constant variance, and autocovariance that depends only on lag, not on time:

$$\rho(u) = \frac{\gamma(u)}{\gamma(0)}$$

In practice, most time series are non-stationary (meaning their properties change over time), so why then is stationarity important? Many statistical procedures used in time series analysis rely on stationarity; non-stationary data are often transformed to become stationary. A trend-stationary process is not strictly stationary, but it can be easily transformed into a stationary process by removing the underlying trend, which is solely a function of time. The same holds true for a stationary process with an added cyclical component.

The core idea is that it's much easier to model dynamic behavior over time if the statistical properties do not change; oscillations happen around the same level, the amplitude does not change too much (in other words, the probability distribution of X_t is the same as the distribution of X_{t+h}); such models are well understood. Algorithms are likely to yield better predictions if we apply them to stationary processes, because we do not need to worry about, for example, concept drift between our training and test sets.

If we are dealing with a process that does not adhere to those characteristics, we can either try to capture them directly or transform it in such a manner that it can be considered stationary.

So, how can we turn a non-stationary series into a stationary one? We transform the data. Common approaches to transformations are as follows:

- **Differencing:** Taking the difference between each observation and the previous observation.
- **Detrending:** Removing the trend by fitting a model and subtracting.
- **Log transformations:** Stabilizing the variance of time series, which can often bring us closer to stationarity.
- **Power transforms:** Raising the value of observations to a power (greater than 1) also stabilizes variance. This is usually better tailored to the data, but comes at a computational cost.

While non-stationary time series are typically more challenging to forecast, particularly with statistical models, machine learning models such as gradient boosted decision trees and deep learning models have a large number of parameters, which in theory can adapt to any properties of the data. Consequently, deep learning models do not need to assume stationarity; the downside to this is that the models not trained on stationary data may learn less about the inner properties of the DGP. This is another factor that, in practice, we have to experiment with.

Tests for stationarity

While inspecting plots before/after transformation can be helpful to assess the presence of trends or seasonalities (as we did previously with the air passengers dataset), in practice, we need a more formal approach, such as testing a hypothesis. The most popular tests dealing with stationarity are the following:

- **Augmented Dickey-Fuller** (**ADF**)
- **Kwiatkowski–Phillips–Schmidt–Shin** (**KPSS**)
- **Philips-Perron** (**PP**)

In our case, we only look at KPSS.

Applying KPSS

The KPSS test is used to test for stationarity in a time series. It determines how strongly a time series is defined by a stationary process versus a unit root process (non-stationary). The null hypothesis (H0) of the test is that the time series is stationary. The alternative hypothesis (H1) is that the time series has a unit root (i.e., it is non-stationary). We'll try this out on our data:

```
from statsmodels.tsa.stattools import kpss

X = series.passengers.values
result = kpss(X)
print('KPSS Statistic: %f' % result[0])
print('p-value: %f' % result[1])
print('Critical Values:')
for key, value in result[3].items():
    print('\t%s: %.3f' % (key, value))
```

```
KPSS Statistic: 1.651312
p-value: 0.010000
Critical Values:
    10%: 0.347
    5%: 0.463
    2.5%: 0.574
    1%: 0.739
```

Clearly, the air passengers data is not stationary before any transformations. Specifically, our KPSS test result suggests the data is not stationary; the test statistic (1.651312) is much larger than all critical values, with a p-value at `0.01`. This tells us the statistical properties of air passengers are changing over time, likely (from visualization) due to trends or seasonality. Transformations are, therefore, necessary before further analysis or modeling.

We can see how transformations change this. First, we'll create a KPSS and plotting function:

```
series.set_index('date', inplace=True)

def kpss_test(series):
    result = kpss(series.dropna())
    return f'KPSS Statistic: {result[0]:.3f}, p-value: {result[1]:.3f}'

def plot_transformation(original, transformed, title):
    plt.figure(figsize=(12, 6))
    plt.plot(original, color=custom_palette[0], label='Original')
    plt.plot(transformed, color=custom_palette[1], label='Transformed')

    original_kpss = kpss_test(original)
    transformed_kpss = kpss_test(transformed)

    plt.title(f'{title}\nOriginal: {original_kpss}\nTransformed: {transformed_
kpss}')
    plt.legend()
    plt.tight_layout()
    plt.show()
```

Starting with differencing, which has a few options, such as which periods to shift the values from. We'll just use the default (previous value). You can see that after differencing; the transformed series has a much lower KPSS statistic of `0.024` and a higher p-value of `0.100`. We can interpret this as evidence for rejecting the null hypothesis in favor of the alternative, meaning the series is stationary, although you should notice that there is significant variance in the data.

Starting with differencing, we shift the values by one period (the default) to remove trends. The following code shows how to difference the series and plot the transformation:

```
diff_series = series.diff().dropna()
plot_transformation(series, diff_series, 'Original vs Differenced Series')
```

The resulting plot (*Figure 2.24*) compares the original airline passengers series with its differenced version.

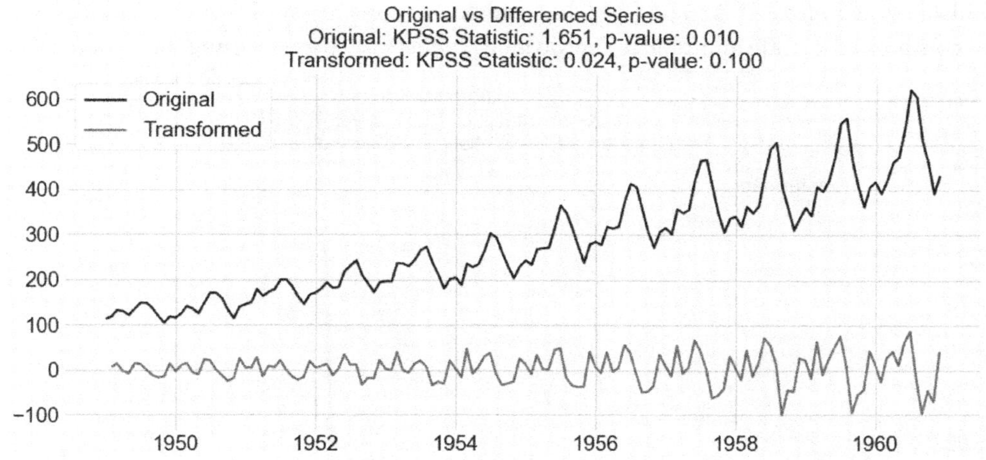

Figure 2.24: Airline passengers' series vs. differenced series

We simply extracted the trend using SciPy's signal. You could detrend an OLS regression model, which is useful if you need to predict later on. For example, say you were validating a model, having the true trend would be a form of data leakage in testing, so you would need to predict the trend for "testing data":

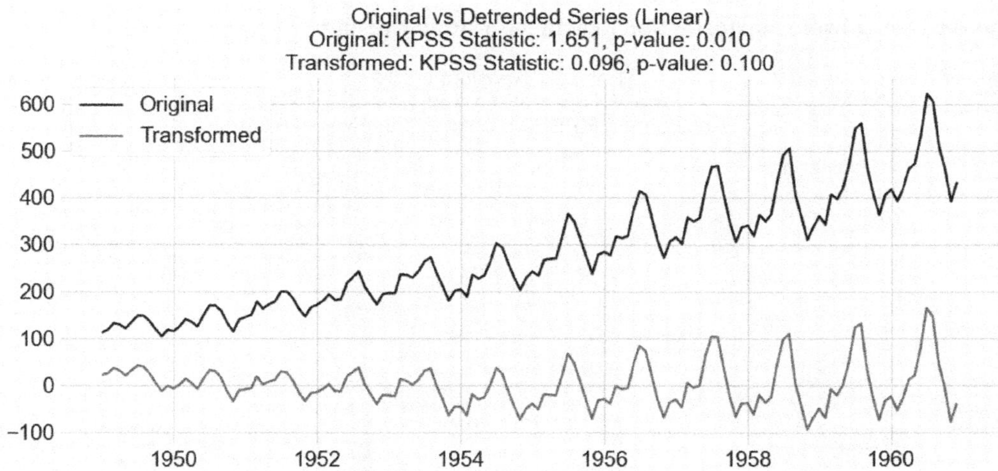

Figure 2.25: Airline passengers original vs. detrended series

Although the differenced data looks quite different, the result is similar: we obtain a much lower KPSS statistic of `0.096` (though not as low as with differencing) and a matching p-value of `0.100`, suggesting stationarity. However, as with the previous detrending method, we still observe a lot of variance in the data.

Economists often address this issue using log transformations. By applying the logarithm to the values, we compress the scale and reduce the variance, effectively transforming the data to a log scale:

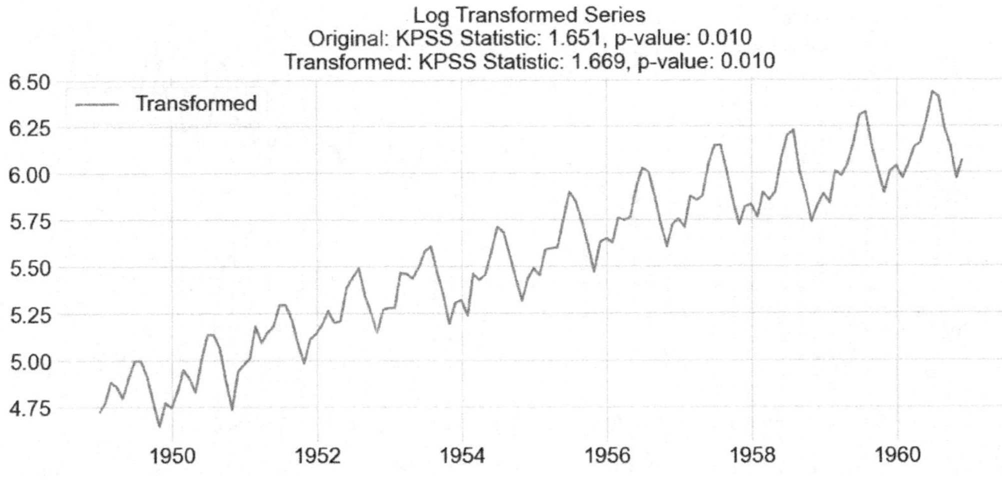

Figure 2.26: Airline passengers log-transformed variance

Looking at the right axis, we can see that the passengers' values change to lower values, and the variance in the seasonal waves is not as pronounced. This doesn't do much for stationarity. We can also use the Box-Cox power transformation to achieve a similar result:

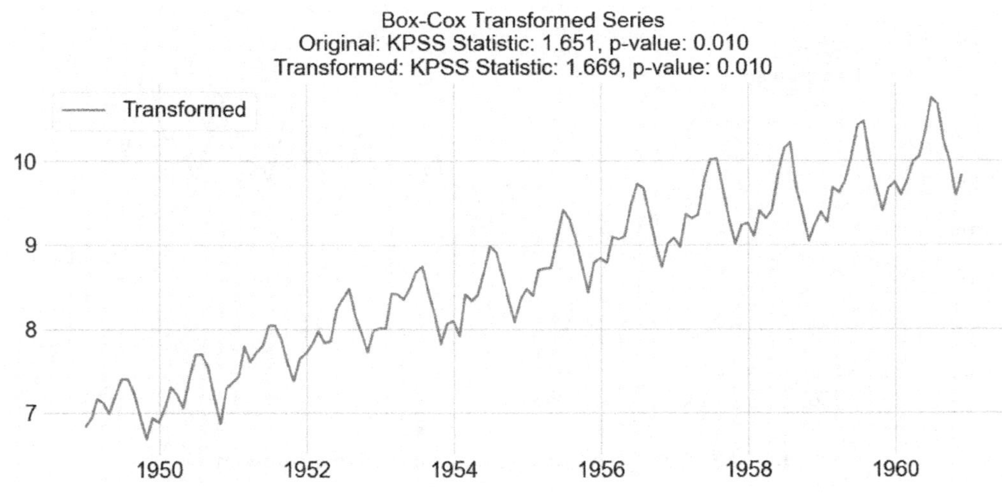

Figure 2.27: Airline passengers Box-Cox transformation

The results aren't too different visually, but there are differences; the log may appear more similar. The scale is slightly different.

Now, let's combine the variance transformation and trend removal, in that order, to account for the changing variance over time properly. Here, we apply the Box-Cox transformation and then detrend, but you can try any combination:

```
bc_series, lambda_param = stats.boxcox(series.values.flatten())
bc_series = pd.Series(bc_series, index=series.index)

bc_detrend_series = pd.Series(signal.detrend(bc_series.values.flatten()),
index=bc_series.index)

plt.figure(figsize=(12, 6))
plt.plot(bc_detrend_series, color=custom_palette[2], label='Transformed')
original_kpss = kpss_test(series)
transformed_adf = kpss_test(bc_detrend_series)
plt.title(f'{'Box-Cox and detrended Transformed Series'}\nOriginal: {original_
kpss}\nTransformed: {transformed_kpss}')
plt.legend()
plt.tight_layout()
plt.show()
```

Here's the output:

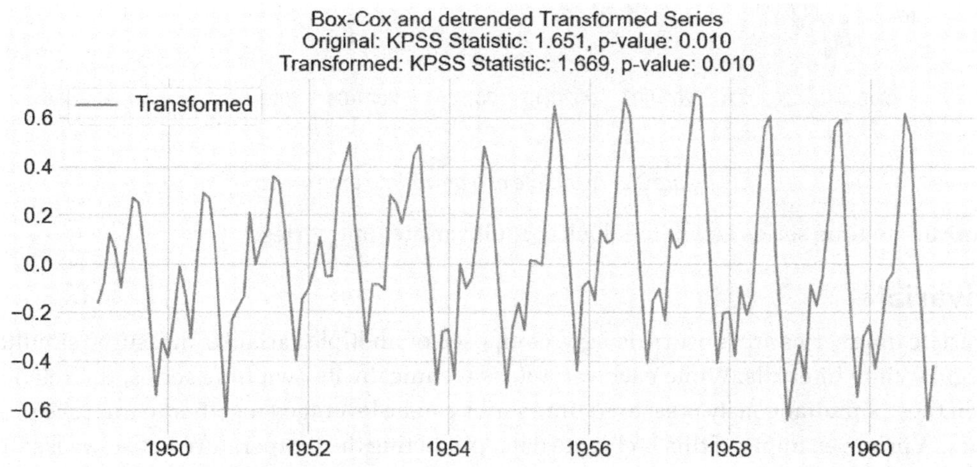

Figure 2.28: Airline passengers power transform stabilizing variance

Hmm, not stationary, but the data oscillates around a constant mean (approximately zero) with no obvious trend. This is caused by the KPSS being sensitive to seasonal fluctuations, which are not removed by these transformations. We actually want to model this remaining data. In addition, there is heteroscedasticity despite the Box-Cox transformation. We could use a different stationary test or different transformations. The main point here is that we should be careful when relying on tests, especially for automated decision-making.

Time series structures

Finally, we should discuss different time series structures. Like all data, time series come in a wide variety of structures. We tend to deal with two types of time series: univariate and multivariate, and extensions on these (i.e., panel data).

Univariate

A univariate time series is the simplest structure, comprising a single variable (sometimes referred to as a component) measured at regular intervals of time. We commonly see this in stock prices, temperatures, and sales figures. Most time series with a single dependent variable can be treated as univariate, which can be advantageous when additional information (i.e., news sentiment) serves to decrease the quality of predictions.

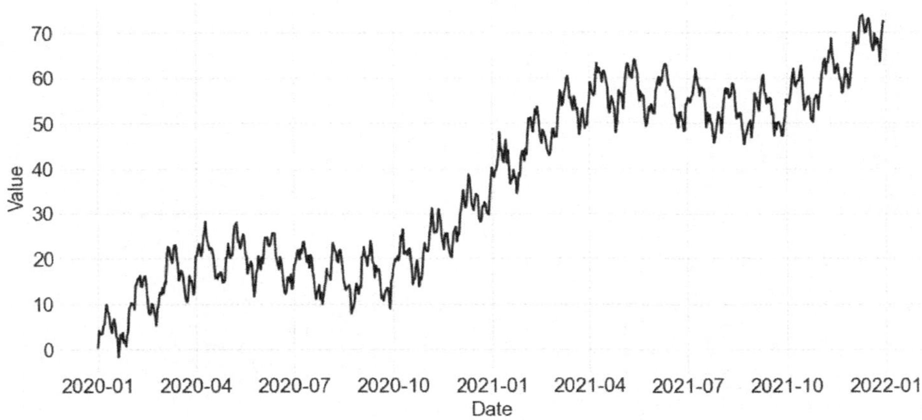

Figure 2.29: Example of a univariate time series

With univariate time series down, let's look at multivariate time series.

Multivariate

Multivariate time series are, unsurprisingly, composed of multiple variables measured simultaneously at regular time intervals. While each variable is technically its own time series, the relationships between these simultaneously measured time series can be leveraged together to improve predictive accuracy. A prime example of this is climate data; predicting the temperature in two weeks would be improved if we used temperature, cloud data, precipitation, and wind speed together. We might, in the case of multivariate time series, wish to predict multiple dependents/targets, or use the additional variables as exogenous variables/covariates, meaning variables that support the model's predictions. An obvious example would be using price and promotional conditions to predict sales of coffee.

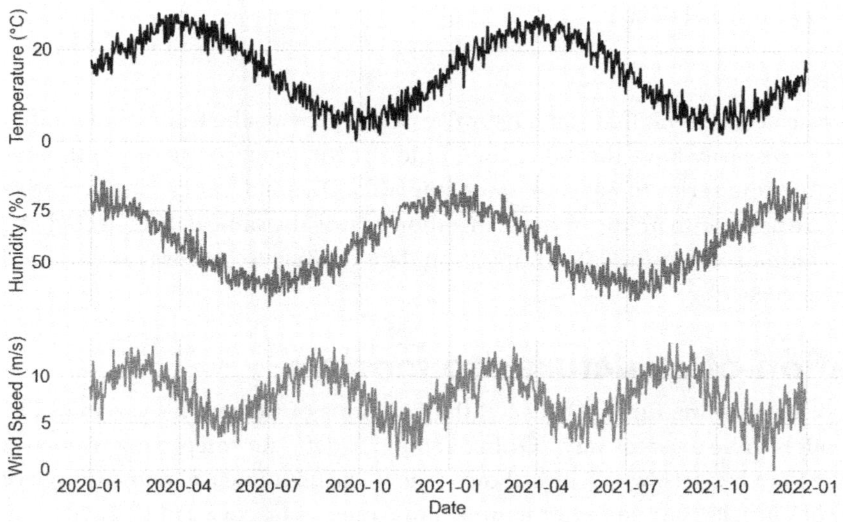

Figure 2.30: Example of a multivariate time series

We will move on to multivariable time series next.

Multivariable

Occasionally, we may also see the term *multivariable time series,* which refers to a scenario where multiple variables are measured at different time intervals. Analyzing and forecasting multivariable time series requires specialized techniques to account for the differing time steps and relationships between variables. We do not discuss this in this book; typically, if a variable is not in the same interval, we can up- or down-sample the data.

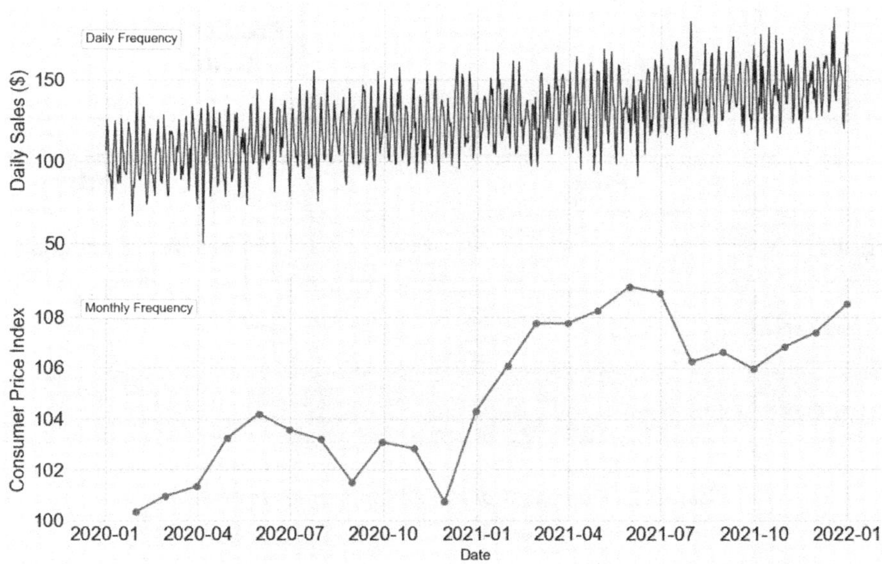

Figure 2.31: Example of a multivariable time series with different frequencies

Next, we will look at panel data.

Panel data

It is worth noting at this point that some neural architectures can be trained on multiple time series, which do not necessarily have the same time index or intervals/frequency, although typically we would expect the time series to have the same intervals. Demand forecasting is a prime area where such datasets are used, with many products in a shop having the same data collected. A good example of such data can be seen in the M5 competition data on Kaggle (`https://www.kaggle.com/c/m5-forecasting-accuracy`).

Application of structures to models

We find that distinguishing further between the functions of time series variables by how they are utilized in models to be a useful way to think about our data. An interesting framework is given by the forecasting library Darts, which also allows us to make use of neural networks out of the box. The Darts framework breaks time series structures into targets, past covariates, and future covariates.

After a time index, the first element of a time series to consider is our dependent variable, which is often referred to as a target or Y. This is the variable(s) we want to forecast. In the Darts framework, we can then break exogenous variables (any variable that is used to inform model building) down into past covariates and future covariates. Past covariates are covariates that a model uses only as a means to inform the initial model build; this is used for model builds that do not have covariates supporting the forecast build. For example, **Long Short-Term Memory** (**LSTMs**) and N-BEATS only use past covariates. If we were using one of these models to predict sales, we might use price to help, but they would not be able to use the prices we plan to use in the future. Future covariates would take the past data for a variable, and the expected or planned future of the variable; in this case, past prices and the planned prices would be one variable.

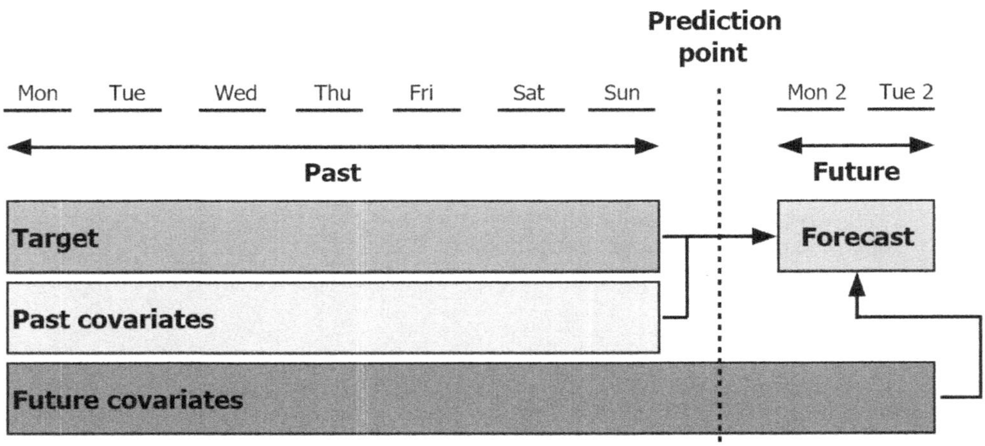

Figure 2.32: Framework of variable application to modeling

Future covariates can be extremely powerful when we have a reasonable expectation that covariates will have a significant influence on future values of the target.

An additional type of covariate to consider, which we often see in the panel form, are static covariates, which provide time-independent information about a target. Static covariates include, for example, product IDs, location IDs, region names, department names, population data, or average temperature of a region, which do not change over the course of the series. They are particularly useful when building models with multiple series, as incorporating time-independent information helps models to contextualize the underlying series being modeled. This allows the model to attribute parameters that can affect predictions of the target variable, which may enhance our forecasting accuracy and our confidence in it.

Summary

This was a bit of a whirlwind tour for time series, and there is far more to learn than we can cover in a single chapter.

We have covered the basic structure of the time series and how we can infer a stochastic process from it, which enables us to make forecasts. We also looked at the fundamental components of time series, and learned how and why we break down our data, and how that can help us to understand it. Finally, we looked at the data structures that time series come in. We covered the definition and components of time series data, including trend, seasonality, and residuals, along with a discussion on handling missing data and irregularities in time series. We explored different time series structures, such as univariate, multivariate, and panel data, and introduced decomposition techniques for understanding underlying patterns in time series. We also looked at stationarity in time series analysis and methods for achieving it, or not, followed by an overview of autocorrelation and partial autocorrelation functions for identifying temporal dependencies. Finally, we briefly touched on time series feature engineering.

While this chapter provided an introduction to time series analysis, there's far more than we could fit in here. We encourage you to explore this further, especially in areas such as advanced decomposition methods, handling complex seasonality, and specialized techniques for specific types of time series data. As we stated earlier, for those who want a more in-depth introduction, we recommend Rob Hyndman and George Athanasopoulos's *Forecasting: Principles and Practice*, third edition (`https://otexts.com/fpp3/`).

Recommended reading

- *Brockwell and Davies, Time Series: Theory and Methods*
- *Shumway and Stoffer, Time Series Analysis and Its Applications*
- *Durbin and Koopman, Time Series Analysis by State Space Methods*
- *Just about anything written by Rob Hyndman*: `https://robjhyndman.com/`
- *Ross Ihaka, Time Series Analysis:* `https://www.stat.auckland.ac.nz/~ihaka/726/notes.pdf`

- *Forecasting: principles and practice (third edition), by Rob Hyndman and George Athanasopoulos:* `https://otexts.com/fpp3/`
- *Time Series Analysis Handbook, by Benjur Emmanuel L. Borja, Sebastian C. Ibañez, and Maria Eloisa M. Ventura:* `https://phdinds-aim.github.io/time_series_handbook/Preface/Preface.html`
- *Time Series Analysis and Its Applications With R Examples, by Shumway and Stoffer (2017)*
- *Elements of Forecasting, by Diebold (1997)*

Join our community on Discord

Join our community's Discord space for discussions with the authors and other readers:

`https://discord.gg/VXpz9YPUTW`

3

Evaluating Time-Series Models

"There are two kinds of forecasters: those who don't know, and those who don't know they don't know" —John Kenneth Galbraith

Once you have built a model, how do you know if it is any good? On the surface, this appears to be a simple question; surely the answer is that the model performs well in the future? But let's say we want to predict the sales of a soft drink such as IRN-BRU (forgive my Scottishness); we could simply select any modeling approach, pass through all available data, and then predict forward into the future forever. Job done! But what happens when unexpected events occur, such as governments globally restricting movement, disrupting supply chains, and drastically changing consumer demand? Even subtler events happen daily that can shift behavior patterns and, consequently, a forecast's usefulness. While no evaluation strategy will allow you to predict a global shutdown perfectly, proper evaluation methodologies provide crucial insights into how your models performed historically, offering statistical inference about their generalizability to future conditions.

By the end of this chapter, you will have learned methods for evaluating forecasting models, along with why the optimal evaluative approach is not always practical. You will also have learned about error metrics, which we use to measure the performance of model forecasts. Error metrics are not one size fits all, and we explain how they can be useful under different conditions. Related to this, we also introduce baseline model forecasts, which we should aim to beat if we are to use more sophisticated methods. This leads us to *relative error metrics*, which utilize forecasts in the error metric calculations. We also discuss evaluating groups of time series. All of this is vital to understanding how to evaluate your model's ability to forecast. Finally, we discuss data leakage, an unfortunate phenomenon that makes predictions appear better than they are.

In this chapter, we will cover the following key topics:

- Why we evaluate
- Partitioning data
- Error metrics
- Length of evaluation
- Data leakage

Proper evaluation separates useful models from those that merely appear impressive in controlled settings but fail when confronted with real-world complexity.

Why we evaluate

So, how can we evaluate time-series modeling performance in a way that helps us to choose the best model? The *no free lunch* theorem tells us that when averaging over all possible problems, all algorithms perform similarly. An algorithm that performs well on one type of problem will perform less well on another; *there is no perfect method* for statistical modeling. If there is not one perfect method, our question becomes: which approaches work best (**model assessment**) and are practical options?

To do this, we follow a structured evaluation process. This helps ensure that model comparisons are fair and that the conclusions we draw about performance are meaningful. The core steps are as follows:

1. Data partitioning: How we split data appropriately.
2. Forecasting: Training algorithms to build models that generate predictions.
3. Error metrics: Selecting appropriate metrics requires understanding the calculations for comparing forecast predictions to actual data.
4. Error calculation: Applying our chosen metric(s) to obtain a value we can use to understand or compare the accuracy of our model predictions.
5. Statistical testing: We can use statistical testing when we have many comparisons to make, which is useful when we have distributions of error metrics.
6. Model selection: Selecting the approach or model to take forward from the options available to use.

> Steps 5 and 6 are optional. We don't require significance testing when deploying models, as we can just select a "best" model. If we combine model predictions (ensembling), we don't need to select at all (Hewamalage, Ackermann, and Bergmeir (2022)).

Part of evaluating models is about considering the *bias-variance trade-off*, which is crucial to selecting and tuning forecasting models.

When evaluating models, we rely on error metrics (discussed later in the chapter), such as the **mean absolute error** (**MAE**), **mean squared error** (**MSE**), and **root mean squared error** (**RMSE**). How any given model performs depends on the capacity ("memory") and complexity of the model (Banachewicz and Massaron (2022)).

Models capture patterns from data through internal parameters, each with a unique ability to detect specific rules or associations. As a model "learns," it stores these patterns in its "memory." The concept of model capacity or expressiveness is often discussed in terms of bias and variance, which relate to the model's predictions and its underlying complexity.

Supervised models are mathematical functions mapping inputs (observed data) to outputs (predictions). The complexity of these functions determines how flexible a model is in fitting to data. There are two concepts we should consider here:

- **Bias**: Reflects errors in a model's ability to capture patterns within data. High bias often indicates underfitting, with a model failing to capture patterns, while low bias suggests a model is adequately capturing the underlying data structure. This doesn't, however, guarantee optimal forecasts without considering variance.
- **Variance**: Describes how sensitive a model is to fluctuations in training data. High variance indicates excessive complexity, causing models to capture noise and produce unstable predictions. Low variance implies stable predictions, but risks oversimplifying, potentially missing key underlying patterns.

Modeling techniques available today, supported by increased computational resources (e.g., gradient boosting and deep **neural networks** (**NNs**)), make variance the primary concern for most forecasting challenges (Banachewicz and Massaron (2022)). Variance becomes a problem once a model has extracted all useful patterns from its training data; any remaining capacity in the model can lead to memorizing irrelevant data characteristics, or noise. This results in overly specific learning that impairs performance on new data—a phenomenon known as overfitting. Conversely, insufficient complexity can lead to underfitting, where the model fails to capture essential data patterns due to high bias.

This is where **validation** comes in.

Using validation to manage complexity

Validation is critical in distinguishing between a model's ability to generalize and its tendency to underfit or overfit. It establishes a score, or performance (error metric) value, that separates these two aspects of model performance, helping to optimize the balance between capturing genuine patterns and avoiding the memorization of training set idiosyncrasies.

The chart in *Figure 3.1* illustrates the bias-variance trade-off, showing how model bias and variance affect predictive performance in terms of error. Initially, when model complexity is low—such as having too few neurons in an NN—a model exhibits high bias due to the oversimplification and underutilization of available data, resulting in underfitting. As complexity increases, bias decreases; beyond a certain point, memorization of irrelevant data features occurs, increasing variance and overfitting. The crossing point between bias and variance indicates the optimal model complexity, minimizing total predictive error.

Refer to color images on the Graphic bundle or the ebook.

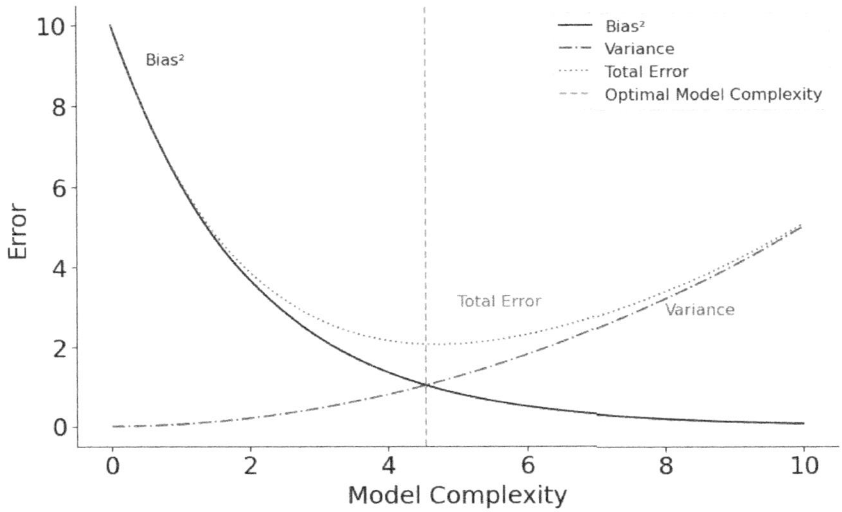

Figure 3.1: Bias-variance trade-off

When plotting measures of loss against the model's learning efforts (i.e., epochs for NNs, or GBDT rounds), a pattern emerges. Typically, training datasets see continuous improvement, while performance on unseen datasets plateaus or degrades after a certain point. This pattern persists across different hyperparameters, data preprocessing techniques, or model types. Although the curves' shapes may vary, there is always an optimal point before overfitting begins. *Figure 3.2* illustrates this concept, depicting training and validation loss over successive iterations of model training. Initially, both losses decrease together, but as training continues, validation loss plateaus and then begins to rise, marking the onset of overfitting. Identifying this inflection point through a robust validation strategy ensures optimal model generalization.

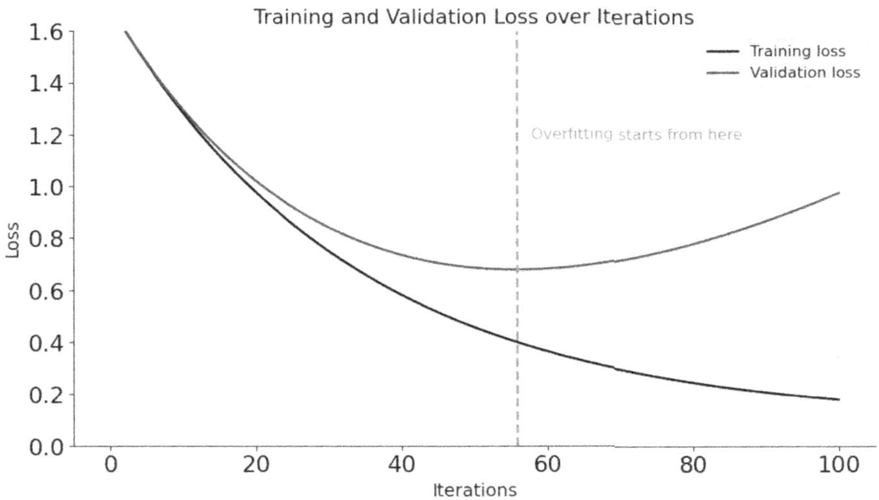

Figure 3.2: Training and validation loss

Overfitting can differ based on model choice, hyperparameter settings, and data preparation methods. This variability underscores the importance of a well-implemented validation strategy to identify the point where overfitting starts.

Importance of proper validation

A robust validation strategy serves as a practical mechanism through which we navigate the bias-variance trade-off. When implemented properly, validation helps us to identify models and/or their hyperparameters that approximate the inflection point, discussed previously, between model bias and variance, to find the signal within our data, rather than noise. Identifying, or approximating, this balance point represents a model's optimal capacity for a specific forecasting task, resulting in predictions that neither oversimplify nor overfit the underlying patterns within data.

Partitioning data

The simplest way to assess a model's performance is to measure its prediction accuracy against data unobserved in model training, known as **out-of-sample** (OOS) testing. We may use OOS testing to quantify how closely the prediction of a model aligns with actual data. There are several approaches that we could use, and it is useful to have an overview of the most common approaches, even though some methods prevail more consistently.

Introducing the dataset

Before moving on to the other topics, we will introduce some data to be used as an example. To keep things simple, we will focus on a single-item univariate time series taken from a subset of the M5 competition data: `FOODS_2_197`. Our data is retail data that captures sales of a single food product, that is, IRN-BRU 450 ml (precise product is not stated), at a single store in Texas, USA, between 2011 and 2017. The series comes from what we refer to as a panel dataset, meaning many time series that are stored in the same table and are related to each other strongly or weakly, that is, products sold at the same store. Toward the end of the chapter, we will focus on rolling error metrics with multiple series from the same dataset to extract model performance across groups of related time series. The precise data, partially wrangled, is stored in the GitHub repository for this book.

In the code, we load our subset of the M5 competition panel dataset, containing multiple series of product sales. We convert the `date` column to a datetime format to enable chronological processing. Then, we implement a function to remove the last 28 days (approximately one month) from each product's time series, as this is not data that we want to use as it is filled with zeros (see the Kaggle link `https://www.kaggle.com/competitions/m5-forecasting-accuracy?` for further information):

```
# Data wrangling
panel_df = pd.read_csv(CFG.data_folder / 'M5_t20_ABC.csv', index_col=False).
drop(columns = 'Unnamed: 0')
panel_df['date'] = pd.to_datetime(panel_df['date'])
panel_df.set_index('date')
```

```
def remove_tail(df, drop_value=28):
    assert 'item_id' in df.columns and 'date' in df.columns
    df_sorted = df.sort_values(['item_id', 'date'])
    def remove_tail_values(group):
        return group.iloc[:-drop_value]
    df_filter = df_sorted.groupby(
        'item_id'
    ).apply(remove_tail_values).reset_index(drop=True)
    return df_filter

panel_df = remove_tail(panel_df, drop_value=28).copy()
```

In this second block, we extract a single product (`FOODS_2_197`) from our panel dataset to focus our analysis. We then generate a time-series plot to visually examine the sales patterns, which helps us get a sense of the underlying structure of our data, that is, trends, seasonality, and potential anomalies, before conducting a formal statistical evaluation:

```
# Single product series

series_df = panel_df.loc[panel_df.item_id == 'FOODS_2_197'].copy()
series_df['date'] = pd.to_datetime(series_df['date'])

fig, ax = darts_plot(series_df, 'date', ['sold'],
                     labels=['Sales'],
                     title='Sales for FOODS_2_197')
plt.show()
```

Here's the output:

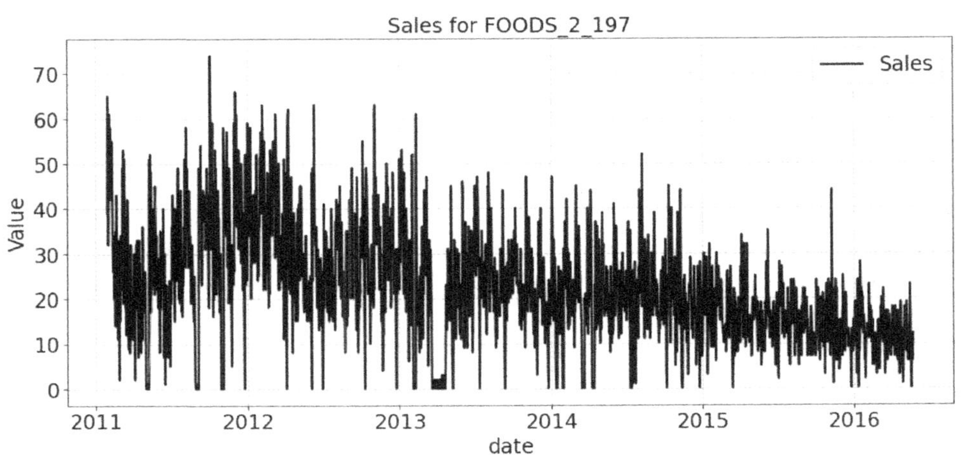

Figure 3.3: Product sales of FOODS_2_197 over time

We can see that the series has a complex pattern, with an obvious downward trend and strong seasonal patterns, although these are difficult to make out at this level. We can also observe quite a bit of volatility. Building on the previous chapter, we will now take a closer look at our data to better understand its structure. What you see here is a common structure that you would see in retail companies; I saw several similar series this morning (at the time of writing) in a completely different dataset in my employer's data!

To analyze our data, we will apply decomposition with a "recent" subset of the data (2015–2016), as not doing so makes it difficult to make sense of the plots.

```
series_df.set_index('date', inplace=True, drop=True)
short_series = series_df['2015':'2016'].copy()
fig = plot_seasonal_decompose(short_series['sold'], model='additive')
plt.show()
```

We now see the data decomposed in the following plot:

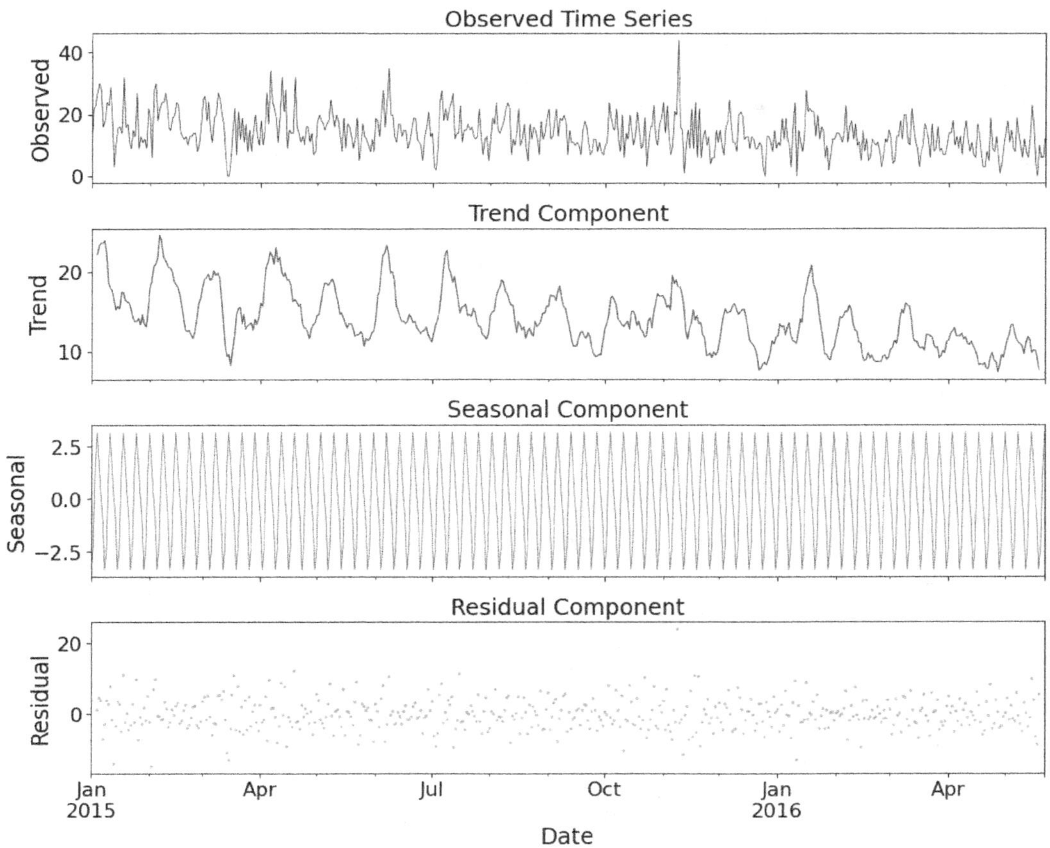

Figure 3.4: Decomposition of the FOODS_2_197 time series

As we saw in the series plot (*Figure 3.3*), the decomposed sales slowly trend down, and there is a strong seasonal component, with both weekly and monthly variation. What is interesting is that despite the obvious negative trend seen in *Figures 3.3* and *3.4*, an **Augmented Dickey-Fuller** (**ADF**) test, conducted next, finds the series to be stationary on both the full and shortened series:

```
from statsmodels.tsa.stattools import adfuller

result = adfuller(series_df['sold'])
print('ADF Statistic: %f' % result[0])
print('p-value: %f' % result[1])
print('Critical Values:')
for key, value in result[4].items():
    print('\t%s: %.3f' % (key, value))
```

```
ADF Statistic: -3.606685
p-value: 0.005635
Critical Values:
    1%: -3.434
    5%: -2.863
    10%: -2.568
```

Our ADF results (statistic: -3.617, p-value: 0.006) are statistically significant, with the ADF value greater than all critical values, suggesting that we should reject the non-stationarity null hypothesis. This clearly demonstrates that the ADF is a poor statistical test to rely on when you have **high volatility**, **complex seasonal** patterns, or **exogenous variables** with a strong impact on the series, as it fails to detect gradual trends when masked by high-frequency seasonal oscillations and volatility patterns.

Of course, we could use a **Kwiatkowski–Phillips–Schmidt–Shin** (**KPSS**) test instead, as discussed in the previous chapter:

```
KPSS Statistic: 4.6110454287271905
p-value: 0.01
Critical Values:
    10%: 0.347
    5%: 0.463
    2.5%: 0.574
    1%: 0.739
```

Unlike the ADF, the KPSS test's null hypothesis is that a time series is stationary. Given our test statistic of 4.611, that is much larger than even the 1% critical value (0.739), and we have statistical significance at p = 0.01, so we can reject the null hypothesis of stationarity. This obviously makes more sense.

ACF and PACF plots

Another feature to look at is the **autocorrelation function** (**ACF**), relating to lags in time, which we can build using the same code as in Chapter 2's notebook. We simply load `plot_acf` and `plot_pacf` from statsmodels and apply the functions, with the specified number of lags (n = 32 is like saying 32 days here as the data is daily observations). This produces plots of the ACF and **partial autocorrelation function** (**PACF**):

```
from statsmodels.graphics.tsaplots import plot_acf, plot_pacf

# Facet plotting
fig, (ax1, ax2) = plt.subplots(1, 2, figsize=(12, 6))

# ACF plot
plot_acf(series_df['sold'], lags=32, ax=ax1, color= '#003DFD')
ax1.set_title('Autocorrelation Function (ACF)')

# PACF plot
plot_pacf(series_df['sold'], lags=32, ax=ax2, color= '#003DFD')
ax2.set_title('Partial Autocorrelation Function (PACF)')
plt.tight_layout()
plt.show()
```

We can see how sales relate to previous points in time.

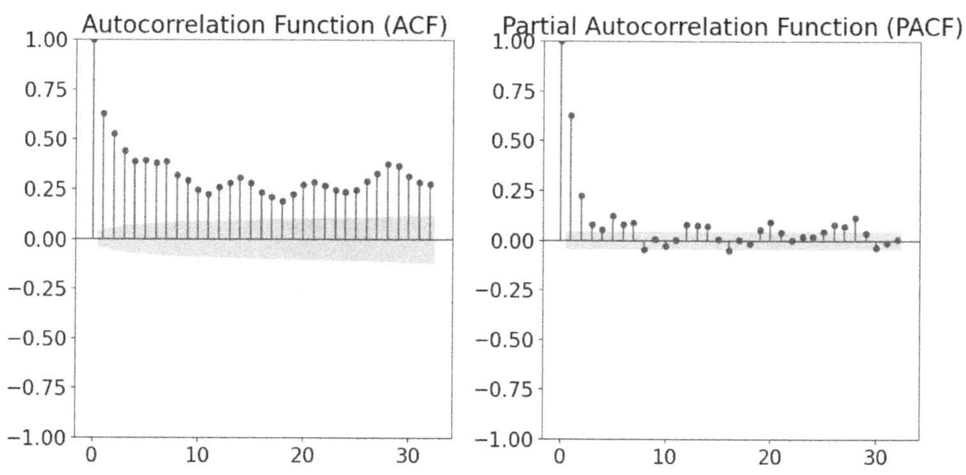

Figure 3.5: Autocorrelations for FOODS_2_197, with white noise threshold

You can observe that with 32 lags, there is regular correlation throughout the series. Values extending outside the blue band (representing the 95% confidence interval) are considered statistically significant. This indicates temporal dependencies in the series, which is worth considering in model builds.

By examining the PACF, which controls for the effects of intermediate lags, we see a distinctive pattern of significant spikes occurring at regular 6-to-7-day intervals. This pattern becomes evident through the prominent spikes at lags 6, 7, 12, 13, 14, 20, and 38, exceeding the significance threshold (blue band). Obviously, this is not an exact 7-day cycle, but this is fairly common for food sales.

Conducting residual analysis

Finally, let's take a look at the residuals, by fitting an OLS linear regression to the data. The code for this is supplied in the GitHub repository.

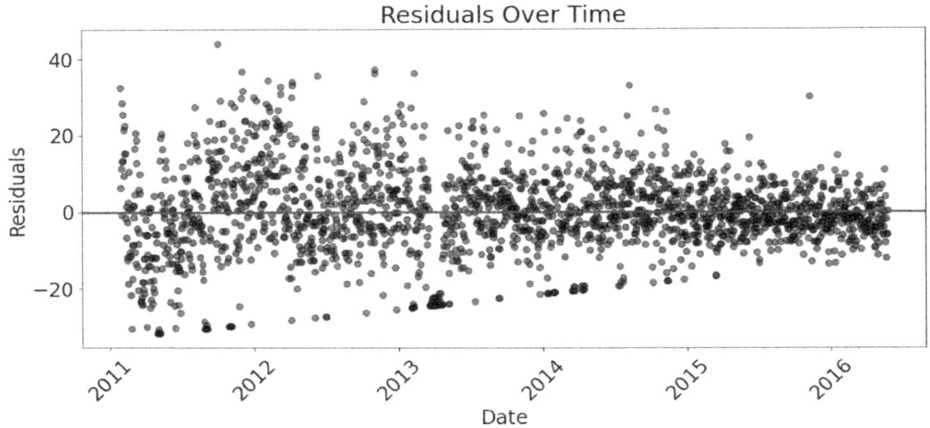

Figure 3.6: Residual plot for FOODS_2_197

The residuals tell us several things. First, they surround 0, which implies that the model is at least going through the middle of the data. But there may be patterns in these residuals, although this is difficult to see with this amount of data, or even the shorter series we made earlier. Being a simple model, it is likely that we are not capturing the full dynamics of the data. Of course, the most obvious pattern is that of heteroscedasticity (non-constant variance in residuals/errors). You can see that the spread of residuals appears to decrease over time, with larger variations in earlier years (2011–2013) compared to later years (2015–2016). This violates an assumption of linear modeling. Clearly, we can do better.

Fixed-origin design

A common approach to portioning time-series data is to use a *fixed-origin* experimental design, under which we simply partition data with one (train-test) or two (train-test-validation) splits based on set points in the dataset.

Train-test split

Under the simplest fixed-origin design, we divide data into two portions: a training and test set, often labeled `y_train` and `y_test`, along with `X_train` and `X_test` when exogenous variables exist. The training data is used to estimate the forecasting method's parameters, and the test data is used to evaluate its accuracy. Because the test data is not used in determining the forecasts, it should provide a reliable indication of how well the model is likely to forecast on new data. However, we have a few rules to follow.

Test sets typically comprise about 20% of your total sample. This size can vary depending on the sample size and the desired forecast horizon (i.e., the number of future time steps to predict, such as 28 days). Ideally, *your test set should contain at least as many observations as the length of this forecast horizon* (Hyndman and Athanasopoulos (2021)).

Unlike traditional approaches to data splitting, we have to *maintain the temporal order* of a time series. As a result, we cannot *randomly stratify* our data, as this would disrupt this order and could result in data leakage; known information influences predictions in the test set, making models appear to perform better than they will on unseen data.

When dealing with time series for regression tasks, with a *fixed-origin* design, we use the earlier data for training, and later (recent) data is "split" off for use as test, or validation, sets. You can see how we do this next, by creating a split point three months before the end of this series:

```
from dateutil.relativedelta import relativedelta
last_date = series_df['date'].max()
split_date = last_date - relativedelta(months=3)

# Create split point
train_df = series_df[series_df['date'] < split_date]
test_df = series_df[series_df['date'] >= split_date]
```

This approach relies on the datetime value, but you could use an ordered numerical index with pandas. It is worth noting here that the Polars library, in many respects, is simpler and faster for working with data bound by time. We have used pandas due to its popularity at the time of writing. Our fixed-origin train-test split is plotted here:

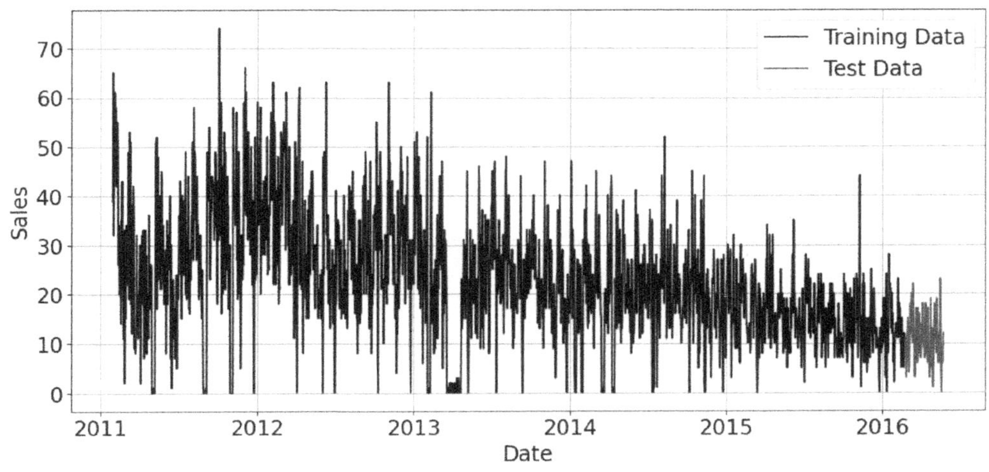

Figure 3.7: Simple train-test split

This approach respects the time-dependent nature of the series. We do, however, still need to be careful about preserving specific characteristics across our splits (analogous to what stratification does for non-time-series data).

This simple approach is useful when we want to use as much of our data as possible when validating a model; for example, say we want to simply pull out the accuracy of a model we use in production, but it is not particularly effective if we want to create a robust estimate of predictive performance (Cerqueira, Torgo, and Mozetič (2020)). The problem with this approach is that comparing multiple models (and/or hyperparameters) with the same split can lead to **sequential overfitting** as we attempt to search for the best score on the test set. Unlike regular overfitting (where a model memorizes training data patterns too specifically), sequential overfitting occurs across the model selection process itself; each iteration of model comparison and refinement against the same test set gradually incorporates information about the test data distribution into the modeling process. Using a test set throughout such a comparative process to guide model selection results in information "about the test set implicitly leaks into the training process" (Lones (2024)), effectively turning the test set into an extension of the training data over time.

In both cases of overfitting, statistical inference—the generalizability of predictions—is reduced.

Train-validation-test split

You can avoid the problems of sequential overfitting by adding another split to make a validation set, which is used to guide training processes, leaving the test set to test the final model. We have done this in the following code by taking the max date of the series, then finding the point three months previous to define the test set, and then another three months back from that point to define the validation set:

```
# Find split points
last_date = series_df['date'].max()
test_split_date = last_date - relativedelta(months=3)
val_split_date = test_split_date - relativedelta(months=3)

# Create train, validation, and test sets
train_df = series_df[series_df['date'] < val_split_date]
val_df = series_df[(series_df['date'] >= val_split_date) & (series_df['date'] <
test_split_date)]
test_df = series_df[series_df['date'] >= test_split_date]
```

This split is visualized using the code given here:

```
# Plot
plt.figure(figsize=(12, 6))
plt.plot(train_df['date'], train_df['sold'],
         label='Training Data', color='black')
plt.plot(val_df['date'], val_df['sold'],
         label='Validation Data', color='#b512b8')
plt.plot(test_df['date'], test_df['sold'], label='Test Data', color='#003DFD')
plt.tight_layout()
plt.title('Train-Validation-Test Split')
plt.xlabel('Date')
```

```
plt.ylabel('Sales')
plt.legend()
plt.grid(True)
plt.show()
```

The result is as follows:

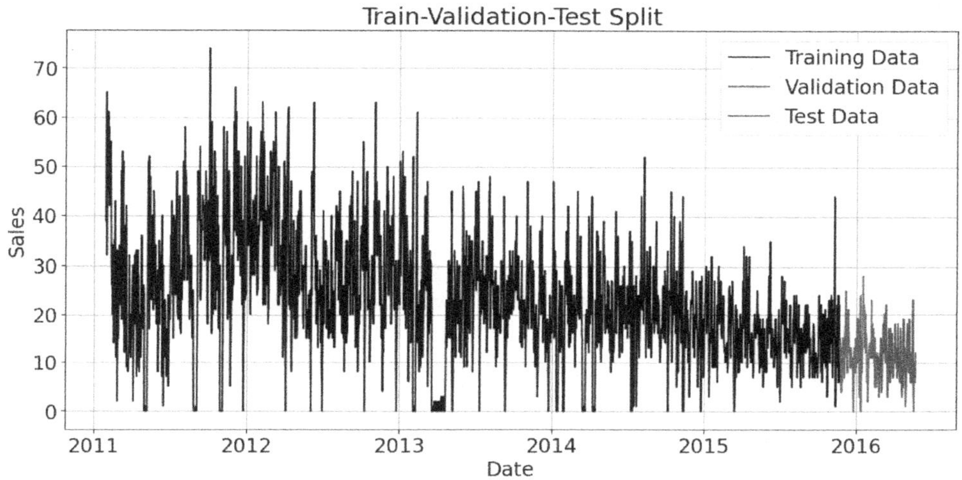

Figure 3.8: Train-validation-test split

The problem with fixed-origin approaches such as these, however, is that a point estimate can change quite dramatically from one model run to another due to the inherent randomness or variability in a model's behavior. For example, as you will learn later, a simple NN, such as a feed-forward network, is **initialized** with weights and trained with an **optimizing algorithm**. Both weights and the optimizer have an element of randomness that leads to different runs of the same NN architecture to converge on different predictions for the same dataset each time it is retrained. This means you might select a model based simply on the random seed. Even a small change to one hyperparameter can produce markedly different results, especially if our data is noisy or volatile.

To address these issues, we can adapt our splitting approach to move across the data.

Shortened data

You will notice with the preceding additional validation split, and with the following back-testing/**cross-validation** (**CV**), that these methods reduce the amount of data we have to create forecasting models. This can be a problem when the most recent data is the most informative; it's important to consider what data is most important when selecting your evaluative approach.

Cross-validation

CV is a good approach for dealing with the stochastic nature of models. Unlike the simple train-test split, k-fold CV iterates over a dataset k times, ensuring each sample has the opportunity to be in both the training and test sets. In each iteration, the data is divided into $k$parts: one for validation and the remaining $k - 1$ for training. You calculate error metrics (such as RMSE) on each validation fold separately, then typically compute their mean or median to get an overall performance estimate. This provides a more robust evaluative assessment of model performance than fixed-origin designs, by giving a distribution of predictive performance across all available data.

When it comes to time-series CV (rolling origin), we are not limited to just non-overlapping folds. Two key approaches are worth considering: expanding windows and rolling windows.

Expanding windows

Expanding-window CV starts with an initial training period that grows over time. For each iteration, you maintain a fixed-size test window that moves forward in time. The process begins by selecting an initial training period (determined either as a proportion of your data or by specifying the number of folds, such as the three folds shown next). As you progress through iterations, the training window expands to include all historical data up to the current test window, while the test window maintains the same size (in our example, three months). This approach progressively increases the training set size as we move through the folds, allowing the model to leverage more historical data with each iteration. It's particularly useful for shorter/smaller time-series datasets (Hewamalage, Ackermann, and Bergmeir (2022)). Think of it as though you are giving your model a growing history lesson with each iteration. You can see how the folds expand in the following figure.

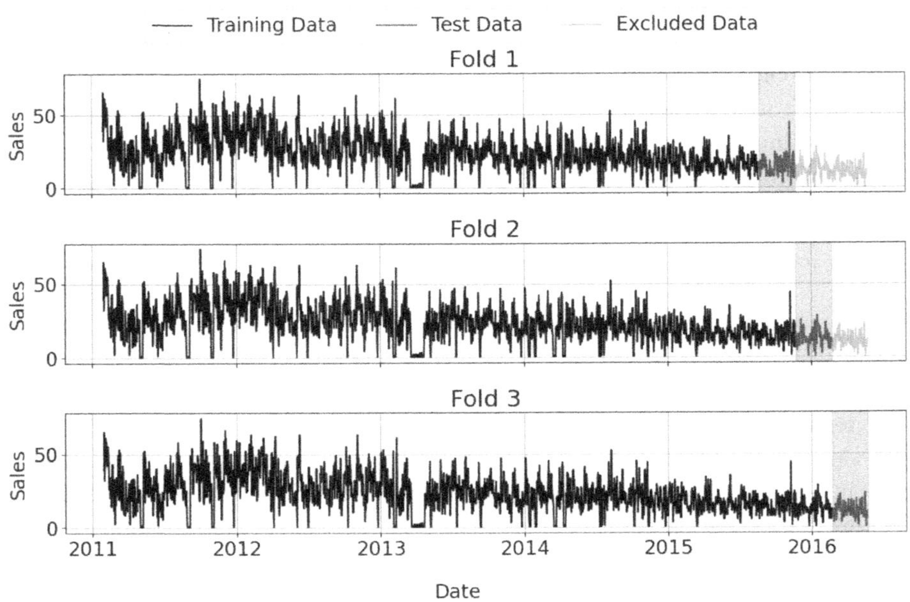

Figure 3.9: Expanding-window time-series CV

The problem with this approach is its known tendency to overestimate validation error. This may occur because the initial folds of the expanding-window approach use less data for training than would be available in production, resulting in poorer performance. When these early, higher-error estimates are averaged with later iterations, it inflates overall error estimates.

There are several options we could apply here:

- Use more folds. The optimal number (in Kaggle competitions) is around 6–7 (Banachewicz and Massaron (2022)), although this is not specific to any one dataset.
- Walk the training set forward with smaller steps than the forecast horizon, that is, one day at a time.
- Create a gap between the training and test sets. This is useful if there is a lag between model builds and predictions being used in production.

Rolling windows

Rolling/sliding-window (also referred to as blocked CV) designs are very similar to expanding windows, except the training set's length remains the same size. Instead, as the forecast origin (start point) moves forward in time, so does the start of the training set. This causes windows to drop older observations while incorporating new data (Cerqueira, Torgo, and Mozetič (2020)). Generally, this is advantageous when we have a series where recent data is considered more important. You can see an example of this in the following figure:

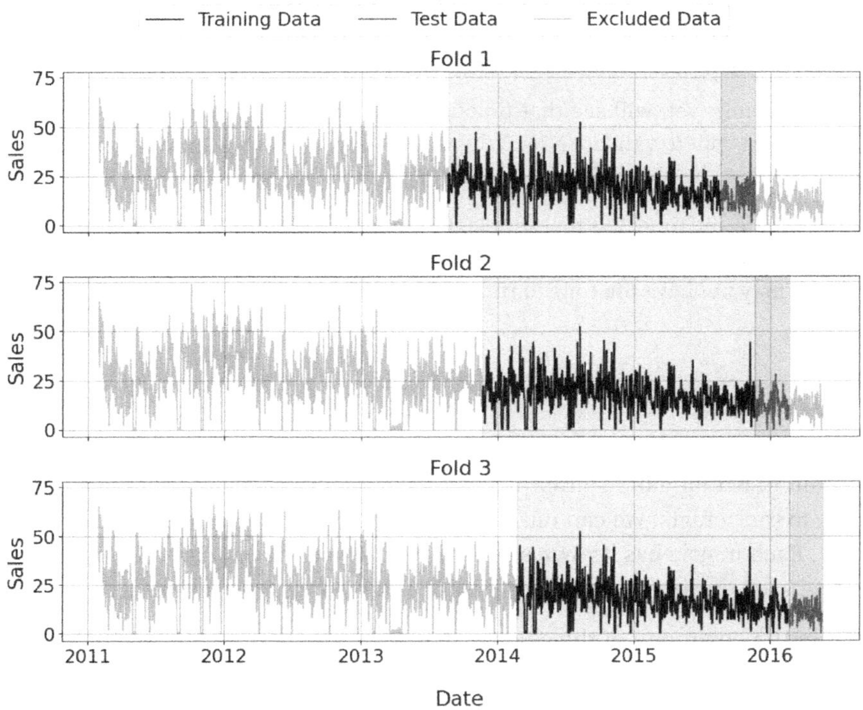

Figure 3.10: Rolling-window time-series CV

For pure autoregressive machine learning models that lack an inherent notion of time beyond the provided window, a rolling-window approach can be beneficial as it prevents the model from being influenced by potentially outdated patterns in the oldest data. Of course, if you have streaming data, this approach of "forgetting" older data in favor of the current window can also be beneficial (Hewamalage, Ackermann, and Bergmeir (2022)). Forecasting competitions and long datasets also seem to benefit from blocked CV designs too (Hidalgo, Maciel, and Barros (2019)). Having said that, it is important to consider a window size, so that you capture important patterns.

Although the recency bias of blocked CV (the tendency to favor more recent observations by excluding older data) helps build good models for many applications, there is no true consensus as to which approach is the best to use. Beyond data recency, one study demonstrated that blocked CV "is applicable to stationary time series" (Cerqueira, Torgo, and Mozetič (2020)). But "when dealing with non-stationary data sets, holdout applied with multiple randomized testing periods" performs better (Cerqueira, Torgo, and Mozetič (2020)). Our choice between expanding and rolling windows can also depend on the forecasting technique being applied. For methods such as **error, trend, seasonality (ETS)** that don't heavily weigh distant past observations, the difference may be minimal.

Ultimately, the selection of a CV method should be guided by the quirks of your data, the forecasting model you're using, and what you're trying to achieve with your forecast. As you will find as the book progresses, often the answer is to run experiments to find the best approach for your specific challenge. In reality, however, we often have to be pragmatic and take the easy-to-compute route. This is especially true of NNs, which may take a long time to train.

Neural network validation

Commonly, you will see that practitioners use a simple training-test or training-validation-test split to validate NNs. Why would they use such a simple approach that may disadvantage the model for selection?

Compute and time. It is that simple. Running CV with CatBoost or XGBoost takes moments compared to doing the same for NNs. If you have several thousand or million models, you may not have the time to run CV for model comparison and/or multiple sets of HPs.

Error metrics

Once we have trained a model and made our predictions, within our evaluative design, we will want to compare them to actual data, so that we can get some understanding of how close our models' predictions are to the actuals. We can measure performance with a number of metrics; we can even create our own. Each metric has its own strengths and weaknesses. It is crucial to select appropriate error measures based on your data characteristics and business requirements.

Remember from the beginning of the chapter that when evaluating forecasts, you need to consider bias and variance. These components are crucial because they lead to different business decisions. For example, an energy company typically sees consistent seasonal patterns in line with minimal trend, while a new company might experience rapid growth and less predictable patterns.

Selecting an appropriate error metric depends on understanding what each metric actually *rewards*. Every point forecast is really a one-number summary of some predictive distribution we have in our heads (or in our model), and different metrics incentivize different summaries. Forecasting literature calls this **elicitation**. A given metric *elicits* a specific functional of a distribution (Gneiting, 2011; Kolassa, 2020). While it sounds abstract we've used a few examples make concrete:

- **Squared error metrics** (MSE, RMSE) are minimized in expectation by the *mean* of a future distribution. So they reward models that predict the conditional mean. Useful for planning decisions, but sensitive to outliers.
- **Absolute error metrics** (MAE, and pinball loss more generally) are minimized in expectation by the *median* (or a specific quantile, for pinball loss). More robust to outliers, but for sparse data when your median is zero, a MAE-optimal forecast is also zero. Usually not what you want if you're deciding whether to stock a product.
- The **MAPE** elicits what Gneiting calls the $(-1) - median$ (Gneiting, 2011). The only definition of this quantity is *the thing MAPE rewards*. Nobody has given a convincing reason why we should care about it, and it may not even exist when your data includes zeros. This is the reason MAPE is so problematic; computational issues with zero denominators are the most visible symptom.

There isn't a *best* evaluation metric that looks at a point forecast from all angles. Each metric rewards a *different* number. Kolassa draws an analogy: evaluating a single-point forecast with multiple metrics is like estimating someone's height and then checking it against a bathroom scale (Kolassa, 2020). The number you gave is a height; the tool expects a weight; neither is wrong on its own, after all, height is somewhat related to weight, but the pairing is nonsensical.

For symmetric distributions, the mean and median coincide, so MSE and MAE agree on the optimal forecast. For skewed distributions, common with intermittent demand, they can differ substantially. A metric that rewards the median may lead you to select a model that forecasts zeros for a slow-moving product; that product goes unstocked, and you miss sales. This is why understanding the distribution of your data matters, and why your metric should align with your business objective. A recent survey listed 38 error metrics (Hewamalage, et al., 2022), each suited to different scenarios; we'll cover the most common here.

Intrinsic metrics

You'll encounter intrinsic error metrics more than extrinsic ones (which we'll discuss later). These are measures you may have heard of, such as the MAE, MSE, RMSE, MASE, and the controversial **mean absolute percentage error** (**MAPE**). They are intrinsic metrics because they evaluate a forecast using only predicted and actual values and don't rely on any external benchmark or reference.

Before we go into the specifics of selecting metrics, let's understand how they work.

Absolute error

Errors in forecasting can, of course, be positive or negative, with models under- or over-predicting. If we calculate the error with the sign (+/-), errors cancel each other out, and we build an understanding of the summated bias of a model. Using AE $|e_t|$ prevents cancellation, ensuring errors accumulate over the validation set.

AE can be defined as follows:

$$AbsoluteError = |e_t| = |y_t - \hat{y}_t|$$

Where:

- y_t is the actual observed value at a time point t
- $\hat{y}_t$ is the predicted value at a time point t
- e_t is the error (difference between actual and predicted) at a time point t

Error metrics that use AEs (e.g., MAE or MASE) are optimizing for the median. This makes them less sensitive to outliers compared to other metrics, such as SE metrics. You can see this in the following chart, which shows consistent growth or error with under- (minus values) or over-forecasting (positive values):

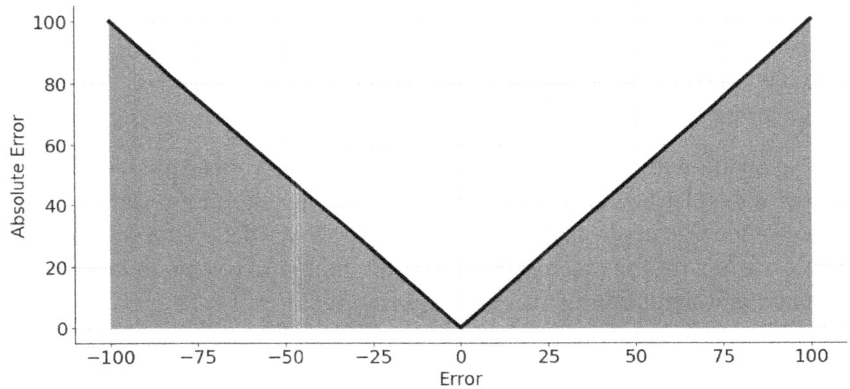

Figure 3.11: AE weighting plot

From a business perspective, AE-based metrics return values in the same units as our target variable (e.g., sale units, kilograms, or miles per hour), making error interpretation intuitive for stakeholders. However, AE's scale dependency requires understanding context—an error of 25 units is negligible when forecasting values between 7,500 and 10,000, but substantial when values range between 75 and 100. In other words, the MAE is great for a task such as model selection, even with highly variable data, but not so great for comparing models across datasets with multi-series comparisons; in that case, scale-independent metrics are more appropriate. Another issue is that selecting the median forecasts can be problematic when you have sparse data and a zero forecast is a legitimate outcome.

Absolute percentage error (APE)

If ever an error metric caused pain, it was this one. Business stakeholders love percentages, because they seem so intuitive and easily interpretable. APE metrics take the scales of one or multiple time series and convert them into percentages of the actual value. This sounds intuitive, but it introduces serious mathematical issues that make APE unreliable for evaluation. If you don't already hate measures such as APE, let me teach you the way!

$$AbsolutePercentageError = |e_t/y_t| \times 100\% = |(y_t - \hat{y}_t)/y_t| \times 100\%$$

Where:

- y_t is the actual observed value at a time point t
- $\hat{y}_t$ is the predicted value at a time point t
- e_t is the error (difference between actual and predicted) at time point t

To demonstrate this issue, we'll create some pseudo-data. Feel free to code along. We build some actuals and build "predictions" of two models we want to compare, using only three values for each. Obviously, this is not realistic, but it allows us to visualize the issue in the calculations of APE. See this data plotted in *Figure 3.12*.

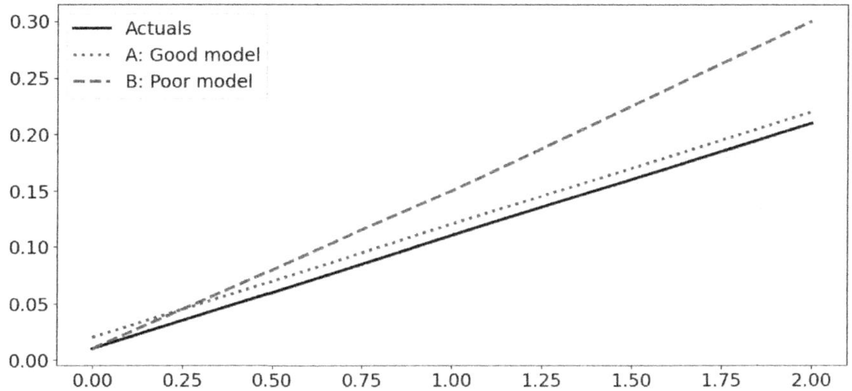

Figure 3.12: Comparison of true values and predictions (A versus B) with MAPE

Clearly, model A (dotted line) is better than model B (dashed line), as its predictions are closer to the actuals (solid line). We'll confirm this by calculating the error, that is, how far each set of predictions is from the actuals, with the MAPE:

```
def mape(true_values, predicted_values):
    return np.mean(np.abs((true_values - predicted_values) / true_values))*100

mape_A = mape(yt_actuals, yt_A)
mape_B = mape(yt_actuals, yt_B)

print(f'Model A has a MAPE of {mape_A}')
print(f'Model B has a MAPE score of {mape_B}')
# Output:
Model A has a MAPE of 37.95093795093795
Model B has a MAPE score of 26.596681096681102
```

How is it possible that what is clearly a poorer model results in a lower error score than one whose predictions are mostly closer to the actuals? The answer is that APE scores like MAPE give disproportionate weight to data points near zero, making poor models look better than they are.

A common workaround is to shift both sides by adding a constant (typically 1) before computing the MAPE:

```
#| label: model comparison with shifted MAPE
#| message: false
#| echo: true
#| eval: true

mape_A = mape(yt_actuals + 1, yt_A + 1)
mape_B = mape(yt_actuals + 1, yt_B + 1)

print(f'Model A has a shifted MAPE of {mape_A}')
print(f'Model B has a shifted MAPE score of {mape_B}')
```

Boomtown! Our rankings are now *correct*. Or maybe not... What does it actually *mean* that forecast A is better than B after we've added some arbitrary number to everything? Our shifted metric isn't MAPE anymore; it's something else, and we don't have a clear definition of what that something else rewards. We've patched a symptom without fixing its cause. MAPE's real problem is that it elicits a functional we have no good reason to care about, and no amount of shifting changes that (Kolassa 2020). You'll see us use this shift later in the chapter for practical comparisons (it's common enough in industry), but be aware of what you're giving up in interpretability.

If we look at the weighting of error, we can see another issue.

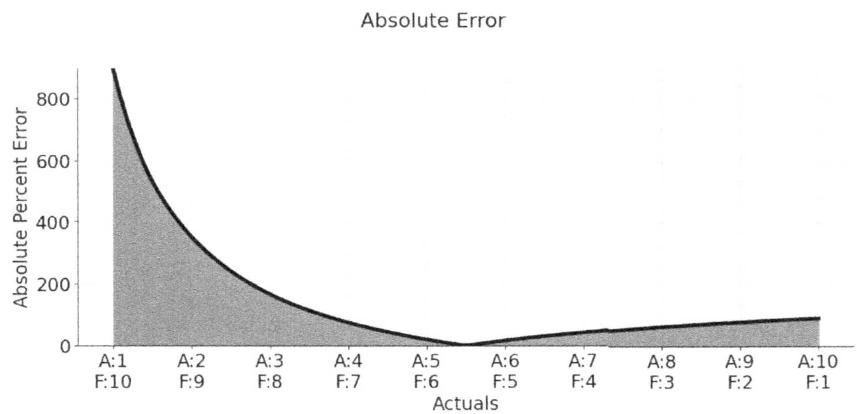

Figure 3.13: APE error weighting plot. Adapted from Manu (2022)

We can see here the dark side of the moon, with a clear asymmetry, which is common to APE-based metrics. When you're dealing with small actual values (look at the left side of the graph), even over-forecasting sends your errors shooting up an exponential path. This is not bad if you want to catch poor forecasts. But the main issue is the downplaying of error, or under-forecasting, which biases model selection toward models that predict lower values. This asymmetry is caused by the denominator in APE being actual values. When actual values are small, even a moderate over-forecast may result in large percentage errors.

There is more to make this a terrible measure—zero values! APE cannot handle zero values because dividing by zero creates infinite values, which is just the cherry on the cake. You can, of course, ignore them, but this kind of hack means we are not representing the data properly. Similarly, using a shift on both sides (actuals and forecasted values) is possible but is a little hacky, and it fails the address the underlying problem.

Another APE approach is the **weighted APE (WAPE)**, which aggregates the AEs and actual values separately before performing the division. Here are some points to note:

- If a single actual value is zero, it doesn't cause a division-by-zero error because the denominator is the sum of all actual values
- Even if the sum of actual values in the denominator is zero (which would be rare in most practical scenarios), it would result in an undefined value for the entire dataset, rather than causing issues with individual data points

However, even with this, if we change the scale, for example, from grams to kilograms, or Celsius to Kelvin (related units), the APE score can change drastically. Here's an example with Celsius to Kelvin, with the MAE, which has scale invariance, for comparison:

Metric	Celsius	Kelvin
MAPE	5.11%	0.46%
WAPE	4.67%	0.46%
MAE	1.40	1.40

Viewing this in a plot shows the issue more clearly.

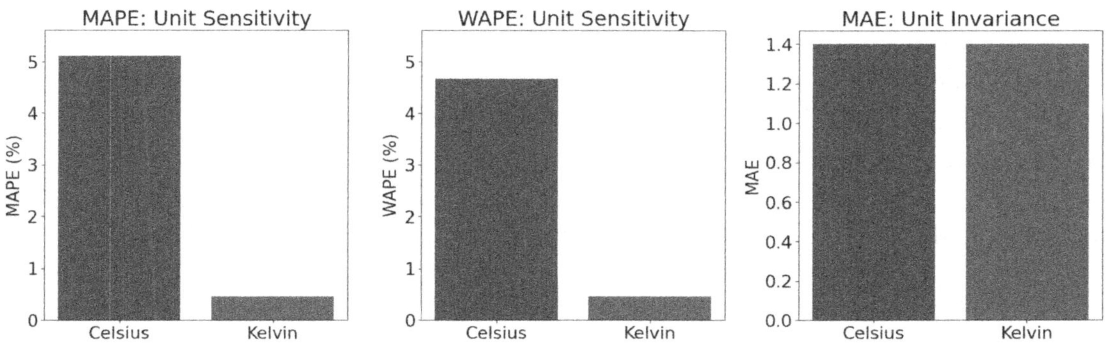

Figure 3.14: APE unit change comparison plotted

Let's summarize:

- APE is undefined for zeros
- It is asymmetrical in its weighting of over- and under-forecasting
- It has a bias (in error representation) toward over-forecasting
- Values close to zero can make poor models look good
- It is inconsistent across scales

We are not saying abandon APE values altogether; WAPE can be useful for large datasets with multiple time series. But these limitations should highlight its careful use. We would suggest using alternative metrics. In scenarios where AEs or costs (e.g., inventory management) matter more than percentage-based errors, we would suggest looking toward measures based on the AE or SE. These can, however, have their own limitations.

Symmetric Mean Absolute Percentage Error (sMAPE)

sMAPE was proposed as a fix for MAPE, particularly its tendency to favor under-forecasting and its tendency to blow up when actuals are zero. The idea is to replace an actual value in the denominator with an average of the actual and the forecast:

$$\text{sAPE}(\hat{y}_t, y_t) = \frac{|\hat{y}_t - y_t|}{\frac{(|\hat{y}_t| + |y_t|)}{2}}$$

sMAPE is then the mean of sAPE values across time steps. It's bounded between 0 and 2 (or 0% and 200%), and it doesn't blow up at zero actuals. We define $\{sAPE\}(0, 0) = 0$. Seems reasonable right? But it has some serious problems.

Despite its name, sMAPE isn't symmetric in a way that matters. Swapping forecast and actuals gives the same sAPE (fine, but rarely useful). However, over-forecasting and under-forecasting by the same amount give *different* sAPE values. With an actual of 10, forecasting 5 (under by 5) gives sAPE = 66.7%, while forecasting 15 (over by 5) gives sAPE = 40%. So sMAPE actually penalizes under-forecasting more heavily; it doesn't fix MAPE's asymmetry, it just reverses it.

It gets worse. For count data (which is what you have in retail, supply chain, and most demand forecasting), sMAPE always elicits integer forecasts. And for sufficiently intermittent data, where more than half of observations are zero, an sMAPE-optimal forecast is a flat zero regardless of how large the nonzero values might be (Kolassa 2026). A rational forecaster evaluated on sMAPE has every incentive to round their predictions to integers, or just predict zero for slow-movers.

sMAPE was used in the M3, M4, NN3, and NN5 competitions, and appears throughout forecasting literature. But the functional it elicits serves no useful purpose: for high-volume data, it approximates the mean (in which case, just use RMSE), and for low-volume count data, it produces nonsense. It is, as Kolassa puts it, either unnecessary or dangerous (Kolassa 2026). You should treat it the same way we treat the MAPE: understand it, because you'll encounter it in literature (and industry), but don't use it for your own evaluations if you can avoid it.

Squared error

The SE takes error values and squares them, which, as you can see in the following formula, means it does not have the problem with division that the APE values have.

$$SquaredError = |e_t^2| = (y_t - \hat{y}_t)^2$$

Why would you do this? Here are a number of reasons:

- Squaring values makes them positive, which prevents the errors from canceling each other out.
- It causes errors to increase rapidly the further we get away from 0.
- As with the AE, and not the APE, the sensitivity of the SE is symmetrical in under- and over-forecasting. This makes it differentiable and useful for optimization as well as forecast error evaluation.

You can see the effects of this on how error is weighted in the following figure, with it weighting higher errors more heavily than lower forecasting errors:

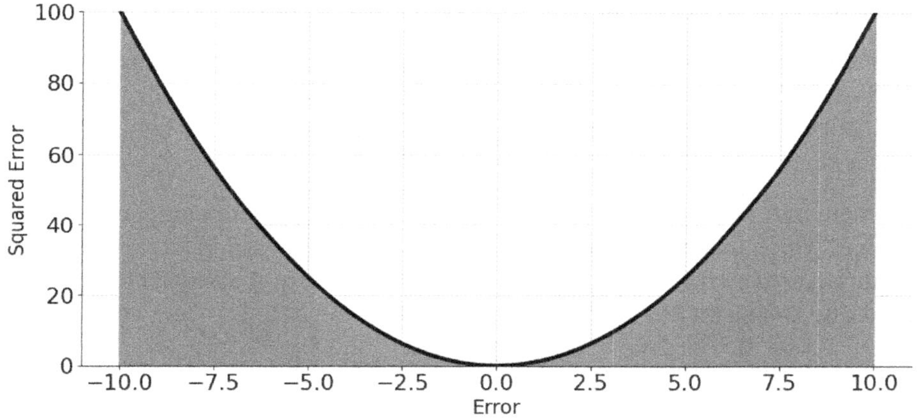

Figure 3.15: SE weight plot

Let's see how the SE compares to the AE by creating some pseudo-data with a forecast value of 5, and "predictions" of 0 to 10:

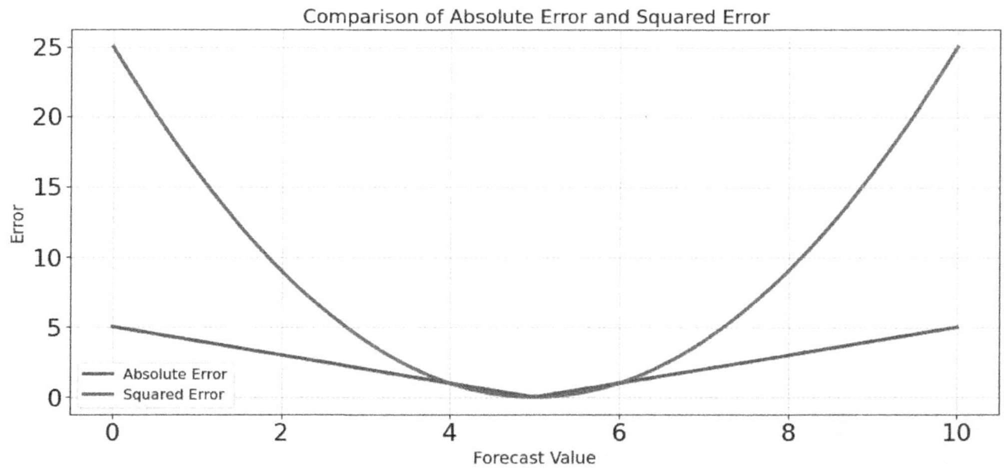

Figure 3.16: Comparison of AE and SE weightings

You can immediately see the difference in the weighting of error under each calculation. Clearly, the SE penalizes higher errors more than lower ones, which is very useful when you're interested in getting the error inside a tolerable margin, rather than looking for the absolute lowest value. In terms of optimization of CV, it is handy for optimizing/tuning hyperparameters. The problem is that it can create absurd values if you have poor forecasts and/or outliers, which, while making bad models easy to spot, can cause errors that require handling in your code.

Unlike the APE, SE metrics (i.e., MSE and RMSE) are not as directly interpretable to non-technical stakeholders, since the values are squared and don't maintain the original units of measurement. This makes them more useful to data scientists and statisticians, though experienced business stakeholders may still understand their significance.

Often, we use the RMSE, which rescales the values to the original units while still penalizing large errors more heavily. However, it doesn't represent a direct ratio between predicted and actual values like percentage-based metrics do, making it mathematically sound but sometimes less intuitive in business discussions.

There is another issue, which is that minimizing the SE implicitly assumes that errors follow a Gaussian distribution, which may not be the case. Errors are, in our experience, rarely Gaussian. Unlike some normalized metrics, the MSE is scale-dependent. This means that if you multiply the actual and forecast values by a constant, the SEs scale quadratically.

This is demonstrated in the following code. You can see how scaling up values affects the MSE:

```
# Define data
actual = np.array([100, 150, 200, 250, 300])
forecast = np.array([105, 160, 190, 240, 310])

# Define error function
def mse(actual, forecast):
    return np.mean((actual - forecast)**2)

# Calculate MSE
mse_original = mse(actual, forecast)
mse_scaled = mse(actual * 10, forecast * 10)

print(f"MSE (original): {mse_original:.2f}")
print(f"MSE (scaled): {mse_scaled:.2f}")
print(f"Ratio of scaled to original MSE: {mse_scaled / mse_original:.2f}")
```

```
MSE (original): 85.00
MSE (scaled): 8500.00
Ratio of scaled to original MSE: 100.00
```

Variations

Most error metrics are derived from the AE, SE, and APE, such as the MAE and MSE, which are uncomplicated applications of the AE and SE. The MAE provides an average of absolute differences between predictions and actual values, retaining the original unit of measurement. The MSE, on the other hand, averages squared differences, penalizing larger errors more heavily but resulting in squared units.

Variations in metrics often come about because we have to model many time series with different scales in the same dataset (i.e., panel dataset). This is where APE metrics rear their heads because of their scale-independence, which is supposed to give an intuition of overall performance between models and across the whole set. If you want to avoid the MAPE (which you do), you can look at *scaling* or *normalization*.

Normalized errors in forecasting tackle the challenge of scale. The process typically involves dividing the error by a measure of the data's scale. For example, the **normalized root mean squared error (NRMSE)** is calculated by dividing the RMSE by either the range or the mean of observed values, yielding a dimensionless metric that can often be expressed as a percentage:

$$\mathrm{NRMSE} = \mathrm{RMSE}/(\max(y) - \min(y)) = \sqrt{\left(1/n \sum_{i=1}^{n} (y_i - \hat{y}_i)^2\right)}/(\max(y) - \min(y))$$

While normalization facilitates cross-dataset error comparisons, it's not without its problems. The data distribution, particularly outliers, can cause issues. When using a range for normalization, extreme outliers will artificially deflate the NRMSE, potentially obscuring poor-performance models. In these cases, where outliers are present, we can consider employing a more robust denominator for the division, such as an interquartile range, mean, or median.

Scaled errors are typically done by using the predictions of an extrinsic forecasting model, such as a naive model or a linear regression, and come under the umbrella of extrinsic metrics, which we will discuss next.

Extrinsic metrics

Sometimes you'll look at a metric and you'll ask yourself, "Is this MAE of 5.4 good or bad?" If you don't have the history of the error metric for that dataset to look up, it is difficult to gauge. Extrinsic metrics give us a tool by which to judge performance, by scaling model error through division by the error of a benchmark model, such as a naive forecast. You can also use more complex models, depending on the comparison requirements, such as when the previous model was built. Often, the goal with extrinsic measures is to quantify the value added of more sophisticated models relative to similar models, providing a standardized measure of performance improvement.

There are different approaches to contextualizing your model's performance:

- **Relative error** is similar to comparing your test scores to a class average; it's directly tied to how others (or, in this case, your benchmark model) performed. It's intuitive but can be fickle, changing dramatically if you switch benchmark models. We can define the relative error as follows:

$$RE_t = \frac{e_t}{e_t^b}$$

Where e_t represents your model's error at time t and e_t^b is the error of your benchmark method at time t. The most common of these are the following:

- RelMAE - $\text{RelMAE} = \frac{\text{MAE}_{\text{model}}}{\text{MAE}_{\text{benchmark}}}$
- RelMSE - $\text{RelMSE} = \frac{\text{MSE}_{\text{model}}}{\text{MSE}_{\text{benchmark}}}$
- RelRMSE - $\text{RelRMSE} = \frac{\text{RMSE}_{\text{model}}}{\text{RMSE}_{\text{benchmark}}}$
- RSE - $\text{RSE} = \frac{\sum_{t=1}^{n}(y_t-\hat{y}_t)^2}{\sum_{t=1}^{n}\left(y_t-y_t^{\text{benchmark}}\right)^2}$

- **Absolute scaled error** is more like grading on a curve that's set by a simple passing strategy (our naive forecast). It's more stable across datasets, making it a go-to for comparing diverse time series. Common versions of these are as follows:

  - MASE - $\text{MASE} = \frac{\frac{1}{h}\sum_{t=1}^{h}|y_t-\hat{y}_t|}{\frac{1}{T-1}\sum_{t=2}^{T}|y_t-y_{t-1}|}$
  - MdMASE - $\text{MdMASE} = \frac{\text{Median}(|y_t-\hat{y}_t|)}{\frac{1}{T-1}\sum_{t=2}^{T}|y_t-y_{t-1}|}$
  - RMSSE - $\text{RMSSE} = \sqrt{\frac{\frac{1}{h}\sum_{t=1}^{h}(y_t-\hat{y}_t)^2}{\frac{1}{T-1}\sum_{t=2}^{T}(y_t-y_{t-1})^2}}$

  Where:

  - y_t: Actual at time point
  - $\hat{y}_t$: Predicted value at time point
  - h: Forecast horizon length
  - T: Length of training set
  - n: Number of observations

Let's have a look at how we might use them. Here we take the last year of `series_df` and the last three months for the test period. We create two simple models for comparison:

- Naive forecast: We take the last value of the training data and predict it forward (flat-line prediction). This is often used as a baseline benchmark.
- Seasonal naive forecast: We use a rolling projection of the last 28 days of the training data to capture weekly seasonality patterns.

Here's what the visualization for these forecasts would look like:

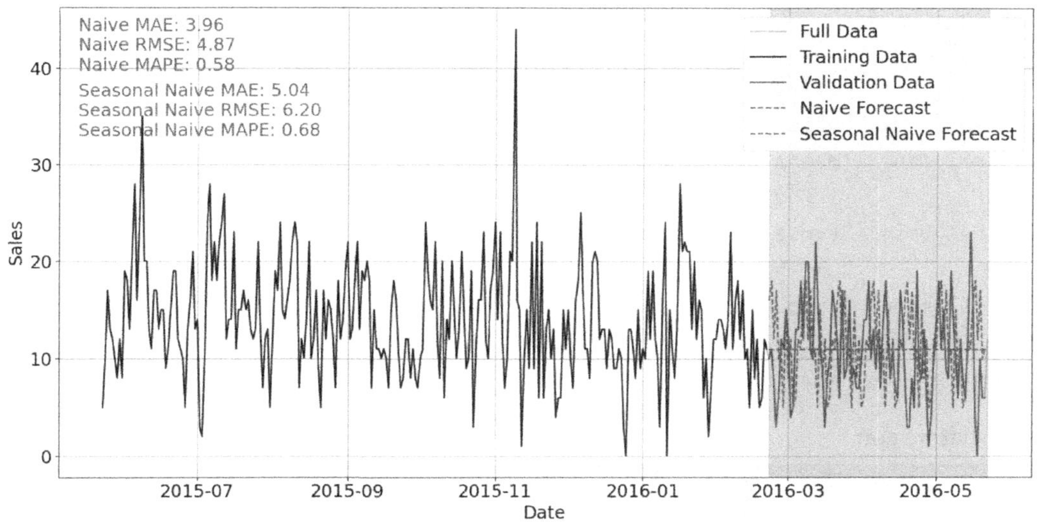

Figure 3.17: Three-month validation: Naive versus seasonal naive forecast

Looking at the plot, we can see that despite the seasonal naive model attempting to capture the weekly patterns in sales data, the simpler naive forecast actually performs better on all three metrics: MAE, RMSE, and MAPE (shifted +1). You can see that the flat naive model is actually the better fit. This illustrates that complex models don't always outperform simpler ones. This is often the main point in using relative metrics, to measure whether a more complex model can beat a simpler one. Of course, you can use a more complex model as your base in a relative error metric. In this instance, though, we want to see how much better or worse the simple naive version is compared to the seasonal one.

We can calculate scaled and relative metrics directly with NumPy and/or use `sklearn.metrics`, such as `mean_absolute_error`, `mean_squared_error`, and `mean_absolute_percentage_error`. See the following code:

```
def calculate_metrics(actual, forecast):
    mae = mean_absolute_error(actual, forecast)
    rmse = np.sqrt(mean_squared_error(actual, forecast))
    mape = mean_absolute_percentage_error(actual+1, np.array(forecast)+1)
    return mae, rmse, mape

def calculate_scaled_metrics(actual, forecast, training_data):
    in_sample_mae = np.mean(np.abs(np.diff(training_data)))
    if in_sample_mae == 0:
        return np.inf, np.inf
    errors = np.abs(actual - forecast)
    squared_errors = (actual - forecast) ** 2
    mase = np.mean(errors) / in_sample_mae
    rmsse = np.sqrt(np.mean(squared_errors)) / in_sample_mae
```

```
    return mase, rmsse

def calculate_relative_errors(actual, forecast, baseline_metrics):
    mae, rmse, mape = calculate_metrics(actual, forecast)
    rel_mae = (mae / baseline_metrics[0]
               if baseline_metrics[0] != 0 else np.inf)
    rel_rmse = (rmse / baseline_metrics[1]
                if baseline_metrics[1] != 0 else np.inf)
    rel_mape = (mape / baseline_metrics[2]
                if baseline_metrics[2] != 0 else np.inf)
    return rel_mae, rel_rmse, rel_mape
```

Here's what we got:

```
Metric            Value
MAE              5.0444
RMSE             6.1968
MAPE             0.6811
MASE             0.9227
RMSSE            1.1335
Relative MAE     1.2753
Relative RMSE    1.2720
Relative MAPE    1.1754
```

So, what do these metrics mean? Let's go through them one by one:

- **MAE = 5.0444:** Means the seasonal naive forecast is, on average, five units away from the actual sales
- **RMSE = 6.1968:** Similar interpretation to the MAE but the value is slightly higher due to how the calculation changes the values.
- **MAPE = 0.6811:** A 68.11% error relative to the actuals. Of course, you can see that there is both under- and over-forecasting. This does not appear to be representative of the error.
- **MASE = 0.9227:** An MASE less than 1 indicates that this forecast is performing better than the average naive forecast. With the value being close to 1, we are only seeing a slight improvement in forecasting.
- **RMSSE = 1.1335:** Being above 1 suggests that when considering SEs, the seasonal naive forecast is performing slightly worse than the naive forecast.
- **Relative MAE = 1.2753:** We can interpret this as the seasonal naive forecast deviating from the actuals by more than 27.53%, in terms of the MAE, than the naive model.
- **Relative RMSE = 1.2720:** Similarly, the RMSE for the seasonal naive forecast is 27.2% higher than the naive forecast in terms of the RMSE.

- **Relative MAPE = 1.1754:** With the MAPE, we are 17.54% higher than the MAPE for our naive baseline model.

Generally, for comparison, relative metrics work well, but they need to be contextualized by descriptions and intrinsic error metrics to give a good understanding. They become particularly valuable when comparing more sophisticated models. For instance, if we developed an ARIMA and an NN model, simple baseline relative metrics would quickly tell us which one provides the most improvement over our baseline, potentially saving implementation and computational resources.

Grouped error metrics

What do we do when we have more than one time-series model to track? What if we have 100 series? A common question is how we build metrics when we build multiple models, such as with a panel dataset. With such data, we can use one model, for example, a whole store, or multiple, for example, each product sold in a store. If we have multiple models, we might want to group the errors. A fatal mistake here would be to aggregate the error metrics for each individual model. There are many reasons to avoid doing this, but the main one is that an RMSE of 23.46 and 104.34 are very different, so aggregating them together would distort the true performance by effectively hiding the distribution of errors. For example, a number of good forecasts can hide poor forecasts and vice versa, depending on the proportion of good and bad forecast and their scales. If you normalized or scaled the error, this would reduce the impact, but it does not eliminate the issue.

The optimal approach is to pool all actuals and predictions of your grouped dataset before building metrics such as the RMSE and MAE. The fundamental calculations (AE, SE, and APE) remain the same, but their application changes. Of course, when scales differ significantly between series, we should probably use normalization or scaling. We can still use scale-independent metrics such as the MASE.

The following code demonstrates the idea with the MAE, RMSE, and MAPE using the previous code:

```
ProductA - MAE: 1.00, RMSE: 1.00, MAPE: 0.07
ProductB - MAE: 1.20, RMSE: 1.26, MAPE: 0.05
ProductC - MAE: 1.00, RMSE: 1.00, MAPE: 0.19

Pooled metrics - MAE: 1.07, RMSE: 1.10, MAPE: 0.11
```

Multivariate forecasting

We don't deal much with multivariate time-series forecasting in this book. Again, the calculations of the AE, SE, and APE remain similar for these types of forecasts, but they're applied to each predicted variable. We could do this separately and take the average, or pool them much like we do for grouped individual models. There are more sophisticated approaches, such as geometric or harmonic means, but the use of them depends on your data (e.g., sensitivity to outliers).

Normalization is also crucial for assessing the error of multivariate models. So, another approach is to use distance metrics such as Euclidean or Manhattan norms. For example, the **Euclidean norm** (L2 norm) for the RMSE in multivariate forecasting is calculated as follows:

$$\text{RMSE} = \sqrt{\frac{1}{T}\sum_{t=1}^{T} \| \mathbf{y}_t - \hat{\mathbf{y}}_t \|^2}$$

Where y_t and $\hat{y}_t$ are vectors of actual and forecasted values across all variables at time t. Similarly, the **Manhattan norm** (L1 norm) for MAE is as follows:

$$\text{MAE} = \frac{1}{T}\sum_{t=1}^{T} \| \mathbf{y}_t - \hat{\mathbf{y}}_t \|_1$$

Which to use and when

Choosing the right error measure isn't obvious. It depends on the characteristics of your data, your forecasting goals, and your business context. If your series contains zero or near-zero values, avoid percentage-based errors, and certainly don't use the MAPE (or sMAPE) for anything other than reporting to stakeholders, if you have to.

- **Single versus multiple metrics**: While practitioners debate this, using multiple measures can provide a more comprehensive view—just understand what each one emphasizes. Squared-error metrics (MSE and RMSE) penalize large errors heavily, while absolute error metrics (MAE) treat all errors more equally but may underperform with sparse data. Personally, I like a relative metric with a seasonal naive base. It provides a great way of seeing whether you are beating the simplest approach, and how much by.
- **For large panels**: The RMSSE provides a single aggregated performance measure for the entire panel. Alternatively, when you need to investigate individual series performance or compare models, calculate error metrics for each series, then scale those distributions (e.g., 0–1 range). This makes relative performance across thousands of series interpretable for stakeholders—similar to the MAPE's intuitive percentages, but without the drawbacks.

Length of evaluation

Earlier, we mentioned that the length of the validation period should be as long as the forecast horizon, the amount of time you want to predict. This is best practice; however, you may need to vary your forecast horizon for business reasons. The length of your forecast should be long enough to see beyond variability in your data, but not so long that it provides no additional detail. Longer horizons can have diminishing returns, as the following graph demonstrates. It shows that accuracy on a test set can increase as the validation length increases, but only to a point:

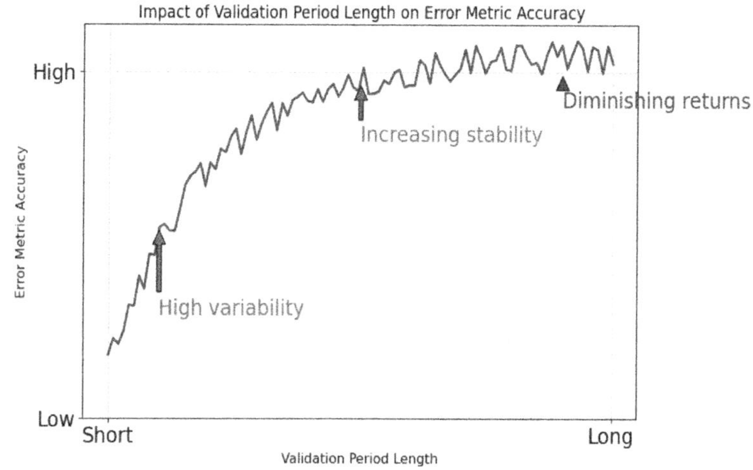

Figure 3.18: Impact of validation length on test set accuracy

Using our example series, we test this logic next, with a simple naive forecast. You can see that changing the validation length has a substantial impact on errors in the test set:

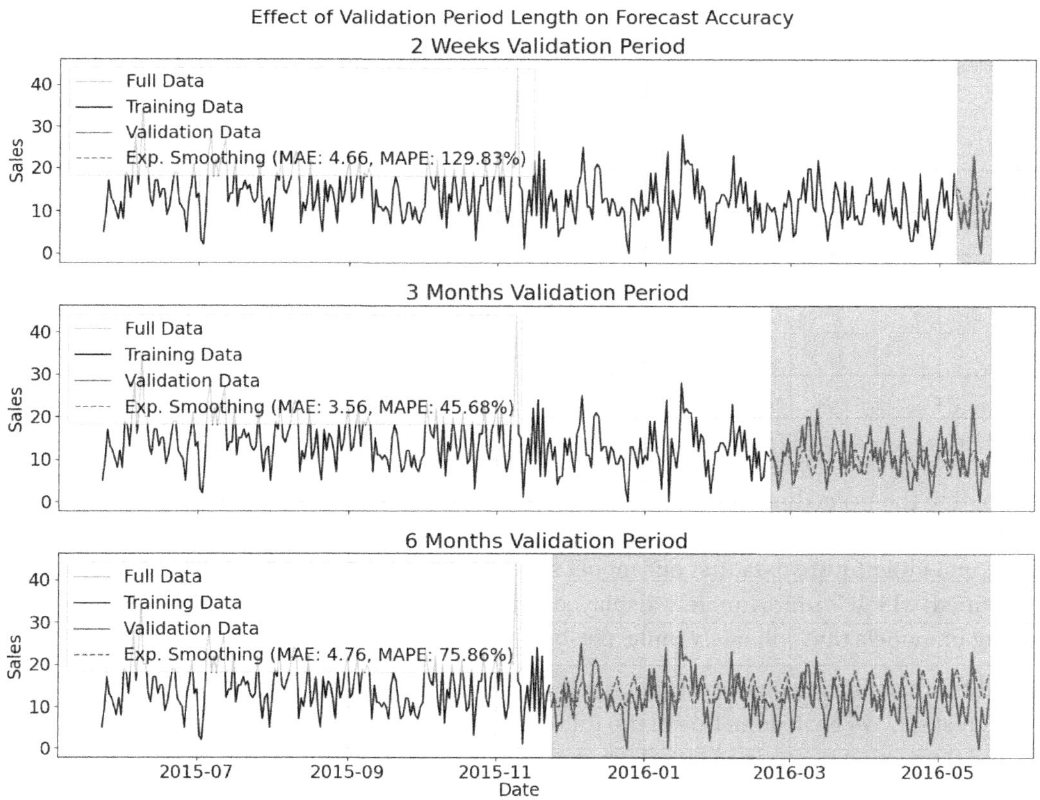

Figure 3.19: Applied validation length changes on test set accuracy

You can see that the forecast is overly affected by the short and long windows, which both make performance worse. Building models with short data is similarly problematic. Our suggestion is to test different lengths and see how well they work for your use cases.

Data leakage

Last, but by no means least, is **data leakage**. Data leakage occurs when information from outside the training set is inadvertently used in model development, leading to overly optimistic performance metrics and poor generalization of models when applied in production. In traditional machine learning, controlling for this is fairly simple; many of the same rules apply in time-series forecasting. However, there are additional complexities for you to navigate.

Transformation-based leakage

The most common sources of data leakage come from transformations of data—that is, Box-Cox, min-max scaling, detrending (with a model), and so on—before you have applied your data splits.

If you are doing backtesting (expanding or rolling windows), you will need to think about how to apply these to each fold without causing leakage.

To make this concrete, let's look at a common implementation mistake. The following example demonstrates a leakage-prone workflow, where scaling is applied before the train–test split. Although this may appear harmless, it introduces future information into the training process.

```
# PROBLEMATIC APPROACH - Leakage!
from sklearn.preprocessing import MinMaxScaler

# Scaling entire dataset before splitting
scaler = MinMaxScaler()
series_df['sold_scaled'] = scaler.fit_transform(series_df[['sold']])

# Then splitting (too late, leakage already occurred)
train_df = series_df[series_df['date'] < split_date]
test_df = series_df[series_df['date'] >= split_date]
```

If we applied the preceding code, the data in our test split would influence the scaling parameters. This distorts model performance because the scaler computes min and max values from our entire dataset, including future data that our model shouldn't have access to during training. Evaluation is undermined, which is unfortunately displayed in artificially improved performance metrics, despite resulting in models that will likely underperform in production.

To avoid this, we need transformations, such as scaling, to be applied to training data, to maintain the OOS test set. We have done this in the following code by fitting and transforming on the training data, and then using the trained transformer on the test without fitting:

```
# CORRECT APPROACH
from sklearn.preprocessing import MinMaxScaler
```

```
# First split data
train_df = series_df[series_df['date'] < split_date]
test_df = series_df[series_df['date'] >= split_date]

# Then transform each set separately
scaler = MinMaxScaler()
train_df['sold_scaled'] = scaler.fit_transform(train_df[['sold']])

# Use same scaler for test data (fit only on training data)
test_df['sold_scaled'] = scaler.transform(test_df[['sold']])
```

The principle of splitting data before wrangling applies to all transformations. If you want to remove outliers? You should only use the training data to identify and remove them. Need to apply Box-Cox or log transformations? Find the optimal lambda parameter using only training data. When doing backtesting with expanding or rolling windows, this becomes even more critical. You need to apply these transformations independently for each fold, using only the data available at that point in time. It's undeniably tedious and requires careful programming, but it's essential for valid evaluation.

One of the most obvious examples of leakage is demonstrated by when we leak information via de-trending (discussed in *Chapter 2*). In the following graph, we can see the impact of leaked compared to unleaked detrending on ARIMA model predictions:

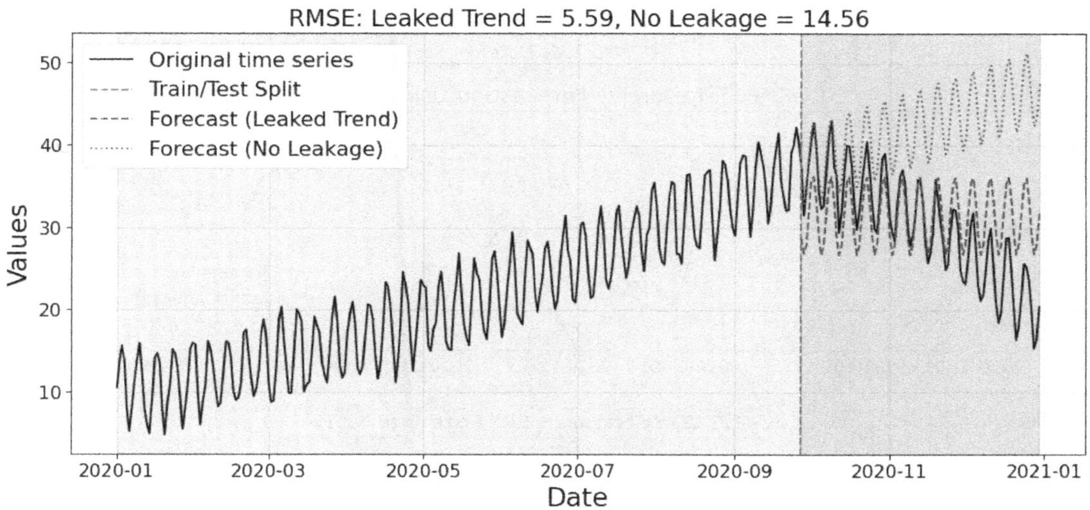

Figure 3.20: Impact of splitting before and after data transformation

You can clearly see that we have brought information about the trend back into the predictions by not splitting the data and applying the trend forward. Our "wrong" detrending produces a flat trend test that artificially improves the forecast. While the correct approach is worse, it preserves the natural behavior of the trend up to the point of the split, resulting in a more realistic evaluation of the model's performance.

Lookforward bias

Another source of data leakage comes from features that are strongly correlated with variable(s) (y) we are trying to predict. It is called lookforward bias. This can come from lagged features or, in the specific case of NNs such as **feed-forward neural networks** (**FNNs**), window embeddings. These kinds of features need to be built using the predictions of the model as we predict recursively. It's easy to mistakenly build these features using the actual data and leave them in your dataset when making predictions, which would cause data leakage. It is also worth mentioning that we can get lookforward bias in panels from cross-sections of data that may inform our covariates.

Let's see how this type of leakage affects performance, undermining evaluation. We use a linear model here with lagged features. In the wrong way, we simply shift values to create lagged features, which we use in the training and test sets. The right way is to do this only for the training set and then to build the lagged features in the test set as we make our predictions; this is called recursive forecasting:

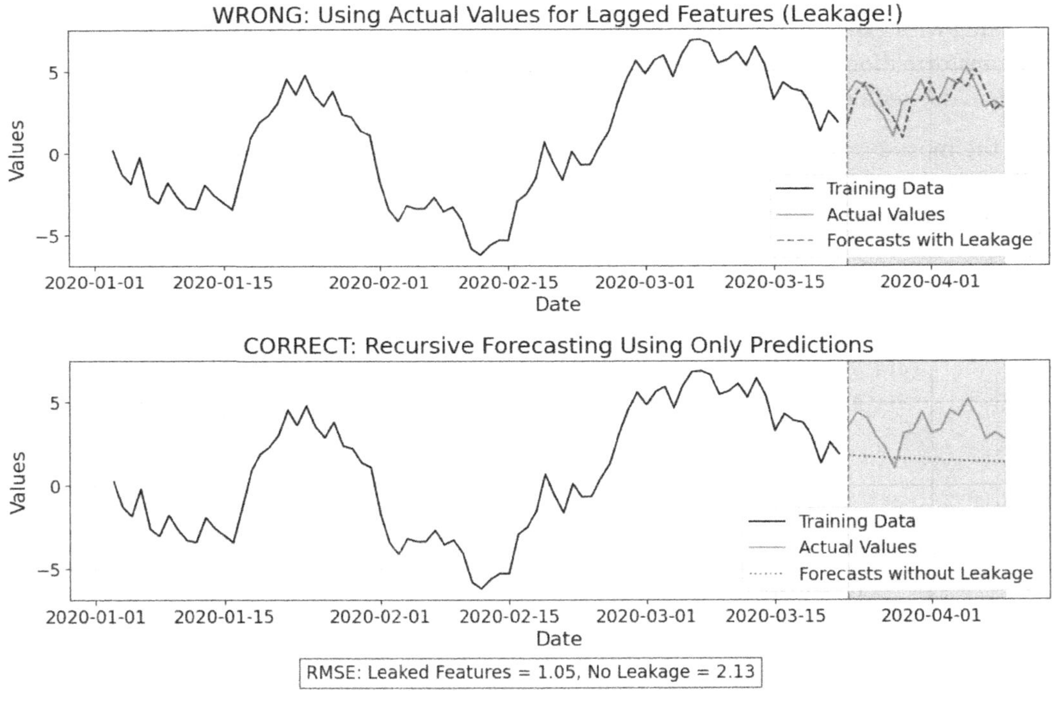

Figure 3.21: Impact of lookforward bias via lagged features

Clearly, using actual values as lag features results in artificially accurate forecasts that won't hold up in production. The "wrong" approach shows suspiciously good tracking of the test data, while the correct recursive approach shows the natural divergence we'd expect in real forecasting.

Panel data leakage

It is also worth mentioning that we can get lookforward bias in panel data from cross-sections of data that may inform our covariates. When you are working with multiple related time series (such as products in the same store), leakage can occur across series. For example, if you aggregate sales across all products to create a store-level feature, then split each product's time series individually, you might inadvertently include future information from one product in the features for another. This is also known, ironically, as a *golden feature*. We have (wrongly) built this in the following graph. The correct approach is to ensure your splits maintain temporal consistency across all related series:

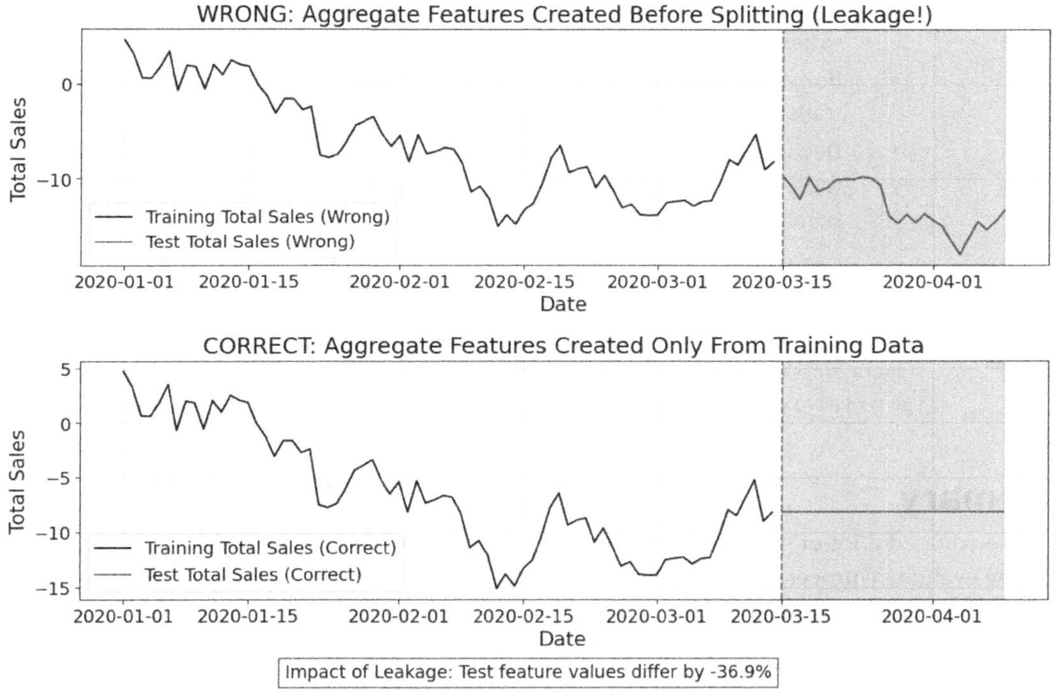

Figure 3.22: Impact of lookforward bias via golden feature

It is clear how the incorrect approach allows actual future sales from all items to leak into the features, while the correct approach maintains proper temporal separation. While performance may appear worse, this true performance is crucial for selecting the appropriate model, which is the whole point of these evaluations: to compare multiple models/hyperparameters.

Practical tips to prevent data leakage:

- Avoiding leakage requires vigilance at every step of your forecasting workflow:
- Always split data before any transformations or feature engineering
- Apply preprocessing separately to training and test sets
- For transformations that require parameters (such as scaling and Box-Cox), fit them only on the training data
- In recursive forecasting, use only previous predictions, never actual values
- For cross-sectional or panel data, ensure temporal consistency across all time series
- Be particularly careful with expanding-window or rolling-window validation—transform each fold independently
- Document your preprocessing pipeline carefully to avoid accidental leakage
- When in doubt, ask: "Would I have access to this information in a real forecasting scenario?"

Applying these precautions is necessary for an honest evaluation. Models that perform well with leaky features often collapse when deployed in production, leading to disappointed stakeholders and potentially costly business decisions. Taking the time to prevent data leakage will save you time later and help you select robust and reliable models for forecasting.

Summary

We have covered a lot of ground here, and we hope that you can now see that understanding and applying evaluative processes appropriately is the bedrock of good forecasting practice. Selecting the best model and tuning models requires careful splitting of data, to avoid information leakage, although often we are constrained by the reality of computational resources and time. But data splitting is just one aspect of evaluation; how you compare the performance of models by understanding the metrics and how to use them is critical.

We examined how evaluation practices have evolved from simple train-test splits to more sophisticated CV techniques, such as expanding and rolling windows. We also explored intrinsic error metrics, including the AE and SE, along with their various derivations.

We discussed the controversial nature of percentage-based errors such as the MAPE and considered alternative approaches that address their well-known limitations. In addition, we introduced extrinsic metrics such as relative error and absolute scaled error, which provide a broader context to assess model performance.

We then reviewed approaches for evaluating grouped time series and offered brief insights into multivariate forecasting evaluation. Finally, we emphasized the importance of evaluation period length on error metrics and highlighted the critical issue of data leakage in time series forecasting.

Recommended reading

- Banachewicz, Konrad, and Luca Massaron. 2022. *The Kaggle Book: Data Analysis and Machine Learning for Competitive Data Science. 1st ed.* Birmingham: Packt Publishing.
- Cerqueira, Vitor, Luis Torgo, and Igor Mozetič. 2020. "*Evaluating Time Series Forecasting Models: An Empirical Study on Performance Estimation Methods.*" Machine Learning 109 (11): 1997–2028. `https://doi.org/10.1007/s10994-020-05910-7`
- Hidalgo, Juan I. González, Bruno I. F. Maciel, and Roberto S. M. Barros. 2019. "*Experimenting with Prequential Variations for Data Stream Learning Evaluation.*" Computational Intelligence 35 (4): 670–692. `https://doi.org/10.1111/coin.12208`
- Lones, Michael A. 2024. "*How to Avoid Machine Learning Pitfalls: A Guide for Academic Researchers.*" Patterns. `https://doi.org/10.1016/j.patter.2024.101046`
- *Forecasting: Principles and Practice* by Hyndman and Athanasopoulos `https://otexts.com/fpp3/accuracy.html`
- Gneiting, T. 2011. "*Making and Evaluating Point Forecasts.*" Journal of the American Statistical Association. `https://doi.org/10.1198/jasa.2011.r10138`.
- Hewamalage, H., et al. 2022. "*Forecast Evaluation: A Data-Driven Approach.*" `https://doi.org/10.1109/TKDE.2022.3214571`.
- Kolassa, S. 2020. "*Why the 'Best' Point Forecast Depends on the Error or Accuracy Measure.*" International Journal of Forecasting. `https://doi.org/10.1016/j.ijforecast.2019.02.017`.
- Kolassa, S. 2026. "*The Symmetric Mean Absolute Percentage Error: Unnecessary or Dangerous.*" Forecasting. `https://doi.org/10.3390/forecast8020024`.

4

PyTorch Fundamentals

Having examined how to evaluate forecasting models, we now move from assessment to construction. The previous chapter established how we quantify predictive reliability; this one establishes the elements of neural networks required to build models capable of achieving it.

Such models require understanding the mathematical principles that guide the construction of neural network algorithms and the tools that bring those principles to life. For this reason, our next step is to learn the basic building blocks of PyTorch that are fundamental to the deep-learning approaches explored throughout the book.

PyTorch provides a transparent environment for experimentation, allowing us to construct, test, and adapt neural networks with the same iterative precision that we apply to statistical models. Its dynamic computation graphs, tensor abstractions, and automatic differentiation (Autograd) capabilities make it an excellent tool for developing forecasting architectures that can scale - from simple **feedforward networks** (**FFNs**) to complex global models such as N-BEATSx and N-HiTS.

This chapter provides a brief but functional foundation in PyTorch. You will learn how tensors encapsulate numerical data, how computational graphs define the mathematical flow within neural networks, and how gradient propagation enables 'learning'. We will also introduce PyTorch Lightning, a structured interface that streamlines training workflows and removes redundant code, allowing us to focus on model design rather than orchestration.

In this chapter:

- Introduction to PyTorch
- Working with tensors
- Understanding computational graphs
- A brief look at autograd
- Practicing PyTorch fundamentals

In essence, this chapter helps us establish the computational grammar required for modern neural forecasting.

Introduction to PyTorch

PyTorch is an open-source machine learning library, created by Meta, that is primarily used to develop deep learning models. It is based on Torch, a scientific computing framework, and provides a flexible and dynamic way of building and training models. What makes it popular is that the defining characteristics of PyTorch are in its "*Pythonic*" design, which you can learn about by importing '`this`' into python:

```
# The zen of python by Tim Peters
import this
```

PyTorch uses Python as its primary interface, which makes it intuitive and easy to learn for developers who are already familiar with the language. This is in contrast to TensorFlow, which is more focused on creating a static computation graph that is compiled and optimized before the model is run.

Its dynamic computation graph allows for considerable flexibility in building models, as developers can modify graphs on-the-fly during training. This makes it easier to experiment with different architectures and to build models that can adapt to varying input sizes and shapes. PyTorch also provides us with a wide range of tools and libraries for building and training deep learning models, including support for distributed training, data loading, and preprocessing, and visualization.

Overall, PyTorch's Pythonic design and dynamic computation graph make it a popular choice for researchers and developers who value flexibility and ease of use. Vox populi does not always equal vox Dei, but at the time of this writing (2026) it is hard to argue with the rising popularity of PyTorch and TensorFlow losing the "market" share.

This is the moment to take PyTorch for a test drive - there are three options:

- Using Kaggle notebooks: `https://www.kaggle.com/code`
- Using Google Colab: `https://colab.research.google.com/`
- Installing it on your own machine: `https://pytorch.org/get-started/locally/`

Whichever option suits you best, going forward we shall assume you have access to an environment where PyTorch is working. Personally, I use the latest stable version of PyTorch (2.10.0) with CUDA support, installed in a virtual environment managed with `poetry` (instead of `anaconda`). If you do not have a CUDA-supported GPU, you should install a CPU version. Using the PyTorch install page (`https://pytorch.org/get-started/locally/`), you can select through the options for local installs.

In the remainder of this chapter, we give an overview of the main concepts in PyTorch: this should give you a basic idea how to start moving, but is not by any means a substitute for practice that gives familiarity. Our exposition follows roughly the official tutorials order: `https://pytorch.org/tutorials/beginner/basics/`.

Working with tensors

The fundamental building blocks of deep learning libraries like PyTorch are tensors (different from the strict mathematical definition). A tensor is a multi-dimensional array, a structured container for holding data with n-dimensions, and some metadata that describes the tensor.

Our reason for using tensors is that we can represent virtually any form of numerical data within them; typically, data within them is formatted in floating point formats, such as float32, or integers. This data can range from a single value to a complex multidimensional sequence. This allows tensors to encode both model parameters and the data passed between layers, whether our data represents a single scalar, multidimensional time-series or something more complex.

While you will often see the data formats in PyTorch tensors as float32, other formats are possible: `https://pytorch.org/docs/stable/tensors.html`

Our choice is dependent on data, modeling approach, and model size. A large floating point number format i.e., float64 will allow greater precision within a model build, but it comes with increases in processing time and memory costs. However, you should be wary of going lower as dropping the floating points available to values i.e. float16 may reduce processing time, and model size, but will also impact model accuracy. A smaller float format can also cause issues with numerical stability due to the low-precision arithmetic.

At this point you're probably noticing that tensors aren't significantly different from `NumPy` arrays, so what separates the two? Well, tensors are specialized for use with GPUs, which means they can parallelize the great many calculations we need to run, in machine learning processes, across multiple cores (CPU or GPU). In other words, tensors speed up processing of matrix computations. Having said that, tensors in PyTorch use CPUs by default.

While PyTorch defaults to CPU execution, its strength lies in seamless acceleration on GPUs. To ensure reproducibility and scalability, it is good practice to detect and explicitly set the computation device at the start of a notebook or training script:

```
import torch

device = "cuda" if torch.cuda.is_available() else "cpu"
print(f"Using device: {device}")
model.to(device)
```

For large-scale forecasting tasks, where we may be training across thousands of series, GPU acceleration can significantly reduce the time spent on matrix operations. Conceptually, GPUs act as a wide vector processor, applying the same computation across multiple data points simultaneously, which is ideally suited for the highly parallel nature of neural architecture training.

You'll often see tensors referred to by the number of dimensions (rank) they have, although it's also common to see terms like scalars, vectors, and matrices used for lower dimension tensors. Generally time series data is often found between 2- to 4-dimensions, depending on the model being applied, but for more complex modeling approaches, like hierarchical forecasting we can see higher dimensions.

Refer to color images on the Graphic bundle or the ebook.

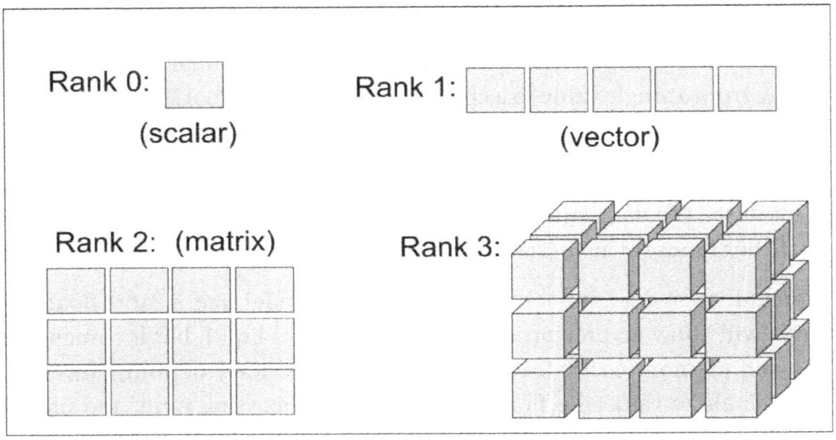

Figure 4.1: Tensor Diagram

Stride

In addition to the data, processor, and structure of tensors, there is a critical piece of metadata called **stride**, which helps map the logical representation of a tensor to its physical location in memory. In PyTorch we use a linear/contiguous memory layout, with each element of a tensor being stored in a memory location immediately adjacent to the previous element, which is efficient for accessing elements in a sequence, often the case in tensor operations. Let's look at a tensor where the elements are 32-bit integers:

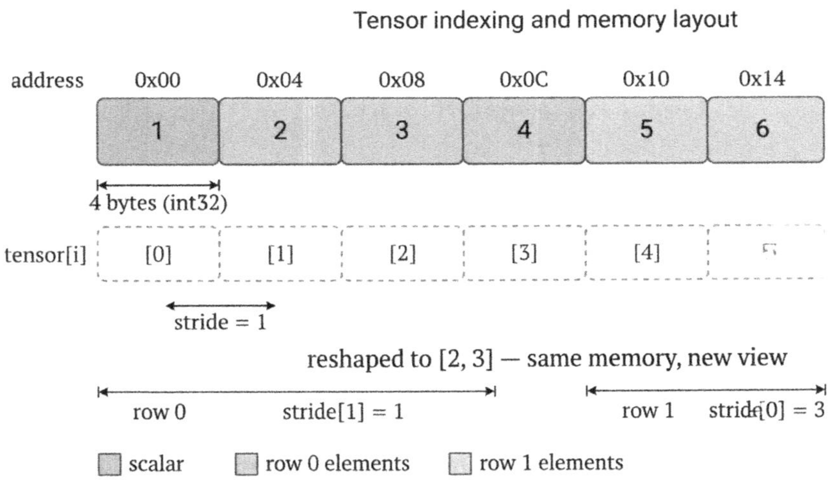

Figure 4.2: Tensor indexing

Figure 4.2 represents a linear array in memory where each slot is 4 bytes wide. The first element of the tensor goes into the first slot, the second element into the second slot, and so on. The contiguous memory layout is the first half of the story; we also need to keep track of the tensor's dimensions.

Strides serve as a bridge between a tensor's logical representation and its physical memory location. To find an element at a specific index, such as tensor[0, 1], in other words the second element of the first row, we use strides in a calculation that involves the tensor's dimensions. This internal mechanism allows PyTorch to map high-level tensor operations to low-level memory manipulations efficiently.

When accessing an element at a particular index, strides help pinpoint its exact memory location. The method involves multiplying each index by its corresponding stride and summing these products. This is vital for operations like slicing or reshaping, enabling efficient memory usage without duplicating data.

Creating a view of a tensor, for example, by extracting a row or column, doesn't generate a new tensor. Instead, we create a new "window" to view the same underlying data. Which is efficient but comes with caveat **modifying the view also alters the original tensor**. To break the link to an original tensor's memory we have to make a copy of a view.

Strides are versatile enough to handle non-contiguous data layouts as well. If we want to extract a column from a 2D tensor, the elements aren't adjacent in memory. In this case, we can use a stride greater than one to jump to the next element, effectively creating a non-contiguous view.

In PyTorch's architecture, the logical tensor, and its physical data, known as **Storage**, are decoupled. This allows multiple tensors to share the same storage but have different sizes, strides, and offsets.

When performing operations like `torch.mm` — PyTorch's matrix multiplication function — the framework employs a two-level dispatch mechanism. The first level determines the device (CPU, CUDA, etc.) and layout (strided, sparse, etc.), while the second level dispatches based on the data type. This modular design supports various tensor types, including specialized ones like XLA or MKL-DNN tensors.

When we're interested in extending PyTorch tensors, we need to decide which of the three key parameters—device, layout, and dtype—to extend. These parameters define the tensor's characteristics and functionalities. Alternatively, a wrapper class around PyTorch tensors can be created, especially if there's no need to pass the tensor through the autograd backward pass.

Basic tensor construction

It's time to get into some coding, you can just follow along, but we recommend playing with tensor manipulation yourself to help develop your understanding of the constructions and operations discussed here.

You can see in the following code that building simple tensors with torch constructors is easy. All we do is provide a number, list, or array to `torch.tensor`, with a `dtype`, in this case `torch.float32`.

```
import torch

scalar = torch.tensor(42, dtype=torch.float32)

vector = torch.tensor([1, 2, 3, 4, 5], dtype=torch.float32)

matrix = torch.tensor([
```

```
    [1, 2, 3],
    [4, 5, 6],
    [7, 8, 9]
], dtype=torch.float32)

tensor_3d = torch.tensor([
    [[1, 2], [3, 4]],
    [[5, 6], [7, 8]],
    [[9, 10], [11, 12]]
], dtype=torch.float32)

print(scalar)
print(vector)
print(matrix)
print(tensor_3d)
```

```
tensor(42.)
tensor([1., 2., 3., 4., 5.])
tensor([[1., 2., 3.],
        [4., 5., 6.],
        [7., 8., 9.]])
tensor([[[ 1.,  2.],
         [ 3.,  4.]],

        [[ 5.,  6.],
         [ 7.,  8.]],

        [[ 9., 10.],
         [11., 12.]]])
```

Of course usually we want to create tensors with a few more numbers than we are willing to type ourselves, for specific purposes like initializing model weights, or creating placeholders etc. In which case `torch` comes with some handy factory functions, such as `torch.rand`, `torch.randn` or `torch.zeros`:

```
torch.manual_seed(42)

randn_tensor = torch.randn(3,2)

print(f'data type \n', type(randn_tensor))
print(f'within tensor data type \n',randn_tensor.dtype)
print(f'tensor values \n', randn_tensor)
```

```
data type
```

```
 <class 'torch.Tensor'>
within tensor data type
 torch.float32
tensor values
 tensor([[ 0.3367,  0.1288],
        [ 0.2345,  0.2303],
        [-1.1229, -0.1863]])
```

You can see in the example we create a two dimensional tensor (matrix) with 3 rows and 2 columns, populated with data taken from a random normal distribution. The tensor we have created also has a float32 format. The additional function applied here `torch.manual_seed()` allows us to use any number to help make results reproducible; you don't have to use 42! 42 is just a commonly used value, referencing Hitchhiker's Guide to the Galaxy. Regardless, setting a random seed is important for research purposes, as we commonly initialize the weight tensors of models that act as the starting point of a model build, making seeds vital for reproducibility between runs and devices.

Let's quickly sidetrack into operations, it's always useful to see how we can manipulate tensors with the functions that are for building neural networks. We'll look at three common operations: **tensor x tensor** (element-wise multiplication, where corresponding elements are multiplied together), **matrix multiplication** (the dot product operation), and **addition**. Element-wise multiplication produces a tensor of the same shape with each position multiplied independently, while matrix multiplication follows linear algebra rules and often changes dimensions.

Element-wise multiplication produces a tensor of the same shape with each position multiplied independently, while matrix multiplication follows linear algebra rules and often changes dimensions. Element-wise multiplications can be performed using either `.mul()` or the * operator, and matrix multiplication using either `.matmul()` or the @ operator. We'll be using the method syntax here for clarity:

```
print(f'tensor x tensor \n', randn_tensor.mul(randn_tensor), f'\n')
print(f'matrix multiplication \n', randn_tensor.matmul(randn_tensor.T))
print(f'add 1 \n', randn_tensor.add_(1))
```

```
tensor x tensor
 tensor([[0.1134, 0.0166],
        [0.0550, 0.0531],
        [1.2608, 0.0347]])

matrix multiplication
 tensor([[ 0.1300,  0.1086, -0.4021],
        [ 0.1086,  0.1080, -0.3062],
        [-0.4021, -0.3062,  1.2955]])
add 1
 tensor([[ 1.3367,  1.1288],
        [ 1.2345,  1.2303],
        [-0.1229,  0.8137]])
```

Some other useful operations are concatenation and simply clearing the numbers of a tensor to zero:

```
zero_tensor = torch.randn(3,2).zero_()
concat_tensor = torch.cat([zero_tensor, randn_tensor], dim = 1)

print(f'zero tensor \n', zero_tensor)
print(f'concatonated tensors \n', concat_tensor)
```

```
zero tensor
 tensor([[0., 0.],
        [0., 0.],
        [0., 0.]])
concatenated tensors
 tensor([[ 0.0000,  0.0000,  1.3367,  1.1288],
        [ 0.0000,  0.0000,  1.2345,  1.2303],
        [ 0.0000,  0.0000, -0.1229,  0.8137]])
```

In PyTorch we can apply inplace and functional operations to tensors. Inplace operations modify tensors directly and have a trailing underscore (i.e., `a.mul_(b)` would change value a inplace after multiplying by `b`). All inplace operations in PyTorch use this naming convention. The functional equivalent would act on a copy rather than modifying the original tensor. In the previous example we created a 3x2 tensor with random values via `torch.randn(3,2)`, we then used the inplace operation `zero_()`, which has an underscore appended to its name. We also wanted to demonstrate method chaining of operations. The final operation we demonstrated is concatenation using torch.cat, which concatenates a list of tensors along a specified dimension. The easiest way to remember is that dimension 0 refers to row-wise operations, while dimension 1 refers to column-wise operations.

Of course, this is a small amount of what PyTorch has to offer for manipulating and calculating data. A short investigation into the PyTorch documentation can take you a long way and can help you understand what PyTorch is doing, mostly under the hood, when modeling with neural networks. You can also read this (`https://machinelearningmastery.com/manipulating-tensors-in-pytorch/`) excellent article by *Adrian Tam* on *Machine Learning Mastery* for slightly more depth on tensor manipulation.

At this point we know how to construct a tensor, but you'll also need to convert data from DataFrame. The simplest approach is to pull the values of the columns you're interested in. If you have a single column, this will just create a 1-dimensional tensor, while additional columns will create a 2-dimensional tensor.

Let's look at a practical example using the airline passengers dataset. We'll demonstrate how to convert `pandas` DataFrame columns into PyTorch tensors, creating both a univariate tensor (single column) and a multivariate tensor (multiple columns).

This is a common pattern you'll use when preparing time series data for modeling.

```
# Load dataset
data = pd.read_csv(CFG.data_folder / 'passengers.csv')

# Add a count column
data['count'] = np.arange(1, len(data) + 1)
tensor_1_col = torch.tensor(data['passengers'].values,
                            dtype=torch.float32)
tensor_2_col = torch.tensor(data[['passengers', 'count']].values,
                            dtype=torch.float32)

#print(f'simple univariate tensor', tensor_1_col)
#print(f'simple multivariate tensor', tensor_2_col)
```

It's worth noting at this point that while the general set of the tensors tends to be the same for most conversions of DataFrame, the specific configuration of the tensors and how we load them vary greatly, for example RNNs require us to create look-back sequences/windows. We will discuss this in *Chapter 5*.

Managing memory well is crucial when working with neural architectures. Every operation can allocate new memory for its results. For example, `B = A + B` redirects tensor `B` to a new location containing the sum of tensor `A + B`, abandoning its old memory address. We can verify this simply by using `id(B)`, which returns an object's address in memory. After an operation, `id(B)` will be in a new location from before.

```
torch.manual_seed(42)

A = torch.randn(3,2)
B = torch.ones(3,2)
print(f'Original location in memory', id(B))

B = A + B
print(f'New location in memory', id(B))
```

```
Original location in memory 1966217801520
New location in memory 1966217798320
```

This can be problematic, as we want to avoid unnecessarily allocating memory, which is wasteful given that we are often faced with large datasets and frequent updates in machine learning. Secondly, parameters are often referenced in multiple places, so inconsistent updates across references can cause issues.

PyTorch provides in-place operations to address these issues. Using slice notation like `Y[:] = <expression>`, we can direct an operation's output to override an existing array in-place. For example, we can reuse `B`'s memory for the sum of `A` and `B`:

```
A = torch.randn(3,2)
B = torch.ones(3,2)
print('Location in before operation', id(B))

B[:] = A + B
print('Location in after operation', id(B))
```

```
Location in before operation 1966217802240
Location in after operation 1966217802240
```

This retains `B`'s original memory address without allocation. The key is that in-place updates keep memory addresses unchanged, ensuring consistency across references and preventing leaks.

Understanding computational graphs

Neural networks are fundamentally mathematical equations into which we feed data that is encapsulated within tensors. These equations are calculated through computational graphs with autograd. **Neurons** within artificial neural networks essentially act as different parts of an equation, which are mathematically 'connected'. We use computational graphs to build these functions and keep track of how values within a network both relate to each other and change, via our calculations with automatic differentiation (autograd). The reason for PyTorch's dominance in the field of deep learning is in part due to its dynamic approach in constructing computational graphs and its application of autograd. Don't worry if you don't understand everything, we'll discuss neural networks in the next chapter.

Computational graph, also referred to as a **directed acyclic graph** (DAG), captures sequences of operations performed on tensors, with nodes representing tensors and edges symbolizing the mathematical operations transforming one tensor to another. The graph structure is a mental model that helps us visualize and understand the sequence of operations that tensors undergo, and importantly, how gradients flow backward during training. In PyTorch this graph is constructed dynamically, built as operations are executed. This contrasts with TensorFlow which utilizes a static graph that is defined and optimized before models run. This difference gives PyTorch an edge in flexibility, allowing for more intuitive debugging and accommodating models with variable input lengths and conditional operations, such as those found in recurrent neural networks. For practitioners, like ourselves, this means a more iterative and experimental approach to model design, which helps to facilitate rapid prototyping and research development in general.

To bring the idea of a computational graph to life, we have created a diagram of an FFN, which is a type of computational graph representing all operations of the FFN as a graph structure in the context of executing and differentiating a neural network. This is essentially a visualization of the graph; this one has an input node, three hidden nodes, and one output node.

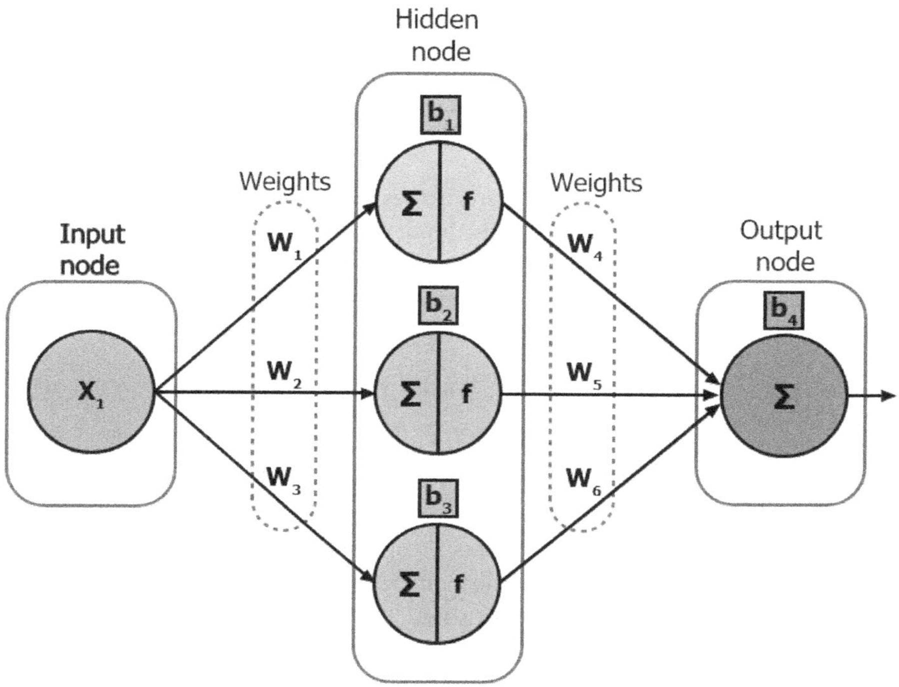

Figure 4.3: Basic Feedforward network (FNN) diagram

What this diagram represents is the basic structure of a FFN network in which we have an input layer, with three hidden layers and an output layer. You can see these are connected with weighted pathways and biases symbolizing the mathematical transformations between layers.

In a computation graph, nodes represent operators, essentially the mathematical operations we perform during both forward and backward passes. The only exception is for user-defined variables, which we also include in the graph for clarity. Although these variables (i.e. x, $w1$, $w2$, $w01$, $b1$, and $b0$) help us understand the graph, they are not inherently part of it. The variables , and are created as a result of mathematical operations, whereas variables a, $w1$, $w2$, $w3$, and $w4$ and are initialized by the user. Since they are not created by any mathematical operator, nodes corresponding to their creation are represented by their name itself. This is true for all leaf nodes in the graph.

We can represent the calculations up to the hidden layer in this simple form:

$$h_1 = w_1 \cdot x_1 + b_1$$
$$h_2 = w_2 \cdot x_1 + b_2$$
$$h_3 = w_3 \cdot x_1 + b_3$$

We would usually have an activation function applied at each node, for example, ReLU or sigmoid, but to keep this example simple we have removed it.

The product of these hidden layer nodes is then passed on to an output layer, during which the outputs are multiplied by three more weights, and have a bias value added. This is represented mathematically as:

$$y = w_4 \cdot h_1 + w_5 \cdot h_2 + w_6 \cdot h_3 + b_4$$

This whole calculation process is fairly simple and makes up most of what we refer to as a **forward pass**; the output of one layer of a neural network is the input of the next layer and so forth. The final aspect of this forward pass is calculating **loss**, which is the outcome of measuring the difference between a network's calculated output and 'actual' values. In the following formula we've simply calculated loss as mean squared error (MSE), but we can use many different **error metrics**.

$$L = (y_{\text{true}} - y)^2$$

Once we have a value for loss, we can calculate gradients, which are partial derivatives of the loss function with respect to each parameter (weights and biases). A derivative measures the rate of change in a function as its inputs are varied infinitesimally. For neural networks, derivatives of a loss function tell us how rapidly loss changes with small adjustments to weights and biases, helping optimization algorithms to select parameter updates that most effectively reduce loss. Let us visualize the relationship between a quadratic function and its derivative.

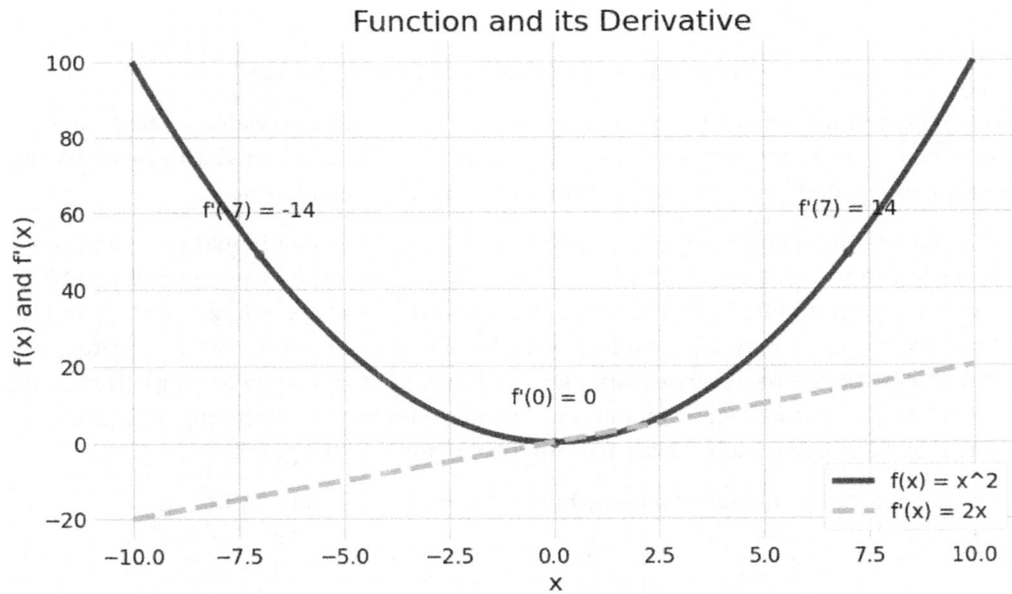

Figure 4.4: Function and derivative

The graph depicts a quadratic function (blue) and its rate-of-change/derivative (orange), which is the slope of the tangent at a point along the curve. The Red dots highlight specific points of interest. At *x=0*, the slope is zero—indicating the curve's lowest point, critical in neural network optimization for finding the minimum loss. Steeper slopes at *x=-7* and *x=7* show rapid changes in the function, guiding the adjustments of parameters (like neural network weights) to reach the minimum point efficiently during training.

We find these rates of change in calculations by making use of the chain rule of calculus during a **backward pass** through a neural network; iterating this process is what is known as backpropagation.

Chain rule revisited

The chain rule underlies backpropagation in neural networks. Consider a simple feed-forward neural network with a single hidden layer. The output is a function of the hidden layer, which itself is a function of the input x. Mathematically, this relationship can be expressed as $y=f(g(x))$. Using the chain rule allows us to decompose the derivative of with respect to into a **product** of simpler derivatives:

$$\frac{dy}{dx} = \frac{dy}{dh} \times \frac{dh}{dx}$$

Each layer's transformation can be considered a function:

- **Forward pass:** we compute outputs by applying these functions sequentially
- **Backward pass:** we use the chain rule to compute gradients for each parameter, starting from the output layer and moving backward through the network

Backpropagation calculations

The backward pass uses the chain rule to calculate gradients, which is essential for backpropagation, where an optimization algorithm updates each network parameter. Once the gradients are obtained, the next step is to update each parameter using an optimization algorithm, such as **stochastic gradient descent** (**SGD**) or one of its variants. The update follows a simple rule: adjust each parameter in the direction that reduces the loss. For a parameter with learning rate η, this looks like:

$$w_i = w_i - \eta \frac{\partial L}{\partial w_i}$$

The intuition here is that the gradient tells us how loss (difference between network output and actual values) changes with respect to a parameter, so we move the parameter in the opposite direction (hence the minus sign) to push the loss toward its minimum. We control the size of each step with a learning rate. Under our defined network, the gradients we need end up looking something like this:

$$\begin{aligned}
\frac{\partial L}{\partial w_{o1}} &= \frac{\partial L}{\partial y} \cdot \frac{\partial y}{\partial w_{o1}} = 2(y - y_{\text{true}}) \cdot h_1 \\
\frac{\partial L}{\partial w_{o2}} &= \frac{\partial L}{\partial y} \cdot \frac{\partial y}{\partial w_{o2}} = 2(y - y_{\text{true}}) \cdot h_2 \\
\frac{\partial L}{\partial w_{o3}} &= \frac{\partial L}{\partial y} \cdot \frac{\partial y}{\partial w_{o3}} = 2(y - y_{\text{true}}) \cdot h_3 \\
\frac{\partial L}{\partial b_o} &= \frac{\partial L}{\partial y} \cdot \frac{\partial y}{\partial b_o} = 2(y - y_{\text{true}}) \\
\frac{\partial L}{\partial w_1} &= \frac{\partial L}{\partial y} \cdot \frac{\partial y}{\partial h_1} \cdot \frac{\partial h_1}{\partial w_1} = 2(y - y_{\text{true}}) \cdot w_{o1} \cdot x \\
\frac{\partial L}{\partial w_2} &= \frac{\partial L}{\partial y} \cdot \frac{\partial y}{\partial h_2} \cdot \frac{\partial h_2}{\partial w_2} = 2(y - y_{\text{true}}) \cdot w_{o2} \cdot x \\
\frac{\partial L}{\partial w_3} &= \frac{\partial L}{\partial y} \cdot \frac{\partial y}{\partial h_3} \cdot \frac{\partial h_3}{\partial w_3} = 2(y - y_{\text{true}}) \cdot w_{o3} \cdot x \\
\frac{\partial L}{\partial b_1} &= \frac{\partial L}{\partial y} \cdot \frac{\partial y}{\partial h_1} \cdot \frac{\partial h_1}{\partial b_1} + \frac{\partial L}{\partial y} \cdot \frac{\partial y}{\partial h_2} \cdot \frac{\partial h_2}{\partial b_1} + \frac{\partial L}{\partial y} \cdot \frac{\partial y}{\partial h_3} \cdot \frac{\partial h_3}{\partial b_1} = 2(y - y_{\text{true}}) \cdot (w_{o1} + w_{o2} + w_{o3})
\end{aligned}$$

Gradients on the right side of our equations can all be computed directly since the numerators are explicit functions of the denominators. While notation may make the mathematics look complicated, calculations are fairly intuitive. Although this is straightforward for a single-layer network, it scales in complexity with network depth, rendering manual gradient computation impractical. Who wants to spend precious hours calculating derivatives anyway? Luckily for us, computational graphs facilitate automation of this process, allowing for efficient, architecture-agnostic gradient computation and enabling us to focus on network design and optimization without the arduous task of manual gradient calculations.

We can reduce the reasoning behind computational graphs to five points:

- **Backpropagation** - It provides a structure on which to apply the chain rule and compute gradients. Starting from the loss node, gradients are backpropagated to earlier nodes, effectively calculating how much each parameter contributed to the loss.
- **Efficiency** - Graph structures allow for optimized tensor operations. Since the graph captures dependencies, it's easier to parallelize operations and reduce computational overhead.
- **Memory Management** - Intermediate values can be discarded after use, reducing memory usage.
- **Debugging and Visualization** - The graph can be inspected to debug model architecture, understand tensor transformations, and optimize performance.
- **Dynamic vs. Static** - PyTorch's dynamic graph is particularly useful for models where the architecture changes conditionally (e.g., RNNs with varying sequence lengths), as it allows for more flexibility.

A brief look at autograd

PyTorch's autograd translates the intricate mathematical calculations of neural networks into computable gradients. Autograd's value lies in handling the chain rule automatically, transforming this repetitive, and cumbersome task of computing derivatives across complex networks into an elegant and precise process.

> Autograd does this while avoiding the direct construction of potentially massive Jacobian matrices (TensorFlow) in favor of a dynamic computational graph that efficiently computes Jacobian-vector products (JVPs) during the backward pass. This approach is anchored in the chain rule, as we have discussed, utilizing backward nodes that contain the JVP versions of derivatives for forward operations, negating the need for explicit matrix construction.

The forward pass of a network, as detailed in our section on computational graphs, unfolds tensors through layers of operations within the graph, with each layer applying its mathematical influence on the data within the tensors. The backward pass, initiated by calling `.backward()`, is where autograd comes into action. It retraces the forward path, utilizing gradients of the loss function to impart learning signals to each parameter through the chain rule:

$$\frac{\partial L}{\partial x} = \frac{\partial L}{\partial y} \cdot \frac{\partial y}{\partial x}$$

Here, represents loss, is the model's output, and represents the parameters we seek to optimize. The partial derivative encapsulates gradient of loss with respect to output, while holds gradients of output with respect to parameters.

Autograd's application of the chain rule within the DAG allows for localized computation of gradients, which it then aggregates to form gradients of higher-order tensors. For each operation that a tensor undergoes, autograd maintains a `grad_fn`, a function that computes the derivative of that operation. For each operation encountered, autograd knows the specific JVP, which it applies during the backward pass to synthesize the necessary gradients. The `grad_fn` function is essentially the embodiment of the JVP for an operation; this approach sidesteps the computational burden of full Jacobian matrices.

In PyTorch, every tensor operation adds to a tape, which is a record that Autograd uses to compute gradients. The backward pass peels back this tape layer by layer, applying the JVPs in reverse sequence. This is the heart of backpropagation—the process which informs how we adjust parameters to minimize loss. We can illustrate this with an example showing how autograd computes gradients for a composition of logarithm and sine functions:

```
# Define operations of forward pass
def forward_pass(x):
    # Operation 1 - Logarithm
    log_x = torch.log(x)
    # Operation 2 - sine
    sin_log_x = torch.sin(log_x)
    return sin_log_x

# Instantiate x tensor, setting requires_grad=True in order to track
computations
x = torch.tensor([2.0], requires_grad=True)

# Forward pass to compute output
```

```
y = forward_pass(x)

# Backward pass, to compute gradient of y with respect to x
y.backward()

# Gradient stored in x.grad
print(f"Gradient of y with respect to x is: {x.grad}")
```

Gradient of y with respect to x is `tensor([0.3846])`

In this code, autograd integrates the logarithmic and sine functions into the computational graph. Upon invoking `y.backward()`, it employs the JVPs of these operations to calculate gradient of the loss with respect to our input tensor x. This gradient is stored in `x.grad`, which is the culmination of complex computations rendered accessible by autograd's design.

Autograd's utility is further exemplified in how it handles complex functions. Rather than constructing voluminous Jacobian matrices, it computes Jacobian-vector products (JVPs)—a far more memory-efficient approach. This method showcases autograd's computational efficiency and adaptability.

Whether dealing with a recurrent neural network or a feedforward model, autograd empowers us to focus on model design and problem-solving. Manual gradient calculation has become unnecessary, replaced by the dynamic computational pathways of autograd in PyTorch's computational graphs.

Practicing PyTorch fundamentals

At this point you should have a reasonable understanding of the fundamentals of neural networks and PyTorch; we can reduce down what we have learned to the following:

- Data in PyTorch is stored in arrays of n-dimensions, that we call tensors
- Tensors make efficient use of CPU and GPU, allowing rapid calculations
- Neural networks are fundamentally a series of mathematical operations we represent in a 'computational graph'
- The movement of data through these operations is the forward pass that creates a loss value
- Loss metrics are used to measure gradients of loss, with respect to weights, using the chain rule
- Loss gradients are used by optimization algorithms to adjust weights and biases in order to improve the accuracy of network outputs

Now is a good time to consolidate your new knowledge, and build our understanding of PyTorch and Lightening. We have built a `jupyter` notebook with the basic tensor operations; from creation and addition to matrix multiplication. Playing with the values and the functions is a great way to develop an intuition for PyTorch functions, which makes algebra and calculus trivial. This allows data scientists, like us, to focus on model design.

In what remains of this section we are going to build some simple neural networks; we'll discuss these more in the next chapter. Here you are simply looking to add depth of understanding for the fundamentals we have already discussed. First, let's expand on our previous gradient calculation example, and apply it to what would be a single layer of a neural network. Here, we define some tensors: input, weights, actual values to predict, and the output layer:

```
# Definitions - input data, weights, and 'true output' for loss calculation
input_tensor = torch.tensor([[1.0, 2.0, 3.0]], requires_grad=True)
weights = torch.tensor([[0.1, 0.2, 0.3], [0.4, 0.5, 0.6]], requires_grad=True)
true_output = torch.tensor([[0.7, 1.5]])

# Forward pass - Matrix multiplication (input with weights), using ReLU
activation
layer_output = torch.mm(input_tensor, weights.t()).relu()

# Calculate loss - MSE
predicted_output = layer_output
loss = (true_output - predicted_output).pow(2).mean()

# Backward pass with autograd to compute gradients
loss.backward()

# dict of tensors and gradients
nn_basic_results = {
    "Input Tensor": input_tensor,
    "Weights": weights,
    "Layer Output after ReLU": layer_output,
    "Loss (MSE)": loss,
    "Gradient with respect to Input Tensor": input_tensor.grad,
    "Gradient with respect to Weights": weights.grad
}
# print result
nn_basic_results
```

Our code here simulates the single layer of a neural network, with predefined inputs and weights, which are utilized in `layer_output` after applying the ReLU activation function to the matrix multiplication of our input tensor and the transposed weights. The loss of the MSE is simply calculated by comparing the network's predicted output from the layer against our true output.

We have also printed out the gradients here, which indicate how much each element of our input tensor and weights tensors contributed to our MSE. This is all calculated automatically in backward pass with autograd, which succinctly shows us everything we have done.

Pure PyTorch neural network

Taking this one step further we can build a full neural network to demonstrate the anatomy of a typical PyTorch model. In PyTorch, constructing a neural network usually begins with defining a class that extends torch.nn.Module, which acts as the blueprint for your model. In the **init** method, you initialize the layers and parameters of a model. For instance, in our SimpleNN class, we define self.weights as a parameter, which will be learned and adjusted during our training process.

In the forward method, where data processing occurs, we apply the network to an input (x), this definition describes how input data flows through a network. In our example, x is passed through a linear transformation (via matrix multiplication with self.weights) and then through a ReLU activation function.

Once your model class is defined, the next step is to set up a training loop, under which a model is trained iteratively over your data. Each iteration (or epoch) involves processing input data through a model, calculating a value of loss, and updating the model's weights accordingly. In our case, we calculate MSE, which is a common choice for regression tasks.

As we have already seen, the backward pass that updates the model's weights is simply called by `loss.backward()`. Then, an optimizer, such as SGD, is used to update the weights of the model based on these gradients.

Although we can manually enter data, we often use a `DataLoader` class to handle the loading and batching of inputs. This simplifies feeding data into a model, and can be easily customized for different data processing needs.

```
# Define neural network
class SimpleNN(torch.nn.Module):
    def __init__(self):
        super(SimpleNN, self).__init__()
        self.weights = torch.nn.Parameter(torch.tensor([
            [0.1, 0.2, 0.3], [0.4, 0.5, 0.6]]))

    def forward(self, x):
        # Forward pass through layer and ReLU activation
        return torch.mm(x, self.weights.t()).relu()

# Instantiate neural network
model = SimpleNN()

# Define optimizer
optimizer = torch.optim.SGD(model.parameters(), lr=0.01)

# Define tensors for input and output
input_tensor = torch.tensor([[1.0, 2.0, 3.0]], requires_grad=True)
```

```
true_output = torch.tensor([[0.7, 1.5]])

# Wrap tensors in a DataLoader
train_loader = DataLoader(TensorDataset(input_tensor, true_output),
                          batch_size=1)

# Training Loop
for epoch in range(1):
    for batch in train_loader:
        # Separate data into inputs and outputs
        inputs, targets = batch
        # Zero gradients
        optimizer.zero_grad()
        # Forward pass
        predicted_output = model(inputs)
        # Compute Loss
        loss = torch.mean((targets - predicted_output).pow(2))
        # Backward pass
        loss.backward()
        # Update weights
        optimizer.step()
```

This sequence of defining the model, setting up the training loop, calculating loss, performing backpropagation, and updating parameters forms the core structure of neural network modeling in PyTorch. It's a structured yet flexible framework that lays the groundwork for a wide array of neural architectures that we can extensively build on and adapt.

Lightning NN

We can simplify our coding of the neural network and its training by using PyTorch Lightning (or simply Lightning). Lightning provides us with a streamlined approach to building and training neural networks, allowing us to focus on the core aspects of our models while it handles the underlying complexities. It organizes our code into logical segments, along with simplifying model definition, data handling, training, and optimization. Using a Lightning-based structure makes code more readable and maintainable, along with significantly reducing the repetitive code (boilerplate), i.e., training loops, that is typically associated with PyTorch. While you'll probably not notice it if you use it as default, Lightning saves us a lot of time on complex projects, where managing training loops and device transfers can become a nightmare.

Lightning adheres to best practices, ensuring that your models are both robust and efficient. It's also designed to scale easily, simplifying any transition to multi-GPU or TPU training, which is crucial for handling large-scale datasets and models. Although it takes a structured approach, the library is highly flexible, and allows us to override default behaviors to suit our needs.

Below we have redefined the SimpleNN as a lightening module. In this version the model is still defined as a class, but this time it inherits from `L.LightningModule`, a Lightning class that wraps around torch.nn.Module. Which provides the lightening structure, neatly compartmentalizes a neural network and its training into distinct methods, each serving a specific purpose:

- **Forward** - Similar to the forward pass in PyTorch, Lightning's forward method defines how data flows through your network. A key advantage here is that Lightning automatically manages device transfers (GPU or CPU) for tensors.
- **Training step** - The training_step method in Lightning is where the action happens in each training iteration. You define how your model processes input data and computes the loss within this method. Lightning's framework takes care of the backward pass and the optimization step, automating away a large chunk of code.
- **Configure optimizers** - Optimization strategies, a crucial part of neural network training, are defined in the configure_optimizers method. This method is isolated from the rest of your training code, meaning you can easily adjust your optimization strategy—changing the learning rate or switching optimizers—without needing to modify other code.
- Data Handling - Data handling in Lightning is also usually accomplished through DataLoaders, but there are dedicated methods like train_dataloader that we can use.

Let's see how this looks in practice with our simple neural network:

```
import lightning as L

class SimpleNN(L.LightningModule):
    def __init__(self):
        super(SimpleNN, self).__init__()
        self.weights = torch.nn.Parameter(torch.tensor([
            [0.1, 0.2, 0.3], [0.4, 0.5, 0.6]]))

    def forward(self, x):
        # Forward pass
        return torch.mm(x, self.weights.t()).relu()

    def training_step(self, batch, batch_idx):
        # Forward pass with loss calculation
        x, true_output = batch
        predicted_output = self(x)
        loss = torch.mean((true_output - predicted_output).pow(2))
        return loss

    def configure_optimizers(self):
        # Optimization with SGD
        return torch.optim.SGD(self.parameters(), lr=0.01)
```

```
# Instantiate neural network
model = SimpleNN()

# Define input and output tensors
input_tensor = torch.tensor([[1.0, 2.0, 3.0]])
true_output = torch.tensor([[0.7, 1.5]])

# Wrap data in a DataLoader
train_loader = DataLoader(TensorDataset(input_tensor, true_output),
                          batch_size=1)

# Define trainer
trainer = L.Trainer(max_epochs=1)

# Fit model
trainer.fit(model, train_loader)
```

In contrast to the more explicit and detailed training loop in pure PyTorch, Lightning's structure abstracts the loop into trainer.fit(model) call. This internally manages the steps we manually wrote in the pure PyTorch implementation, like the forward pass, loss computation, backward pass, and optimizer step, making the overall process more streamlined.

Our Lightning implementation, therefore, improves upon the pure PyTorch structure by providing a more organized, readable, and maintainable framework. It allows for focusing more on model development and less on the underlying mechanics of training a neural network, which is particularly beneficial as models and training processes become more complex.

While Lightning offers advantages in code organization and scalability, it hasn't completely replaced vanilla PyTorch for a few good reasons. Firstly, it adds an additional abstraction layer, which makes debugging trickier when things go wrong — you're troubleshooting both your model logic and Lightning's framework. When we are experimenting with new models, training procedures or custom backward passes, Lightning's structure can lack flexibility and may require workarounds for non-standard training loops.

Performance overhead is worth considering too. Lightning's abstractions around logging, callbacks, and framework operations can slow training if misconfigured. This can occur with small batch sizes or fast-iterating models, where Lightning's per-step overhead becomes significant relative to actual computation. Some implementations report 20-30% longer training times compared to vanilla PyTorch. Proper configuration minimizes this, but it's something to watch when training time is critical or you're running extensive hyperparameter searches.

That said, for production code, multi-GPU training, and complex projects managing multiple experiments, Lightning's benefits typically outweigh these drawbacks. Many practitioners prototype with vanilla PyTorch to understand their models, then transition to Lightning when project complexity demands better organization.

Summary

In this chapter we explored PyTorch's fundamentals and applied Lightning to enhance neural network training, validation, and testing workflows. We have learned how we encapsulate data in tensors, facilitating efficient computation on CPUs and GPUs. You have also been introduced to the computational graph, which dictates the sequence of operations in neural architectures. As a result, we came to understand this sequence informs forward pass calculations.

Critical to neural network modeling is computing loss gradients using the chain rule, which allows optimization algorithms to adjust network parameters (weights and biases) toward improved predictive accuracy. Finally you learned about the basic coding structure for pure PyTorch and Lightning, with the latter highlighted for its ability to streamline and automate training workflows, allowing us to focus on model design rather than procedural code. In the next chapter you will learn about the powerful but 'simple' neural network design: the feedforward network (FFN/MLP), and how it can be used for forecasting.

Recommended reading

- Dive into deep learning, `https://d2l.ai/index.html`
- Neural Networks and Deep Learning, `http://neuralnetworksanddeeplearning.com/`
- Overview of PyTorch Autograd Engine, `https://pytorch.org/blog/overview-of-pytorch-autograd-engine/` PyTorch internals `http://blog.ezyang.com/2019/05/pytorch-internals/`
- PyTorchTutorials, `https://pytorch.org/tutorials/beginner/blitz/autograd_tutorial.html`

Join our community on discord

Join our community's Discord space for discussions with the authors and other readers:

`https://discord.gg/VXpz9YPUTW`

5

Simple Neural Architecture

"Deep learning is going to be able to do everything."—Geoffrey Hinton

As we continue our journey into time series forecasting with PyTorch, we are ready to delve into neural networks and build our own. Compared to traditional statistical methods like generalized linear models, neural networks are more complex modeling structures — a complexity that grants them remarkable flexibility and power. This is not as daunting as it seems; neural architectures are, in reality, combinations of mathematical operations arranged in networks, and as we'll see in later chapters, even these networks can be combined into greater structures. Think of these mathematical components as building blocks — like Lego pieces — that we can connect to create intricate and versatile structures, such as a castle or, if you prefer, a section of a brain. This composability is precisely why **Deep Learning** (**DL**) refers to a family of learning algorithms, spanning recurrent to attention-based architectures among others, rather than a single method.

Our primary goal in this chapter is to develop our understanding of how neural architectures are structured and function. The fundamentals we will learn throughout this chapter and the book regard time series forecasting, but may be applied more widely to other modeling challenges. Once we pass the initial theory, we'll start to build a neural network with PyTorch and Lightning, which will give greater depth to the features of PyTorch we learned in the previous chapter.

In this chapter:

- Basic structures of neural networks
- Artificial neurons
- Neural nets
- From scalars to matrices
- Building a neural network

We will kick off this chapter by familiarizing ourselves with the basic components and functions of neural nets, and learn how they allow us to model and replicate almost any pattern. After this, we'll proceed to building a simple **Feedforward Neural Network** (**FFN**), also known as **Artificial Neuron Net** (**ANN**), using PyTorch. Once our model is complete, we will evaluate the performance of our *models*. This initial model-building exercise is designed to familiarize you with PyTorch, so don't worry too much when we touch on some topics briefly; we will delve much deeper into these, i.e. activation function, in the next chapter.

Basic structures of neural networks

Artificial neurons and neural networks draw from two mathematical foundations we'll build on throughout this section: linear functions and activation functions.

Artificial neurons, and consequently **Neural Networks** (**NN**), were initially conceptualized to mimic the structure and interactions between neurons in mammalian brains. The degree to which they successfully do this is up for debate, but what artificial neurons do well is replicate the general modularity of biological neurons while being excellent processors of data. The earliest attempts at what we would today call a neural network started in 1795 with Gauss and Legendre's linear networks — a *shallow learning* approach that Gauss used to re-discover Ceres. The major leap toward deep learning didn't come until the late 1950s with Rosenblatt's development of the **perceptron**, a more direct attempt to model biological neurons. Perceptrons combine Gauss and Legendre's linear networks with an output threshold function (**Heaviside/step**) — later called an **activation function** — to perform classification tasks.

We can define a perceptron as a type of artificial neuron that takes several binary inputs and produces a single binary output. Each input has an associated weight reflecting its importance in determining the output, and the perceptron fires (outputs 1) when the weighted sum of inputs exceeds a threshold value. In biological neurons, inputs from neighboring cells gradually depolarize the neuron; when cumulative depolarization surpasses a threshold, the neuron fires an action potential and propagates the signal forward. Artificial neurons work analogously — weighted inputs accumulate, and the activation function determines whether and how the neuron fires.

We can write out the perceptron's rule in terms of bias (b):

$$\text{output} = \begin{cases} 0 & \text{if } \mathbf{w} \cdot \mathbf{x} + b \leq 0 \\ 1 & \text{if } \mathbf{w} \cdot \mathbf{x} + b > 0 \end{cases}$$

Refer to color images on the Graphic bundle or the ebook.

Breaking this down into its components:

- **Input:** An input vector **x**, representing a data point with its features.
- **Weights:** A weight vector **w**, where each element reflects the importance of a particular feature.
- **Bias:** b is a scalar that shifts the decision boundary, adjusting for situations where data points are not centred around the origin.
- **Dot product:** $\mathbf{w} \cdot \mathbf{x}$ computes the weighted sum of input features; a measure of how well the input aligns with the weight vector.
- **Activation function:** The perceptron uses a step function. When $\mathbf{w} \cdot \mathbf{x} + b > 0$, the output is 1; otherwise 0, creating a decision boundary that separates the input space into two classes.

As a visual aid, refer to this diagram:

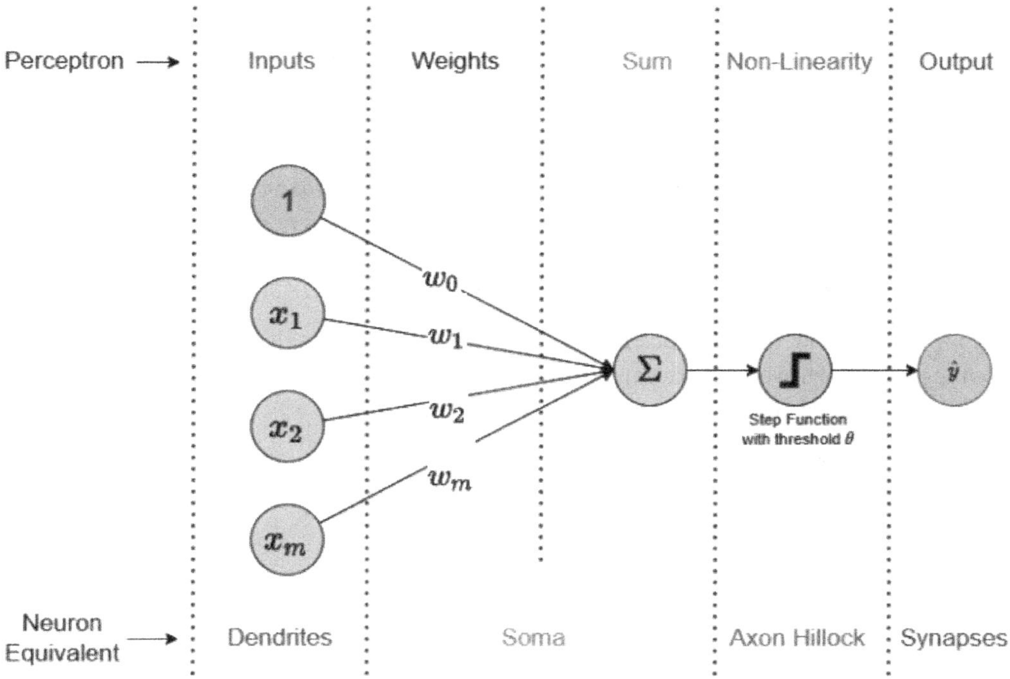

Figure 5.1: Diagram of perceptron structure showing inputs, weights, bias, and step activation function

Strictly speaking, perceptrons are classifiers due to their employment of the unit step activation function, which limits them to binary outputs. This is handy for linearly separable classification problems, but not regression. Now, perceptrons may not seem useful for forecasting where you want a float rather than a class, but they are the precursor to artificial neurons that form the building blocks of neural networks.

> Confusingly, the term *perceptron* can be used interchangeably with *artificial neuron*, which can have various activation functions. However, *true* perceptrons are generally associated with the Heaviside/unit step function for their activation, which is not strictly differentiable. A true perceptron is, therefore, a binary classifier, while artificial neurons employ a broader range of (generally) differentiable activation functions.
>
> It is also worth noting that neural networks can also be referred to as **Multi-Layer Perceptrons (MLPs)**. Again, some distinction is useful: the first MLP, also suggested by Rosenblatt, was a composition of multiple perceptrons and is more commonly referred to as an extreme learning machine. Later developments of the MLP architecture utilized artificial neurons with activation functions that were/are *differentiable*. Regardless, people can, and do, call neural networks MLPs, which I personally think is unhelpful, which is why we use the term neural network, or the more general architecture.

To fully appreciate the mechanics of a perceptron, it is best — as with most of data science — to build and train one, which we'll do in Python directly for simplicity. Our `true_perceptron` class has three methods. `__init__` constructs the object with an `input_dim` matching the feature vector, a `learning_rate` (default `0.1`) controlling the step size for weight updates, and a weights array initialized to zero with length `input_dim + 1` — the extra element holds the bias. `predict` computes the perceptron's output for a given input, calculating and applying the unit step function. `train` iteratively updates weights based on prediction errors, following the perceptron learning rule, for a set number of epochs (default `20`) — each epoch being a full pass through the training data.

For every input-label pair, the perceptron computes a prediction and calculates the error as the difference from the true label. Weights are adjusted in proportion to the learning rate, the error, and input features; bias is adjusted similarly but without the input scaling.

Learning rate determines the magnitude of weight and bias updates at each iteration over a batch of training data. Higher learning rates mean faster learning but risk overshooting the optimal solution; lower rates are slower but typically more precise.

- Weight updates

  $$\mathbf{w}_{\text{new}} = \mathbf{w}_{\text{old}} + \alpha \cdot \left(\text{target} - \text{prediction}\right) \cdot \mathbf{x}$$

  Where,

  - $\mathbf{w}_{\text{new}}$ and $\mathbf{w}_{\text{old}}$ are updated and previous weight vectors, respectively
  - α is the learning rate
  - $\left(\text{target} - \text{prediction}\right) \cdot \mathbf{x}$ is the error scaled by the input

- Bias update

  $$b_{\text{new}} = b_{\text{old}} + \alpha \cdot \left(\text{target} - \text{prediction}\right)$$

  Where, b_{new} and b_{old} are updated and previous bias terms respectively.

  Bias is updated like weights but without input scaling.

In the following example, we use **Stochastic Gradient Descent** (**SGD**), updating parameters after each individual training example. By contrast, batch gradient descent updates parameters once per epoch, after all data has passed through, and mini-batch gradient descent splits the data into small batches, updating parameters after each batch. Each such update is an iteration; a full pass over the dataset is an epoch.

```
class true_perceptron:
    def __init__(self, input_dim, learning_rate=0.1):
        self.weights = np.zeros(input_dim + 1)  # +1 for the bias
        self.learning_rate = learning_rate

    def predict(self, inputs):
        # Unit step activation function
        summation = np.dot(inputs, self.weights[1:]) + self.weights[0]
        return 1 if summation > 0 else 0

    def train(self, training_inputs, labels, epochs=20):
        for _ in range(epochs):
            for inputs, label in zip(training_inputs, labels):
                prediction = self.predict(inputs)
                self.weights[1:] += self.learning_rate * (
                    label - prediction) * inputs
                self.weights[0] += self.learning_rate * (
                    label - prediction)
```

Now we have built our perceptron, we can finally train it. In the following code, we create a simple scenario where the perceptron is given binary pairs in an array, and should learn to return a `1` only when both inputs are `1`. This input-output relationship is an `AND` function. We can verify if our perceptron has learned the pattern from the four inputs by providing it with some unseen data, and comparing its predictions to the expected output (`actuals`):

```
# Create training inputs for AND function
training_inputs = []
training_inputs.append(np.array([1, 1]))
training_inputs.append(np.array([1, 0]))
training_inputs.append(np.array([0, 1]))
training_inputs.append(np.array([0, 0]))

# Labels for AND function
labels = np.array([1, 0, 0, 0])

# Create a Perceptron object
perceptron = true_perceptron(2)
```

```
# Train perceptron
perceptron.train(training_inputs, labels)

# Test perceptron
# Output should be 1
inputs = np.array([1, 1])
print(perceptron.predict(inputs))

# Output should be 0
inputs = np.array([0, 1])
print(perceptron.predict(inputs))
1
0
```

As we can see, our perceptron easily learned associations between paired input values and labels. We can visualize the decision boundary by plotting the data points and coloring the regions corresponding to each class – this code computes predictions over a grid and uses `contourf` to display the boundary:

```
def plot_decision_boundary(perceptron, inputs, labels):
    plt.style.use('fivethirtyeight')

    # Scatterplot of data points
    df = pd.DataFrame(np.concatenate(
        [inputs, labels.reshape(-1,1)], axis=1
    ), columns=['x', 'y', 'label'])
    sns.scatterplot(x='x', y='y', hue='label', style='label', data=df)

    # Generate grid over input space to plot colour of classification
    xmin, xmax = plt.xlim()
    ymin, ymax = plt.ylim()
    xs = np.linspace(xmin, xmax, 30)
    ys = np.linspace(ymin, ymax, 30)
    Y, X = np.meshgrid(ys, xs)

    # Compute predictions over grid
    predictions = np.array(
        [perceptron.predict(np.array([x, y]))
        for x, y in zip(np.ravel(X), np.ravel(Y))]
    )
    predictions = predictions.reshape(X.shape)
```

```
    # Show boundary line
    ax = plt.gca()
    ax.contourf(X, Y, predictions,
                colors=['#D55E00', '#0072B2'],
                alpha=0.2, levels=range(-1, 2))
    plt.show()

# Call decision boundary function with trained perceptron
plot_decision_boundary(perceptron, training_inputs, labels)
```

The result:

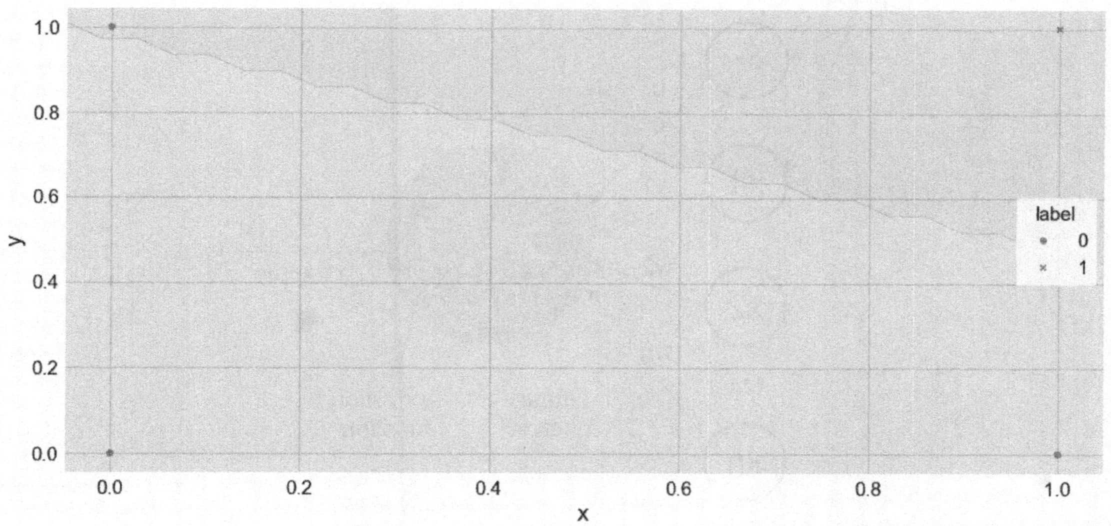

Figure 5.2: A scatter graph of a perceptron

As shown in *Figure 5.2*, the perceptron successfully learns a decision boundary that separates the input space into two classes. This makes it a useful tool for classification tasks.

However, this same mechanism also highlights a key limitation. Because the perceptron relies on a unit step activation function, its outputs are restricted to binary values (0 or 1). This means it cannot produce continuous predictions.

When we apply the perceptron to regression data, where the goal is to predict a range of values, it fails to capture the underlying trend. Instead, its predictions collapse into near-constant outputs, which appear as an almost flat line in a scatter plot.

This limitation stems from the perceptron's use of a unit step activation function, which restricts outputs to binary values. To model continuous relationships, we need a more flexible formulation—one that allows for smooth, differentiable outputs.

Artificial neurons

The key difference between perceptrons and artificial neurons lies, as we briefly mentioned, with **differentiable** activation functions. These introduce non-linearity into the neuron's output, enabling it to learn complex patterns from input data. We use artificial neurons as the smallest computational units within neural architectures, combining them in various ways to approximate more complex, non-linear functions. By selecting different activation functions, artificial neurons can adapt to different tasks, including regression.

Before delving into activation functions and combining neurons, let's redefine the perceptron as an artificial neuron in a little more detail.

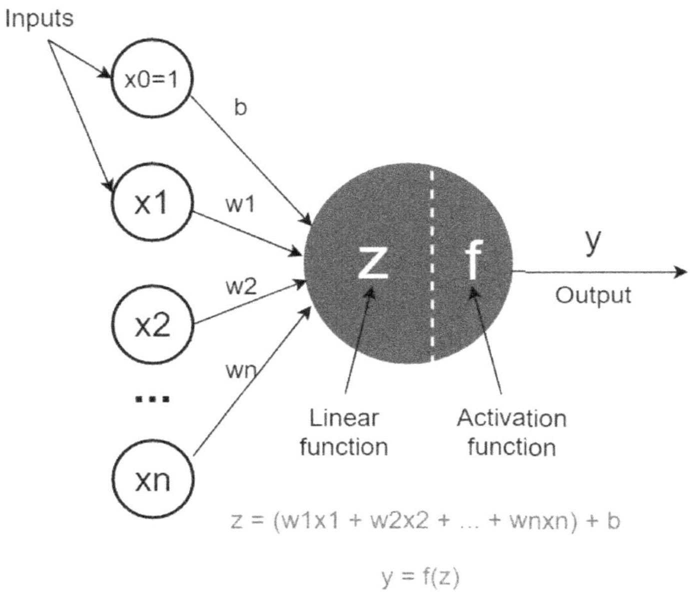

Figure 5.3: Artificial Neuron Diagram

The bias parameter is sometimes denoted as w_0, similar to how the intercept in a linear model is often denoted as β_0. The input to the bias always takes a value of 1.

Let's break down *Figure 5.3*:

- **Inputs:** Artificial neurons receive multiple inputs, represented as $x1, x2, \ldots, xn$. These can be past values of our dependent variable or features of a time series, such as lags, weather data, or sales figures. In a network, inputs can also be the outputs of other neurons.
- **Weights:** Each input to a neuron is assigned a weight, represented as $w1, w2, \ldots, wn$. Initially, weights are random values (although different approaches exist) that are adjusted during a learning process to improve model predictions.
- **Bias:** The bias term b is added to a weighted sum of the inputs. Bias helps improve the predictions of artificial neurons by accounting for situations where the data points are not centered around the origin. Bias determines if, and by how much, a neuron will propagate its output.

- **Linear function**: The weighted sum of inputs $\sum_{i=1}^{n} w_i \, x_i$is combined with the bias term $b$to form the linear function.
- **Activation function**: An activation function, f, is applied to the linear function, z. For regression tasks, activation functions are required to be differentiable and continuous, such as a sigmoid, or ReLU function.
- **Output**: Applying an activation function to a linear function results in the artificial neuron's output, y. This output can be used for tasks like forecasting future values of a time series or performing regression on continuous data.

As you can surmise, the neuron is made of two mathematical functions: the linear and activation functions.

Starting with the linear function expressed using **w**, **x**, and b:

$$z = \mathbf{w} \cdot \mathbf{x} + b$$

Expanding this into summation form:

$$z = \sum_{i=1}^{n} w_i \, x_i + b$$

where z is the weighted sum of inputs plus bias, w_i is the weight assigned to the i-th input, x_i is the i -th input, and b is the bias term.

> In most diagrams of neurons and perceptrons, the bias parameter is often hidden to simplify images, but it is always applied in calculations.

We can think of z with the same intuition as linear regression — weights act like regression coefficients and bias acts as an intercept with a constant input of 1.0, though this too is weighted. Our model inputs are scaled by these weights, which determine the importance of each input within the neuron. Higher weights translate simply as an input having greater significance. When we calculate the weighted sum of inputs, we multiply each input by its corresponding weight and sum them, then add the bias. This ensures the neuron can still activate when all inputs are zero — in that case $z = b$.

When we combine neurons together, weights, and biases represent the strength of inputs and connections between neurons, determining the complexity, and sensitivity of each individual neuron and the network as a whole. We optimize these parameters during the learning (training) phase via backpropagation, gradually improving the network's predictions. The number of weights required depends on the number of inputs — *a neuron with three inputs needs four weights*: one per input plus one for the bias. Initial weight values are set via *initializations*, commonly small values near zero, though numerous methods have been developed to improve training stability (see note below).

In the context of forecasting, the trained weights and biases of the artificial neurons and networks provide valuable information about the relationships and patterns present in the input data. This information allows the network to make predictions or forecasts based on the learned patterns. Different initialization methods can impact the training process and the final model's performance. By carefully selecting these initial values, a network can converge to an optimal solution efficiently, ultimately leading to more accurate forecasts.

We mentioned that there are different ways to initialize a neural network; we have listed a few below just to give an idea of how they are applied.

- **Glorot initialization**: Commonly used with RNNs and LSTMs, as its designed to handle the problems known as vanishing and exploding gradients, which can occur from repeated multiplication of weight matrices. Glorot initializations draw the initial weights from a Gaussian distribution with zero mean and variance that is dependent on the number of input and output connections to each neuron.
- **Orthogonal initialization**: Also useful for building RNNs and LSTMs, the orthogonal approach helps to preserve gradient magnitudes during backpropagation through time, which can also help avoid vanishing or exploding gradients. It sets the neurons initial weights to be an orthogonal matrix.
- **N-BEATS initialization**: N-BEATS is a recent model entry into the forecasting space, that uses fully connected layers with gating mechanisms. Don't worry if that doesn't make sense we will cover this is detail later. The authors of N-BEATS proposed an initialization that sets the weights of fully connected layers to be drawn from a uniform distribution with a mean of zero and a standard deviation of $\sqrt{\frac{2}{n}}$, where n is the number of input connections to each neuron.
- **Temporal Fusion Transformer (TFT) initialization**: TFT is another contemporary neural architecture for forecasting, which is a variation of transformer architecture (discussed later). The proposed method of initialization for TFTs suggests weights be drawn from a truncated normal distribution with a mean of zero and a standard deviation of 0.02.

The second mathematical function of the artificial neuron is the activation function, which adds non-linearity to the neuron's output. It does this by operating on the linear function to modify its output, enabling the neuron to learn from non-linear relationships in data. Without an activation function, the neuron would be limited to modeling linear relationships— inadequate given the complex patterns commonly found in real data. The choice of activation function therefore significantly affects a neuron or network's performance.

To build intuition, consider how biological neurons work: inputs from neighboring cells gradually depolarize a neuron, and when cumulative depolarization surpasses a threshold, it fires an action potential, propagating the signal forward. Artificial neurons work analogously—weighted inputs and bias accumulate into z, and the activation function determines whether and how the neuron fires.

This is easiest to see with a unit step function. Setting a threshold at 0, z must surpass this for the neuron to activate, outputting 1; if not, the neuron remains inactive and outputs 0. The bias term influences how easily this threshold is reached.

Taking our linear function $z = w \cdot x + b$, and expanding:

$$z = (w1 \cdot x1 + w2 \cdot x2 + \cdots + wn \cdot xn) + b$$

If we have a bias set to -2, and we're using a step activation function (with the threshold at 0), for our neuron to fire, and output a 1, the weighted sum of inputs needs to exceed +2. In mathematical terms, this looks as follows:

$$z > 0$$

$$(weightedsumofinputs) > -(-2)$$

With $b = -1$, only $+1$ is needed:

$$z > 0$$

$$(weightedsumofinputs) > -(-1)$$

Now we have a full understanding of the perceptron and artificial neuron, we can write the complete mathematical expression:

$$y = f\left(\sum_{i=1}^{n} w_i\, x_i + b\right)$$

In summary, we can interpret this linear function as a sum of weighted inputs plus bias, which we then apply an activation function to. Let us employ this for modeling the regression data that stumped our perceptron. Below we have modified our Perceptron class to include a parameter that allows us to specify an activation function, along with changing the train method to accept continuous input and output from the perceptron. Everything else remains the same.

```
class Diff_Perceptron:
    def __init__(self, input_dim, learning_rate=0.1,
                 activation_function=lambda x: x):
        self.weights = np.zeros(input_dim + 1)
        self.learning_rate = learning_rate
        self.activation_function = activation_function

    def predict(self, inputs):
        summation = np.dot(inputs, self.weights[1:]) + self.weights[0]
        return self.activation_function(summation)

    def train(self, training_inputs, labels, epochs=20):
        for _ in range(epochs):
```

```
        for inputs, label in zip(training_inputs, labels):
            summation = np.dot(
                inputs, self.weights[1:]) + self.weights[0]
            prediction = self.activation_function(summation)
            error = label - prediction
            self.weights[1:] += self.learning_rate * error * inputs
            self.weights[0] += self.learning_rate * error
```

Now we are simply going to apply two common activation functions: the identity function, which is to say just the linear function with no activation function, and the **Rectified Linear Unit** (**ReLU**) function, to the regression data. Don't worry if you don't fully understand ReLU, that is something for the next stage of our learning (*Chapter 6*). All that you need to understand here is that an activation function that is linearly separable allows us to make predictions for regression.

```
# set seed
np.random.seed(42)

# Instantiate Perceptron class with linear activation
nn_linear = Diff_Perceptron(1, activation_function=lambda x: x, learning_rate=
0.001)

# Train Perceptron
nn_linear.train(x_train_reshaped, y_train, epochs=200)

# Some activation functions to test
relu = lambda x: np.maximum(0, x)
sigmoid = lambda x: 1 / (1 + np.exp(-x))
tanh = lambda x: np.tanh(x)

# Instantiate Perceptron class with ReLU activation
nn_relu = Diff_Perceptron(1, activation_function=relu, learning_rate= 0.00001)

# Train Perceptron
nn_relu.train(x_train_reshaped, y_train, epochs=1000)

# Scatterplot of training data
plt.scatter(x_train, y_train, color='#0F95D7', label='Training data')

# Lineplot of predictions for linear activation
y_pred_linear = [nn_linear.predict(np.array([x])) for x in x_train]
plt.plot(x_train, y_pred_linear, color='#FF2700', label='Linear Activation')
```

```
# Lineplot of predictions for ReLU activation
y_pred_relu = [nn_relu.predict(np.array([x])) for x in x_train]
plt.plot(x_train, y_pred_relu, color='#5FA613', label='ReLU Activation')
plt.xlabel('X values')
plt.ylabel('Y values')

plt.title('Single-Layer Neural Network with Different Activation Functions')
plt.legend(title='Data', loc='best')

plt.show()
```

The output:

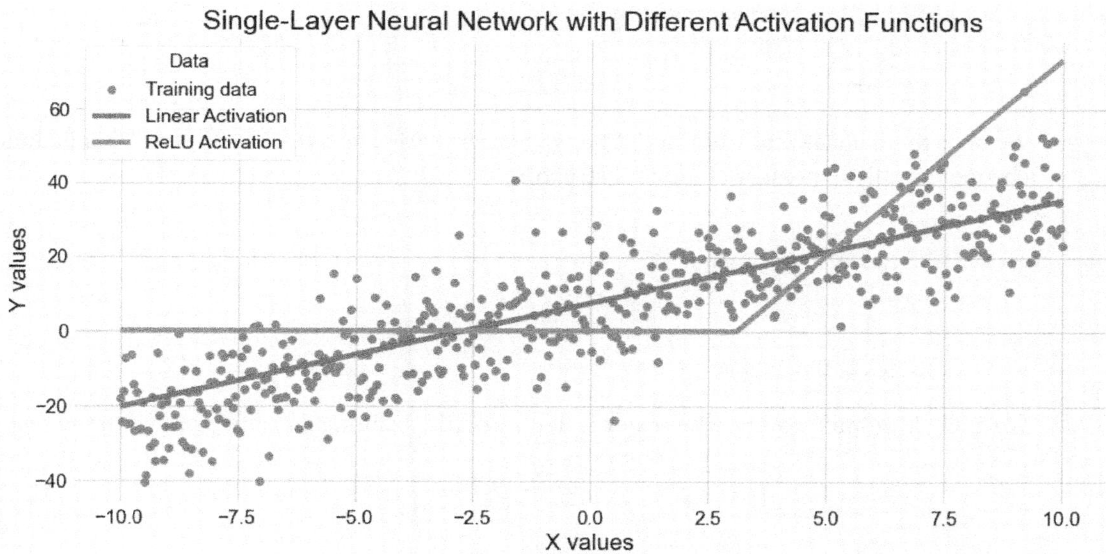

Figure 5.4: A neural network with different linear functions

Not bad! The ReLU activation isn't as good as the simple linear activation, simply because the data is a linear construct with noise. Regardless, both do a much better job than the unit step activation of the true perceptron. Now what happens if we up the complexity of the data? Let us attempt to model the *Airline Passengers* data.

To build a model of time series data with our neuron we need some features to help capture the sequential nature of the data, so we create `1:12` lag vectors of the dependent (passenger numbers), which we can interpret as the number of passengers from the previous month to year. We do this with a minmax data transformation, and use `200` epochs.

We apply the following process:

1. We load the dataset, convert the date column to datetime, and set it as the `index` for time series handling:

```
# Set seed
np.random.seed(42)

# Load dataset
data = pd.read_csv(CFG.data_folder / 'passengers.csv')

# Preprocess data
# Convert 'date' to datetime
data['date'] = pd.to_datetime(data['date'])
# Set 'date' as the index
data.set_index('date', inplace=True)
```

2. We split the dataset into training (60%), validation (20%), and test (20%) sets. This is done chronologically to preserve temporal structure.

```
# Sequential split
train_size = int(len(data) * 0.6)
val_size = int(len(data) * 0.2)
train, val, test = data[:train_size], data[train_size:train_size + val_
size], data[train_size + val_size:]
```

3. We apply Min-Max scaling. The scaler is fit only on the training data and applied to the rest.

```
# Scaling: Min-max
scaler = MinMaxScaler()
train_scaled = scaler.fit_transform(train)
val_scaled = scaler.transform(val)
test_scaled = scaler.transform(test)
```

4. We convert the time series into supervised learning format using `12` lagged observations to capture yearly seasonality:

```
# Prepare data with lags
def prepare_data(data, max_lag=12):
    dataX, dataY = [], []
    for i in range(max_lag, len(data) - 1):
        dataX.append(data[i - max_lag:i, 0])
        dataY.append(data[i, 0])
    return np.array(dataX), np.array(dataY)

lags = 12
```

```
trainX, trainY = prepare_data(train_scaled, lags)
valX, valY = prepare_data(val_scaled, lags)
testX, testY = prepare_data(test_scaled, lags)
```

5. We define the `ReLU` activation function to introduce nonlinearity:

```
# Initialize model with ReLU activation function
def relu(x):
    return np.maximum(0, x)
```

6. We create a differentiable perceptron using the specified input size, learning rate, and activation function:

```
diff_perceptron = Diff_Perceptron(
    input_dim=lags, learning_rate=0.001, activation_function=relu)
```

7. We train the model iteratively and monitor validation loss. Training stops if no improvement is seen for `10` consecutive epochs:

```
# Early stopping parameters
best_val_loss = float('inf')
patience, wait = 10, 0

# Train model with early stopping
for epoch in range(200):
    diff_perceptron.train(trainX, trainY, epochs=1)
    val_predictions = diff_perceptron.predict(valX)
    val_loss = mean_squared_error(valY, val_predictions)

    if val_loss < best_val_loss:
        best_val_loss = val_loss
        wait = 0
    else:
        wait += 1
        if wait >= patience:
            print(f"Early stopping on epoch {epoch}")
            break
```

8. We generate predictions for the training, validation, and test sets:

```
# Generate test predictions
scaled_predictions = diff_perceptron.predict(testX)

# In-sample predictions
train_predictions = diff_perceptron.predict(trainX)
```

```
unscaled_train_predictions = scaler.inverse_transform(train_predictions.
reshape(-1, 1))
train_pred_index = data.index[
    lags:lags + len(unscaled_train_predictions)]

# Validation predictions
val_predictions_plot = diff_perceptron.predict(valX)
unscaled_val_predictions = scaler.inverse_transform(val_predictions_plot.
reshape(-1, 1))
val_pred_index = data.index[
    train_size + lags:train_size + lags +
    len(unscaled_val_predictions)
]
```

9. We convert predictions back to the original scale for interpretation:

```
# Inverse Scaling
unscaled_predictions = scaler.inverse_transform(
    scaled_predictions.reshape(-1, 1))
test_index = data.index[-len(unscaled_predictions):]
```

The plot is produced after *step 9* (inverse scaling), using outputs from *step 8* (predictions):

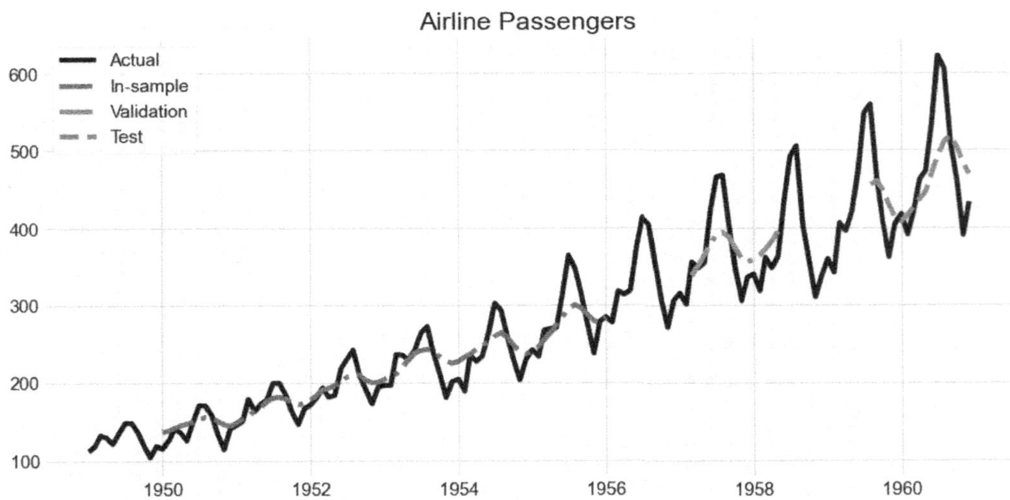

Figure 5.5: Modeling the Airline Passengers dataset

This is a good time to apply some of the error metrics we discussed in *Chapter 3*. We simply calculate our error metrics directly by using our back-transformed predictions against actuals:

MAE	41.13
MSE	3066.11
RMSE	55.37
MAPE	8.30%
sMAPE	8.49%

Table 5.1: Error metrics

The accuracy isn't great for such regular data — an RMSE of 55.37 can definitely be improved on. The in-sample and validation predictions (gaps caused by lags), show us that the model struggles earlier in the series, with less pronounced peaks and troughs. While the model gradually adjusts the data becomes more structured. Adding features, adjusting the learning rate, or applying a log transformation (which drops error) all help. But regardless, we can do much better with this dataset — and we will.

Obviously, we need to up our modeling game. As we are now armed with knowledge of neural architecture building blocks, we are ready to discuss networks!

> There are a large number of metrics that we can use for assessing forecasting accuracy, each with its own positives and negatives. Commonly we see **MAPE** used in business, as people often find it useful to deal with percentages. However, in practice **MAE**, **MSE**, and **RMSE**, and the relative versions (using naive models), tend to help us create and select better models, while percentage-based metrics tend to underemphasize error, for either over-forecasting (**sMAPE**), or under-forecasting (**MAPE**).

Neural nets

While we've delved into the inner workings of an individual artificial neuron, it's important to note that the sophistication of these units is typically not sufficient for modeling complex data types, especially when we're dealing with intricate time series data. A single neuron, despite its impressive ability to model linear and certain non-linear relationships, often falls short when it comes to capturing the intricacies of real-world data. This is where neural networks come in: in these architectures, we can stack, layer, and combine neurons to build complex, interconnected webs, like a section of the digital brain. These structures, which we refer to as **Artificial Neural Networks** (**ANNs**), have a remarkable ability to model complex, multi-dimensional, non-linear data. According to the Universal Approximation Theorem, neural networks can model virtually any non-linear relationship. Without getting into the mathematical details, which are beyond the scope of this book, the theorem states that a simple neural network with a single hidden layer can approximate any continuous function, provided the layer contains an adequate number of neurons.

It helps to think of each neuron in a network as a guard, each keeping an eye on a specific pattern or region of feature space. The area each neuron must cover is determined by the total number of neurons — decrease them and each guard oversees a larger area, reducing the network's ability to accurately approximate underlying patterns. Increase them and each neuron focuses on a smaller area, allowing the network to piece together a more precise approximation. It's like constructing a Lego castle: the more pieces we have, the more precise and intricate the result. The theorem comes with one small caveat though — a neural network's activation functions must be non-linear. And while billion-parameter architectures exist, scaling a simple network to that size for a 730-period time series with three features isn't guaranteed to get results: we risk severe overfitting, and the computational cost in time and processing would be considerable.

When we start to assemble these neurons together, we create a structure known as an Artificial Neural Network, also called a **Multi-Layer Perceptron** (**MLP**). We can think of these a being built of a series of neurons, neatly organized in layers, with each layer communicating with the next through a web of connections. While this might seem like a complex setup, it's essentially a community of neurons working in concert. Each neuron in a layer connects to every neuron in the next layer, allowing for a rich exchange of information. This extensive interconnection is why an MLP is also referred to as a **Fully Connected Neural Network** (**FCNN**) or a **Densely Connected Neural Network.** Neurons within these layers operate independently of each other, without sharing connections. Think of each layer as a team in a company, with each neuron being a team member. Team members work independently, but their outputs inform the work of the next team up the project ladder.

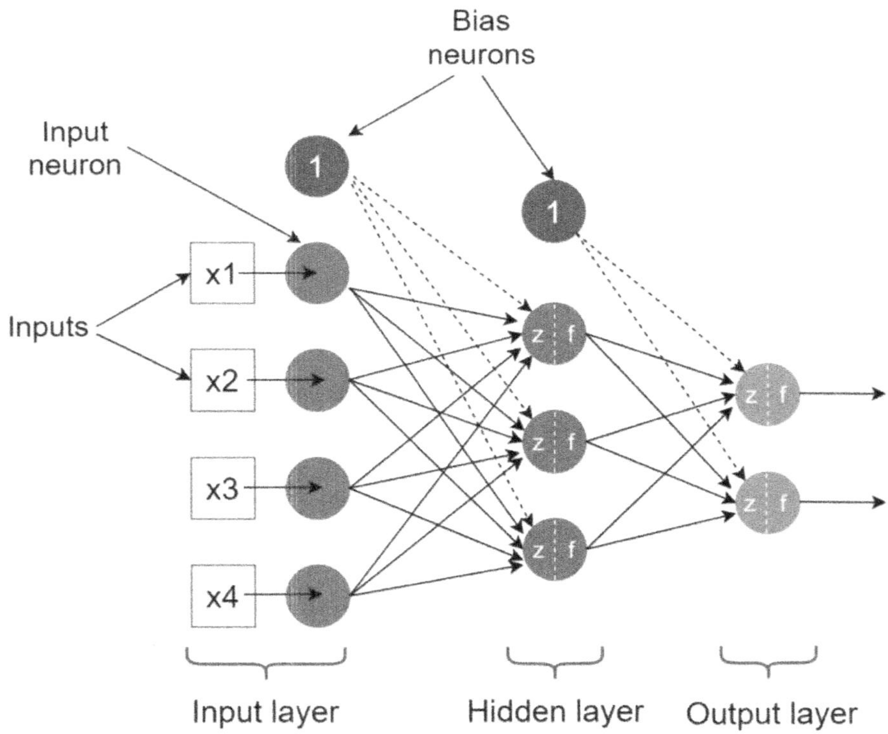

Figure 5.6: Simple ANN

The diagram depicts a simple neural network with an input layer, a single hidden layer, and an output layer — a **Shallow Neural Network**. Adding hidden layers produces a **Deep Neural Network.**

Depth (number of hidden layers) in the hidden layer/s sitting between input and output is often a hyperparameter to tune; more layers can allow intricate pattern recognition, but increase training time and the risk of overfitting. We define a hidden layer as any layer between input and output layers, with no direct connection to either raw data or output predictions.

In terms of implementation, in PyTorch we can choose different types of input layers, such as *Fully Connected*, *Convolutional*, or *Recurrent*, depending on our data and modeling needs. Fully Connected layers connect each input to each node, Convolutional layers handle spatial data like images, and Recurrent layers deal with sequential data like time series. The input layer type also dictates the shape of your input data.

Recurrent layers expect input data with three dimensions: (`batch_size`, `sequence_length`, `number_of_features`). Here, `number_of_features` should match the number of neurons in an input layer, `sequence_length` refers to the number of time steps, and `batch_size` is the number of sequences in a single batch.

Following the input, we have the hidden layer sandwiched between the input and output layers. A hidden layer is any layer with no direct connection to either the raw input data or the final output — it exists purely to transform representations between layers. The number of hidden layers, another hyperparameter, determines a network's depth. More layers mean a network can model more intricate relationships, but at the cost of increased training time and potential overfitting.

Lastly, the output layer makes the network's final prediction. Its size is determined by the problem we're solving; a single dependent variable needs one output neuron, while predicting two stock prices together would require two.

How this all works together is really just an expansion of the mathematics we discussed earlier, but instead of using single values (scalars) we use vectors and matrices. When expanding to multiple layers we simply increase our weight matrices and bias vectors to match the hidden layers. We aren't going to discuss this further here as PyTorch handles it for us, so long as input and output dimensions match our needs. If you'd like to explore how the mathematics develops, we've provided links in the further reading section.

From scalars to matrices

For a single neuron with two inputs, our linear function is simply:

$$z = w_1 x_1 + w_2 x_2 + b$$

When we stack three such neurons into a hidden layer, we can express all three calculations at once using matrix multiplication:

$$\mathbf{z} = \mathbf{W}\mathbf{x} + \mathbf{b}$$

where **W** is a weight matrix (3 neurons, 2 inputs each), **x** is our input vector, and **b** is a bias vector of length 3. Each row of represents one neuron's weights — so matrix multiplication is simply running all neurons in a layer simultaneously rather than one at a time.

When reading about deep learning you may see different conventions regarding how layers are counted, which can be a little confusing. There are three main approaches...

- Count only hidden layers, ignoring input and output layers. This is the most common approach.
- Count all the layers in a network, including input and output layers.
- Only count layers with weights, which translates to all layers except the input layer.

Before building our network, it's worth understanding how training works. We split data into three sets: a **training set** used to fit the model, a **validation set** used to monitor performance during training and guide decisions like early stopping, and a **test set** held back entirely for a single final evaluation once training is complete. Using the validation set for final evaluation would give an overly optimistic picture, as training decisions will have been influenced by it.

There are different methods to train networks depending on the architecture, but the fundamentals are best understood from the perspective of a **Feed-Forward Neural Network** (**FFNN**). The term 'feed-forward' describes how data flows through the network — in one direction, from the input layer, through the hidden layers, to the output. While travelling through intermediate layers there's no rerouting or backtracking; feedback only comes into play once data has reached the output layer. Many networks you'll encounter in this book will be feedforward; others, predominantly sequential models like RNNs, allow information to flow backward.

The training process that allows neural networks to learn is an iterative process with three main features:

- Forward propagation
- Loss calculation
- Backward propagation

These processes guide networks toward minimizing prediction error.

Forward propagation

The initial step in our training process is forward propagation, also referred to as a forward pass. This process describes the forward movement of data through a network, where information starts at the input layer and moves forward through the hidden layers, finally ending at the output layer. We can think of the process like a chain of dominos, as they cascade—falling dominos as the calculations move forward, with the output of each neuron being the input to the following layer. We can break the forward propagation/pass down into steps we have covered in this chapter.

Before we start the training process, we have to initialize our weights and biases with some small values, as we discussed earlier. It is these values that we seek to adjust via the training process. The training process begins by feeding data into the input layer, which simply funnels data into our network; there is no computing yet.

As data moves forward into hidden layers, each layer performs two calculations: the linear function which is passed to the activation function for introducing non-linearity. The output of this is then passed on to the next hidden layer or to the output layer. If the information is propagated to the output layer, this layer will summate its inputs and then output predictions.

Loss calculation

Once we've generated predicted values, we can compare them to actual data to assess a model's performance using a loss function (MSE, MAE, Huber, etc.) to generate a score representing a model's error. Ideally, we want to see a loss score of zero, indicating a perfect prediction, but you should not focus too much on getting loss to zero as this can indicate overfitting. When a model overfits, it becomes so attuned to the specific nuances and noise in training data that it performs poorly on unseen data, meaning an overfit model is unable to generalize.

Overfitting is particularly problematic as models become more complex. To some extent you can interpret model complexity as memory capacity; the more capacity a model has, the greater its likelihood of overfitting. In the case of forecasting, that could mean a model would not create useful forecasts. Luckily, it's relatively easily avoided with various strategies:

- **Data**: We must ensure we have enough varied data, allowing models to learn general patterns instead of focusing on specific instances
- **Regularization**: Adds a penalty to the loss function, proportional to the sum of model coefficients, which discourages the model from assigning too much importance to any single feature
- **Dropout**: Randomly disabling neurons during training encourages a network to distribute knowledge across all neurons, rather than concentrating too much processing on a few
- **Early stopping**: Stopping the training process when a neural network's performance on a validation set stops improving or plateaus, even if it continues to improve on the training set

Backpropagation

The final process of training is backward propagation, or the backward pass. Just as the name implies, this phase involves the backward flow of information, starting from the output layer and moving in reverse through the hidden layers toward the input layer. This process is akin to retracing our steps along the domino line, adjusting each domino (each neuron's weight) based on how much it contributed to any errors in the output. This process begins by calculating differences between a network's predicted output and the actual target values using a loss function, symbolized as $L(y, \hat{y})$, where y is the actual target value and $\hat{y}$ is the predicted output.

The essence of backpropagation is to calculate the gradient of the loss function with respect to a network's weights and biases, using the chain rule. This gradient provides a measure of responsibility for each neuron regarding the total error. Effectively, it tells us how much a slight adjustment in each weight or bias affects the overall error. The gradient calculation and subsequent parameter update together constitute the backward pass; the two are inseparable in practice. From this, the backpropagation algorithm adjusts weights and biases in the direction that opposes or reduces the error, typically using an optimization algorithm like gradient descent.

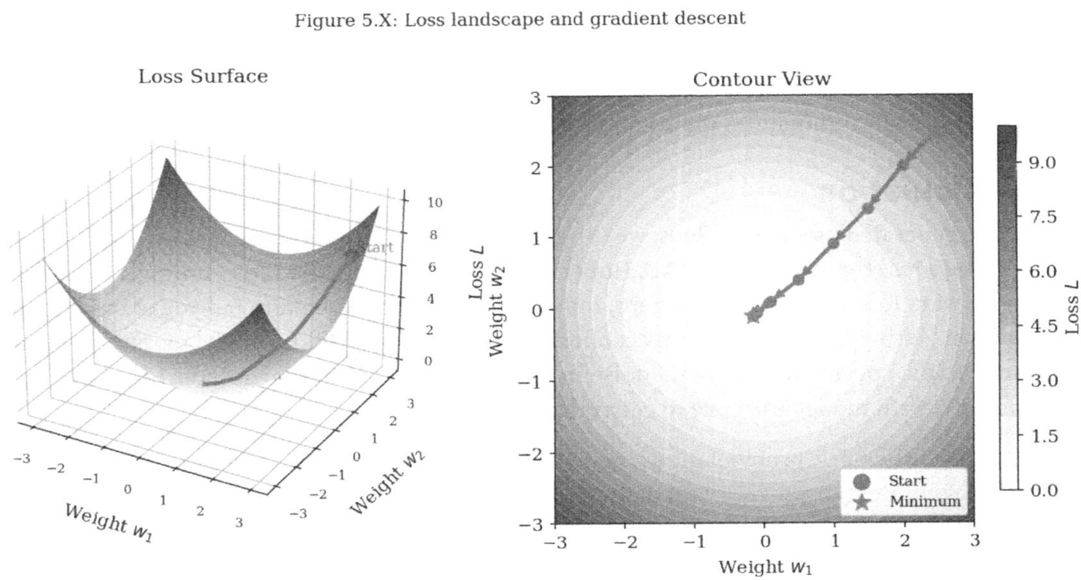

Figure 5.7: Gradient descent

There are, of course, various optimization algorithms which apply different strategies and techniques for updating parameters, such as using momentum, adaptive learning rates, or some combination of both. But in general, they all rely on the gradient of the loss function to determine the direction in which parameters should be updated. We will discuss how they do this in the next chapter.

Building a neural network

It's time to bring everything we've covered together, and more. Our previous models were simple perceptrons and neurons, for which we did not fully control the training process. In training a neural network we add some complexities, like backpropagation.

First, we simply load the libraries we will need.

```
import torch
from torch import nn
from torch.optim import Adam
from sklearn.model_selection import train_test_split
from sklearn.metrics import mean_squared_error
```

As we know, time series data has temporal dependencies. FFNs like the one we are creating treat data as if each observation is independent from every other, which isn't accurate for our data. To overcome this, we can use window embeddings, which create sequences of consecutive data points used as inputs for our model. In the code below, we define a `create_time_windows` function to build our time window embeddings. We apply this to our splits of the Airline Passengers dataset with a window size of `12`, as the cyclic patterns are clearly annual. This creates two new datasets: `inputs`, where each is a sequence of 12 time steps, and `targets`, which is the next time step (13th). The window slides forward one place for the inputs, and the next value becomes the target.

It is worth noting that these splits are not randomly selected — time series data is sequential, so randomly shuffling would break temporal order and allow future data to leak into training. Instead we split chronologically with the earliest portion for training, middle for validation, and most recent held back for testing. A validation set is used throughout training to monitor performance each epoch and inform decisions like early stopping. Because training decisions are influenced by it, validation sets cannot give unbiased assessments of generalization. This is why the test set is used once, after training is complete.

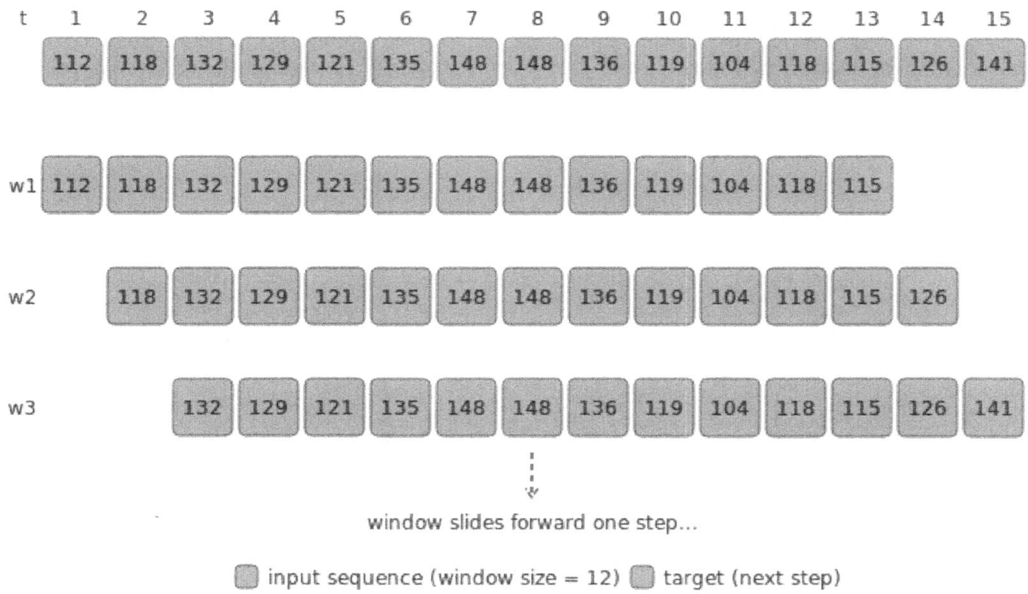

Figure 5.8: Window Embeddings

Let's work on an example to create some time window embeddings:

```
def create_time_windows(data, window_size):
    inputs = []
    targets = []
    for i in range(len(data) - window_size):
        inputs.append(data[i:i+window_size])
        targets.append(data[i+window_size])
    return np.array(inputs), np.array(targets)

window_size = 12 # good parameter to play with to understand effect
```

```
train_inputs, train_targets = create_time_windows(train_scaled, window_size)
val_inputs, val_targets = create_time_windows(val_scaled, window_size)
test_inputs, test_targets = create_time_windows(test_scaled, window_size)
```

We then take these and convert them into simple one-dimensional tensors for the FFN. This will convert them from 3D arrays to 1D tensors, with the following dimensions:

- **Batch size/sample number.** Each sample is a time window of lagged values.
- **Window size**, length of each time window (`n = 12`).
- **Feature dimension**, as we have only one feature (passengers), this dimension is `1`.

In PyTorch nomenclature, this gives us a `[BatchSize, SequenceLength, FeatureDimension]` tensor, a typical format for sequential data. We can verify this by calling `.shape` on the tensor objects.

```
# Convert numpy arrays to PyTorch tensors
x_train = torch.FloatTensor(train_inputs)
y_train = torch.FloatTensor(train_targets)
x_val = torch.FloatTensor(val_inputs)
y_val = torch.FloatTensor(val_targets)
x_test = torch.FloatTensor(test_inputs)
y_test = torch.FloatTensor(test_targets)

# Print shapes
x_train.shape, y_train.shape, x_val.shape, y_val.shape, x_test.shape, y_test.
shape
(torch.Size([74, 12, 1]),
 torch.Size([74, 1]),
 torch.Size([16, 12, 1]),
 torch.Size([16, 1]),
 torch.Size([18, 12, 1]),
 torch.Size([18, 1]))
```

Once our data is prepared, we're ready to build our model. Below we build a `ff_network` class, a feed-forward neural network (MLP), which forms the backbone of our time series forecasting model, consisting of an input layer, an activation function, and an output layer.

The input layer, `nn.Linear(input_dim, hidden_dim)`, handles data shaped by our window size, encapsulating past time steps to forecast the future. Its output is transformed by a ReLU activation function, `nn.ReLU()`, introducing non-linearity to capture complex patterns before passing into an output layer, `nn.Linear(hidden_dim, output_dim)`, responsible for our passenger predictions. As we've discussed, the model learns by minimizing discrepancies between predicted and actual data, adjusting weights in its linear layers accordingly.

The `forward` function dictates the path of data through our network. Starting with the weighted sum of inputs computed by `self.fc1(x)`, data passes through `self.relu(out)`, eliminating negative values and retaining positive ones. Finally, `self.fc2(out)` computes another weighted sum from the hidden layer's transformed data, producing our predictions.

The next section of code covers our model instantiations and training loop. We begin by instantiating a feed-forward neural network with an input dimension that matches the window size, then we specify the number of units in our hidden layer (dimension). We also have a single neuron in the output dimension (which is default), as we are doing regression with a single dependent variable (passengers). We define MSE as our loss function and use Adam as our optimizer with a learning rate of `0.0001`. Adam is a good default choice, as it adapts learning rates during training, making it more forgiving than SGD without requiring careful tuning of the learning rate. We will discuss optimizers in more detail in the next chapter.

Our hidden layer size and learning rate parameters were selected after some manual experimentation. Selecting optimal parameters, loss function, and optimizer are generally best found as part of a structured optimization process. We suggest experimenting with these **hyperparameters** and graphing their impact on model performance to help develop your understanding and decision-making.

Overfitting is a significant issue, especially when we don't use cross-validation, so we implement early stopping. Our early stopping has a patience variable (set to `20`), which determines the number of non-improving epochs a model will tolerate before stopping, unless training reaches the maximum specified epochs (`n = 2000`). We also create variables to track the best loss and log training and validation losses in lists so we can plot training performance.

An **epoch** is one complete pass through the entire training dataset. During an epoch, our model is set to train mode, and a forward pass is performed, pushing training data through the model to obtain predictions. Our selected loss function is then applied to calculate differences between the predictions and actuals, with the result appended to a list of `train_losses`.

Next, we perform a backward pass and update model parameters with our optimizer. PyTorch then evaluates performance on the validation set by switching to evaluation mode and disabling gradient computation. This validation loss is calculated and logged. When validation loss improves, the model's state is saved. If no improvement is observed for the patience number of epochs, training stops early to prevent overfitting.

```
np.random.seed(42)

# Initialize FFN
input_dim = 12
hidden_dim = 100 # Tunable
output_dim = 1 # We are predicting a single output
model = ff_network(input_dim, hidden_dim, output_dim)

# Loss and optimizer
criterion = nn.MSELoss()
```

```
optimizer = Adam(model.parameters(), lr=0.0001)


# Early stopping parameters
best_val_loss = float('inf')
patience, wait = 10, 0

# Lists to hold loss values for plotting
train_losses = []
val_losses = []

# Best model checkpoint
best_model_state = None

# Training loop with early stopping
for epoch in range(2000):
    model.train()
    optimizer.zero_grad()
    # Flatten time window dimension
    outputs = model(x_train.view(-1, window_size))
    loss = criterion(outputs, y_train)
    loss.backward()
    optimizer.step()

    # Store train loss
    train_losses.append(loss.item())

    model.eval()
    with torch.no_grad():
        val_outputs = model(x_val.view(-1, window_size))
        val_loss = criterion(val_outputs, y_val)

    # Store validation loss
    val_losses.append(val_loss.item())

    if val_loss.item() < best_val_loss:
        best_val_loss = val_loss.item()
        wait = 10
        # Save best model's state
        best_model_state = model.state_dict()
    else:
        wait += 1
```

```
        if wait >= patience:
            print(f"Early stopping on epoch {epoch}")
            break

# Load best model's state
if best_model_state:
    model.load_state_dict(best_model_state)

# Output:
Early stopping on epoch 1645
```

Congratulations, we have just trained our first neural network! The model trained on my computer in a few seconds. The first thing we want to look at is the training and validation loss. Although we could have used PyTorch's TensorBoard, we simply stored the data to graph training and validation loss over each epoch to check for overfitting.

A decreasing training loss with a stable or decreasing validation loss indicates a model is learning well. A sign of overfitting would be validation loss starting to increase, or simply staying flat, while training loss continues to decrease. This translates as the model memorizing training data and would mean that it is unlikely to generalize well on unseen data.

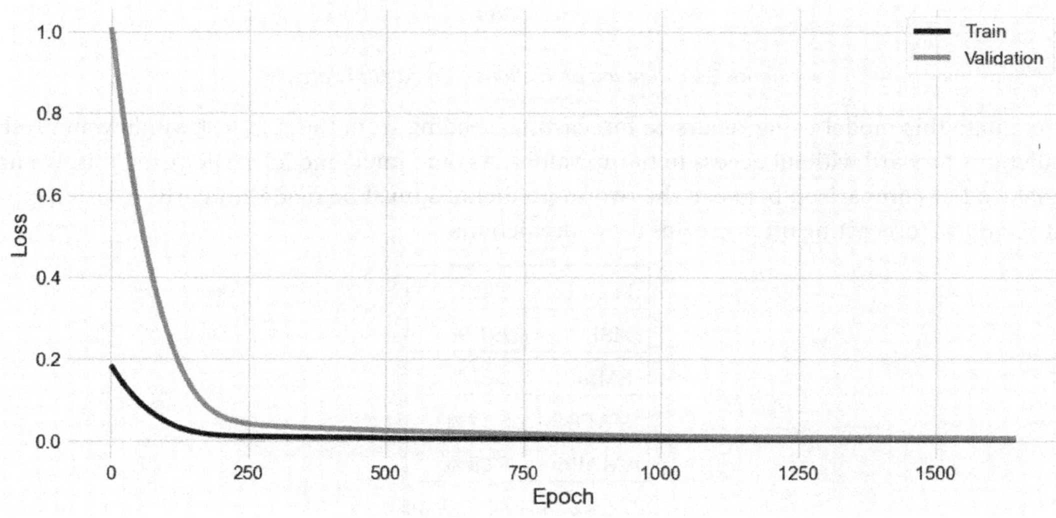

Figure 5.9: Graphing out loss values

Graphing out the loss values shows that we are not overfitting, as training and validation losses do not cross over. We can also see a large initial drop in loss, which indicates that the model quickly learned to reduce prediction errors between training and validation data. After this, the model runs with small but continuous improvements to reach a maximum training epoch, just short of **400**. A flat tail appears at the end of our graph, where training and validation loss essentially stops; this suggests our model may have reached its limit in terms of what it can learn from the dataset.

Let's have a look at the performance on the actual data. We have graphed out the resulting model's predictions below, which are not bad. Something to note here is that although the predictions of the FFN do not look much better (`RMSE = 91.095`) than the single neuron model (`RMSE = 99.89`). And you can pretty much guarantee we could beat this with an ARIMA model.

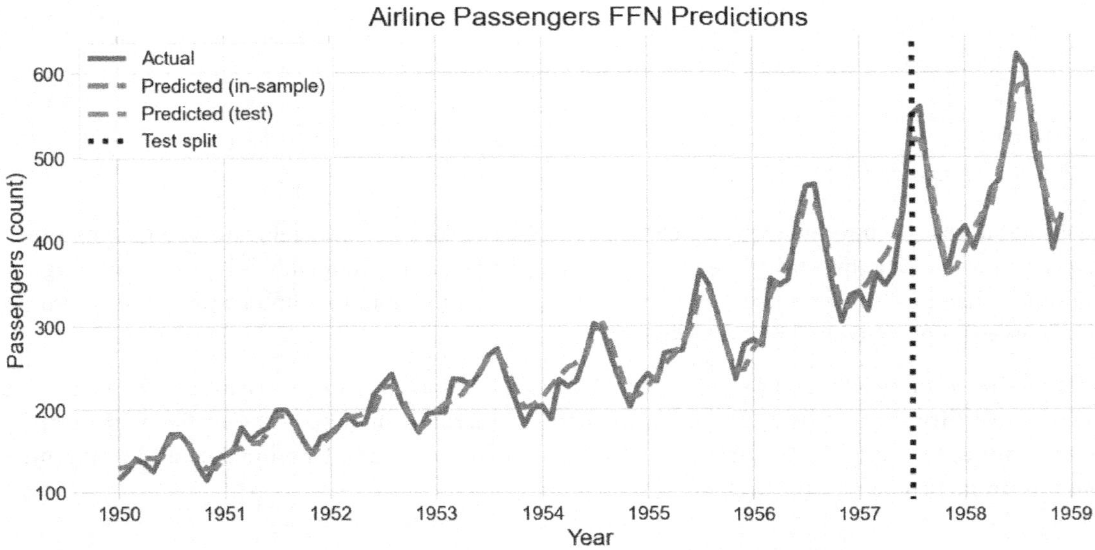

Figure 5.10: Running predictions on an actual dataset

We evaluate this model using recursive forecasting, seeding from the first test window and rolling predictions forward without access to future values. As our simple model earlier used this, we need to make a fair comparison between the two so predictions must be made similarly. It also happens that recursive forecasting often creates *better* predictions.

MAE	23.91
MSE	791.9
RMSE	28.14
MAPE	5.17%
sMAPE	5.08%

Table 5.2: Metrics for evaluation

Our FFN (`RMSE = 28.14`) is a significant improvement over the single neuron models(`RMSE = 55.37`), which is what we'd hope to see from a more sophisticated architecture. That said, you can pretty much guarantee we could still beat this with an ARIMA model on data this regular. In the next chapter we will look at how to improve on this further.

Summary

Excellent work! Once again we have encountered a lot of new terminology. We have covered the perceptron and how it led to the artificial neuron which forms the basis of today's neural architectures. We now know how these work mathematically with a weighted linear summation of input values, and an activation function to add non-linearity.

We also know because of the Universal Approximation Theorem, that we can stack artificial neurons together to approximate almost any data structure. This led us to learning about feedforward networks, and how to train them with forward propagation of information, loss calculations, and backward propagation to adjust weights and biases.

Finally we built our first neural network for forecasting, and made predictions on the Airline Passengers dataset, which did alright for a simple model. But this could be a lot better. In the next chapter we will take this knowledge further, and optimize our model builds and training with Lightning!

Recommended reading

- Hornik, K. 1991. "Approximation capabilities of multilayer feedforward networks." *Neural Networks*. `https://doi.org/10.1016/0893-6080(91)90009-T`.
- Kratsios, A. 2021. "The universal approximation property: Characterizations, extensions, and applications." *arXiv*. `http://arxiv.org/abs/2102.10917`.
- LeCun, Y., Bengio, Y., and Hinton, G. 2015. "Deep learning." *Nature*. `https://www.nature.com/articles/nature14539`.
- Rosenblatt, F. 1958. "The perceptron: A probabilistic model for information storage and organization in the brain." *Psychological Review*. `https://psycnet.apa.org/record/1959-09865-001`.
- Schmidhuber, J. 2022. "Annotated history of modern AI and deep learning." `https://people.idsia.ch/~juergen/deep-learning-history.html`

6

Optimization

"If he [Thomas Edison] had a needle to find in a haystack, he would not stop to reason where it was most likely to be, but would proceed at once with the feverish diligence of a bee, to examine straw after straw until he found the object of his search... Just a little theory and calculation would have saved him ninety percent of his labor." — Nikola Tesla

As we have seen, neural network architectures are highly flexible structures with vast *parameter spaces* that allow them to capture intricate patterns within datasets. However, realizing each architecture's full potential requires thoughtful consideration of hyperparameter optimization — the process of finding the optimal *design* of a model relative to our data, including selecting activation functions, regularization methods, and learning rates, all of which govern how a model trains and compresses data into function. This process is both critical and challenging due to the complex interplay of architectural decisions, training algorithms, and hyperparameters. For instance, choices made in selecting the training algorithm for *simple* FFNs/MLPs profoundly impact model sensitivity and forecasting accuracy. In this chapter, we will build on our understanding of FFNs/MLPs by exploring neural network optimization and the processes that allow us to extract the best performance from our models.

We can tailor neural network hyperparameters to fit different time series datasets, but additional care in architecture saves time, compute, and improves performance. You could, of course, undertake a brute-force approach, testing every permutation of design, but an intelligent approach, as Tesla suggests, is usually more efficient. Focused experimentation aligns more closely with contemporary optimization practices across machine learning. The optimization of neural architectures applies similar principles to the approaches we take for tuning hyperparameters for algorithms like boosted decision trees, although it's not unreasonable to suggest that neural networks tend to be more sensitive to tuning processes than other modeling approaches; haphazard optimization easily triggers overfitting, underfitting, and convergence to local minima.

A note on terminology

Optimization algorithms are used in training networks via backpropagation. When we talk of optimization here, we are discussing the overall process of optimization, which includes selecting elements of network design and training parameters.

In this chapter:

- Optimization costs
- Optimization of neural networks
- Understanding activation functions
- Gradient descent, loss functions, and regularization techniques

Optimization costs

"I have not failed. I've just found 10,000 ways that won't work." — Thomas A. Edison

While neural networks provide forecasting practitioners with intricate tools for modeling, they come with a large number of options and a parameter space that can dwarf simpler models, which translates into neural networks often consuming significant computational resources and time to train. An excellent study by Nixtla investigated modeling time series data (n = 3,003 series) with different approaches, finding that a simple statistical ensemble (a combination of models) can outperform most individual deep learning models. The statistical ensemble was 25,000x faster to train than a *slightly more accurate* ensemble of deep learning models (−0.36% sMAPE), which the authors stated took 14+ days to train at a cost of approximately $11,000. This contrasts starkly with the statistical ensemble, which took 6 minutes and cost $0.50 (Nixtla Statsforecast GitHub repository). While the findings can be questioned (see note *Criticism of Nixtla's study*), they aptly demonstrate that we must be careful when applying deep learning methods, and think about how best to train neural architectures efficiently. In situations where modeling and training is well planned, deep learning architectures can outperform other modeling approaches; a study by Zalando's data science team demonstrated that they achieved their best results for a panel of demand data (retail time series) with over 1 million series for approximately $275 USD (Kunz et al. (2023)). To get the best out of neural architectures, we need to think about optimization intelligently. This involves selecting the right activation functions, loss functions, and initialization methods, as well as tuning hyperparameters and using techniques like batch normalization and regularization.

Criticism of Nixtla's study

It is worth pointing out that Nixtla's study was conducted using the M3 dataset, which consists of 3,003 time series that were sampled at yearly, quarterly or monthly frequency, with 40 to 120 observations per series. Such time series structures disadvantage more complex neural architectures, which require significant amounts of data to build models.

In addition, I am skeptical about the cost of running a deep learning ensemble, but then again I don't use cloud processing because it is ridiculously expensive. If you have to do a lot of processing, I personally recommend getting some rack space and plugging in some graphic cards; it is more cost effective in many cases

The most important component of any model is your data, and the quality of that data. Neural networks and other approaches do not magically find patterns; the quality of your data not only affects how well a model fits, but how quickly it trains. It is not uncommon to encounter time series data that are slow to train. Commonly you'll find this issue with:

- **Low-frequency data:** Sales of pickles in a small village shop, for example, may record only a few transactions each week, making data sparse and infrequent. When combined with complex patterns (which is common), the convergence of neural network training can slow down significantly. These kinds of time series are generally better suited to simpler, linear models, which have far fewer parameters to estimate and do not require the large volumes of data that neural networks need to learn stable representations.
- **High signal-to-noise ratio**: Stock market data is a good example. The underlying patterns are generally obscured by irrelevant and random variations we cannot easily account for in modeling. With such data, models can spend a significant amount of time trying to differentiate between signal and noise.
- **Long sequence dependencies**: Some time series are very long, which while providing useful data can also hinder modeling with neural networks like RNNs and LSTMs.
- **High-dimensional data:** Time series with many features, whether multivariate or multiexogenous-univariate, can be slow to train, especially if some of the additional features are weakly predictive. The additional dimensionality increases model complexity, slowing training and hindering convergence.
- **Sparse or irregular data**: Missing data is always annoying, sparse data is perhaps worse, but irregular data is the worst. Most modeling approaches rely on data that is evenly spaced; neural networks are no different. When data is irregular, the best approach is often to make it regular.

That isn't to say that we can't model such data, but that a good forecast may be slow to achieve with neural networks. This highlights the need for matching data with appropriate models, which we will come back to later when we discuss clustering.

Optimization of neural networks

Optimization with neural networks starts with **Exploratory Data Analysis** (**EDA**) to inform the preprocessing of data, such as handling missing values, creating features, and transforming data. Once your data is prepared, you should focus on optimizing network design, which you can test on training data; this largely boils down to three areas (Chen et al., 2023):

- **Preprocessing:** Missing values, dimensionality reduction, dealing with outliers and leverage, feature selection, decomposition, transformations, scaling and normalization
- **Architecture:** Activation functions, number of nodes, number of hidden layers
- **Training optimizers and hyperparameters:** Learning rate, batch size, regularization, dropout

Refer to color images on the Graphic bundle or the ebook.

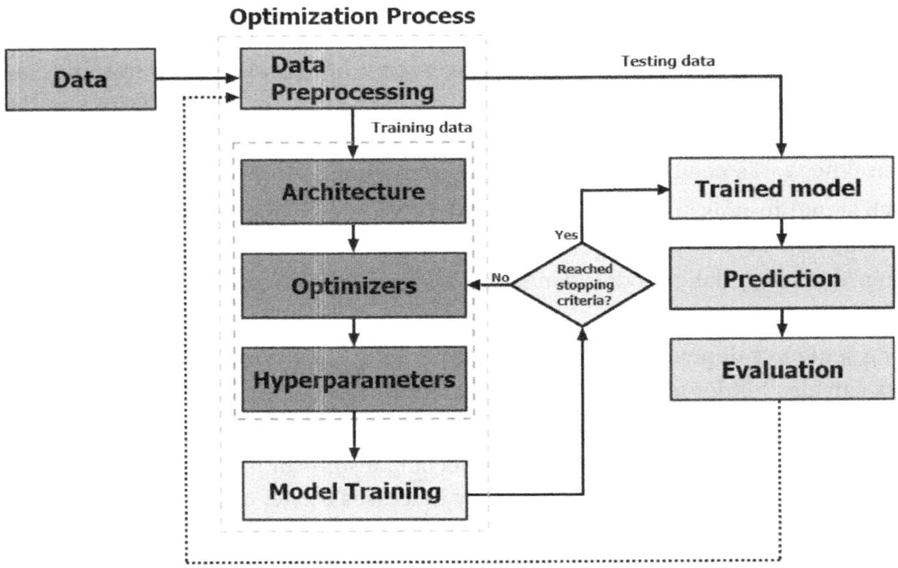

Figure 6.1: Neural architecture optimisation process

Understanding activation functions

Activation functions are crucial parts of neural architecture that determine outputs, accuracy, and computational efficiency (Moon et al., 2019; Ramachandran, Zoph, and Le, 2017). As we already know, without activation functions, networks essentially build linear regressions, which, while excellent tools, have limited power for complex data challenges. As we learned in the previous chapter, at the heart (neuron) of a neural network lies:

$$y = f\left(\sum_{i=1}^{n} w_i\, x_i + b\right)$$

Here f is an activation function, applied to a linear combination of inputs to a neuron: the sum $\sum_{i=1}^{n}$ of weighted w_i inputs x_i, plus bias. Activation functions are mathematical components we use to transform these inputs. What makes them crucial is that they allow networks to transform input vector space (a multi-dimensional space, with each dimension corresponding to an input or feature) into a linearly separable vector space. To do this, activation functions must be non-linear (have multiple degrees). Transforming inputs this way is not simply reshaping, but an *unwarping* which makes underlying patterns in your data easier for a network to learn. This additional complexity enables neural networks to learn and adapt better to prediction errors during backpropagation, resulting in better functional approximations... *in theory*.

Why can't we use a linear activation function?

Activation functions in neural networks, such as the linear function, generate a series of activation values. However, a major drawback of linear activation functions is that they have a constant derivative, which impacts backpropagation. A linear function's derivative (from $f(x) = cx$) implies no relationship with input x, hindering learning, as a constant derivative means that weight adjustments are not influenced by input variations.

If linear functions are used in all layers of a neural network, the resulting linearity would limit a network to simple, linear data patterns, undermining any benefits of deep learning. Specifically, neurons in earlier layers would not add any representational capacity to the interconnecting layers (Kızrak, 2020).

Our first challenge then is to work out which activation functions to choose. This is not a simple task because when we change activation function, we also change the partial derivatives of our neurons, which has differing effects on model builds (Nielsen, 2015). The goal of this section is not to point you to specific activation functions, as that's not practical, but to develop an understanding of how activation functions alter modeling and forecast accuracy.

Taxonomy

There are a large number of activation functions, which are increasing in number with research, but we can generally categorize them as follows:

- **Fixed-shape activation functions** do not change shape during training.
    - **Rectified-based** include the ReLU family of functions (ReLU, Leaky ReLU, etc.).
    - **Classic activation functions** are functions not in the rectifier family, such as sigmoid and tanh.
- **Trainable activation functions** change shape during model training. Their shapes adapt to the training data, although many trainable functions can be approximated by adding more layers with fixed activations to a network.
    - **Parameterized standard functions** start as standard functions but include additional parameters that are learned through training.
    - **Functions based on ensemble methods** are ensembles of functions created by combining several distinct functions, generally via a linear combination.

Trainable activation functions are a relatively recent development. The main idea is that these functions have parameters which can be updated, meaning the function's shape can be adapted via training to better suit patterns in your data. While it can be beneficial to allow a network to learn the best weights, biases, and form of activation function in a given context, the downside of more parameters is *higher computational demand* (Apicella et al. (2021); Bingham (2023)). At the time of writing, trainable activation functions are still largely in the realm of research, as benefits and drawbacks continue to be explored. We briefly cover one (Swish) in this chapter; a more thorough investigation can be found in this chapter's associated notebook.

This chapter will focus on the most popular activation functions, which are currently fixed-shape functions more prominent in industry.

Layer-wise activation functions

While it's possible to apply activation functions to individual neurons, in practice we apply them to entire layers of neurons in a network for *efficiency* and *simplicity*. Applying the same activation function to all neurons in a layer means we can compute a function once for all neurons, rather than individually for each one. This makes it easier to train a network and reduces computational cost, with the additional benefit that the simplicity makes it easier to comprehend what each layer is doing.

Fixed-shape activation functions

The fixed-shape activation functions we cover here are the most commonly used in practice. We start with sigmoid (historically important), move through tanh, and then to ReLU and its variants, which are popular in modern architectures.

Sigmoid

Sigmoid activations are logistic functions that are differentiable at every point, meaning we can find a gradient at any point on a curve. If you recall, the original perceptron used a step/Heaviside function that was not differentiable, which prevented us from optimizing networks with gradient descent via backpropagation, as it relies on a function's derivative. The sigmoid resolved this, allowing us to take any real-valued number and map it as a float between 0 and 1. This made sigmoid a strong choice for problems where predicted probabilities were required, as probabilities lie within the same range.

$$\sigma(x) = \frac{1}{1 + e^{-x}}$$

Furthermore, the rapid change in the sigmoid's values in the input range between `-2` and `+2` helps a network *quickly* adjust its weights during the early stages of training, when it may be far from an optimal solution.

We can visualize sigmoid and its derivative using PyTorch:

```
x = torch.arange(-8.0, 8.0, 0.1, requires_grad=True)

sigmoid = torch.nn.Sigmoid()

plot_activation_fn_and_derivative(
    sigmoid, x, xlabel='x', ylabel='Sigmoid(x)',
    title='Sigmoid Activation Function'
)
```

The sigmoid plot look like this:

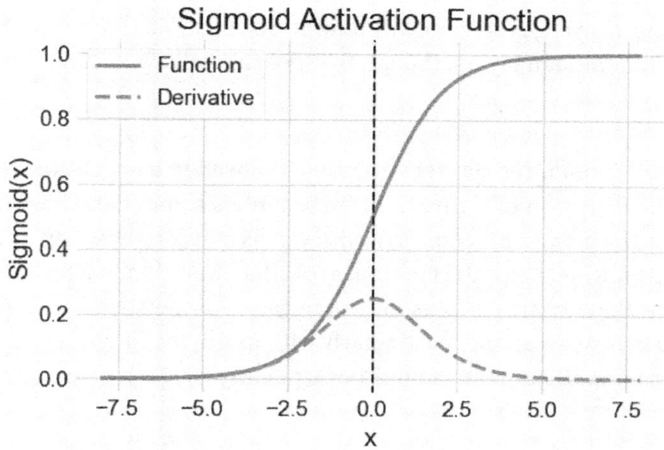

Figure 6.2: Graphs inspired by Manu (2022) - recommended read!

Despite being helpful early on, sigmoids have several drawbacks. You can see in our graph that in the extremes y-values change minimally with respect to x. When training via backpropagation, these small gradients propagate through layers, becoming smaller until learning effectively stops. This becomes more of a problem as layers are added to a network (Bingham, 2023; Nwankpa et al., 2018). This saturating of activation is a phenomenon best known as the vanishing gradient problem. It can slow learning and cause an optimization algorithm to get stuck in local minima, limiting model performance. The related exploding gradient problem, where gradients grow uncontrollably large during backpropagation, is less common with sigmoid (whose derivatives are bounded below 1) but becomes relevant with other architectures and activation functions.

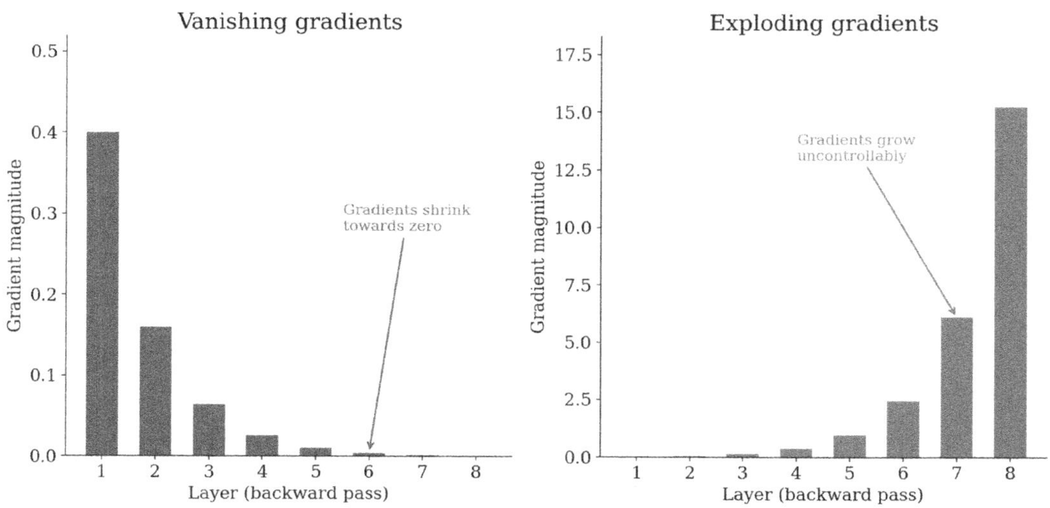

Figure 6.3: Vanishing and exploding gradients

Another issue is that outputs are not centered around zero, as they are squashed between 0 and 1. This makes all the outputs of a hidden layer either positive or negative, which obstructs gradient updates and slows training and convergence. You will tend to see sigmoids in output layers these days, for classification rather than hidden layers, though that doesn't mean you can't use them.

Tanh

To address issues with sigmoid, researchers eventually moved to using tanh; we can think of tanh as a mathematically shifted version of sigmoid. Because tanh is sigmoidal it has similar properties, but improves upon sigmoid by being zero-centered, with values mapping between −1 and 1. Zero-centering facilitates backpropagation and helps the function avoid the saturation that leads to vanishing gradients. Using tanh also means that negative inputs map to negative outputs, and near-zero values map close to zero, which results in faster training, particularly with multiple hidden layers. These factors helped it dominate training of neural networks in the 1990s to early 2000s (Jagtap and Karniadakis, 2022).

It can be defined as:

$$\sigma(x) = \frac{e^x - e^{-x}}{e^x + e^{-x}}$$

Where,

- $\sigma(x)$ is the output of the tanh activation function
- x is the inputs
- e is the base of the natural logarithm

We can plot tanh and its derivative similarly:

```
y_tanh = torch.nn.Tanh()

plot_activation_fn_and_derivative(
    y_tanh, x, xlabel='x', ylabel='Tanh(x)',
    title='Tanh Activation Function'
)
```

The derived plot:

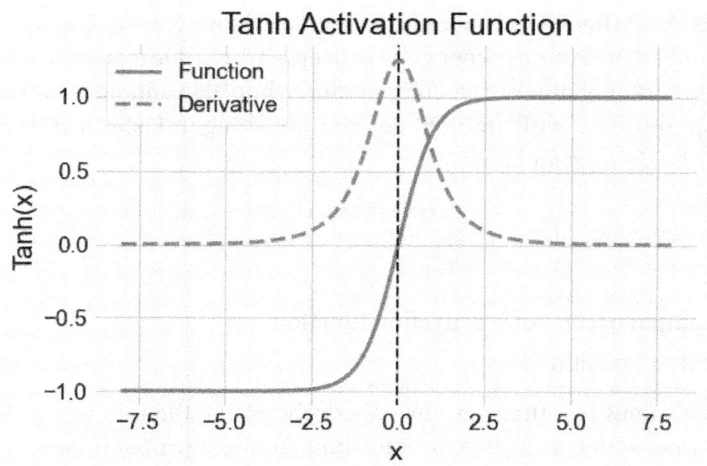

Figure 6.4: The tanh activation function

Vanishing gradients are, however, *still* a problem for tanh, despite the derivatives being steeper than sigmoid's. This is especially an issue as hidden layers increase in a neural network. An additional issue is that tanh only attains its maximum gradient of 1 when the input is 0; for inputs further from zero the gradient falls toward 0, which can produce *dead neurons* — neurons that become effectively inactive because their near-zero gradients result in negligible weight updates during backpropagation (Nwankpa et al., 2018; Szandała, 2021). Despite this, tanh activations were, and are used extensively in recurrent neural networks, just not so much in *deeper* models.

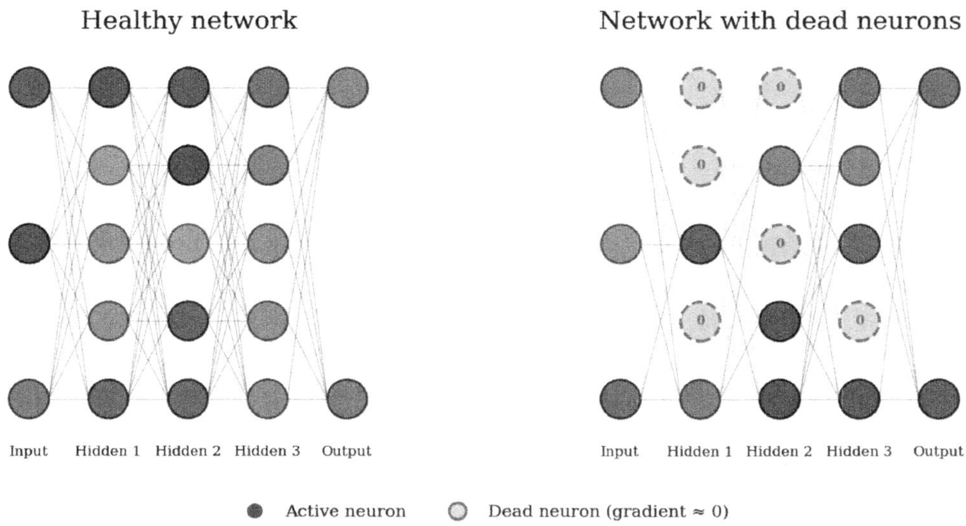

Figure 6.5: Dead neurons in a network

ReLU

The **Rectified Linear Unit** (**ReLU**) activation function was introduced by Nair and Hinton in 2010 Nair and Hinton (n.d.), and has become a cornerstone in deep learning due to its efficiency and effectiveness. As a nearly linear function, it allows fast computation while also adding non-linearity to a network, which facilitates optimization with gradient descent, resulting in faster learning and outpacing traditional functions like sigmoid and tanh.

$$\sigma(x) = \max(0, x)$$

Where,

- $\sigma(x)$ is the output of the ReLU activation function
- x is the input to function

ReLU rectifies input values less than zero by effectively setting them to zero, while being unbounded on the positive side, which mitigates the vanishing gradient problem seen in earlier (saturating) activation functions. This also adds sparsity to networks by zeroing out negative activations, which reduces space, and time complexity by avoiding complex operations like exponentials and divisions. As a result we require less computational power, and generate more concise models with less noise and usually greater statistical inference.

```
y_relu = torch.nn.ReLU()

plot_activation_fn_and_derivative(
    y_relu, x, xlabel='x', ylabel='ReLU(x)',
    title='ReLU Activation Function'
)
```

The resulting plot is:

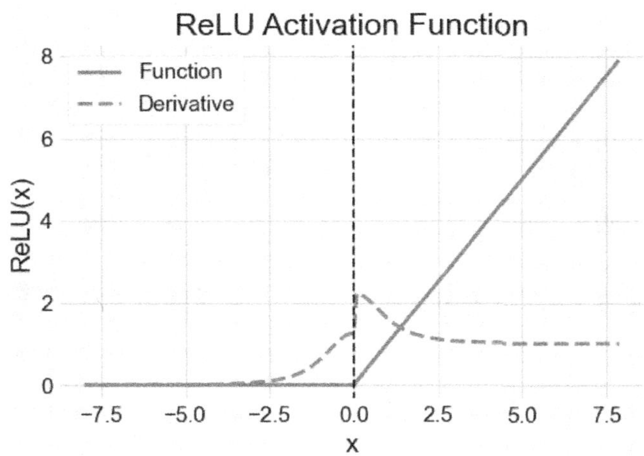

Figure 6.6: The ReLU activation function

To visualize how we can mitigate the *dying ReLU* problem, we use the following code to plot Leaky ReLU:

```
y_leaky_relu = torch.nn.LeakyReLU(negative_slope=0.1)
plot_activation_fn_and_derivative(
    y_leaky_relu, x, xlabel='x', ylabel='Leaky ReLU(x)',
    title='Leaky ReLU Activation Function', ylim=(-2, 6)
)
```

The output of this code is:

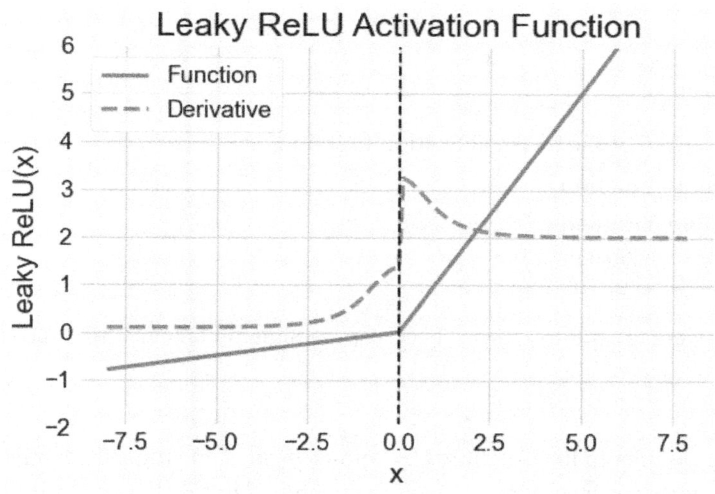

Figure 6.7: The Leaky ReLU activation function

ReLU-activated networks may have some issues. The first is the **dying ReLU problem**, which is caused by setting negative values to 0, resulting in gradients of 0. This makes ReLU *fragile*, leading to 'dead' neurons that give zero activation, hindering learning because there is no weight update during backpropagation (Nwankpa et al., 2018).

It's not hard to imagine this being an issue with sparse demand data, for example. In addition, ReLU has a propensity to overfit compared to sigmoidal functions, although we can mitigate this with techniques like dropout. Despite this, ReLU's characteristics help it handle non-linearity and improve computational efficiency, making it a useful function to apply in the hidden layers of networks you will commonly use for forecasting, such as CNNs, RNNs, TCNs, and transformers.

As with tanh's saturation issue, ReLU has issues with dying neurons; several ReLU variants have been proposed. Leaky ReLU $\sigma(x) = \max(0.01x, x)$ is common, as is its parameterized version (**PReLU**) $\sigma(x) = \max(\alpha x, x)$. In both cases, instead of setting negative inputs to 0, they assign a small value proportional to the input. As a result, gradients do not saturate to 0; if an input is negative, the gradient will be α. In the case of Leaky ReLU, this is a constant value of generally 0.01 or 0.02, with no provision for choosing a value during training. In the case of PReLU, we learn the value of α as we train the weights and biases of our model, which of course adds computational complexity but can improve accuracy.

Trainable activation functions

Trainable activation functions like Swish add an extra layer of complexity to neural networks, which can improve performance (Apicella et al., 2021), or at least that is what is often seen in image studies. Trainable activation functions demonstrate the idea that almost any part of a network can be trained, not just weights and biases.

Swish is a relatively recent activation function, proposed by researchers at Google in 2017 (Ramachandran, Zoph, and Le, 2017).

$$\sigma(x) = x \cdot \frac{1}{1 + e^{-\beta x}}$$

Where,

- $\sigma(x)$ is the output of the Swish activation function
- x is the input to function
- β is a trainable parameter

If we were to set β to 1, Swish becomes the **Sigmoid Linear Unit** (**SiLU**) function.

While Swish appears similar to ReLU it differs by being *self-gated*, meaning it adaptively adjusts its shape based on the characteristics of your data. Specifically, when β is 0, Swish is essentially a linear function $\sigma(x) = x$, but when β is large, Swish approximates ReLU $\sigma(x) = x \cdot (x > 0)$. The trade-off is that this adds an extra parameter to each neuron, increasing computational demand, and risking overfitting.

```
class Swish(nn.Module):
    def __init__(self, beta=1.0):
        super().__init__()
        self.beta = nn.Parameter(torch.tensor(beta))

    def forward(self, x):
        return x * torch.sigmoid(self.beta * x)

swish_fn = Swish(beta=1.0)

plot_activation_fn_and_derivative(
    swish_fn, x, xlabel='x', ylabel='Swish(x)',
    title='Swish Activation Function'
)
```

The resulting plot:

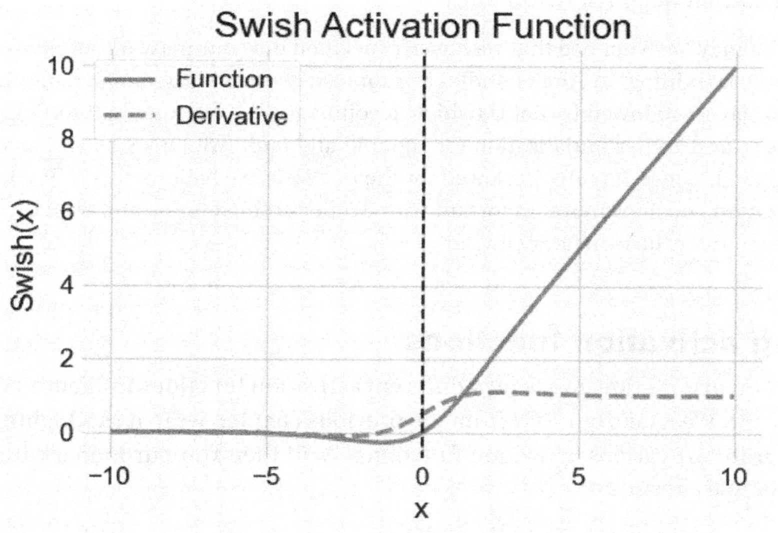

Figure 6.8: The Swish activation function

Similarly to ReLU, Swish is unbounded above, meaning that for large positive values gradients do not become 0. It is also bounded below, with the output tending to *some* constant as values approach negative infinity. This allows Swish to 'forget' very large negative values, which are usually deactivations, which helps regularize a model. Unlike monotonic functions, Swish varies direction in the negative region. This non-monotonicity enhances a model's information storage and discriminative capacity, as its mix of positive and negative derivatives boosts expressivity.

It does this by taking inspiration from LSTM's self-gated function. Like LSTM's sigmoid gates, Swish uses sigmoid to control information flow to the next layer. In activation, a value remains the same, but in deactivation, it's reduced by multiplying with the sigmoid of the value. Self-gating uses one scalar input, unlike multi-gating, which requires two.

Swish is also smooth, unlike ReLU's abrupt change near 0. This smoother transition results in a less oscillatory loss function, promoting faster convergence, and better generalization. A disadvantage here, is that it is more computationally expensive than ReLU, but it remains a useful alternative in hidden layers.

There are a plethora of activation functions in PyTorch we could you that you can look over here: `https://pytorch.org/docs/stable/nn.html`. There is also the option to make your own.

A note on weight initialization

You may have noticed that we haven't specified how our network's weights are initialized before training. PyTorch handles this for us with sensible defaults: Kaiming initialization for layers followed by ReLU (which accounts for the fact that half the inputs are zeroed out), and Xavier initialization for sigmoid and tanh (which keeps variance stable across layers). These defaults work well for the networks we build in this chapter. Weight initialization becomes more important with deeper architectures, and we'll revisit it when we encounter those in later chapters.

Comparing activation functions

Having covered theory, it's time to see how different activation functions influence performance with a basic experiment. We'll take our FFN from the previous chapter, wrap it as a Lightning module, and add a parameter to test various activation functions. We'll then run our model with each activation function and compare results.

First, we'll prepare our data. Let's load the airline passengers dataset, apply a sequential train-validation-test split, scale with min-max normalization, and create windowed input-target pairs:

```
np.random.seed(42)

data = pd.read_csv(CFG.data_folder / 'passengers.csv')
data['date'] = pd.to_datetime(data['date'])
data.set_index('date', inplace=True)

train_size = int(len(data) * 0.6)
val_size = int(len(data) * 0.2)
train, val, test = data[:train_size], data[train_size:train_size + val_size],
data[train_size + val_size:]

scaler = MinMaxScaler()
train_scaled = scaler.fit_transform(train)
val_scaled = scaler.transform(val)
test_scaled = scaler.transform(test)

def create_time_windows(data, window_size):
    inputs = []
    targets = []
    for i in range(len(data) - window_size):
        inputs.append(data[i:i+window_size])
        targets.append(data[i+window_size])
    return np.array(inputs), np.array(targets)
```

```
window_size = 12

train_inputs, train_targets = create_time_windows(train_scaled, window_size)
val_inputs, val_targets = create_time_windows(val_scaled, window_size)
test_inputs, test_targets = create_time_windows(test_scaled, window_size)

x_train = torch.FloatTensor(train_inputs)
y_train = torch.FloatTensor(train_targets)
x_val = torch.FloatTensor(val_inputs)
y_val = torch.FloatTensor(val_targets)
x_test = torch.FloatTensor(test_inputs)
y_test = torch.FloatTensor(test_targets)
```

Then we define our Lightning data module, our FFN architecture (with a configurable activation function), and a trainable Swish class:

```
class TimeSeriesDataModule(L.LightningDataModule):
    def __init__(self, train_data, val_data, test_data, batch_size=64):
        super().__init__()
        self.train_data = train_data
        self.val_data = val_data
        self.test_data = test_data
        self.batch_size = batch_size

    def setup(self, stage=None):
        self.train_dataset = TensorDataset(*self.train_data)
        self.val_dataset = TensorDataset(*self.val_data)
        self.test_dataset = TensorDataset(*self.test_data)

    def train_dataloader(self):
        return DataLoader(
            self.train_dataset, batch_size=self.batch_size, shuffle=False)

    def val_dataloader(self):
        return DataLoader(
            self.val_dataset, batch_size=self.batch_size, shuffle=False)

    def test_dataloader(self):
        return DataLoader(
            self.test_dataset, batch_size=self.batch_size, shuffle=False)

class ffnetwork(L.LightningModule):
```

```
    def __init__(
        self, input_dim, hidden_dim, activation_func,
        output_dim=1, learning_rate=0.0001
    ):
        super(ffnetwork, self).__init__()
        self.fc1 = nn.Linear(input_dim, hidden_dim)
        self.activation = activation_func
        self.fc2 = nn.Linear(hidden_dim, output_dim)
        self.learning_rate = learning_rate

    def forward(self, x):
        x = x.squeeze(-1)
        x = self.fc1(x)
        x = self.activation(x)
        x = self.fc2(x)
        return x

    def training_step(self, batch, batch_idx):
        x, y = batch
        y_hat = self.forward(x)
        loss = nn.functional.mse_loss(y_hat, y)
        self.log('train_loss', loss)
        return loss

    def validation_step(self, batch, batch_idx):
        x, y = batch
        y_hat = self.forward(x)
        loss = nn.functional.mse_loss(y_hat, y)
        self.log('val_loss', loss)

    def configure_optimizers(self):
        optimizer = torch.optim.Adam(self.parameters(), lr=self.learning_rate)
        return optimizer

class swish(nn.Module):
    def __init__(self):
        super().__init__()
        self.beta = nn.Parameter(torch.tensor(1.0))

    def forward(self, x):
        return x * torch.sigmoid(self.beta * x)
```

After setting up our data, model, training process, and activation functions, we define our hyperparameters. Here we have selected some typical values; to get a sense of what these parameters do it is always a good idea for us to test multiple values.

```
input_dim = 12
hidden_dim = 50
output_dim = 1
learning_rate = 0.0001

batch_size = 64
data_module = TimeSeriesDataModule(train_data=(x_train, y_train),
                                   val_data=(x_val, y_val),
                                   test_data=(x_test, y_test),
                                   batch_size=batch_size)

activations = {
    'ReLU': nn.ReLU(),
    'Sigmoid': nn.Sigmoid(),
    'Tanh': nn.Tanh(),
    'LeakyReLU': nn.LeakyReLU(0.01),
    'Swish': swish()
}
```

Next we define a training loop with early stopping (`patience = 10`) that iterates through each activation function:

```
results = {}
for name, activation_func in activations.items():
    print(f"Training with {name} activation...")

    torch.manual_seed(42)
    torch.cuda.manual_seed(42)

    model = ffnetwork(
        input_dim, hidden_dim, activation_func, output_dim, learning_rate)

    early_stop_callback = EarlyStopping(
        monitor='val_loss', patience=10, verbose=False)

    trainer = L.Trainer(max_epochs=2000, callbacks=[early_stop_callback])
    trainer.fit(model, data_module)

    model.eval()
```

```
    last_window = x_val[-1].view(1, -1)
    recursive_preds = []

    for _ in range(len(x_test)):
        with torch.no_grad():
            pred = model(last_window)
        recursive_preds.append(pred.item())
        last_window = torch.cat((last_window[:, 1:], pred.view(1, 1)), dim=1)

    unscaled_predictions = scaler.inverse_transform(np.array(
        recursive_preds).reshape(-1, 1))
    results[name] = unscaled_predictions

    del model
    torch.cuda.empty_cache()
```

Recursive forecasting

The prediction approach we use throughout this chapter is **recursive** (also called **autoregressive**) forecasting. We seed our model with the last known window from our validation set, predict one step ahead, then roll the window forward by appending that prediction and dropping the oldest value. This is repeated for the length of our test set. It's a natural approach for univariate models, but be aware that errors accumulate with each step, a poor early prediction shifts the entire input window, and subsequent predictions compound drift. We'll revisit forecasting strategies in more detail in later chapters, but for now this approach serves us well for comparing architectural choices on a level playing field.

Finally, we'll plot our results:

```
plt.figure(figsize=(CFG.img_dim1, CFG.img_dim2))
for idx, (name, preds) in enumerate(results.items()):
    test_index = data.index[-len(preds):]
    plt.plot(test_index, preds, label=f'{name}',
             color=custom_palette[idx + 1], linewidth=2.0)

plt.plot(data.index, data['passengers'], label='Actual', color=custom_
palette[0], linewidth=2.0)
plt.title(
    'Airline Passengers - Predictions with Different Activation Functions')
plt.legend()
plt.show()
```

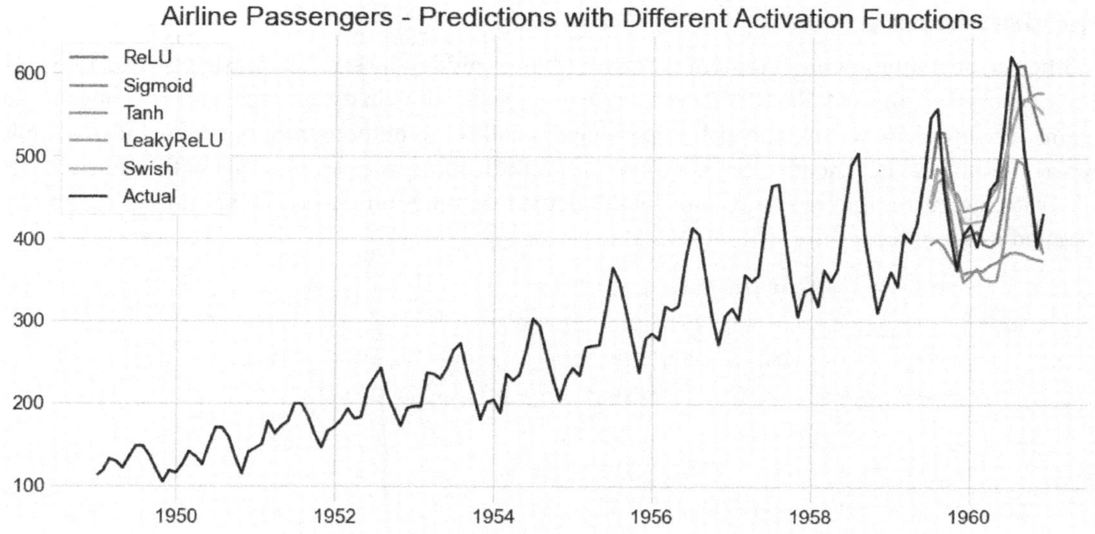

Figure 6.9: Predictions with different activation functions

Our plotted predictions suggest that ReLU appears to help our FFN create a better forecast for this dataset, which is reflected in the metrics we have used here: RMSE (and MSE) but not MAE (Swish does slightly better). ReLU does not achieve the lowest error for all datasets; selection may vary with neural architecture, hyperparameters, time/compute cost and crucially, as always, your data. We demonstrate this further in the notebook with our M5 subset. Activation functions have a significant impact on even the simplest neural network designs, and it is worth conducting experiments before making a choice.

	Activation	MAE	MSE	RMSE	MAPE	sMAPE
0	ReLU	63.202089	5580.584110	74.703307	13.868372	13.237820
1	Sigmoid	94.374972	13336.849200	115.485277	18.443492	21.068519
2	Tanh	66.534065	6673.879197	81.693814	13.464732	14.810079
3	LeakyReLU	66.207058	6830.479637	82.646716	14.515465	13.561442
4	Swish	61.119099	5684.264051	75.394058	13.223443	12.516499

Warning

Airline passengers dataset is unusually clean, with strong trend, regular seasonality, no missing data, and no exogenous variables. We also have not modeled on residuals here, which is often worth considering when building models.

On messier real-world data, the differences between activation functions (and all the other architectural choices in this chapter) can be more pronounced, and rankings may well shift. The chapter notebook extends some experiments to a subset of the M5 competition dataset, but in reality everything is dependent on your data and its quality. One question people rarely ask is: can it even be forecasted? You'd be surprised how often the answer to that question is no.

Hidden layers

Another critical architecture element is the number of **hidden layers**. Obviously you won't be using the 96 layers that an LLM like GPT-3 uses, but extra layers can help disentangle patterns in your data. Again, for simplicity we use the airline passengers dataset to demonstrate here; for more complex experimentation visit the chapter notebook on optimization. We can test this with a few changes to our architecture: making ReLU our default activation function and allowing hidden layers to be changed programmatically.

```
class ffnetwork(L.LightningModule):
    def __init__(
        self, input_dim, hidden_dim, num_layers=1, output_dim=1,
        learning_rate=0.0001, activation_func=nn.ReLU()
    ):
        super(ffnetwork, self).__init__()
        self.layers = nn.ModuleList()
        self.layers.append(nn.Linear(input_dim, hidden_dim))
        for _ in range(num_layers - 1):
            self.layers.append(nn.Linear(hidden_dim, hidden_dim))
        self.layers.append(nn.Linear(hidden_dim, output_dim))
        self.activation = activation_func
        self.learning_rate = learning_rate

    def forward(self, x):
        x = x.squeeze(-1)
        for i in range(len(self.layers) - 1):
            x = self.activation(self.layers[i](x))
        x = self.layers[-1](x)
        return x

    def training_step(self, batch, batch_idx):
        x, y = batch
        y_hat = self.forward(x)
        loss = nn.functional.mse_loss(y_hat, y)
        self.log('train_loss', loss)
        return loss

    def validation_step(self, batch, batch_idx):
        x, y = batch
        y_hat = self.forward(x)
        loss = nn.functional.mse_loss(y_hat, y)
        self.log('val_loss', loss)
```

```
    def configure_optimizers(self):
        optimizer = torch.optim.Adam(self.parameters(), lr=self.learning_rate)
        return optimizer
```

For this experiment we simply run the model with different numbers of layers:

```
results = {}
hidden_layer_configs = range(1, 7)

for num_layers in hidden_layer_configs:
    print(f"Training model with {num_layers} hidden layers...")

    torch.manual_seed(42)
    torch.cuda.manual_seed(42)

    model = ffnetwork(input_dim, hidden_dim, num_layers=num_layers,
                      output_dim=output_dim, learning_rate=learning_rate)

    early_stop_callback = EarlyStopping(
        monitor='val_loss', patience=10, verbose=False)
    trainer = L.Trainer(max_epochs=2000, callbacks=[early_stop_callback])
    trainer.fit(model, data_module)

    model.eval()
    last_window = x_val[-1].view(1, -1)
    recursive_preds = []

    for _ in range(len(x_test)):
        with torch.no_grad():
            pred = model(last_window)
        recursive_preds.append(pred.item())
        last_window = torch.cat(
            (last_window[:, 1:], pred.view(1, 1)), dim=1)

    unscaled_predictions = scaler.inverse_transform(
        np.array(recursive_preds).reshape(-1, 1))
    results[f'{num_layers} layers'] = unscaled_predictions

    del model
    torch.cuda.empty_cache()
```

As we can see in *Figure 6.10*, the number of hidden layers has a significant impact on forecast accuracy, with a few extra layers improving performance before it decreases with four layers. Although 5-layers shows improvement again, in these cases Occam's Razor should be applied, especially if the differences are marginal.

```
plt.figure(figsize=(CFG.img_dim1, CFG.img_dim2))
for idx, (name, preds) in enumerate(results.items()):
    test_index = data.index[-len(preds):]
    plt.plot(test_index, preds, label=f'{name}',
             color=custom_palette[idx + 1], linewidth=2.0)

plt.plot(data.index, data['passengers'], label='Actual',
         color=custom_palette[0], linewidth=2.0)
plt.title('Airline Passengers - Predictions with Different Numbers of Hidden 
Layers')
plt.legend()
plt.show()
```

The result of the code is the following plot:

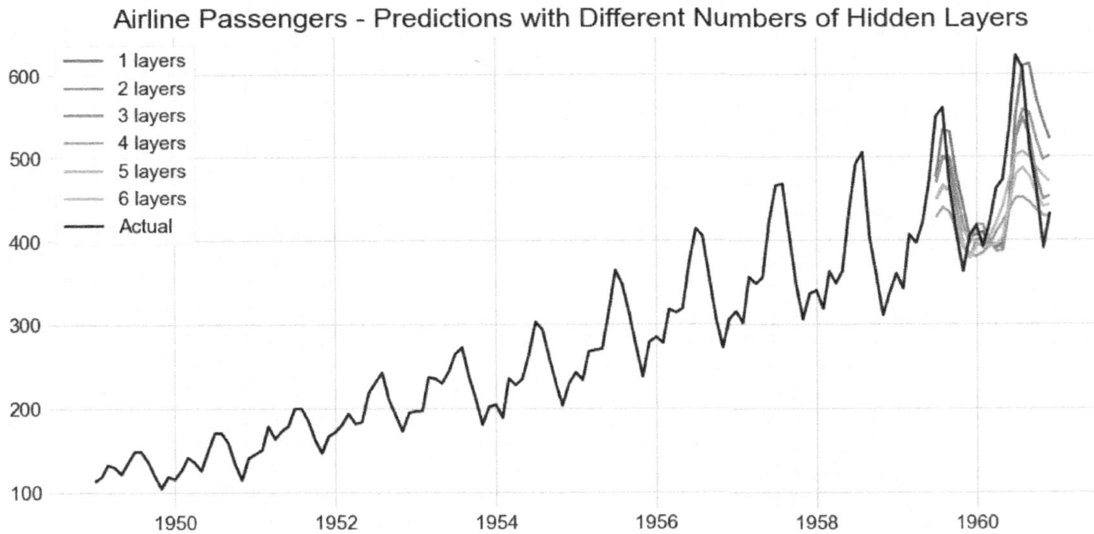

Figure 6.10: Predictions with different hidden layers

We can also see how the number of layers determine performance related scores:

	Model	MAE	MSE	RMSE	MAPE	sMAPE
0	1 layers	63.202089	5580.584110	74.703307	13.868372	13.237820
1	2 layers	45.515614	2869.036939	53.563392	9.332868	9.688681
2	3 layers	56.687399	4082.960045	63.898044	11.984523	12.024014

```
3  4 layers  58.790394  6021.330113  77.597230  11.356693  12.349161
4  5 layers  46.014103  3555.361974  59.626856   9.261752   9.535999
5  6 layers  50.953759  4317.074975  65.704452  10.032113  10.695182
```

You can use a similar approach to work through other architectural parameters, such as nodes (which we have discussed previously) and kernel sizes (specifically for CNNs). Doing so individually gives us some idea of what parameter values we might limit ourselves to, and some idea of their effect on models. The problem with these experiments is that, while tuning individual hyperparameters can help reduce options, it does not always reflect the optimal architectural parameters. This is because the parameters work in tandem with other hyperparameters, such as learning rate, regularization, dropout, iterations, and batch size. Not to mention the random process that affects reproducibility; while writing this I forgot to set the seed, resulting in completely different accuracy metrics for each layer on a second run!

Typically the best approach to finding bounds for hyperparameter tuning is to review the literature—look at papers, GitHub code, Kaggle notebooks, and books—to see what practitioners are using for similar challenges and data patterns. Use rules of thumb from experience and community best practices and test on your own data.

Research around neural network design often emphasizes learning algorithms and architectural design, neglecting the impact of activation functions. But in practice, the choice of activation function greatly influences the performance of neural networks, playing a crucial role in the convergence of learning (Da S. Gomes, Ludermir, and Lima, 2011). Clearly, the number of nodes and layers also have a strong impact on performance, as does the choice of optimization algorithm. These parameters interact with one another, so while testing them individually can help narrow the search space, tuning them together is ultimately what produces the best results.

Gradient descent, loss functions, and regularization techniques

Our next focus for optimization is the selection and tuning of network optimizers, such as **Stochastic Gradient Descent** (**SGD**), Adam, and RMSprop. These optimization algorithms are concerned with minimizing an objective function, typically a loss function (e.g. MSE), computed on your training dataset. Their role is to update model parameters (weights and biases) in a direction that reduces training error. This is not equal to the goal of statistical inference, which is to reduce the generalization error of a model. These goals are to some extent in opposition with one another; optimizing for immediate accuracy can negatively affect long-term predictiveness by causing us to overfit models to the data.

Loss functions

The loss function (also called a **cost function** or **objective function**) is what our optimizer is trying to minimize. Throughout this chapter we use MSE, which is the default for most regression tasks, but it is worth understanding why and when you might choose differently.

$$\text{MSE} = \frac{1}{n}\sum_{i=1}^{n}(y_i - \hat{y}_i)^2$$

MSE penalizes large errors heavily because of the squaring; a prediction that is off by 10 contributes 100 to our loss, while one that is off by 2 contributes only 4. This makes MSE a good choice when large errors are particularly undesirable, which is often the case in forecasting scenarios where a big miss can mean stockouts or excess inventory.

Mean Absolute Error (**MAE**) is a simpler alternative:

$$\text{MAE} = \frac{1}{n}\sum_{i=1}^{n}|y_i - \hat{y}_i|$$

MAE treats all errors proportionally and is more robust to outliers, since it doesn't amplify large deviations. If your data has occasional spikes or anomalies that you don't want to dominate the training signal, MAE can be a better choice. In PyTorch, swapping is straightforward.

```
# MSE loss (default throughout this chapter)
loss = nn.functional.mse_loss(y_hat, y)

# MAE loss
loss = nn.functional.l1_loss(y_hat, y)

# Huber loss, combines MSE and MAE, with a threshold delta
loss = nn.functional.huber_loss(y_hat, y, delta=1.0)
```

Huber loss is a useful middle ground; it behaves like MSE for small errors (below the delta threshold) and like MAE for large errors, giving you sensitivity to small deviations without letting outliers dominate training. In practice, MSE is a safe default for most forecasting tasks, but if you find your model is chasing outliers or producing erratic predictions, switching to Huber or MAE is one of the first things to try.

To illustrate loss optimization, we can represent the underlying function of some data as f and the observed function f_d, which is not as smooth due to the noise inherent in collected data. This is adapted from *Dive into Deep Learning*.

```
def f(x):
    return x * torch.cos(np.pi * x) + 2
def fd(x):
    return f(x) + 0.3 * torch.cos(6 * np.pi * x)

x = torch.arange(0.5, 1.5, 0.01)

plt.figure(figsize=(CFG.img_dim1, CFG.img_dim2))
plt.plot(x, f(x), label='f (generalisation)', color=custom_palette[1])
plt.plot(x, fd(x), label='fd (training)', color=custom_palette[2])
```

```
plt.xlabel('x')
plt.ylabel('Value')
plt.legend()
plt.show()
```

In the following plot we are looking for the minimal value. We can see that the minimum of f_d is shifted to the right of the minimum of f — the function we are actually trying to generalise toward.

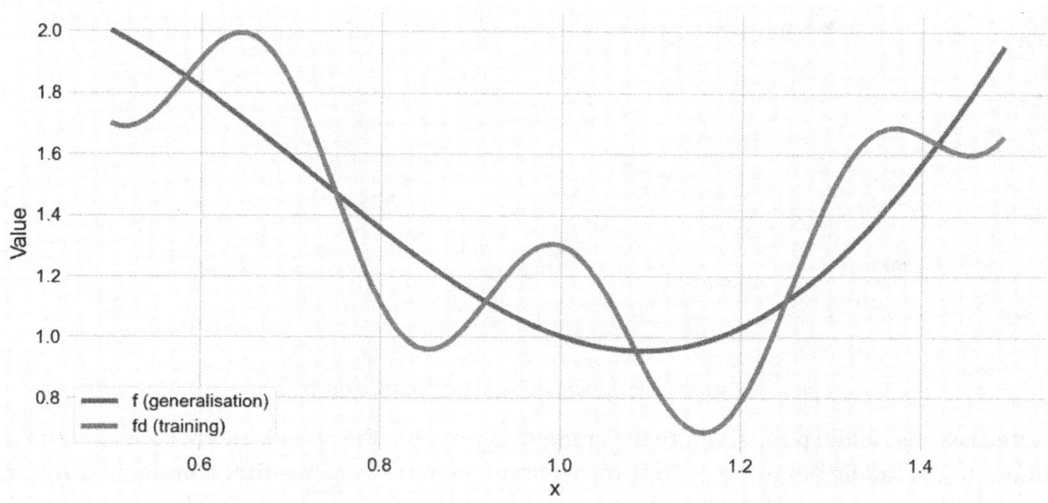

Figure 6.11: Loss optimization

Optimization algorithms encounter a number of issues we should be aware of: local minima, saddle points, and vanishing gradients. Let's imagine we are minimizing the error of a model and we plot it. We can see a *ditch* at around `0.8`, which we refer to as a local minimum, but the lowest value of the function is the global minimum, where we want our optimization algorithm to reach. The problem we encounter when training neural networks is that the loss surfaces we have to navigate often contain many local optima in which optimization algorithms can become stuck. This occurs because optimization algorithms are attempting to minimize/optimize your objective locally instead of globally. There are ways to deal with this, with the simplest approach being to just add some noise to the optimizer's weight changes, which may help an optimizer (i.e. minibatch stochastic) get out of a local ditch/minima.

```
def fd(x):
    return f(x) + 0.3 * torch.cos(4 * np.pi * x)


plt.figure(figsize=(CFG.img_dim1, CFG.img_dim2))
plt.plot(x, fd(x), label='fd (training)', color=custom_palette[2])
plt.xlabel('x')
plt.ylabel('Value')
plt.legend()
plt.show()
```

The resultant plot is:

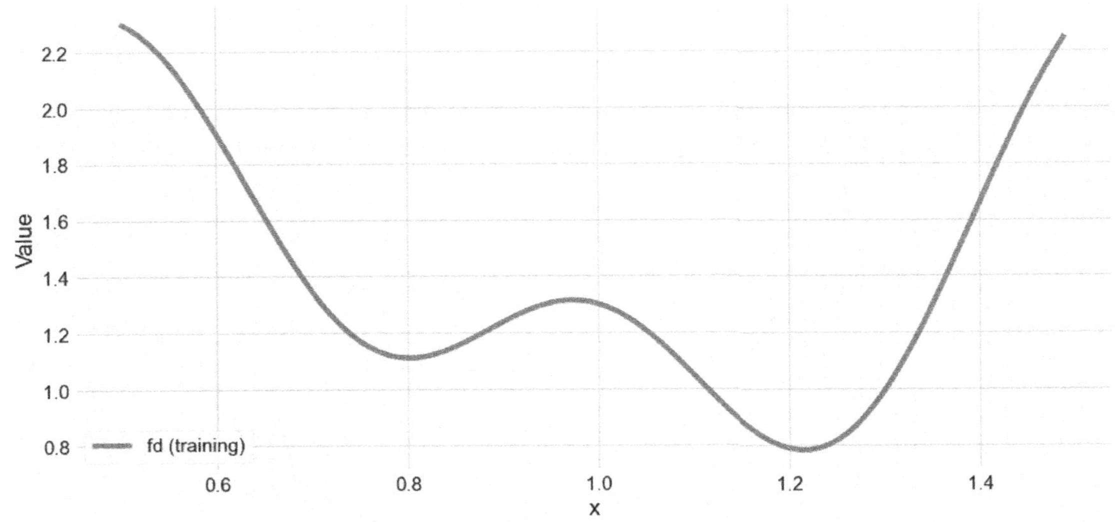

Figure 6.12: Adding noise to overcome a ditch

A related issue is **saddle points**, where the gradient is zero, but the point is neither a maximum nor a minimum. At a saddle point, we see a surface curving upward in some directions and downward in others, which can cause gradient-based optimizers to stall because gradient signals become very small.

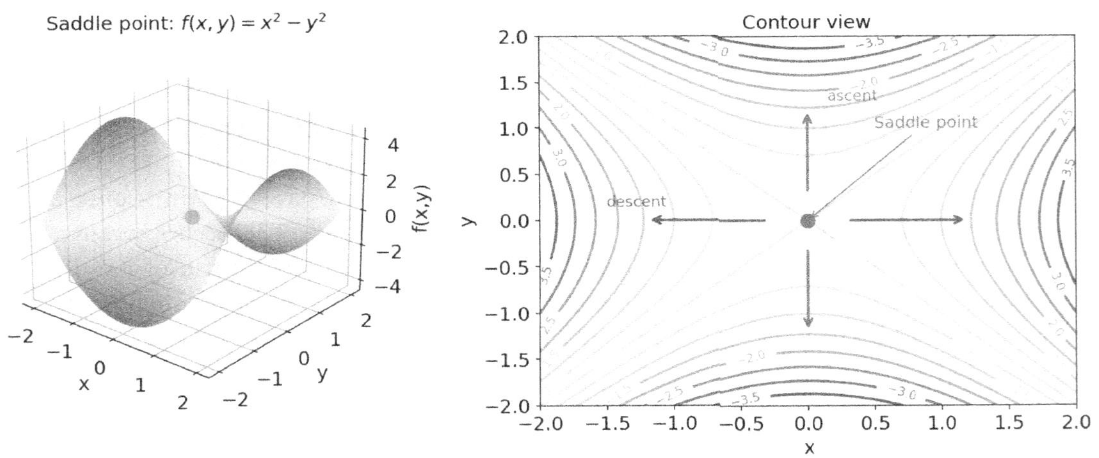

Figure 6.13: Saddle point: the surface curves upward along the x-axis (descent) and downward along the y-axis (ascent), creating a point where the gradient is zero, but no minimum exists

Optimization algorithms are a large research area within deep learning, but in terms of application we typically use those provided by `https://pytorch.org/docs/stable/optim.html`, which includes the most common optimizers. Of these you will find the following used most often:

- **Stochastic gradient descent** (**SGD**)
- **Batch gradient descent** (**GD**)
- **Mini-batch gradient descent** (**MGD**)
- Adam
- Adadelta
- RMSprop
- Adagrad

The gradient descent algorithms require more tuning, typically of the learning rate, than the adaptive optimizers which adjust learning rate intrinsically as part of the optimization process. It's useful to know a little about how these algorithms operate if you want to reduce your options and research time. Gradient descent algorithms are largely separated by the amount of data they use to compute the gradient of a function:

- **GD**, or simply gradient descent, uses an entire dataset to perform parameter updates, making it computationally expensive.
- **SGD** uses *individual samples* to update parameters, which is faster than GD. But because the model is updated after every single sample, this high variance causes large fluctuations in the objective function during training. This can be advantageous for escaping local minima, but can also lead to getting trapped in one. We can tune the learning rate to reduce oscillations in the objective.
- **MGD** computes gradients on small *batches* of data, resulting in less variation of the objective function compared to SGD. It is generally preferred over other gradient descent methods because of its balance between efficiency and convergence stability.
- **Adagrad** adapts learning rates, with smaller updates for frequent features and larger updates for rarer ones. While this allows the optimizer to capture information from rarer features, it can result in vanishing learning rates, where the rate becomes too small too early in training.
- **Adadelta** improves on Adagrad by using a fixed window of past gradients to avoid vanishing learning rates.
- **RMSprop** is similar to Adadelta but uses a different approach to managing past gradients. It adjusts learning rates for each parameter dynamically, dividing the learning rate by a running average of the magnitudes of recent gradients for that weight.
- **Adam** combines the advantages of Adadelta/RMSprop and momentum. It is currently the default optimizer for most forecasting libraries utilizing neural networks. Its advantage is that it works well with sparse data due to its adaptive learning rates, without requiring manual tuning of the global learning rate.

Although Adam is generally your go-to, it is not always the best approach. For example, a linear regression is faster to train with SGD than Adam, and even on more complex datasets you will find that a well-tuned learning rate for a gradient descent algorithm can result in a better model or faster training. This is why it is still worth running experiments with different optimization algorithms.

We can extend our FFN to accept different optimizers as a parameter. Note that we have also added dropout to this version of the network, which is applied after each hidden layer, but it is set to `0.00`, which means it is not used here:

```
from lightning.pytorch.loggers import CSVLogger


class ffnetwork(L.LightningModule):
    def __init__(
        self, input_dim, hidden_dim, num_layers=1, output_dim=1,
        learning_rate=0.0001, dropout_rate=0.00,
        activation_func=nn.ReLU(), optimizer_name='Adam'
    ):
        super(ffnetwork, self).__init__()
        self.layers = nn.ModuleList([nn.Linear(input_dim, hidden_dim)])
        self.dropout = nn.Dropout(dropout_rate)

        for _ in range(num_layers - 1):
            self.layers.append(nn.Linear(hidden_dim, hidden_dim))

        self.layers.append(nn.Linear(hidden_dim, output_dim))
        self.activation = activation_func
        self.learning_rate = learning_rate
        self.optimizer_name = optimizer_name

    def forward(self, x):
        x = x.squeeze(-1)
        for i in range(len(self.layers) - 1):
            x = self.activation(self.layers[i](x))
            x = self.dropout(x)
        x = self.layers[-1](x)
        return x

    def training_step(self, batch, batch_idx):
        x, y = batch
        y_hat = self(x)
        loss = nn.functional.mse_loss(y_hat, y)
```

```
        self.log('train_loss', loss)
        return loss

    def validation_step(self, batch, batch_idx):
        x, y = batch
        y_hat = self(x)
        loss = nn.functional.mse_loss(y_hat, y)
        self.log('val_loss', loss)

    def configure_optimizers(self):
        optimizers = {
            'SGD': torch.optim.SGD(
                self.parameters(), lr=self.learning_rate),
            'Adam': torch.optim.Adam(
                self.parameters(), lr=self.learning_rate),
            'Adadelta': torch.optim.Adadelta(
                self.parameters(), lr=self.learning_rate),
            'RMSprop': torch.optim.RMSprop(
                self.parameters(), lr=self.learning_rate),
            'Adagrad': torch.optim.Adagrad(
                self.parameters(), lr=self.learning_rate),
        }
        if self.optimizer_name in optimizers:
            return optimizers[self.optimizer_name]
        else:
            raise ValueError(
                f"Optimizer '{self.optimizer_name}' not recognized.")
```

Let's run our experiment, training with each optimizer and logging the results:

```
# List of optimiser algorithms
optimizers = ['SGD', 'Adam', 'Adadelta', 'RMSprop', 'Adagrad']
results = {}
for optimizer_name in optimizers:
    print(f"Training with {optimizer_name} optimizer...")

    torch.manual_seed(42)
    torch.cuda.manual_seed(42)

    model = ffnetwork(
        input_dim, hidden_dim, num_layers=2, output_dim=output_dim,
        learning_rate=0.001, optimizer_name=optimizer_name
```

```
)

early_stop_callback = EarlyStopping(
    monitor='val_loss',
    patience=10,
    verbose=False,
    mode='min'
)
# Dictionary for logs to be saved
log_dir = f'lightning_logs/{optimizer_name}'
logger = CSVLogger("lightning_logs", name=optimizer_name)
# Early stopping
trainer = L.Trainer(
    max_epochs=2000,
    default_root_dir=log_dir,
    callbacks=[early_stop_callback],
    logger=logger # CSV logging of training
)
# Train model
trainer.fit(model, data_module)
# Evaluate model with test data and store predictions for plotting
model.eval()
last_window = x_val[-1].view(1, -1)
recursive_preds = []

for _ in range(len(x_test)):
    with torch.no_grad():
        pred = model(last_window)
    recursive_preds.append(pred.item())
    last_window = torch.cat((last_window[:, 1:], pred.view(1, 1)), dim=1)

unscaled_predictions = scaler.inverse_transform(
    np.array(recursive_preds).reshape(-1, 1))
results[optimizer_name] = unscaled_predictions

del model
torch.cuda.empty_cache()
```

To visualize how these optimizers differ in their search behavior, we can plot their paths across a known loss surface. The **Beale** function is a classic test surface with a single global minimum at `(3, 0.5)` and a complex landscape that challenges gradient-based methods. Each panel below shows how a different optimizer navigates from the same starting point `(-1.5, -1.0)`.

In this figure, each panel shows a different optimizer starting from the same point, with the cross marking the global minimum. Notice how SGD barely moves with a small learning rate, while Adam finds its way to a global minimum. Adagrad makes good progress but approaches from a different angle, and Adadelta struggles to escape the starting region.

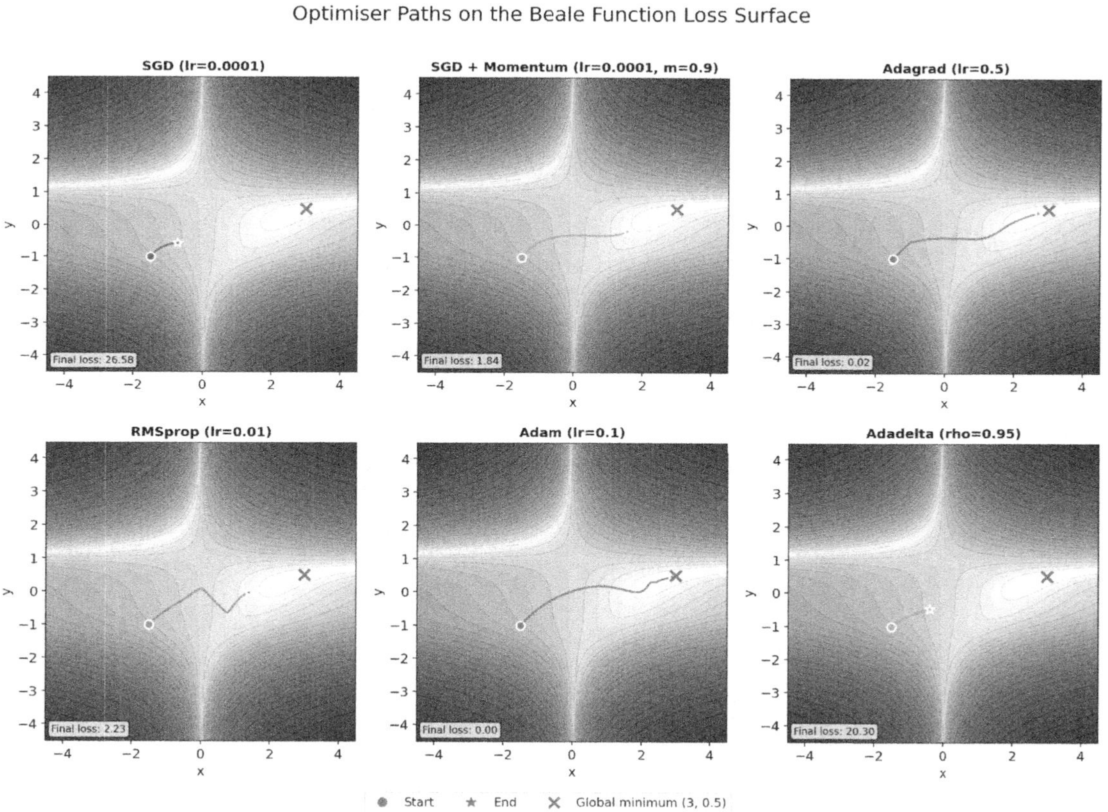

Figure 6.14: Optimiser paths on the Beale function loss surface

A visual inspection shows us something interesting, Adam performs poorly as does SGD and Adadelta, we know performance varies by architecture, and we didn't tune the learning rate, but this is terrible.

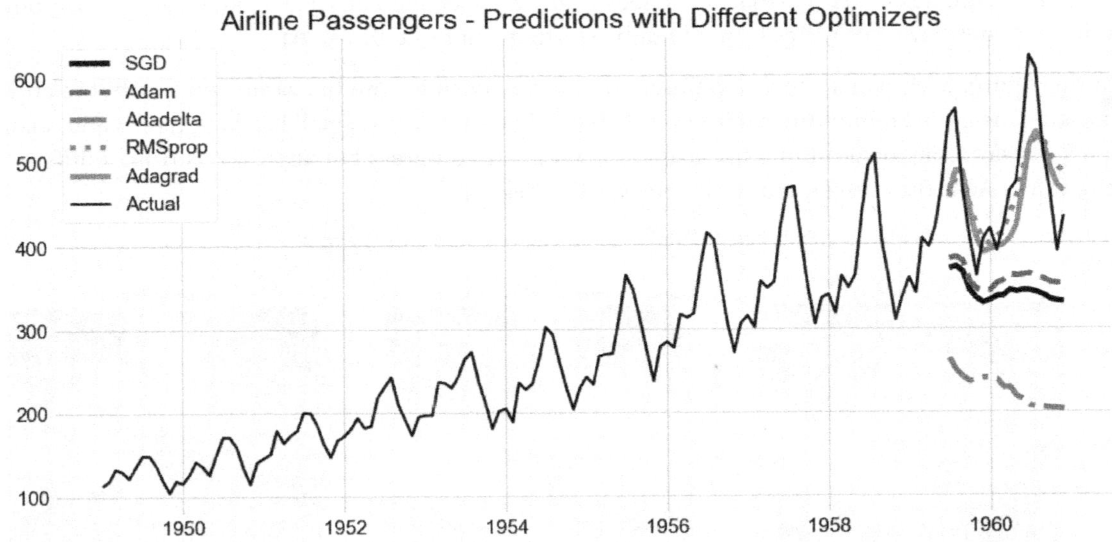

Figure 6.15: Predictions with different optimizers

Why are these optimizers performing so poorly? Let's look at the validation loss.

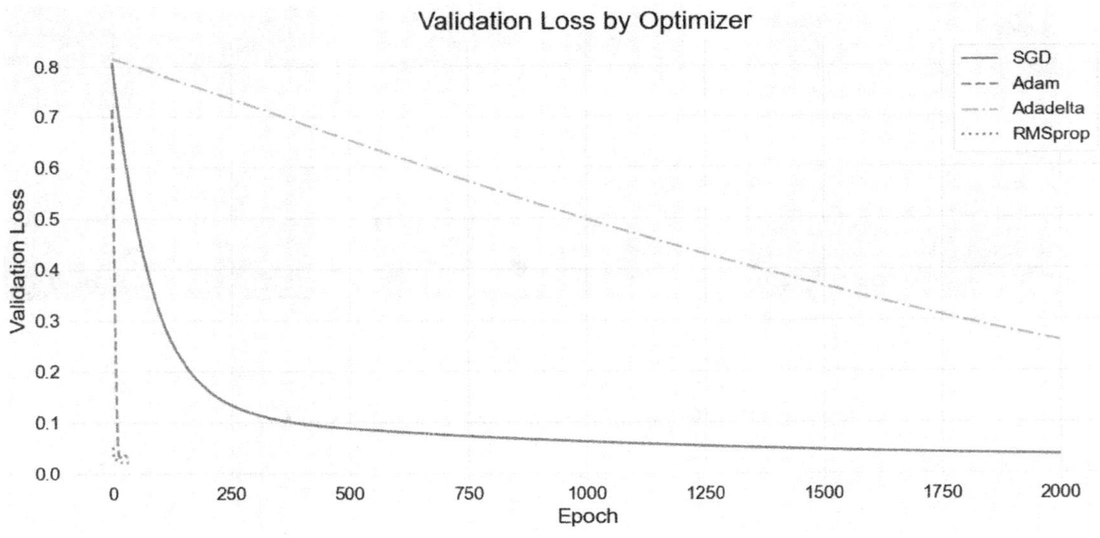

Figure 6.16: Validation loss by optimizer

Each optimiser has very different loss curves over epochs, with both Adam and RMSprop stopping at around 30-40 epochs, clearly getting stuck in a local minima. While this works in RMSprop's favor, it is catastrophic for Adam, which is reflected in the error metrics.

	Model	MAE	MSE	RMSE	MAPE	sMAPE
0	SGD	126.722022	20841.284405	144.365108	25.423805	29.893697
1	Adam	110.533553	16851.939089	129.815019	21.923355	25.382636
2	Adadelta	244.829587	66358.090440	257.600641	50.839653	69.130685
3	RMSprop	46.301069	3253.420483	57.038763	9.593590	9.617378
4	Adagrad	46.235060	2982.060727	54.608248	9.499967	9.728563

Optimizing hyperparameters

Learning rate is the most important hyperparameter for optimizers, as it influences a model's ability to generalize by affecting how thoroughly networks are adapted to the parameter space. It controls the size of steps taken in the direction of the negative gradient during training, scaling the adjustment of network weights with respect to the loss function. Finding the optimal learning rate is vital for creating a good model, as it helps models overcome local minima, saddle points, and vanishing gradients by moving training in a way that avoids these traps (though it will not resolve issues caused by poor data).

For a little more depth on optimization algorithms we recommend reading *Chapter 12* of

`https://d2l.ai/chapter_optimization/index.html.`

Learning rate scheduling

Using a fixed **Learning Rate** (**LR**) for an entire training run is not a good idea. A large LR will make good progress early but may overshoot as your model approaches a minimum, while a rate small enough for fine-grained convergence may be painfully slow at the start of training. LR schedulers dynamically adjust LR rate during training.

The most practical scheduler for forecasting work is `ReduceLROnPlateau`, which monitors a metric (typically validation loss) and reduces the LR when it stops improving. Which compliments early stopping, as the scheduler gives the model a chance to escape a plateau before early stopping terminates training. We can add it to our `configure_optimizers` method:

```
def configure_optimizers(self):
    optimizer = torch.optim.Adam(self.parameters(), lr=self.learning_rate)
    scheduler = torch.optim.lr_scheduler.ReduceLROnPlateau(
        optimizer, mode='min', factor=0.5, patience=5
    )
    return {
        'optimizer': optimizer,
```

```
        'lr_scheduler': {
            'scheduler': scheduler,
            'monitor': 'val_loss'
        }
    }
```

Here, `factor=0.5` halves the learning rate each time the validation loss fails to improve for five consecutive epochs (`patience=5`). Our scheduler and early stopping work together, with the scheduler reducing LR to allow finer convergence; if that still doesn't help, early stopping terminates training.

Other common schedulers include **cosine annealing**, which smoothly decays LR following a cosine curve, and **step decay**, which reduces LR by a fixed factor at set intervals. For most forecasting tasks, `ReduceLROnPlateau` is a practical starting point as it adapts to your data rather than requiring you to guess when to reduce LR.

Before choosing a learning rate, it is worth running a learning rate range test. The idea, proposed by Leslie Smith in 2017, is to train for a short period while gradually increasing your learning rate from a very small value to a large one, and plot loss against learning rate. The optimal starting LR is typically just before the point where loss begins to rise sharply. Lightning makes this simple with Tuner:

```
from lightning.pytorch.tuner import Tuner

trainer = L.Trainer(max_epochs=100)
tuner = Tuner(trainer)

# runs LR range test and updates model.learning_rate
lr_finder = tuner.lr_find(model, datamodule=data_module)

# plot loss curve
fig = lr_finder.plot(suggest=True)
fig.show()

# suggested LR
print(f"Suggested LR: {lr_finder.suggestion()}")
```

This is one of the most useful practical tools for setting an initial learning rate before you start tuning other hyperparameters.

Epochs

The number of times we pass an entire dataset through a neural network during training (epoch) impacts its learning; by affecting the number of opportunities it has to learn patterns from our dataset, which affects the overall accuracy of the model. If we allow a model too few opportunities it can underfit, conversely too many epochs may lead to overfitting which can be overcome with regularization and early stopping, with the latter being especially important.

The following plot shows this trade-off. Early in training (*left* region), both training and validation loss are high, as the model is underfitting. As training continues, both losses decrease. Eventually, training loss continues to fall while validation loss begins to rise, and the gap between them starts widening as the model overfits to the training data. Early stopping halts training at the point where validation loss is minimized.

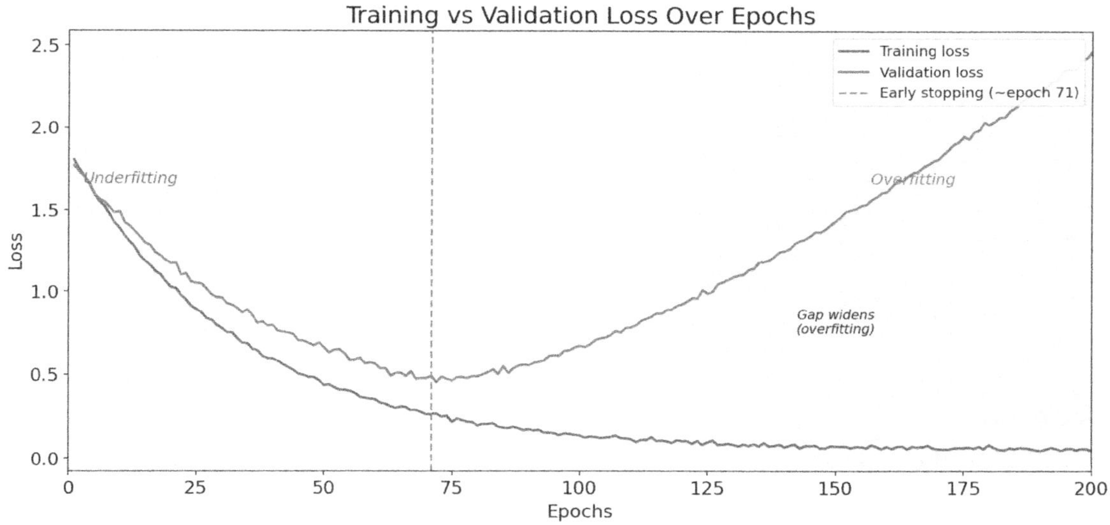

Figure 6.17: Training vs validation loss over epochs

Batch size

Batch size is a critical hyperparameter because it affects the number of training samples used to compute gradients for each parameter update. Generally, finding the optimal batch size is a trade-off between gradient estimation accuracy and computational efficiency. Larger batch sizes generally lead to more accurate gradient approximations across a dataset and more stable convergence, with a smoother optimization trajectory, but they require more memory and can slow each training iteration. Smaller batch sizes allow more frequent parameter updates with less computational cost, though the resulting gradient estimates are noisier. This noise can introduce more variance into the learning process and a more erratic optimization path, which can help optimizers escape shallow local minima.

Mini-batch approaches to training are fairly common because they utilize moderate batch sizes which provide some noise-reduced averaging while allowing reasonably fast iteration times. While there is no universal answer for the ideal batch size, 32 is usually a good starting point for datasets with fewer than 1,000 samples.

The effect of batch size on training dynamics is shown below. Smaller batches (**8**) produce noisier gradient estimates, resulting in more erratic loss curves but faster individual updates. Larger batches (**32** and **128**) give smoother convergence at the cost of slower iterations and higher memory usage.

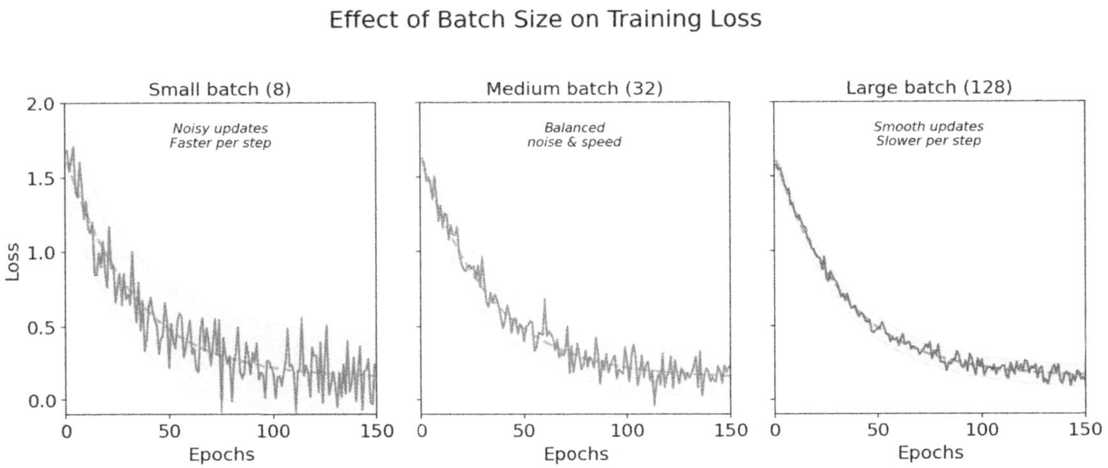

Figure 6.18: Effect of batch size on training loss

Dropout

Dropout is another regularization technique, which helps to prevent overfitting. We simply provide a float value between 0 and 1 (e.g. 0.2 tells a network to drop 20% of the nodes and their connections at each training iteration). This creates *thinned* network architectures; applying this stochastic process promotes distributed and generalized learning across a network, ensuring that no single neuron is solely responsible for predicting specific features. Essentially, dropout breaks down complex co-adaptations among neurons, forcing networks to learn more robust features, which in theory generalize better to held-out data. It is implemented by multiplying neuron activations by a Bernoulli random mask with keep probability p. During inference, dropout is not applied; instead, neuron weights are scaled by p to compensate for the increased number of active neurons and prevent shifts in activation magnitudes.

Figure 6.19 shows how dropout creates a *thinned* network at each training iteration. On the *left*, all neurons are active. On the *right*, dropout with $p = 0.3$ randomly deactivates roughly 30% of hidden neurons (shown with hatched pattern), along with their connections. This forces the remaining neurons to learn more distributed representations.

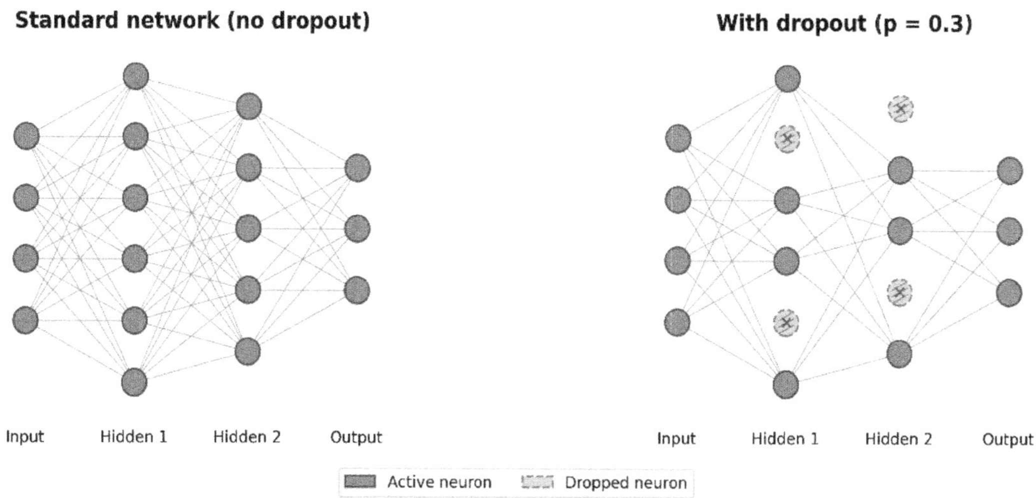

Figure 6.19: Dropout applied to a neural network

In practice, dropout is applied after each hidden layer in the forward pass. This code shows how to add it to our FFN:

```
class ffnetwork(L.LightningModule):
    def __init__(
        self, input_dim, hidden_dim, num_layers=1, output_dim=1,
        learning_rate=0.0001, dropout_rate=0.5, activation_func=nn.ReLU()
    ):
        super(ffnetwork, self).__init__()
        self.layers = nn.ModuleList([nn.Linear(input_dim, hidden_dim)])
        self.dropout = nn.Dropout(dropout_rate)

        for _ in range(num_layers - 1):
            self.layers.append(nn.Linear(hidden_dim, hidden_dim))

        self.layers.append(nn.Linear(hidden_dim, output_dim))
        self.activation = activation_func
        self.learning_rate = learning_rate

    def forward(self, x):
        x = x.squeeze(-1)
        for i in range(len(self.layers) - 1):
            x = self.activation(self.layers[i](x))
            x = self.dropout(x)  # Apply dropout after each hidden layer
        x = self.layers[-1](x)
        return x
```

```
    def training_step(self, batch, batch_idx):
        x, y = batch
        y_hat = self.forward(x)
        loss = nn.functional.mse_loss(y_hat, y)
        self.log('train_loss', loss)
        return loss

    def validation_step(self, batch, batch_idx):
        x, y = batch
        y_hat = self.forward(x)
        loss = nn.functional.mse_loss(y_hat, y)
        self.log('val_loss', loss)

    def configure_optimizers(self):
        optimizer = torch.optim.Adam(self.parameters(), lr=self.learning_rate)
        return optimizer
```

Personally I find dropout to be one of the most effective hyperparameters, in terms of improving statistical inference.

Another common regularization technique is *batch normalization*, which normalizes inputs to each layer to stabilize training. Batch normalization normalizes the inputs to each layer during training, reducing internal covariate shift, which is the change in the distribution of layer inputs as weights are updated. By keeping these distributions more stable, batch normalization allows higher learning rates and makes training less sensitive to weight initialization. It also has a mild regularizing effect, similar to dropout, because the normalization statistics computed on each mini-batch introduce noise. We cover batch normalization in later chapters when we work with deeper architectures where it becomes more important; for the shallow networks in this chapter, dropout and early stopping are generally sufficient.

Weight decay

Weight decay is a form of L2 regularization that penalizes large weights by adding a fraction of the squared weight values to the loss function. This discourages a network from relying too heavily on any single connection, which helps prevent overfitting, particularly when we have limited training data. In PyTorch, weight decay is a parameter of the optimizer rather than something we add to the model architecture:

```
# Adam with weight decay (L2 regularization)
optimizer = torch.optim.Adam(
    self.parameters(), lr=self.learning_rate, weight_decay=1e-4
)
```

```
# AdamW separates weight decay from gradient updates, generally preferred
optimizer = torch.optim.AdamW(
    self.parameters(), lr=self.learning_rate, weight_decay=1e-2
)
```

The difference between Adam with `weight_decay` and `AdamW` is worth knowing; standard Adam applies weight decay to a gradient before adaptive scaling, which couples the two in ways that can be counterproductive. `AdamW` decouples them, applying weight decay directly to weights after a gradient step. In practice, `AdamW` with a weight decay of 0.01 is a sensible default for most forecasting networks. You can include `weight_decay` in your Optuna search space alongside learning rate and dropout to let a tuner find the right balance.

Data splitting for neural networks

When we are preparing time series data for forecasting modeling, specifically training and hyperparameter tuning, we need to take account of how neural networks work compared to other methods, like gradient **Boosted Decision Trees** (**GBDTs**), **Multivariate Adaptive Regression Splines** (**MARS**), **Support Vector Machines** (**SVM**), etc. **Classical Machine Learning** (**CML**) methods, such as **Decision Trees** (**DTs**), are particularly susceptible to overfitting, especially with smaller datasets. Similarly, hyperparameters like maximum depth, which produces *deep trees*, can also lead to overfitting, while low or untuned values may result in underfitting. Sensitivity in ML models is exacerbated by their deterministic nature; although there is some randomness in the models (which we can control), they largely provide similar models when provided with the same HPs and data in training. This, among other factors, makes **Cross-Validation** (**CV**) strategies (e.g. nested CV) critical for hyperparameter tuning. Given that GBDTs are generally quicker to train than neural networks, nested cross-validation is a more practical choice for evaluating model hyperparameters and their impact on performance.

By contrast, neural networks benefit from several forms of implicit regularization, including architectural choices and techniques like dropout and early stopping. They also have a stochastic nature; optimization approaches like mini-batch gradient descent add additional regularization that helps prevent overfitting. Neural networks operate in a continuous optimization landscape, which offers some robustness to minor hyperparameter variations, such as learning rate adjustments. And their capacity to model complex functions allows for decent generalization, even when not perfectly tuned. Early stopping, guided by validation performance, also serves as an adaptive regularization method. This makes neural networks generally less sensitive to hyperparameter fluctuations, reducing the need for exhaustive cross-validation strategies like those used for modeling with GBDTs.

In essence, this means we can split data more simply for neural networks, using a fixed train-validation-test split rather than nested cross-validation. As training neural networks takes longer than other ML methods, this saves some computational resource.

Approaches to splitting

When working with neural networks for time series forecasting, we typically progress through three stages of data splitting, each serving a different purpose:

1. **Training-validation-test split** (hyperparameter tuning). When establishing hyperparameters for neural networks, we use a fixed training-validation-test split rather than nested cross-validation. Instead of splitting training and validation into multiple folds, we reuse the same split for each hyperparameter set. This is more robust than a simple training-test split for hyperparameter tuning while being computationally feasible. For example, with a time series of 730 periods (2 years of daily data) and a forecast horizon of 90 periods, we might split our data into approximately 550-90-90, with the training set comprising approximately 75% of our data.

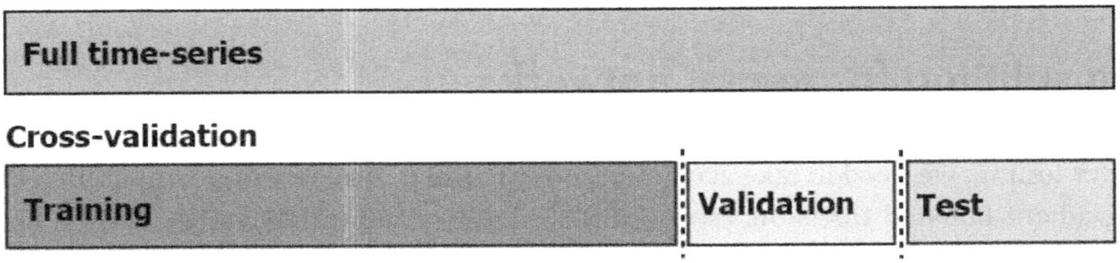

Figure 6.20: Training-validation-test split

2. **Training-validation split (final model training).** After identifying your architecture, optimizer, and hyperparameters, you train a model on a larger portion of the data using a training-validation split. There is no separate test set here because hyperparameter selection is already complete. The validation set provides a final performance metric on unseen data that can be compared against the tuning stage.

Figure 6.21: Training-validation split

3. **Full dataset training and fine-tuning.** Once we are confident that our model and its hyperparameters perform as expected, we can train on the full dataset. At this point, we're assuming that our model is trained with optimal hyperparameters and consequently will have its weights and biases well-calibrated, allowing the model to generalize well when given access to all available data. Alternatively, we can fine-tune; rather than rebuilding from scratch, we take an already-trained model and continue training it on new or additional data (e.g. recent pricing or sales data). This allows us to update our model with all available data without starting over.

While fine-tuning is a useful approach, especially if our data is stable, our models can drift out of sync with the data they are modeling. It is wise to monitor performance carefully when applying this approach over time.

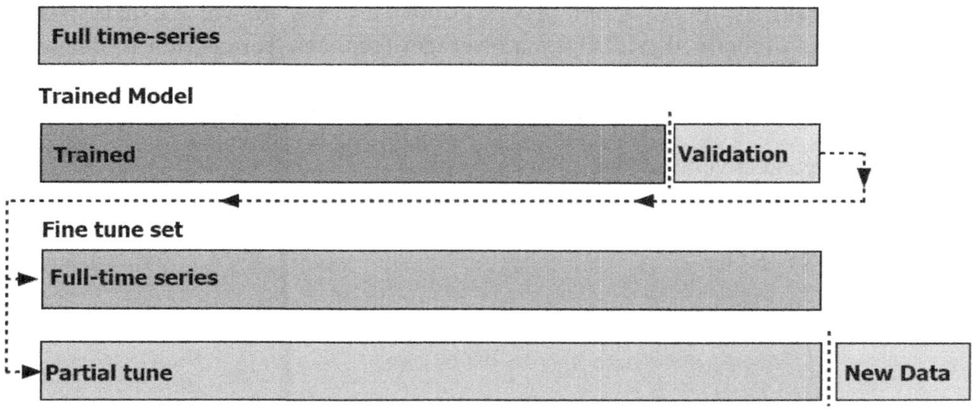

Figure 6.22: Full dataset training and fine-tuning

Hyperparameter tuning

There are various approaches to hyperparameter tuning, with the most common being grid search, Bayesian optimization, and metaheuristics:

- **Grid search** systematically creates and evaluates models for each combination of hyperparameter values specified in a predefined grid. You can do a full grid search (slow!), random search with a selected number of hyperparameters to compare, or more statistical approaches like a Latin hypercube, which attempts to find a plausible set of values within a multidimensional space.
- **Bayesian optimization** models the performance as a function of hyperparameters using a surrogate model, generally a *Gaussian Process*. It selects new hyperparameter values to test, balancing exploration of new areas and exploitation of known good areas.
- **Metaheuristics** make use of heuristic algorithms to explore the search space. Some common approaches include genetic algorithms, particle swarm optimization, and ant colony optimization.

While these are all decent approaches, metaheuristics tend to be more useful when we are conducting a large number of comparisons (Tani and Veelken, 2024). Grid search has its place too, especially when we want to create generalized but tailored default hyperparameters for large groups of data. However, the tried and tested way to find optimal hyperparameters for individual models is often Bayesian optimization.

Optuna (`pip install optuna`) is an excellent tool for this. It uses an objective function to guide hyperparameter optimization by evaluating performance. Within the objective function, we define a search space with upper and lower bounds for the hyperparameters we want to investigate, along with the model training and evaluation process. Unlike grid search, which exhaustively evaluates all combinations in a predefined grid, Optuna leverages feedback from earlier trials to efficiently navigate the search space, focusing on promising regions and avoiding the combinatorial explosion of exhaustive methods. We set up an Optuna experiment with three components:

1. Define an objective function:

```
import optuna

# Define objective function
def objective(trial):
    # Define model and hyperparameters
    # Train and evaluate model
    # Return evaluation metric
```

2. Create a study object:

```
# Create a study object
study = optuna.create_study(direction='minimize')
```

3. Run optimization process:

```
# Run optimization process
study.optimize(objective, n_trials=20)
```

Under an objective function, we instantiate and fit models with the bounds and categories we wish to investigate. When selecting bounds, you specify a method for selecting values: for example, you can use an integer suggestion, specify categorical values, or use a log-uniform distribution for Optuna to sample from on each trial. After setting up the parameters, you build predictions and score them against an evaluation metric returned by an objective function. Crucially, when we create a study, we instruct Optuna to maximize or minimize an evaluation metric; for forecasting, we generally select minimize. Finally, we run our study by calling the `.optimize()` method, which takes our objective function and the number of trials allowed. Here is a more complete example:

```
def objective(trial):
    learning_rate = trial.suggest_float('learning_rate', 0.0001, 0.1, log=True)
    dropout_rate = trial.suggest_float('dropout_rate', 0.0, 0.3)
    hidden_dim = trial.suggest_categorical('hidden_dim', [6, 10, 20, 40])
    num_layers = trial.suggest_int('num_layers', 1, 6)

    model = ffnetwork(
        input_dim, hidden_dim, num_layers=num_layers, output_dim=output_dim,
        learning_rate=learning_rate, dropout_rate=dropout_rate
```

```
    )

    early_stop_callback = EarlyStopping(
        monitor='val_loss', patience=10,
        verbose=False, mode='min'
    )
    trainer = L.Trainer(
        max_epochs=100, callbacks=[early_stop_callback],
        logger=False, enable_checkpointing=False
    )

    trainer.fit(model, data_module)

    model.eval()

    last_window = x_val[-1].view(1, -1)
    recursive_preds = []
    for _ in range(len(x_test)):
        with torch.no_grad():
            pred = model(last_window)
        recursive_preds.append(pred.item())
        last_window = torch.cat((last_window[:, 1:], pred.view(1, 1)), dim=1)

    recursive_preds = np.array(recursive_preds).reshape(-1, 1)
    y_test_np = y_test.numpy().reshape(-1, 1)
    mse_val = np.mean((recursive_preds - y_test_np) ** 2)

    trial.set_user_attr("MSE", mse_val)

    return mse_val

study = optuna.create_study(direction='minimize')
study.optimize(objective, n_trials=20)

# Extract best results
print(f"Best hyperparameters: {study.best_params}")
print(f"Best MSE: {study.best_value:.4f}")
```

You could, of course, run hyperparameter tuning on everything at once, but the feasibility of this really depends on the number of models you have to train. The notebook for this chapter conducts experiments on a wider range of data for a subset of the M5 dataset, using Nixtla's MLP in `NeuralForecast`. Running this will give you a sense of the time involved in hyperparameter tuning. You'll tend to find that developing good subject knowledge and testing on your own data is the best approach to learning and understanding.

Summary

We have covered optimization in general in this chapter, and it is important to highlight that there is some nuance in applying optimization to different neural architectures. For example, with dropout we apply it to nodes in any layer of an FFN/MLP, but with LSTMs we typically apply dropout only when we have more than one layer. You will develop your understanding by experimenting with network designs and building experiments. Of course, for those that just want to forecast, we can use Darts or Nixtla's libraries, but these do have set design choices, even as they make your task easier.

For those building neural architectures from scratch, we would suggest you optimize in the following order:

1. Architectural elements
2. Optimization algorithm
3. Hyperparameters

This ordering follows the chain of dependency: architectural choices (activation functions, layer counts, node counts) constrain which optimizers make sense, and both architecture and optimizer constrain hyperparameter space worth searching. There is of course crossover between these elements, and separating out experiments prevents one from properly searching multidimensional space, so if you have time and compute, do everything (within reason)!

Recommended reading

- Apicella, A., Donnarumma, F., Isgrò, F., and Prevete, R. 2021. "A Survey on Modern Trainable Activation Functions." *Neural Networks*, 138: 14–32. `https://doi.org/10.1016/j.neunet.2021.01.026`.
- Bingham, G. 2023. "Optimizing Neural Networks Through Activation Function Discovery and Automatic Weight Initialization." *arXiv*. `http://arxiv.org/abs/2304.03374`.
- Chen, C., Lai, J., Chang, Y., Lai, C., and Pai, P. 2023. "A Study of Optimization in Deep Neural Networks for Regression." *Electronics*, 12 (14): 3071. `https://doi.org/10.3390/electronics12143071`.
- Da S. Gomes, G. S., Ludermir, T. B., and Lima, L. M. 2011. "Comparison of New Activation Functions in Neural Network for Forecasting Financial Time Series." *Neural Computing and Applications*, 20 (3): 417–39. `https://doi.org/10.1007/s00521-010-0407-3`.
- Jagtap, A. D., and Karniadakis, G. E. 2022. "How Important Are Activation Functions in Regression and Classification? A Survey, Performance Comparison, and Future Directions." *arXiv*. `http://arxiv.org/abs/2209.02681`.

- Kızrak, A. 2020. "Comparison of Activation Functions for Deep Neural Networks." *Medium*. https://medium.com/data-science/comparison-of-activation-functions-for-deep-neural-networks-706ac4284c8a.
- Kunz, M., et al. 2023. "Deep Learning Based Forecasting: A Case Study from the Online Fashion Industry." *arXiv*. http://arxiv.org/abs/2305.14406.
- Manu, J. 2022. *Modern Time Series Forecasting with Python*. 1st edition. Packt Publishing Limited.
- Moon, J., Park, S., Rho, S., and Hwang, E. 2019. "A Comparative Analysis of Artificial Neural Network Architectures for Building Energy Consumption Forecasting." *International Journal of Distributed Sensor Networks*, 15 (9). https://doi.org/10.1177/1550147719877616.
- Nair, V., and Hinton, G. E. 2010. "Rectified Linear Units Improve Restricted Boltzmann Machines."
- Nielsen, M. A. 2015. "Neural Networks and Deep Learning." http://neuralnetworksanddeeplearning.com.
- Nixtla. n.d. "Statsforecast/Experiments/M3 at Main · Nixtla/Statsforecast." *GitHub*. https://github.com/Nixtla/statsforecast.
- Nwankpa, C., Ijomah, Winifred., Gachagan, A., and Marshall, S. 2018. "Activation Functions: Comparison of Trends in Practice and Research for Deep Learning." *arXiv*. http://arxiv.org/abs/1811.03378.
- Ramachandran, P., Zoph, B., and Le, Q. V. 2017. "Searching for Activation Functions." *arXiv*. http://arxiv.org/abs/1710.05941.
- Smith, L. N. 2017. "Cyclical Learning Rates for Training Neural Networks."
- Szandała, T. 2021. "Review and Comparison of Commonly Used Activation Functions for Deep Neural Networks." *Bio-Inspired Neurocomputing*, 903: 203–24. https://doi.org/10.1007/978-981-15-5495-7_11.
- Tani, L., and Veelken, C. 2024. "Comparison of Bayesian and Particle Swarm Algorithms for Hyperparameter Optimisation in Machine Learning Applications in High Energy Physics." *Computer Physics Communications*, 294: 108955. https://doi.org/10.1016/j.cpc.2023.108955.
- Zhang, A., Lipton, Z. C., Li, M., and Smola, A. J. 2023. *Dive into Deep Learning*. https://d2l.ai/chapter_optimization/index.html.

Join our community on discord

Join our community's Discord space for discussions with the authors and other readers:

https://discord.gg/VXpz9YPUTW

7

Conformal Prediction

"Uncertainty is an uncomfortable position. But certainty is an absurd one." — Voltaire

All our predictions miss the mark to some degree, and that is fine - the old adage about all models being wrong is a cliché, right alongside the correlation and causation one. But it points to an important question, which is *how wrong are we?* Translating to slightly more formal language (this is a book dealing with machine learning, after all), it is useful to think about forecasting in terms of point and interval forecasting: the former predicts a specific value, while the latter quantifies the error or uncertainty around it.

Under classical time series modeling, the interval aspect is fairly straightforward to handle. ARIMA-class models rely on parametric assumptions, so deriving a confidence interval around point forecasts amounts to manipulating conditional distributions. With more general types of classical time series models (e.g., non-Gaussian state space models), there are no closed-form formulas, but sampling distributions can be obtained using **Markov Chain Monte Carlo** (**MCMC**). Bayesian methods capture both parametric and sampling approaches as special cases (after a fashion). If these approaches fail, we could always bootstrap the data, adjusting to preserve temporal dependence as in block bootstrapping.

While each of those approaches achieves some degree of success in different applications, there is always some fragmentation to the overall setup. This problem was exacerbated with the advent of machine learning methods, and deep learning afterwards—quantification of uncertainty in regression problems (of which, time series is a special form) became something of an afterthought in most applications—but the practical issues that motivated the study of forecast uncertainty did not go away. As you will find in practice, uncertainty is an important factor in decision-making for many business problems.

As in every good story, cometh the hour, cometh the man: the year was 1998, and the men were Alex Gammerman, Vladimir Vovk, and Vladimir Vapnik. By modifying the original **Support Vector Machine (SVM)** algorithm, they demonstrated a method of assigning confidence scores to predictions of a classifier. Their paper *Learning by transduction* (Gammerman, Vovk, and Vapnik 2013) laid the foundation on which conformal prediction was built.

Conformal Prediction (CP) is a method for constructing prediction intervals in machine learning, moving from point to interval forecasts. Under mild assumptions, conformal intervals are guaranteed to contain the true value of the next observation with a certain probability: this feature makes CP particularly useful in situations where reliable predictions with uncertainty bounds are required. In this chapter, we introduce CP as a general approach to prediction uncertainty quantification for time series, which will be useful throughout the remainder of the book.

In this chapter, we will learn:

- Uncertainty quantification
- Understanding conformal prediction
- Conformalised regression

Uncertainty quantification

Inherent to any forecast predictions is an element of uncertainty as a consequence of factors that are unknown to us, along with difficulties in measuring and modeling our data. In general, we can say that any model we build is done so with incomplete data to create approximations of reality, which will produce predictions that are influenced by many sources of uncertainty.

When you create a supervised regression model, you will separate data into exogenous variables (X) and dependent variables (Y), so you can regress exogenous variables, such as price, on dependent values, like sales. You are attempting to capture patterns in training data, when passing this data through models, that may be used to predict Y (sales in this case) for new points X. If you have X price of £0.6, your model should predict a value of sales ($\hat{Y}$), based on the learned model parameters; this is a point prediction. While useful, point predictions do not provide us with any understanding of potential distances between themselves and actual values (Y). This is a fundamental problem from a business perspective, as modeled predictions can be terrible, which can undermine confidence in modeling systems. This is especially true where modeled predictions vary from run to run, as most Gradient-Boosted Decision Trees (GBDTs) and neural networks do when we do not control the seeding of random processes.

Refer to color images on the Graphic bundle or the ebook.

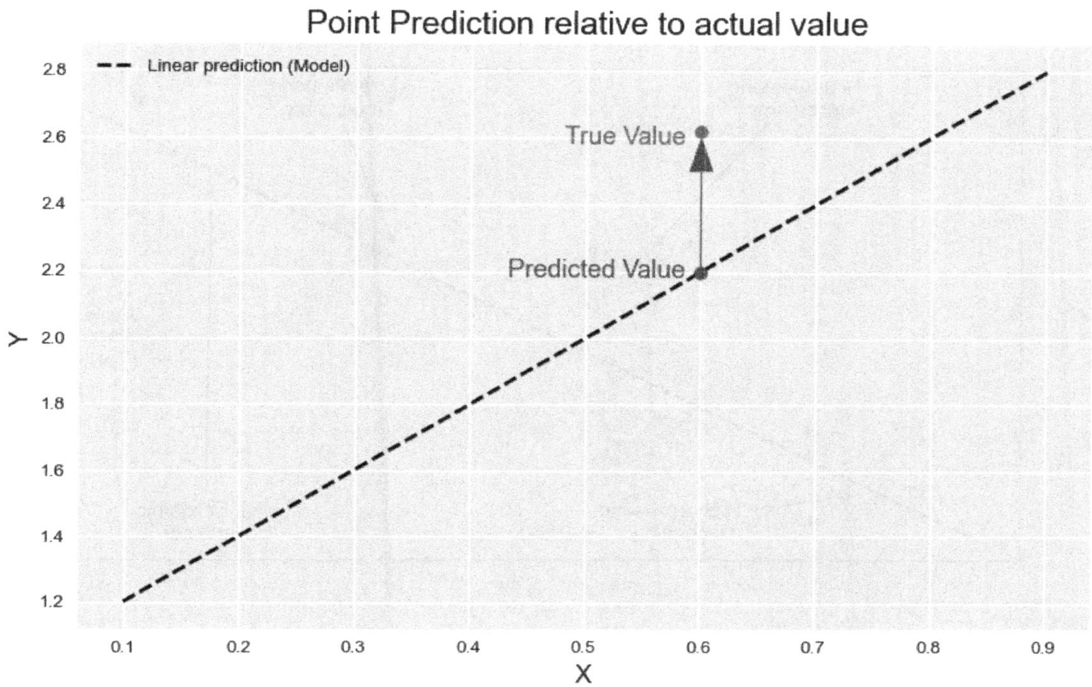

Figure 7.1: Point prediction relative to actual value

In *Figure 7.1*, the vertical arrow shows the distance (error) between a model's prediction and a true observed value.

Quantifying the likelihood of predictions being correct, or not, guides our judgments about the degree to which a model and its predictions may be trusted. Of course, this is relative to our tolerance of risk, and what to do if a model does not meet expectations.

Uncertainty Quantification (**UQ**) aims to help us estimate the probability of predictions actually being observed. We can broadly categorize uncertainty as either **aleatoric**—caused by intrinsic randomness and variability within data, as a feature of the systems we investigate, i.e., stock market prices—or **epistemic**—coming from incomplete knowledge/understanding.

These concepts are demonstrated in the following diagram (inspired by Manokhin and Sudjianto 2023):

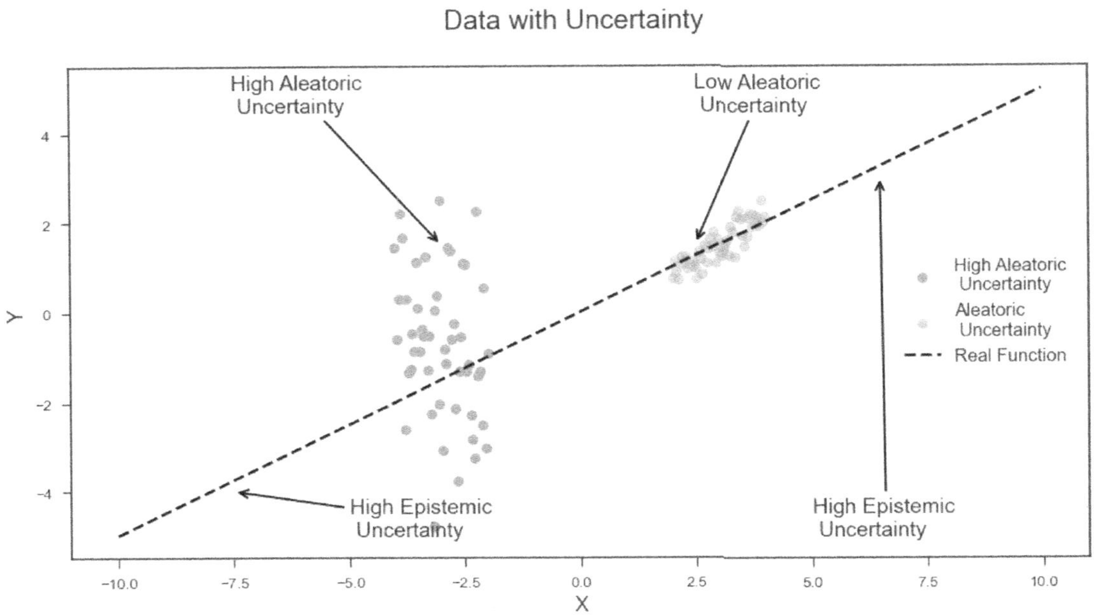

Figure 7.2: Aleatoric and epistemic uncertainty

In *Figure 7.2*, the *left* cluster exhibits **High Aleatoric Uncertainty** due to pronounced randomness from a linear model, while the *right* has **Low Aleatoric Uncertainty** due to the regularities in data. Additionally, we encounter voids with no data, indicating **High Epistemic Uncertainty.** This is where we lack observations, and our knowledge of the underlying function is incomplete.

There are four properties we should consider when building prediction intervals:

- **Validity:** Our goal in using uncertainty quantification is to create valid prediction regions, such as intervals for regression or sets for classification, that guarantee our probabilistic statements are accurate. For example, if we have a prediction interval with 80% coverage, it means that there must be at least an 80% probability that the actual value falls within that interval. This is crucial for effectively quantifying uncertainty, as failing to meet this requirement undermines the fundamental purpose of the prediction.
- **Verifying coverage:** We need to consider robustness when creating prediction intervals from finite datasets. Due to the limited nature of our datasets, we often require prediction intervals that maintain coverage validity even with a small dataset; this property is known as **finite-sample coverage validity**.
- **Efficiency:** Efficiency refers to how tight a prediction interval is. The tighter the interval, the more efficient it is. For example, a prediction of 100 units ± 5 is more efficient than 100 units ± 50, because a narrower interval gives us a tighter interval for decision making. When assessing intervals, it's important that we account for both coverage validity and efficiency.

- **Model-agnostic and distribution-free:** An effective interval predictor should not rely on the particularities of the point predictor or data distribution. It should be able to work with any model, without requiring an understanding of the model's internal mechanisms.

With these properties in mind, we can now consider the general approaches to uncertainty quantification:

- Statistical: Relies on probability distributions to model data and prediction uncertainties.
- Bayesian: Relies on prior knowledge about data to update beliefs about prediction uncertainties.
- Fuzzy logic: Represents uncertainty using membership functions and fuzzy sets, allowing partial set membership.

While all useful, they have their place and problems. Bayesian statistics, which is commonly discussed for UQ, is useful when one has thoroughly investigated the data/topic to be modeled, but its reliance on prior distributions is a problem for us when we are modeling hundreds, or more, items. It is simply not practical to properly investigate priors at scale, and not doing so often results in oversimplified or poorly selected priors, and therefore poor UQ. There is also the issue with Bayesian methods of being computationally expensive.

Fuzzy logic simply lacks the precision needed for many business tasks, such as forecasting. Generally, for business applications, statistical methods (i.e. regression analysis, confidence intervals, and Monte Carlo simulations) strike a better balance between efficiency and accurate uncertainty representation. Statistical approaches are more useful, but rely on parametric assumptions and computationally intensive methods like Monte Carlo bootstrapping.

A newer non-parametric statistical approach to UQ is **conformal prediction**, which tends to perform well in business contexts due to its ability to provide precise probabilistic intervals that are distribution-free, model-agnostic, and generally computationally efficient (Manokhin and Sudjianto 2023):

- It is non-parametric and distribution agnostic saving the need to investigate every model's data set fully.
- It provides probabilistic predictions with confidence metrics, supporting informed decision-making.
- It produces valid prediction regions, enabling rigorous uncertainty assessment (Angelopoulos and Bates 2023).
- It maintains validity with finite datasets, ensuring reliable coverage even with limited data.
- It is robust to imperfect and noisy data, making it suitable for real-world applications.
- It is computationally efficient, avoiding the overhead of many traditional uncertainty methods.
- It is applicable to most supervised learning models, making it widely usable across tasks.

With these advantages in mind, we can now introduce conformal prediction more formally.

Understanding conformal prediction

Conformal prediction has a few requirements, the most important being that data is *exchangeable*. A sequence of random variables is exchangeable if their joint distribution is invariant to any permutation of the indices; the order does not matter for the joint distribution, but unlike IID the variables can be dependent. This is a slightly weaker assumption than the standard IID imposed in most machine learning.

A prediction region is said to be valid if a true value falls within the region at least $(1 - \alpha)\%$ of the time. The efficiency of a prediction region is its size. For classification, this is the number of classes in the set and for regression, the interval width. Under exchangeability, split conformal prediction gives us a finite-sample guarantee:

$$P\big(Y_{n+1} \in C(X_{n+1})\big) \geq 1 - \alpha$$

where $C(X_{n+1})$ is the prediction region for a new test point X_{n+1}, and α is the desired miscoverage rate. If we set $\alpha = 0.1$, our intervals will contain the true value at least 90% of the time.

In its purest form, conformal prediction is designed for an online (*transductive*) setting. This means that once a label is predicted in a supervised learning scenario, its true label is revealed before the next prediction is made. The underlying model can therefore be retrained with this newly acquired data point, and the next prediction is made using a calibration set containing $n + 1$ data points, where the previous model had n.

Intuition behind conformal prediction

Consider the concept of point prediction we discussed earlier. Instead of providing a single value, we can use historical data to predict a range or interval of values. We then assign a probability to this interval, indicating the likelihood of the actual value(s) we are trying to predict falling within it. For example, in our earlier regression problem, we used historical pairs of X and Y values to create a model, which allowed us to predict values of Y given a chosen value of X. Under uncertainty quantification with conformal intervals, we simply create a range of plausible values by using actual data to gauge the fit of our predictions. Any newly predicted points may align, or not, with our plausible range of previously observed values. Put simply, conformal intervals (or sets when talking about classification) help us find values that sufficiently conform with expectations based on our historical data, while values that do not conform are flagged as outliers.

To build a set of plausible values, we need a function that can measure how well a single value fits within a group of values. This is relatively trivial, as we can simply use distance metrics like **Absolute Error** (**AE**) to measure the non-conformity of a value. Of course, we could use many other measures (i.e. squared error, standardized residuals, Huber loss, etc.). The point is to apply a function that takes a single point and a group of points as input and calculates the distance/non-conformity between them.

To make this salient, we'll apply a simple regression, with most points falling onto a linear path, then use the resulting model to predict a couple of points.

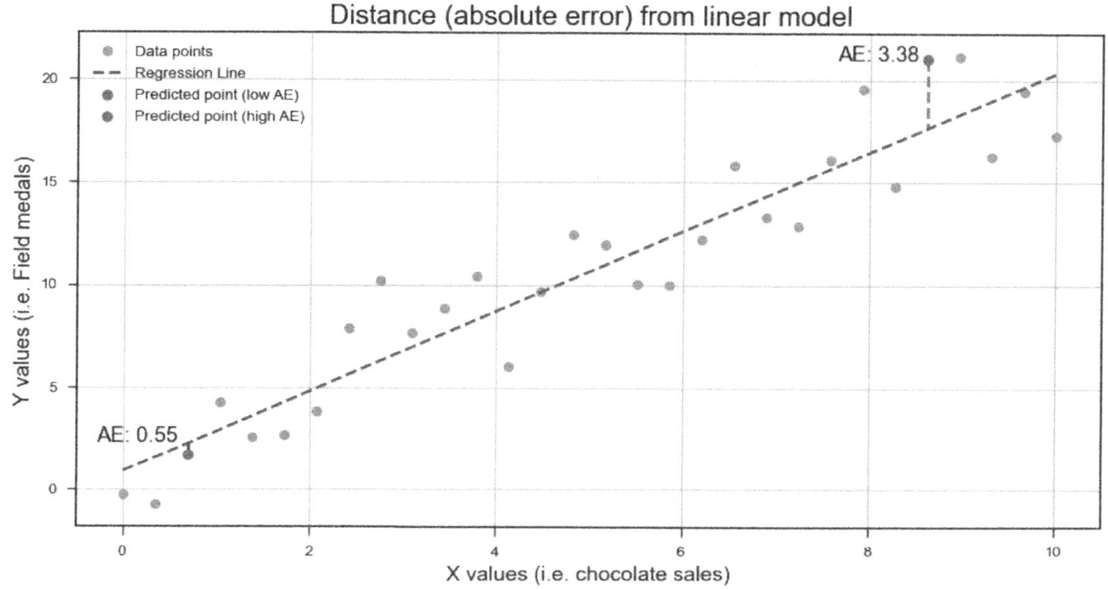

Figure 7.3: Distance (absolute error) from a linear model

In *Figure 7.3*, two predicted points are highlighted: one close to the regression line with low absolute error (**AE=0.55**), and one far from it with high absolute error (**AE=3.38**).

You can see the conformity of our predicted points to the interval/set (all points here) varies; the predicted point closest to the line appears to conform better to the entire data set, while our other prediction stands out in distance from most of the other points in the data set. Of course, distance doesn't just apply to the predictions, but to all the points within the interval, which all exhibit differing values of non-conformity. Any function we select should measure these differences. A simple approach is to fit a linear OLS model and use it to measure absolute error ($AE = y - \hat{y}$) as our non-conformity metric (you can use other metrics). As you can see in our example, the predicted point closest to the line has the lowest absolute error of the two predictions.

Simply establishing a difference between values and modeled expectations is not sufficient for determining conformity. Being a relative concept, conformity depends on the degree data in a set *conforms*. We resolve this by ranking values within a set by their respective distance (non-conformity); in essence, we create a *ranking/conformity ladder*, ordering conformity values from higher (low AE) to lower (high AE). Finally, we scale these ranking values to a standard 0 to 1 range.

To demonstrate this, we take a simple regression dataset, highlight one data point (diamond box), and observe where it sits on the conformity ladder. We then shift that same point to an extreme position and see how its ranking changes and how the rankings of all other points shift as well.

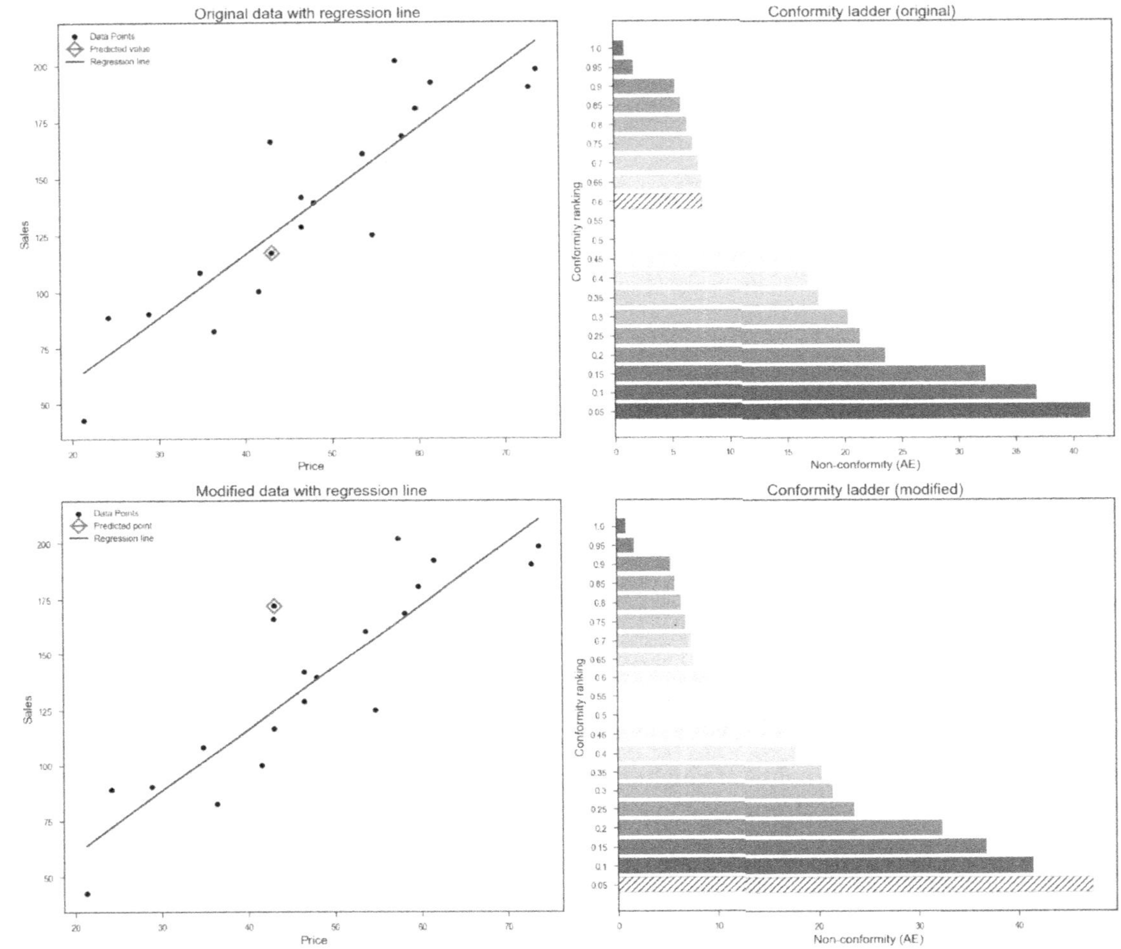

Figure 7.4: Conformity ranking ladders for original and modified data

In *Figure 7.4*'s top row, a data point (*diamond box*) sits near the regression line and ranks highly. In its bottom row the same point is shifted to an extreme position, dropping its ranking, and shifting all others.

Our ladder graphs above show how distance (non-conformity metric) from the simple regression (*solid* line) affects the position (conformity value) of our predicted values within the set, highlighting relative conformity. By scaling the values we have created a conformity δ; delta) for what we consider to be a sufficient threshold of conformity, defining which values from our data are accepted, or not, into a conformal interval based on a probability range with coverage. Specifically, when we set a δ threshold to 0.8, it will include values within the top 80% of our ranking ladder, with values outside (higher) than this considered non-conforming.

Now that we have the intuition, we can express it formally. The notation for this may look dense, but each piece maps directly to what we've already discussed: B is our set of data points, F is our non-conformity function (absolute error in our examples), and Γ is a prediction set, a collection of values that conform sufficiently.

Formally, when given a confidence threshold (δ) and a set of paired training values $(x_0, y_0), \ldots, (x_{n-1}, y_{n-1})$, with a non-conformity function (F), a conformal prediction set consists of all plausible values for which the fraction of points conforming better than the test point is less than our confidence delta (adapted from Rad 2023).

We begin by defining the set of observed examples, including a candidate value for the test point:

$$B \triangleq \{(x_0, y_0), \ldots, (x_{n-1}, y_{n-1}), (x_n, y)\}$$

Next, we compute nonconformity scores, which measure how well each point aligns with the rest of the data.

$$f_i \triangleq F\big(B, (x_i, y_i)\big), \quad \forall i \in \{0, \ldots, n-1\}$$

$$f_n \triangleq F\big(B, (x_n, y)\big)$$

Using these scores, we define the prediction set as the collection of values that conform sufficiently to the observed data:

$$\Gamma(x_0, y_0, \ldots, x_{n-1}, y_{n-1}, x_n) = \{y \in Y: \big|\{n \mid i = 1, \ldots, n: \hat{f}_i < \hat{f}_n\}\big|/n < \delta\}$$

Finally, this construction satisfies the following coverage guarantee:

$$P(y_n \in \Gamma) \geq \delta$$

Under the assumption of exchangeability, this ensures that the probability of an actual value falling within our constructed interval is at least 80%. We can, therefore, state with 80% confidence that the actual value lies within our predicted interval. This is referred to as an 80% prediction interval. We are acknowledging the possibility of error in our predictions and quantifying that by stating there is less than a 20% chance of being incorrect, considering all plausible values that have the possibility of being assigned.

A prediction set consists of all plausible values that, when assigned to new data points, will make them conform sufficiently with the previously established points within the set. We define a sufficient conformity threshold δ on the standardized rankings of our conformity ladder, i.e. $\delta = 0.8$ means 80%, which will include values in the set below this threshold (adapted from Rad 2023).

A note on notation

Throughout the intuition sections above we have used δ for our confidence level, following Rad (2023) — so $\delta = 0.8$ means we want 80% coverage. In conformal prediction literature and in libraries like MAPIE, you will commonly see α for *miscoverage* rate, where $\alpha = 1 - \delta$. A $\delta = 0.8$ confidence level is the same as $\alpha = 0.2$. We switch to α notation from the applied sections onward, as this is what the code expects.

Improving conformal intervals

Up to here, we have learned about full conformal prediction, which is of little benefit to us for two reasons:

1. It is computationally expensive! Because whenever a new data point is tested we have to refit the underlying model for all points in the set used to build our non-conformity values.
2. When we move a validation point with an unknown Y from a conforming position to an extreme non-conforming position, it distorts the fit of the underlying model. This impacts the non-conformity values, which consequently changes the values in our conformity ladder, impacting our measurements of plausible values.

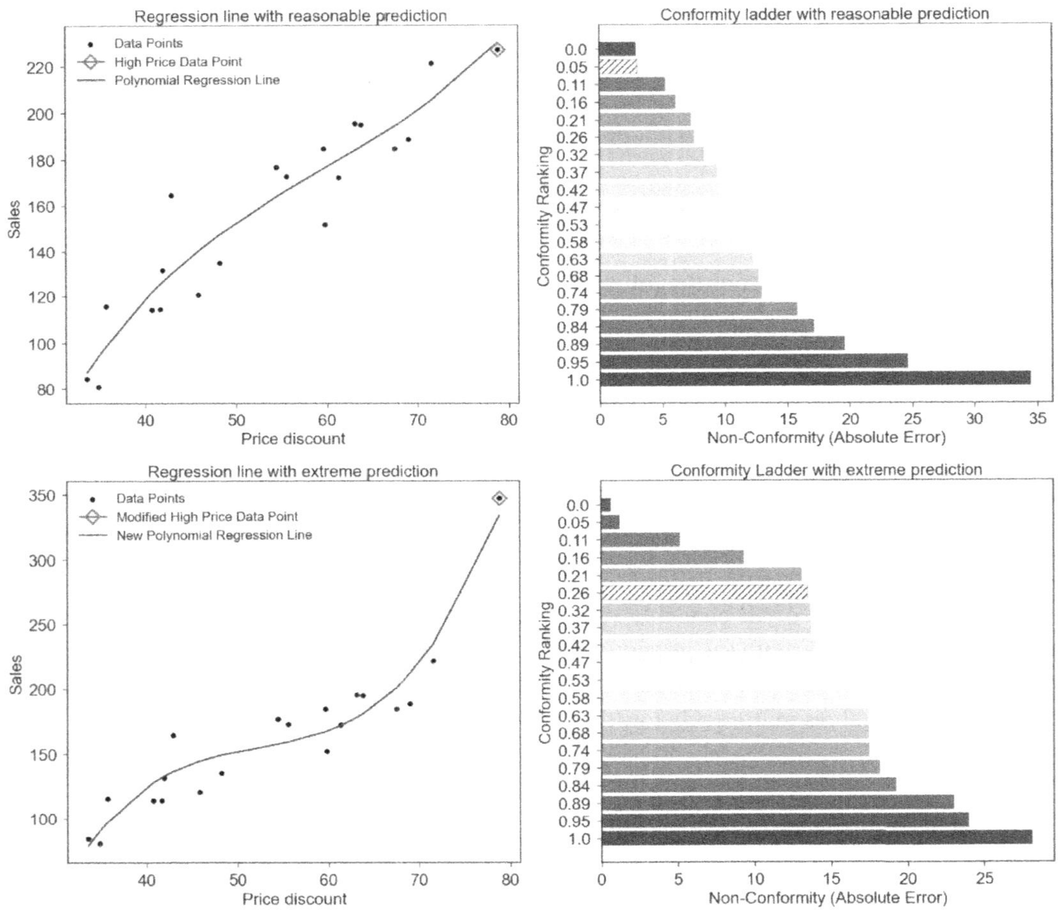

Figure 7.5: The refitting problem in full conformal prediction

Top row: a data point (diamond box) sits at a reasonable position with a stable polynomial fit. Bottom row: the same point is shifted to an extreme position, distorting the fit and reshuffling the conformity ladder.

In our code, we use a raw rank convention where higher scores mean *less* conforming (higher error), which is the inverse of our earlier ladder plots. Both conventions are common; what matters is consistency within each calculation.

While the computational overhead in this example is small, refitting the model underlying conformity metrics makes the full approach scale poorly with larger datasets and more complex models. Of course, we only need to refit because we are using the whole set for building the model, including the validation point(s). Being data scientists, we know a few ways around this:

- **Exclude the validation/test point when building the model:** Excluding the validation point from the build is a straightforward no-go, as it often leads to values in our set having lower non-conformity values (AE here) than validation points, increasing a validation point's non-conformity. This makes the prediction set too narrow.
- **Split conformal methods:**
    - Machine learning practice for building models is to do a training-test split (don't fix what isn't broken!). All we do here is hold more points from the set we use for model fitting; also, we call the splits training and calibration sets.
    - We use a sample of values from our set to fit the underlying model; once this is built, we use the remaining values to construct a conformity ladder to which we apply our confidence threshold.

The split method, as we now know, measures conformity based on one fit of the underlying model, which means we can move validation points wherever we choose without impacting the underlying model and non-/conformity measures, which as a bonus reduces computational cost.

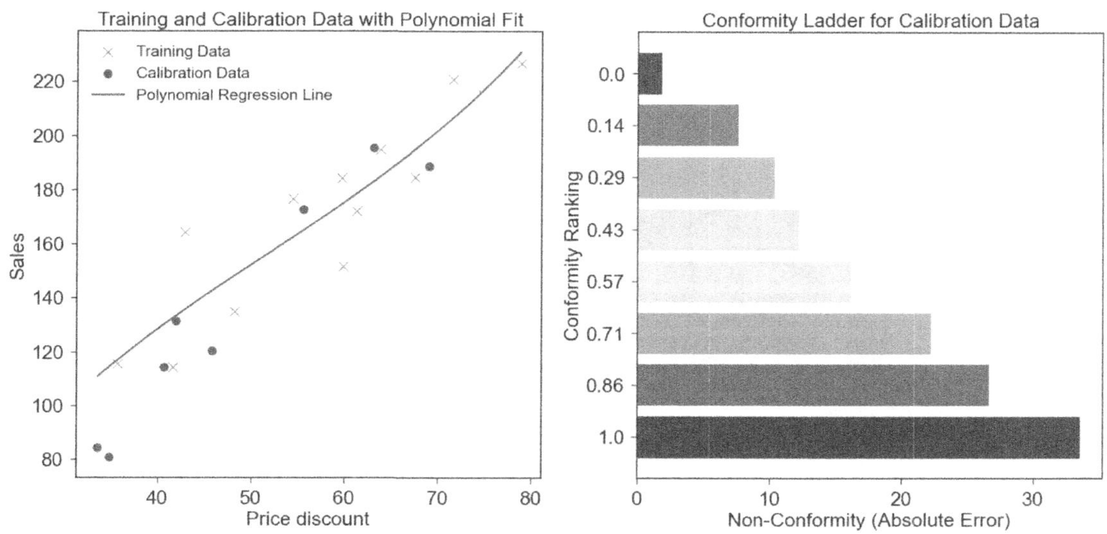

Figure 7.6: Split conformal method

In this image, the training data (crosses) is used to fit the polynomial regression, while calibration data (circles) is held out to compute nonconformity scores and build the conformity ladder.

Conformal intervals with quartiles

Now let's say we have a validation point with an unknown value. If we use an 80% prediction interval, this means there is an 80% probability that an interval will contain the true value around that point; specifically, the interval consists of all plausible values that, if assigned to the test point, will put it within the top 80% of the conformity ladder.

How can we construct this interval? Rather than attempting a brute force approach and trying to generate all plausible values, we can take an intelligent approach, which is allowed for two reasons:

- We only need to fit our model once with a split approach
- There is a monotonic relationship between the distance of the prediction point and the model, meaning that as a point moves away from a model, its non-conformity increases, dropping its conformity ranking

Using these properties we can find the upper and lower bounds of an interval, that deem acceptable. We only need to create subintervals of the rankings from our calibration set, which we use for the values in the conformity ladder.

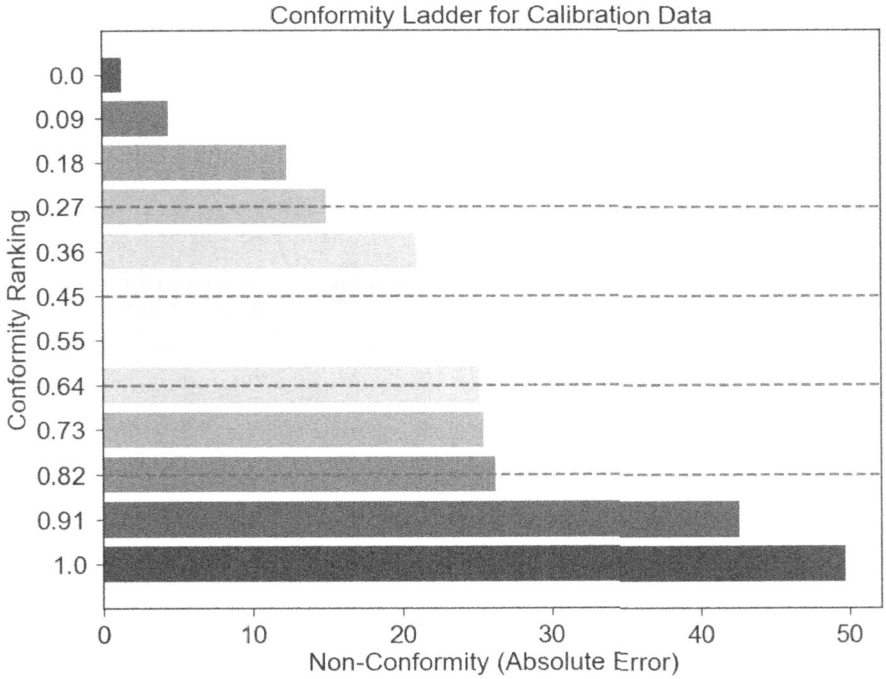

Figure 7.7: Conformity ladder divided into quartile sub-intervals

The dashed lines in *Figure 7.7* mark the boundaries at confidence levels 0.2, 0.4, 0.6, and 0.8.

We built quartiles by simply selecting four points from our ladder to create five sub-intervals of the non-conformity score. These values correspond with confidence levels: 0.2, 0.4, 0.6, and 0.8. You can see that each sub-interval of conformal ranking (Y) corresponds to a range of non-conformities on the X. You can see that the highest probability corresponds to non-conformity scores between 0 and 6.39, the next interval scores are 6.4 and 8.62, etc. As the sub-intervals correspond to XY pairs, we can simply build a validation point with our exogenous (X) variable, get our prediction ($\hat{Y}$), and using the non-conformity sub-intervals, we can see where the prediction falls.

According to exchangeability, we should have equally probable sub-intervals, with the true unknown value for any validation point equally likely to fall within any of the sub-intervals. As the example we used has five intervals, there is a *0.2* probability of a true value falling within any one of these intervals; in referring to *0.8* probability we are saying we will accept a value falling between 0 and 23 AE, which corresponds to 80% of our conformity ladder.

Our prediction intervals' width relative to confidence levels are discrete values determined by non-conformity scores and the size of a calibration dataset. Take our four calibration points, which restrict us to intervals of 20%, 40%, 60%, and 80%, and nothing else. As our confidence levels are not arbitrary, they are intrinsic and we denote them as δ_i.

$$\delta_j = \frac{j}{n+1} \quad \text{for } j = 1, \dots, n$$

Where the calibration set size is n.

As non-conformity sub-intervals are equally probable, we can use the non-conformity scores as quantiles, which divide a probability distribution into equal-probability intervals.

$$q^{\text{th}} \text{ Quantile} > q \times 100\% \text{ of All Possible Values}$$

Mathematical notation

We now switch to the standard α-based notation used in the literature and in libraries like MAPIE (recall from our earlier note that $\alpha = 1 - \delta$).

Prediction intervals for a new test point with prediction $\hat{y}$ are constructed by finding a non-conformity quantile q at a desired confidence level and applying it symmetrically:

$$C(\hat{y}) = \left[\hat{y} - q_{\lceil (n+1)(1-\alpha) \rceil / n}, \quad \hat{y} + q_{\lceil (n+1)(1-\alpha) \rceil / n}\right]$$

where q_k is the k-th smallest non-conformity score from the n calibration points, with $k = \lceil (1-\alpha)(n+1) \rceil$.

Given the following quantiles:

$$q_{1/4} = 6.39, q_{2/4} = 8.62, q_{3/4} = 13.99, q_{4/4} = 22.40$$

These values represent the quantiles of a distribution, specifically, the values below which a certain fraction of the data falls. For example, $q_{1/4} = 6.39$ means that 25% of the data points are below this. More generally, we index quantiles as:

$$q_{j/n} \quad \text{for } j = 1,2, \dots, n$$

Where j denotes the rank, and n is the total number of calibration points (here $n = 4$).

These quantiles correspond directly to intrinsic confidence levels δ_i, where each level represents the proportion of conforming points retained. To relate the position of a given quantile to the desired confidence level, we can express this relationship as:

$$\frac{q_{(j/n)}}{q_{((n+1)\delta'/n)}}$$

We now walk through the calculation using the δ-based confidence levels from our intuition section. Here S is our non-conformity score function (absolute error, in our case), and S^{-1} maps a non-conformity score back to a distance in the original Y scale. So if $S^{-1}(0.31) = 0.2$, it means a non-conformity score of 0.31 corresponds to a deviation of 0.2 units from the point prediction:

1. Select a confidence level.

$$\delta i = 0.4\ (40\%)$$

2. Determine calibration size.

$$n = 4$$

3. Apply the quartile formula, which measures non-conformity for the desired δi

$$q\left(\frac{(n+1)\delta'}{n}\right) : q\left(\frac{(4+1)\cdot 0.4}{4}\right) = q\left(\frac{2}{4}\right)$$

4. This tells us that the confidence level of 0.4 corresponds to the specific quantile 2 (which we labeled B), which has a corresponding nonconformity quantile of:

$$q\left(\frac{2}{4}\right) = 0.31$$

5. The function S^{-1} maps this non-conformity score back to the corresponding value in the original scale. Suppose $S^{-1}(0.31)$ which gives us a specific value, say:

$$\Delta y = 0.2$$

6. The prediction interval is then constructed using this non-conformity value applied as the upper and lower bounds from the point prediction of the model:

$$\{\hat{y} - \Delta y_{\max}, \hat{y} + \Delta y_{\min}\} = \{\hat{y} - S^{-1}(0.31), \hat{y} + S^{-1}(0.31)\}$$

7. We apply this to our example:

$$\{2.5 - 0.2, 2.5 + 0.2\} = \{2.3, 2.7\}$$

The coverage probability for the specified confidence interval should be at least equal to the requested δi.

Arbitrary confidence levels

We can't simply take two intervals and work out the value between our established intervals, as we cannot be confident that the upper and lower intervals have equal values in the non-conformity space. Instead, we select the highest sub-interval, so say we have intervals of 40 and 60%, but we want to use $\delta = 0.5$ we select the subinterval for 60%.

$$\{\hat{y} - S^{-1}\left[q_{\lceil (n+1)\delta/n \rceil}\right], \hat{y} + S^{-1}\left[q_{\lceil (n+1)\delta/n \rceil}\right]\}$$

This expands quartiles to a ceiling while removing the desired intrinsic value from our formula, creating a wider confidence interval than originally desired. This protects us from violating the coverage validity. But compared to the desired probability, how much higher is the actual probability?

In this upper bound, we can focus on sub-intervals 3/4 and 4/4 only. The worst-case scenario occurs when we want to use a confidence level (δ) that is just above an intrinsic confidence level, for example 60.1%. This forces us to extend a prediction interval by almost an entire sub-interval to reach the next intrinsic confidence level.

The probability associated with one full sub-interval is $\frac{1}{n+1}$ as there are $n + 1$ equiprobable sub-intervals. Including an entire sub-interval, which we need to do when the requested Δ is just above an intrinsic confidence level, would increase the coverage probability by $\frac{1}{n+1}$. Thus, the upper bound for the requested Δ is:

$$\Delta + \frac{1}{n+1}$$

The upper probability bound found in the literature may not be the most precise. When adding $\frac{1}{n+1}$ to the requested Δ, it assumes Δ is only slightly above one of the intrinsic confidence levels.

$$\delta \leq P(y^{\text{true}} \in \text{Interval}) \leq \delta + \frac{1}{n+1}$$

However, consider another scenario where Δ is just below one of our intrinsic confidence levels, for example, 79.9%. Reaching the next intrinsic width (80%) does not require a complete sub-interval adjustment. Instead, we only need a small adjustment to reach the next intrinsic level. We can modify the right-hand side of the inequality to create a more precise new upper bound that is smaller than or equal to the old one but never greater.

$$\delta \leq P(y^{\text{true}} \in \text{Interval}) \leq \frac{\lceil (n+1)\delta \rceil}{n+1}$$

Approaches to splitting

We have discussed using a single split to train the underlying model to calibrate conformal rankings. But as is true in most statistical learning procedures, there are approaches to splitting and training data that can add robustness to our UQ.

- Bootstrapping
- Cross-conformal – essentially K-fold splitting
- Jackknife

Under cross-conformal, data is split into sections, i.e. $k = 4$. The first three pieces are combined to train an underlying model, and non-conformity scores are computed for the remaining section. We repeat this process for each fold, ensuring all data is used as a calibration set, which gives us non-conformity scores for the entire dataset and allows us to compute quantiles.

We can take this idea further using a jackknife method, which is in essence leave-one-out validation, where n models are trained, each with $n - 1$ data points. There are variations on the jackknife approach: jackknife+ and jackknife-minimax, with the former providing better coverage, while the latter is more conservative and quicker to run.

Jackknife+-after-Bootstrap combines the **jackknife+** and **bootstrap** approaches, which involve resampling the training set with replacement, fitting regression functions on the bootstraps, and computing predictions on the complementary sets. This is an ensemble methodology that aggregates predictions to compute conformity scores, with reduced computational cost compared to jackknife+.

Conformalized Quantile Regression (CQR) is particularly effective for heteroscedastic data. It uses quantile regressors to estimate conditional prediction bounds directly, rather than assuming a symmetric interval around the point prediction. Residuals from these regressors are then used to create guaranteed coverage intervals. CQR adapts to your data's variability, providing accurate interval widths where they matter.

See `https://mapie.readthedocs.io/en/latest/theoretical_description_regression.html` for more details on these approaches.

Having read this, you might think which approach is best. Single split is the least computationally intensive method, but it can result in higher variance in prediction sets and suboptimal data use since it ignores variance from model refits. This works better with large data sets. Of course, jackknife is computationally expensive. We would recommend using it when you need high precision or model refitting is not expensive for you.

We typically find the cross-conformal approach (CV and other resampling methods) strikes a good balance between single split and jackknife methods.

Conformalized regression

Our primary goal in modeling uncertainty for regression tasks is to take our pre-trained models, e.g., RNN, GBDT, etc., and use them to forecast point predictions, then add uncertainty context to these. We could code this directly in Java, C, etc., or simply use a library like MAPIE or conformal-tights, which can both wrap around models whether they are PyTorch, TensorFlow, or scikit-learn.

The process for regression is straightforward — fit, predict, compute residuals, quantile, interval. We'll apply this directly to time series in the next section:

1. Fit model on training data.
2. Make prediction on calibration data.
3. Compute the absolute errors made by the model on calibration data.
4. Get the $(1 - \alpha)$ quantile from the distribution of absolute errors obtained at the previous point.
5. Make prediction on test data.
6. Compute left (right) end of the interval by subtracting (adding) the quantile obtained at point 4 to the predictions obtained at point 5.

Conformalized forecasting

So far, we have discussed regression modeling outside of time in order to build intuition and understanding. But to make conformal intervals applicable to time series data, we need to consider a critical issue: exchangeability. Time series data is not inherently exchangeable because the order of our points matters; observations at t are typically dependent on observations at $t - 1, t - 2, \ldots$ in ways that permutation would destroy.

Several strategies exist to recover valid conformal intervals despite this. The common thread is constructing residuals that are approximately exchangeable, either by weighting recent observations more heavily, by using out-of-bag errors from bootstrap ensembles, or by aggregating across multiple models.

To get around this, we can use weightings relative to temporal relevance, error estimation from bootstrapped models, and aggregation of predictions.

In a simple weighting approach we calculate a prediction interval based on windows of observations, i.e. a fixed window giving temporal relevance to recent observations. Of course, while this ensures predictions resemble recent data, it might miss key factors like seasonality or exogenous variable influence like price.

A better approach is to create an ensemble of multiple predictions, which can capture different aspects and trends within our data, and to use aggregations of the ensemble to reduce model-specific errors. Several methods have been proposed to achieve this:

- **Simple weighted** follows a standard conformal prediction split and assigns weights to predictions based on temporal proximity, giving more influence on recent observations - (Angelopoulos and Bates 2023).
- **Ensemble Bootstrap Prediction Interval** (**EnbPI**) uses multiple bootstrap samples to train multiple versions of a model. It then combines predictions from these models and uses errors from a model's out-of-sample predictions to adjust prediction intervals (Xu and Xie 2023a).
    - There is an adaptive version (**ACI**) that updates the intervals online.
- **Sequential Conformal Prediction** extends conformal prediction to sequential settings, updating intervals as new data becomes available (Xu and Xie 2023b).

- **Hopfield Network** (**HopCT**) applies a Hopfield network architecture to extract features from a time series and compute prediction errors. Errors are used to weight time-series values based on similarity to past errors. It is primarily used with multivariate data (Auer et al. 2023).
- **Bellman Conformal Inference** (**BCI**) formulates a stochastic control problem that minimizes interval lengths while ensuring target coverage levels. Use dynamic programming to solve this problem, adjusting the trade-off between interval length and miscoverage. Mostly used for multistep forecasting, making it unsuitable for recursive modeling (Yang, Candès, and Lei 2024).

Working with EnbPI

EnbPI is a common approach at this time, and is what we will be using in this example; however, we do recommend trying different approaches and libraries (i.e., PUNCC). EnbPI works with the following process:

1. Select a bootstrap ensemble estimator.
2. Fit B ensemble estimators by bootstrapping from the training data B times (generally 10–20). Each estimator is trained on a different bootstrap sample, resulting in B unique models.

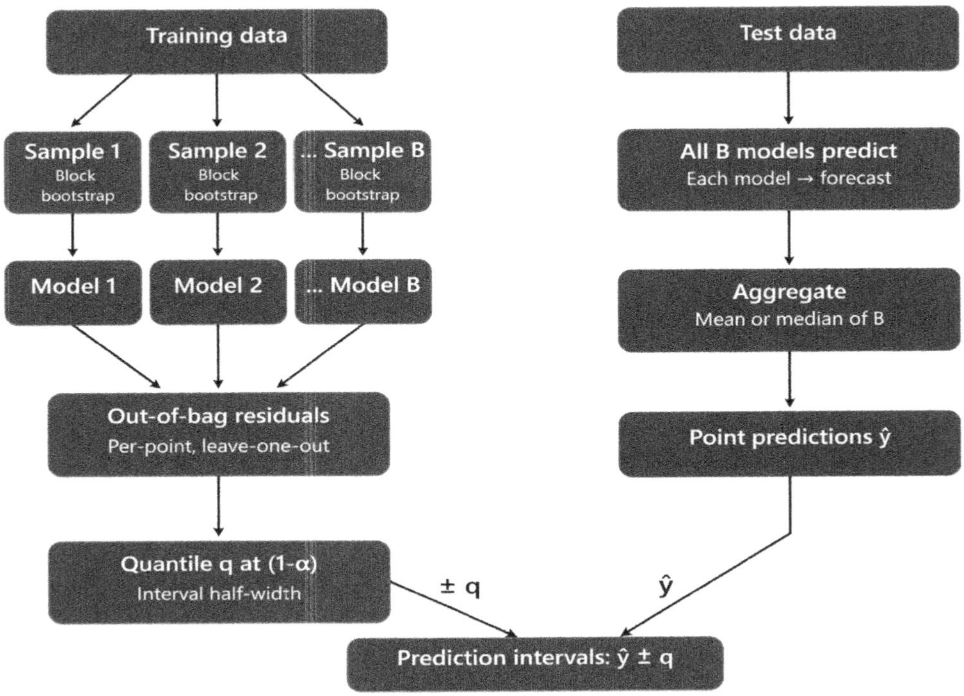

Figure 7.8: EnbPI process

In *Figure 7.8*, training data is bootstrapped **B** times to create **B** models. Each model produces predictions, which are later aggregated.

3. Compute residuals for each time point ($t = 1, \dots, T$) using estimators that did not include time point t in their training. These out-of-sample errors serve as a measure of prediction variance.

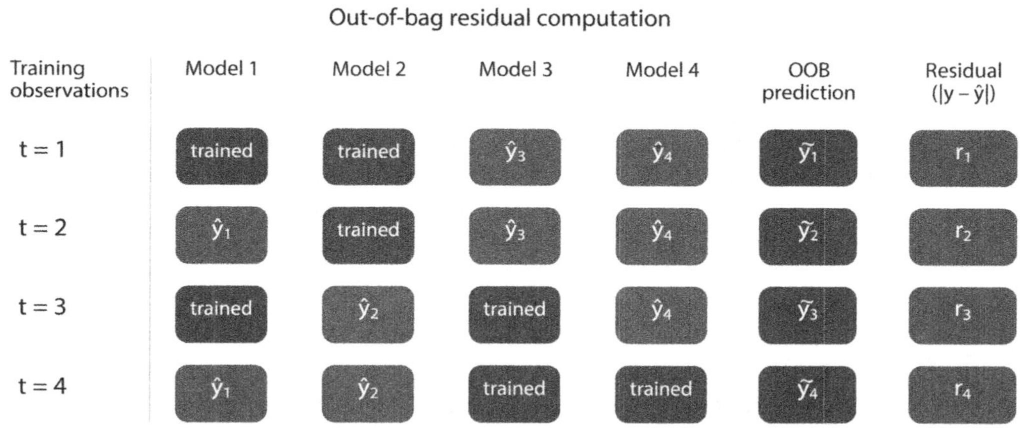

Figure 7.9: Out-of-bag residual computation

For each training observation, only models that did not include it in their bootstrap sample contribute predictions. The aggregated prediction is compared to the true value to produce a residual.

4. Collect all errors into a single array.
5. Use an ensemble estimator to generate and aggregate (mean or median) predictions for test data.
6. Construct prediction intervals using aggregated predictions and a chosen confidence level delta.
7. Finally, select quantiles from the distribution of out-of-sample errors:

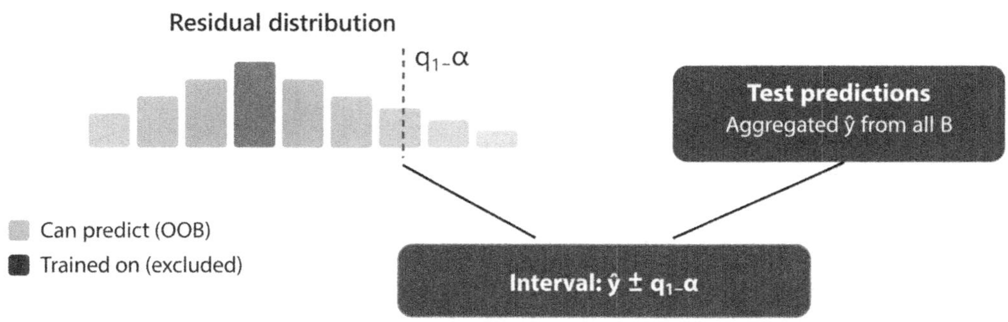

Figure 7.10: From residuals to prediction intervals

The distribution of out-of-bag residuals provides the quantile that sets the interval half-width. Combined with aggregated point predictions from the full ensemble, this produces the final prediction intervals.

The key mechanism in EnbPI is how it computes out-of-bag residuals. For each training observation, only models that did *not* train on that observation produce predictions. These predictions are aggregated, and the absolute error against the true value becomes the residual for that time point. Because each residual comes from an out-of-sample prediction, these residuals are approximately exchangeable even though the underlying time series is not.

EnbPI achieves approximate exchangeability by computing residuals only from models that did not train on the corresponding time point. Because each residual comes from an out-of-sample prediction, these residuals are approximately exchangeable even though the underlying time series is not.

Under EnbPI, we need to be cautious when using larger training sets. While they can improve coverage guarantees, they can also slow down residual updates if the model isn't refitted. Balancing the number of training samples is crucial to fitting a model and updating prediction intervals effectively.

Forecasting libraries

Libraries like Darts and NeuralForecast provide conformal prediction intervals directly in their forecasting pipelines, making the wrapping approach we show here unnecessary for those frameworks. We demonstrate the manual approach to build understanding of what these libraries do. Investigating the GitHub code and reconstructing it can also be worthwhile.

Applying EnbPI to our PyTorch model

We now apply EnbPI to a feedforward network trained on our pseudo sales data. The process involves several steps: loading and preparing the data, defining and training the PyTorch model, wrapping it for compatibility with MAPIE, and finally computing conformal prediction intervals.

We start by loading the data and creating windowed features, which we covered in earlier chapters.

```
import os
import pandas as pd
import numpy as np
import torch
import torch.nn as nn
from torch.utils.data import DataLoader, Dataset
import lightning as L
from lightning.pytorch.callbacks import EarlyStopping
import matplotlib.pyplot as plt
from sklearn.preprocessing import MinMaxScaler

device = torch.device("cuda" if torch.cuda.is_available() else "cpu")

data = pd.read_csv(CFG.data_folder / 'pseudo_sales.csv')
data['date'] = pd.to_datetime(data['date'], format='%d/%m/%Y')
data = data.sort_values('date').reset_index(drop=True)

def add_datetime_features(df, datetime_column):
    df = df.copy()
    df[datetime_column] = pd.to_datetime(df[datetime_column])
```

```
        df['year'] = df[datetime_column].dt.year
        df['month'] = df[datetime_column].dt.month
        df['week'] = df[datetime_column].dt.isocalendar().week.astype(int)
        df['day'] = df[datetime_column].dt.day
        return df

    df = add_datetime_features(data, "date")

    date_col = df['date'].copy()
    target = df['sales'].copy()
    features = df.drop(columns=['sales', 'date']).copy()

    feature_scaler = MinMaxScaler()
    target_scaler = MinMaxScaler()

    features_scaled = feature_scaler.fit_transform(features)
    target_scaled = target_scaler.fit_transform(target.values.reshape(-1, 1))

    def create_time_windows(features, target, window_size):
        inputs, targets = [], []
        for i in range(len(features) - window_size):
            inputs.append(features[i:i + window_size])
            targets.append(target[i + window_size])
        return np.array(inputs), np.array(targets)

    window_size = 32
    inputs, targets = create_time_windows(
        features_scaled, target_scaled, window_size)
```

Next, we create the PyTorch dataset and split it into training, validation, and test sets. Note that we preserve temporal order in the split, which is critical for time series.

```
class TimeSeriesDataset(Dataset):
    def __init__(self, inputs, targets):
        self.inputs = torch.tensor(inputs, dtype=torch.float32)
        self.targets = torch.tensor(targets, dtype=torch.float32)

    def __len__(self):
        return len(self.inputs)

    def __getitem__(self, idx):
        return self.inputs[idx], self.targets[idx]
```

```
train_size = int(0.6 * len(inputs))
val_size = int(0.2 * len(inputs))

train_dataset = TimeSeriesDataset(inputs[:train_size], targets[:train_size])
val_dataset = TimeSeriesDataset(inputs[train_size:train_size + val_size],
targets[train_size:train_size + val_size])
test_dataset = TimeSeriesDataset(inputs[train_size + val_size:], targets[train_
size + val_size:])

train_loader = DataLoader(train_dataset, batch_size=32, shuffle=False)
val_loader = DataLoader(val_dataset, batch_size=32, shuffle=False)
test_loader = DataLoader(test_dataset, batch_size=32, shuffle=False)
```

We define the feedforward network using PyTorch Lightning. This is the same architecture used in *Chapter 6*, with hyperparameters from our Optuna tuning.

```
class FFNetwork(L.LightningModule):
    def __init__(self, input_dim, sequence_length, hidden_dim,
                 num_layers=1, output_dim=1, learning_rate=0.0001,
                 dropout_rate=0.5, activation_func=nn.ReLU()):
        super(FFNetwork, self).__init__()
        self.layers = nn.ModuleList(
            [nn.Linear(input_dim * sequence_length, hidden_dim)]
        )
        for _ in range(num_layers - 1):
            self.layers.append(nn.Linear(hidden_dim, hidden_dim))
        self.layers.append(nn.Linear(hidden_dim, output_dim))
        self.dropout = nn.Dropout(dropout_rate)
        self.activation = activation_func
        self.learning_rate = learning_rate

    def forward(self, x):
        batch_size, sequence_length, input_dim = x.size()
        x = x.view(batch_size, sequence_length * input_dim)
        for i in range(len(self.layers) - 1):
            x = self.activation(self.layers[i](x))
            x = self.dropout(x)
        return self.layers[-1](x)

    def training_step(self, batch, batch_idx):
        x, y = batch
        loss = nn.functional.mse_loss(self.forward(x), y)
```

```
        self.log('train_loss', loss)
        return loss

    def validation_step(self, batch, batch_idx):
        x, y = batch
        loss = nn.functional.mse_loss(self.forward(x), y)
        self.log('val_loss', loss)

    def configure_optimizers(self):
        return torch.optim.Adam(self.parameters(), lr=self.learning_rate)

input_dim = inputs.shape[2]
sequence_length = inputs.shape[1]

torch.manual_seed(3527)

model = FFNetwork(
    input_dim=input_dim, sequence_length=sequence_length,
    hidden_dim=63, num_layers=4, output_dim=1,
    learning_rate=0.000689, dropout_rate=0.170971
).to(device)

trainer = L.Trainer(
    accelerator="auto", devices=1, max_epochs=300,
    callbacks=[EarlyStopping(monitor='val_loss', patience=50, mode='min')],
    logger=True
)
trainer.fit(model, train_loader, val_loader)
```

Our hyperparameters in the code (`hidden_dim=63`, `learning_rate=0.000689`, `dropout_rate=0.170971`) were found using Optuna in *Chapter 6*. (See that chapter's GitHub notebook for tuning code).

MAPIE expects a scikit-learn compatible estimator, so we need to wrap our PyTorch model. The wrapper exposes `fit` and `predict` methods while keeping the underlying model frozen. There is one subtlety here: MAPIE works with 2D arrays (`samples`, `features`), but our `FFNetwork` expects 3D input (`samples`, `window_size`, `features`). The wrapper handles this by reshaping the flattened input back to 3D before passing it to the model.

```
from sklearn.base import BaseEstimator, RegressorMixin

class SklearnWrapper(BaseEstimator, RegressorMixin):
    def __init__(self, model, window_size, n_features, device='cpu'):
        self.model = model
```

```
        self.window_size = window_size
        self.n_features = n_features
        self.device = device
        self.model.to(self.device)

    def fit(self, X, y):
        return self  # model is pre-trained

    def predict(self, X):
        self.model.eval()
        # MAPIE passes flattened 2D (samples, window*features)
        # FFNetwork expects 3D (samples, window, features)
        X_3d = X.reshape(-1, self.window_size, self.n_features)
        X_tensor = torch.tensor(X_3d, dtype=torch.float32).to(self.device)
        with torch.no_grad():
            predictions = self.model(X_tensor).cpu().numpy()
        return predictions.squeeze()

n_features = inputs.shape[2]
wrapped_model = SklearnWrapper(
    model, window_size=window_size,
    n_features=n_features, device='cpu'
)

def extract_data_from_loader(loader):
    inputs_list, targets_list = [], []
    for batch in loader:
        inp, tgt = batch
        inputs_list.append(inp.cpu().numpy())
        targets_list.append(tgt.cpu().numpy())
    return np.vstack(inputs_list), np.concatenate(targets_list)

X_train, y_train = extract_data_from_loader(train_loader)
X_test, y_test = extract_data_from_loader(test_loader)

# Flatten windowed features for MAPIE's sklearn interface
# (samples, window_size, features) → (samples, window_size * features)
X_train = X_train.reshape(X_train.shape[0], -1)
X_test = X_test.reshape(X_test.shape[0], -1)
y_train = y_train.ravel()
y_test = y_test.ravel()
```

With our wrapper in place, we can apply MAPIE's `MapieTimeSeriesRegressor` with the EnbPI method. Block bootstrapping respects temporal structure by resampling contiguous blocks rather than individual observations.

```
from mapie.regression import TimeSeriesRegressor
from mapie.subsample import BlockBootstrap

confidence_level = 0.9  # 90% prediction intervals

cv_mapiets = BlockBootstrap(
    n_resamplings=30, n_blocks=30, overlapping=False, random_state=59
)
mapie_enbpi = TimeSeriesRegressor(
    wrapped_model, method="enbpi", cv=cv_mapiets,
    agg_function="mean", n_jobs=1
)
mapie_enbpi.fit(X_train, y_train)

y_pred_enbpi, y_pis_enbpi = mapie_enbpi.predict(
    X_test, confidence_level=confidence_level, ensemble=True
)
```

With our model configured, we can now fit the EnbPI estimator and generate prediction intervals on the test data.

Evaluating conformal interval quality

Two metrics matter when evaluating conformal intervals: *coverage* (does the interval contain the true value at least $(1 - \alpha)\%$ of the time?) and *width* (how tight are the intervals?). A wide interval that always covers is useless; a tight interval that misses too often is dangerous.

```
y_lower = y_pis_enbpi[:, 0, 0]
y_upper = y_pis_enbpi[:, 1, 0]

in_interval = (y_test >= y_lower) & (y_test <= y_upper)
coverage = float(in_interval.mean())
width = float(np.mean(y_upper - y_lower))

print(f"Target coverage: {confidence_level:.0%}")
print(f"Achieved coverage: {coverage:.3f}")
print(f"Mean interval width: {width:.3f}")
```

The output of this code is:

```
Target coverage: 90%
Achieved coverage: 0.887
Mean interval width: 0.154
```

Our output shows `0.887 coverage` with an `interval` of `0.154`, which is close to what we were looking for. Let's visualize this:

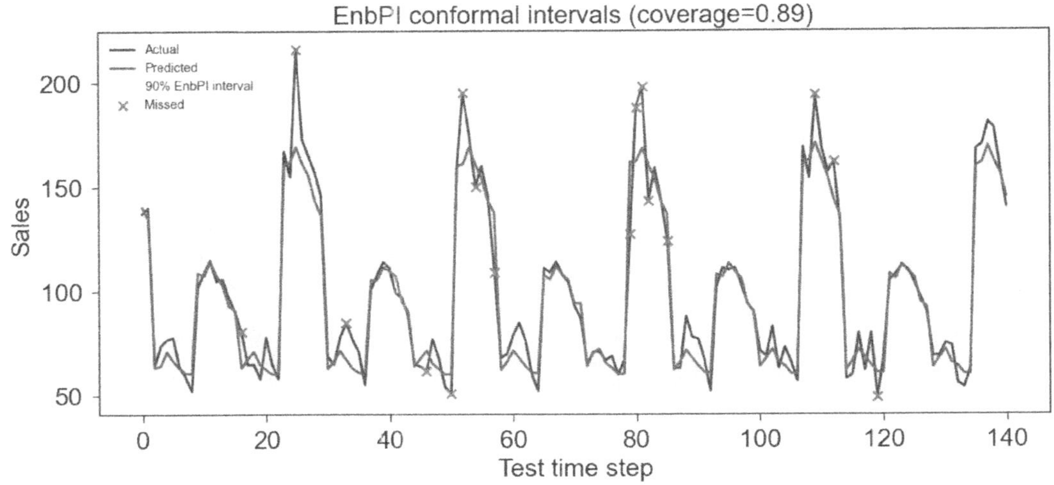

Figure 7.11: Time series plot showing actual vs predicted sales with a shaded 90% conformal prediction interval and x markers for missed points

The top panel shows our point predictions against actuals, with the 90% EnbPI interval shaded. Points falling outside our interval are marked with x. Notice that we miss peak sales, sharp spikes where the model systematically underpredicts. This is a limitation of our underlying model rather than of conformal prediction itself. If our FFN doesn't capture peak behavior $\hat{y} \pm q$, the intervals will also miss these peaks, so they can't adapt to directional bias. . **Conformalized Quantile Regression** (**CQR**) addresses this by allowing asymmetric intervals, and is worth investigating when your model has systematic bias in one direction.

It is also worth noting that our intervals have constant width across all test points. This is because basic EnbPI uses a single quantile from the calibration residuals. **Adaptive variants** (**AEnbPI**) update the residual distribution as new observations arrive, producing intervals that widen or narrow in response to changing conditions; we leave this as an exercise for the companion notebook in the repository.

Summary

In this chapter you have learned that uncertainty quantification is critical for giving us confidence in our point predictions, and is a useful tool for tracking the quality of our forecasts. We introduced conformal intervals and built an intuitive understanding of how error metrics such as absolute error can be used to measure non-conformity to historical data relative to a regression model. You now understand that we rank the errors and standardize them to give us a probability measure that we can apply to our data, one that has guarantees about coverage and validity. Finally, you learned how to apply this with a forecasting model by weighting EnbPI, which we applied to a PyTorch model. This is very much the start of the uncertainty quantification journey, but you now have the tools to break into this fast-moving and fascinating field of research.

Recommended reading

- Angelopoulos, A. N., and Bates, S. 2023. "A gentle introduction to conformal prediction and distribution-free uncertainty quantification." *arXiv*. `https://arxiv.org/abs/2107.07511`.
- Auer, et al. 2023. "Conformal prediction for time series with modern Hopfield networks." *arXiv*. `http://arxiv.org/abs/2303.12783`.
- Gammerman, A., Vovk, V., and Vapnik, V. 2013. "Learning by transduction." *arXiv*. `http://arxiv.org/abs/1301.7375`.
- Kim, B., et al. 2022. "MAPIE: an open-source library for distribution-free uncertainty quantification." *arXiv*. `https://arxiv.org/abs/2207.12274`.
- Manokhin, V. 2022. "MAPIE explained." Toward Data Science. `https://towardsdatascience.com/mapie-explained-exactly-how-you-wished-someone-explained-to-you-78fb8ce81ff3`.
- Manokhin, V., and Sudjianto, A. 2023. *Practical guide to applied conformal prediction in Python*. Packt Publishing. `https://www.packtpub.com/en-gb/product/practical-guide-to-applied-conformal-prediction-in-python-9781805122760`.
- McBurnie Ellis, C. 2022. "Regression prediction intervals with MAPIE." *Kaggle*. `https://www.kaggle.com/code/carlmcbrideellis/regression-prediction-intervals-with-mapie`.
- Rad, S. R. 2023. "Uncertainty quantification using conformal predictors." *YouTube*. `https://youtube.com/playlist?list=PLQkjeU22jN9tpbe7PpoA5nPG43SCgd7Sj`.
- Shafer, G., and Vovk, V. 2008. "Conformal prediction: A gentle introduction." *Journal of Machine Learning Research*. `https://doi.org/10.1561/2200000101`.
- Xu, C., and Xie, M. 2020. "Conformal prediction for time series." *arXiv*. `http://arxiv.org/abs/2010.09107`.
- Xu, C., and Xie, M. 2022. "Sequential predictive conformal inference for time series." *arXiv*. `http://arxiv.org/abs/2212.03463`.
- Yang, I., Candès, E. J., and Lei, L. 2024. "Bellman conformal inference: calibrating prediction intervals for time series." *arXiv*. `http://arxiv.org/abs/2402.05203`

Get this book's PDF version and more

Scan the QR code (or go to `packtpub.com/unlock`). Search for this book by name, confirm the edition, and then follow the steps on the page.

UNLOCK NOW

Note: Keep your invoice handy. Purchases made directly from Packt don't require an invoice.

8

Recurrent Neural Networks

Time-series data is a type of sequential data, and the positions of the values come from the temporal dynamics. Meanwhile, in deep learning, **recurrent neural networks** (**RNNs**) are designed to consume sequential data and predict results sequentially. Not surprisingly, RNNs are good candidates for time-series forecasting models. In this chapter, we take a dynamic system view of time-series data and make a connection between time-series forecasting and RNNs. Following the introduction of RNNs, we introduce our homemade RNN model for forecasting.

In this chapter, we will cover the following topics:

- Time-series data and states
- Introducing RNNs
- Forecasting using RNNs
- Experimenting faster

Time-series data and states

If we were asked to predict the position of a passing car in the next second, we would likely base our prediction on the car's current speed. Naively, we would estimate the speed based on the distance the car travelled in the last second. That being said, without any external aid, we need the time series of the historical positions of the car when we are forecasting the future positions. Mathematically speaking, we need to estimate how the position changes, that is, the velocity of the car, which requires us to know the position history. We also need to estimate how the velocity changes, that is, the acceleration of the car, which will require an even longer history for us to estimate it. And it goes on, as we also need to know how the acceleration changes. In the end, forecasting the position of a car over time relies on understanding the sequence of past positions, especially the position changes through time.

Swing angle forecasting

The car position prediction problem mentioned previously is too naive. In the real world, many time-series datasets show cyclic phenomena. For example, sales for T-shirts may be higher in the summer and lower in the winter. Being able to capture non-linear patterns, such as cyclic patterns, is crucial in time-series forecasting. Another real-world example we can study is a swing.

I'm sure we all played on swings when we were kids. When we are on a swing, we know that we move more slowly when we are higher and faster when we are closer to the ground. Now, think about this question: at a certain time, what is the position of the swing?

To formalize the problem, we look into the angle $\theta(t)$ of the swing at different times, t, as shown in *Figure 8.1*. As a side note, for physicists, it is described by the pendulum model.

Refer to color images on the Graphic bundle or the ebook.

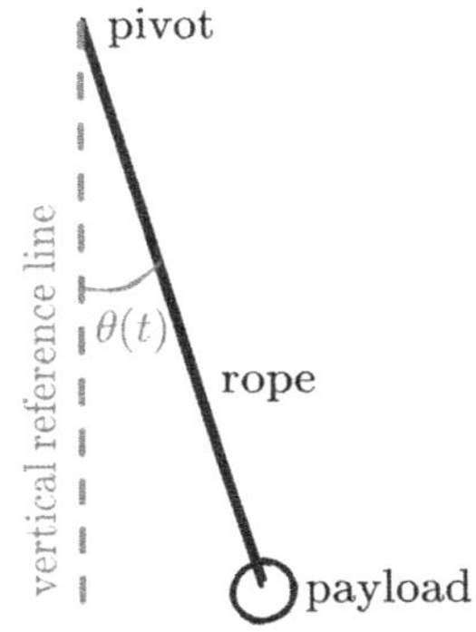

Figure 8.1: Swing setup

The swing oscillates periodically, and the angle $\theta(t)$ cannot be determined only by the previous measurement of position $\theta(t - \Delta t)$ at time $t - \Delta$, without additional assumptions.

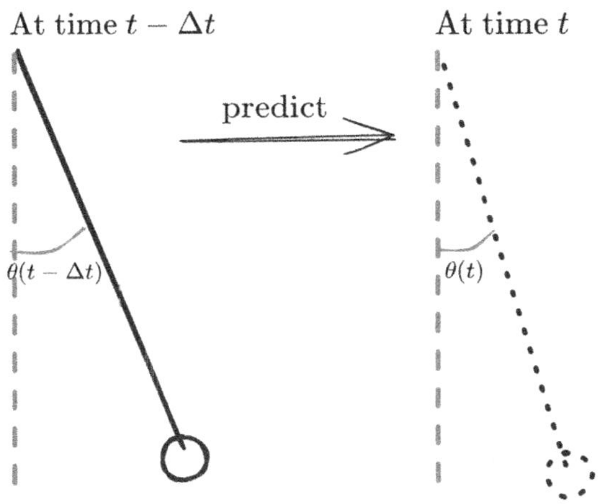

Figure 8.2: Predict swing angle at time t

It is not feasible to predict the angle $\theta(t)$ at time t from the angle $\theta(t - \Delta t)$ at $t - \Delta t$ without any other information or assumptions.

To forecast the angle $\theta(t)$, we need to consider the angles in a certain time window, $[t - \Delta t, t]$. This means we look into the angle history from time $t - \Delta t$ to time t, to understand the dynamic system behind the observations, including speed and acceleration, similar to how we forecast car positions based on past data. To get a rough estimation of the angle, we can apply a first-order thought process by ignoring acceleration: if we know the angle is changing faster, that means the swing is closer to the bottom and the prediction of $\theta(t)$ will be further away from $\theta(t - \Delta t)$. If the angle is changing slowly, our prediction of $\theta(t)$ will be close to $\theta(t - \Delta t)$.

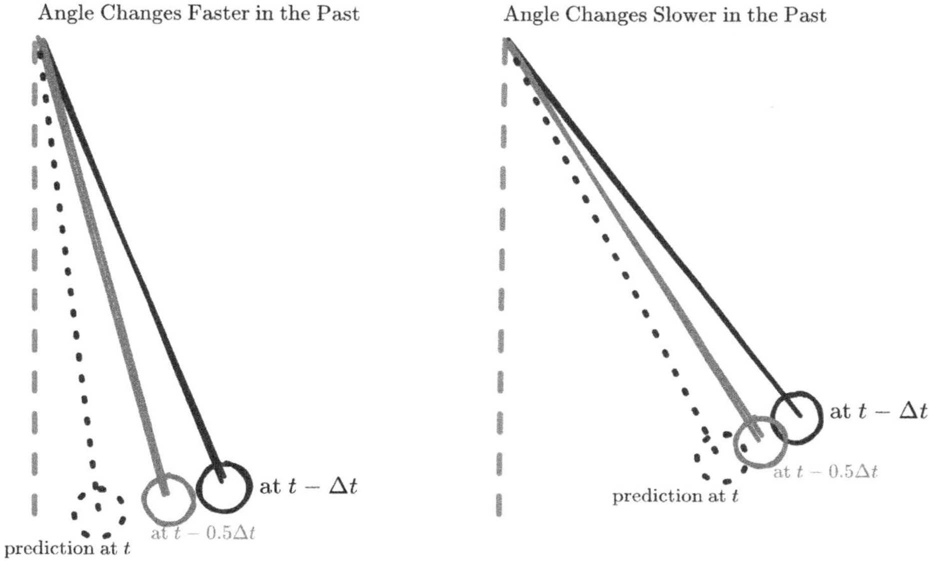

Figure 8.3: Predict swing angle at time t. Intuition of historical angle dependence

The swing angle is an example of a time series generated by the laws of physics. Similar ideas can also be found in other domains, such as socioeconomic data.

Why recurrent models help

There are two crucial properties for a lot of time-series data generated in the physical world and by socioeconomic processes:

- **Sequentiality:** The sequentiality of the data means a lot. In the swing angle forecasting example, the swing angle $\theta(t)$ at time t happens after the swing angle $\theta(t - \Delta t)$. In many situations, the order even has causality implications. Models such as naive MLP do not have the concept of sequentiality; that is, they are sequence-agnostic (Lipton, Berkowitz, and Elkan 2015).
- **Historical sequence dependency**: The second property is that forecasting future values requires a sequence of historical values. Most real-world time series depend on at least a range of past data.

The two properties hint that many time-series forecasting problems require a model that takes in the historical events one by one, sequentially and incrementally. This falls into the sequence model.

Traditional models such as statistical auto-regressive models and models based on the Markov process capture this idea. However, they are limited to less complicated states. In deep learning, we have a class of models called the recurrent neural network, or RNN, that works in a similar fashion, but RNNs can forecast time series with more complicated dynamics.

Introducing RNNs

Without touching the details of the actual model, we can sketch out our grand plan of forecasting based on the sequential modeling that we just discussed. In the following architecture, we use the $h(\cdot)$ function to transit the time-series value from a previous step to the next step. For example, $h(t_i)$ transits $z(t_{i-1})$ at time t_{i-1} to the value $z(t_i)$ at the next time step, t_i. Since the $h(\cdot)$ machine is reused, the architecture can have memory of the past. It is worth mentioning that the $h(\cdot)$ machine is the hidden state.

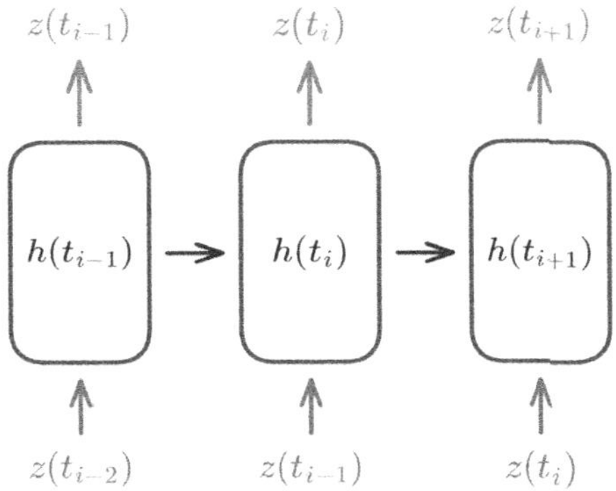

Figure 8.4: Capture historical values sequentially through time

In fact, the structure we designed in *Figure 8.4* is recurrent. To observe the recurrent nature of it, we "fold" the time dimension. In other words, since we reuse the hidden state, $h(\cdot)$, the architecture can be folded into what is shown in *Figure 8.5*. The black line that indicates the state is fed back into the same hidden state, $h(\cdot)$, thus making it recurrent.

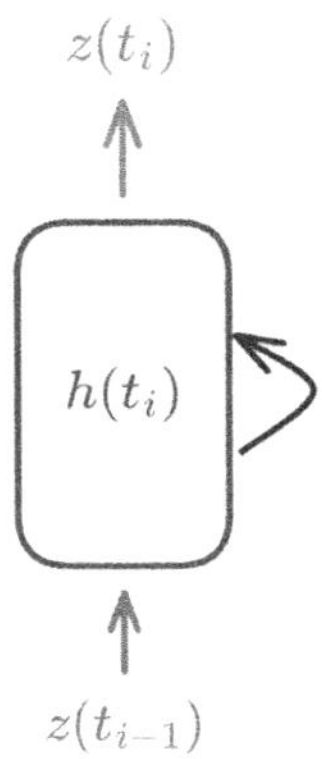

Figure 8.5: Folded representation of an RNN. Green represents the input values, while red represents the output values

In fact, there are different ways to use the hidden state, $h(\cdot)$, recurrently. In general, RNNs are quite flexible in terms of the input sequence and output sequence (Karpathy 2015). In this chapter, we will dive into two different scenarios. The first is the autoregressive architecture shown in *Figure 8.6* (adapted from the paper titled *DeepAR: Probabilistic Forecasting with Autoregressive Recurrent Networks*, available at `https://arxiv.org/pdf/1704.04110`).

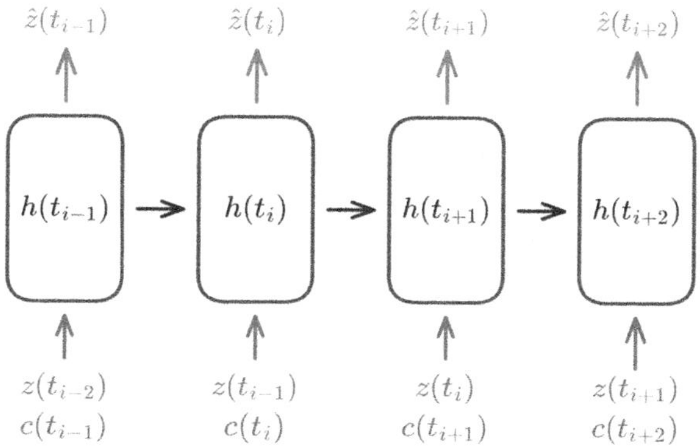

Figure 8.6: RNN autoregressive architecture

We will also look at the encoder-decoder architecture shown in *Figure 8.7* (adapted from the paper *Learning Phrase Representations Using RNN Encoder–Decoder for Statistical Machine Translation*, available at `https://doi.org/10.3115/v1/d14-1179`).

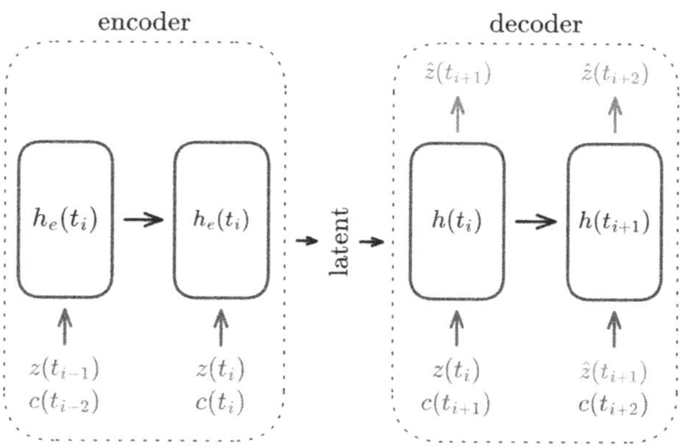

Figure 8.7: Encoder-decoder using an RNN

The autoregressive architecture in *Figure 8.6* captures the temporal dependencies by directly pinning each time step onto each iteration of the RNN. At each iteration (time t_i), the model takes in the previous value in the time series, $z(t_{i-1})$, and if required, also the covariates $c(t_{t_i})$ at time t_i. The output of the iteration for time t_i is the forecast $z(t_i)$.

One of the issues with *Figure 8.6* is that it is time-invariant. For example, if we provided covariates for the past values, we would also have to provide covariates for the future. However, not all time-series data comes with both past covariates and future covariates. The autoregressive architecture becomes inconvenient when the future covariates are missing. The encoder-decoder architecture in *Figure 8.7* is more flexible. We use RNNs to encode the history into the latent space and decode the latent-space representation together with the previous time step and covariates into the forecasts. Since we separate the data input for the history and the future, we don't have to provide future covariates if future covariates are missing.

Simple RNN

The RNN architecture can be traced back as early as 1920 (Schmidhuber 2020). However, Elman RNN and Jordan RNN are the first modern RNN models to be trained using data based on backpropagation, and they are collectively referred to as simple RNNs. Simple RNNs, as the name indicates, have a simpler structure than the **long short-term memory** (**LSTM**) and **gated recurrent unit** (**GRU**) models we will discuss later. Yet, simple RNNs convey the core philosophy of RNNs.

Theoretical background

The hidden state, h, of an Elman RNN has the following form (Bianchi et al. 2017):

$$\mathrm{h}_i = \sigma(\mathrm{W}_i \mathrm{z}_i + \mathrm{W}_h \mathrm{h}_{i-1} + \mathrm{b}) \qquad (8.1)$$

For simplicity, we have used $\mathrm{h}_i \equiv h(t_i)$, $\mathrm{z}_i \equiv z(t_i)$, $\mathrm{h}_{i-1} \equiv h(t_{i-1})$. The hidden state update in *Equation 8.1* can be visualized using the computational graph shown in *Figure 8.8*.

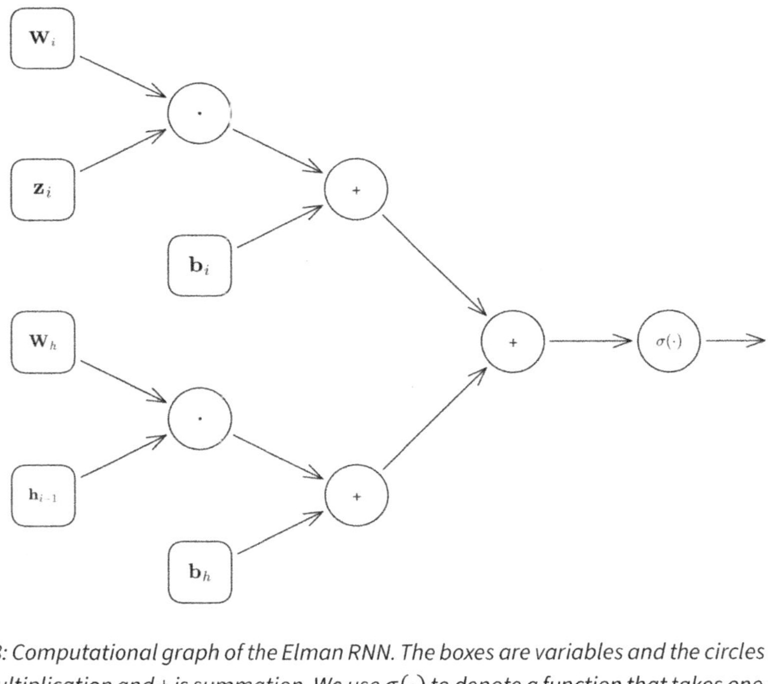

Figure 8.8: Computational graph of the Elman RNN. The boxes are variables and the circles are operators· is multiplication and + is summation. We use $\sigma(\cdot)$ to denote a function that takes one argument

RNNs and differential equations

To get an intuition of how it works, let's simplify *Equation 8.1* and investigate its dynamics. Sherstinsky (refer to *Fundamentals of Recurrent Neural Network (RNN) and Long Short-Term Memory (LSTM) Network* available at `https://doi.org/10.1016/j.physd.2019.132306`) investigated the connections between RNNs and differential equations. The author connects discrete RNN iterations with the continuous dynamics of a function, $h(t)$, and shows that RNNs are special cases of **delayed differential equations (DDEs)**. In DDEs, the rate of change in a function, $h(t)$, denoted by $\frac{\mathrm{d}(t)}{\mathrm{d}t}$, explicitly depends on the history, for example, $h(t-\tau)$, with τ being the time delay.

However, the derivations are lengthy and cluttered. Here, we use Taylor expansion to reestablish the connection between RNNs and differential equations. This will be a sloppy and naive math approach with a lot of assumptions. But it serves the purpose for us to get an intuition of RNNs. For more rigorous derivations, please refer to the Sherstinsky paper mentioned in the previous paragraph.

In the following derivation, we assume that the variables are scalars. First of all, we know that the value of the hidden state at time $t-1$ can be expressed as a Taylor series expansion around t:

$$h(t-1) = h(t) + \sum_{n=1}^{\infty} \frac{h^{(n)}(t)}{n!}(-1)^n$$

Here, $h^{(n)}$ is the nth derivative of $h(t)$. If we assume the higher-order terms do not contribute a lot, we can take the first-order term and plug it back into *Equation 8.1*, assuming σ is an identity function:

$$\frac{\mathrm{d}(t)}{\mathrm{d}t} = W'_i z_i + W_h h(t) + b' \qquad (8.2)$$

Here, $W'_i = W_h^{-1} W_i$ and $b' = W_h^{-1} b$.

Equation 8.2 is a standard first-order differential equation. We know that there may be exploding or shrinking components in the solutions. This hints that if we only use a simple RNN to iteratively predict the far future, we may run into stability issues or get constants.

In this derivation, we have used an identity function for the activation function, σ. A bounded activation function, σ, is often helpful when the value explodes.

Apart from the illustrated connection, RNNs are also very similar to ARMA models when we use an identity function for σ (see *Forecasting Across Time Series Databases Using Recurrent Neural Networks on Groups of Similar Series: A Clustering Approach* by Kasun Bandara, Christoph Bergmeir, and Slawek Smyl. 2017. Available at `http://arxiv.org/abs/1710.03222`).

Implementation

You may imagine that the iterative updates from *Equation 8.1* would lead to diminishing values or exploding extreme values if we iterate long enough. This is one of the issues with RNNs for long-horizon forecasting tasks. To observe this intuitively, we plot out some scenarios of using an already-trained RNN cell for predictions.

We create a simple RNN cell using the following Python code:

```
from functools import cached_property
import numpy as np
import numpy.typing as npt
import pandas as pd

class RNNState:
    """
    Describes a RNN state and computes the history of the states
    based on inputs.

    :param w_h: $W_h$, the weight for $h$
    :param w_i: $W_i$, the weight for $x$
    :param b: $b$, the bias before applying activation
    :param h_init: the initial hidden state
    :param activation: the activation function to be applied.
    """

    def __init__(
        self,
        w_h: npt.ArrayLike,
        w_i: npt.ArrayLike,
        b: npt.ArrayLike,
        h_init: npt.ArrayLike,
        activation: callable = np.tanh,
    ):
        self.w_h = np.array(w_h)
        self.w_i = np.array(w_i)
        self.b = np.array(b)
        self.h_init = np.array(h_init)
        self.activation = activation

    def compute_new_state(
        self, h_current: npt.ArrayLike, z_i: npt.ArrayLike
    ) -> tuple[npt.ArrayLike, npt.ArrayLike]:
        """
        compute_new_state computes a new state

        :param h_current: the current hidden state
        :param z_i: external input
```

```
        :return: the new hidden state, and the difference between new and old
        """
        h_new = self.activation(
            np.dot(self.w_h, h_current)
            + np.dot(self.w_i, z_i) + self.b
        )
        h_delta = h_new - h_current
        return h_new, h_delta


    def states(self, z: npt.ArrayLike) -> dict[str : npt.ArrayLike]:
        """
        states calculates the history of RNN states.

        We designed function to be easily readable by
        computing the values step by step.
        It is not efficient.

        :param z: input values for the RNN
        :return: history of the RNN hidden states
        """
        h_t = [self.h_init]
        t_steps = [0]
        h_t_delta = [np.zeros_like(self.h_init)]
        for t, z_i in enumerate(z):
            h_new, h_delta = self.compute_new_state(
                h_current=h_t[-1], z_i=z_i
            )
            h_t.append(h_new)
            h_t_delta.append(h_delta)
            t_steps.append(t + 1)

        total_time_steps = len(t_steps)

        return {
            **{
                "t": np.array(t_steps),
                "h": np.array(h_t),
                "dh": np.array(h_t_delta),
                "z": np.pad(z, (1,0), constant_values=0)
            },
```

```
            **{
                k: [v] * total_time_steps
                for k, v in self._metadata.items()
            },
        }

    @cached_property
    def _metadata(self) -> dict:
        return {
            "experiment": (
                f"w_h={np.array2string(self.w_h)}; "
                f"w_i={np.array2string(self.w_i)};"
                f"b={np.array2string(self.b)}; "
                f"h_init={np.array2string(self.h_init)}; "
                f"{self.activation.__name__}"
            ),
            "w_h": np.array2string(self.w_h),
            "w_i": np.array2string(self.w_i),
            "b": np.array2string(self.b),
            "activation": self.activation.__name__,
            "initial_state": np.array2string(self.h_init),
        }
```

We feed the RNN with four different scenarios. For simplicity, we choose scalar values for all the variables. For the phenomena we are investigating, we observe similar behaviors for higher-dimension variables. Since W_i and b can be absorbed into other terms and do not change the pattern of the outputs, we always keep $W_i = 1$ and $b = 0$. We will provide the following sets of parameters:

W_h	**Initial State** $h(t = 0)$
0.5	0.5
1.5	1.5
0.5	0.5
.5	.5

Table 8.1: Simple RNN parameters

Using the sets of parameters, we observe the convergence behavior in *Figures 8.9* and *8.10*. In other words, the RNN forgets about the distant histories. *Figure 8.9* shows the simple RNN inference with linear input:

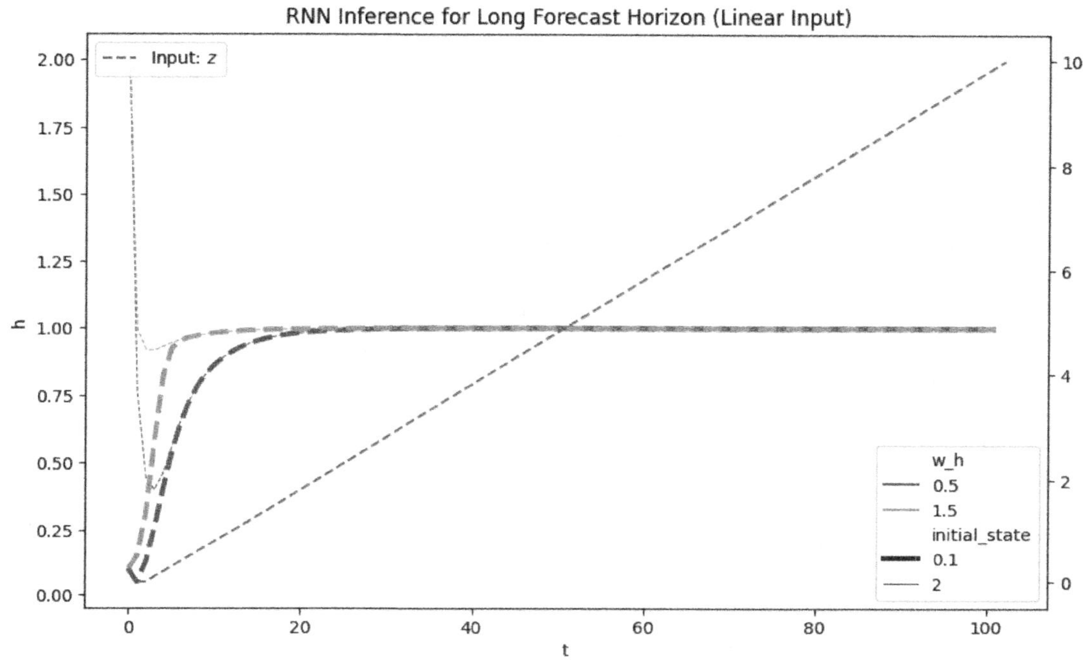

Figure 8.9: Simple RNN inference with linear input

On the other hand, *Figure 8.10* shows a simple RNN with a sinusoidal input.

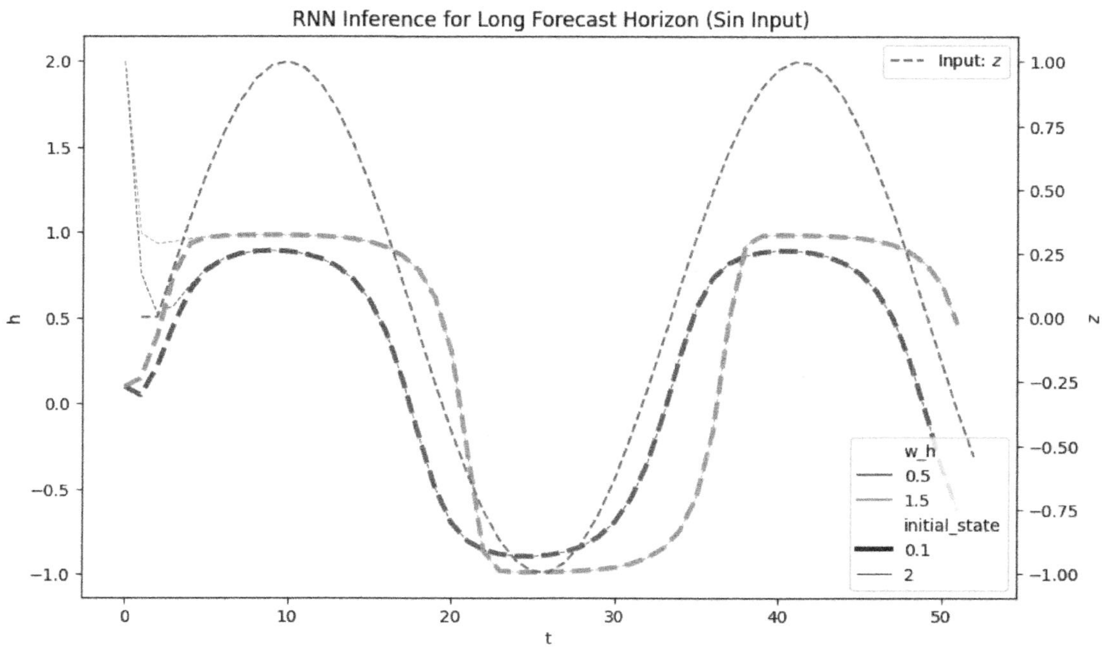

Figure 8.10: Simple RNN inference with sinusoidal input

In *Figures 8.9* and *8.10*, the gray dashed lines show the input value, z. In both scenarios, the far future forecasts are independent of the initial hidden state, h(t = 0). They are dominated by the network parameters and the current step input, z(t). All of the history information is lost for far future forecasts.

The intrinsic problem of simple RNNs for long-term dependencies has been researched thoroughly in the past 30 years (Bengio, Simard, and Frasconi 1994; Hochreiter and Schmidhuber 1996; Hochreiter and Schmidhuber 1997a; Kolen and Kremer 2009; Graves 2013; S. Zhang et al. 2016; Bandara, Bergmeir, and Smyl 2017; Sherstinsky 2020; Zhao et al. 2020). It turns out that deep simple RNNs are also hard to train using data. The problem lies in the backpropagation during training. Brute-force computation of the gradient is expensive. The most popular trick for training RNNs is **backpropagation through time (BPTT)**, which computes the gradients on the unfolded RNN and uses the hidden state at each time step as mediators. Given a loss $L(\{\hat{z}_1, \cdots, \hat{z}_N\}, \{z_1, \cdots, z_N\})$, the gradient with respect to W_h is as follows:

$$\frac{\partial L}{\partial \mathrm{W}_h} = \sum_{i=1}^{N} \frac{\partial L}{\partial \mathrm{h}_i} \mathrm{h}_{i-1}^T$$

Here, $\frac{\partial L}{\partial \mathrm{h}_i}$ contains a term that is proportional to $(\mathrm{W}_h^T)^{N-t}$ (for more details about the derivations, refer to the book *Dive into Deep Learning*, available at `https://d2l.ai/chapter_recurrent-neural-networks/bptt.html`). The exponential form indicates that the hidden state from the distant past can either grow exponentially or diminish, depending on the eigenvalues of W_h^T. This concludes the notorious vanishing gradient problem in RNNs.

Introducing LSTM

The seminal paper *Long Short-Term Memory* (available at `https://doi.org/10.1162/neco.1997.9.8.1735`) by Hochreiter and Schmidhuber introduced the widely used LSTM network, which effectively handles both short-term and long-term interactions.

The hidden state update in an LSTM is defined as follows:

$$\mathrm{h}_i = \sigma(\mathrm{W}_{ih}\mathrm{z}_i + \mathrm{W}_{hh}\mathrm{h}_{i-1} + \mathrm{b}_h) \odot \tanh(\mathrm{c}_i) \qquad (8.3)$$

Here, c_i is the so-called cell state in LSTM, and it is responsible for controlling how information is passed onto the next iteration. Note that the only difference between an Elman RNN and an LSTM is the term $\tanh(\mathrm{c}_i)$, which is multiplied by each element of the Elman RNN hidden state. If the network decides that the cell state has all elements close to 1, our LSTM network reduces to an Elman RNN.

The update rule for c is where the forget and remember come in:

$$\mathrm{c}_i = \mathrm{f}(\mathrm{h}_{i-1}, \mathrm{z}_i) \odot \mathrm{c}_{i-1} + \mathrm{i}(\mathrm{h}_{i-1}, \mathrm{z}_i) \cdot \mathrm{g}(\mathrm{h}_{i-1}, \mathrm{z}_i)$$

Here, $f(h_{i-1}, z_i)$ is the forget mechanism, and $f(h_{i-1}, z_i)$, $i(h_{i-1}, z_i)$, and $g(h_{i-1}, z_i)$ are functions limited to $(-1,1)$. This formalism guarantees that even if our initial cell state, c_0, is large, c_t will eventually become smaller, and the term $\tanh(c_t)$ in *Equation 8.3* will start to suppress the final outputs. See *Figure 8.11* for a better understanding.

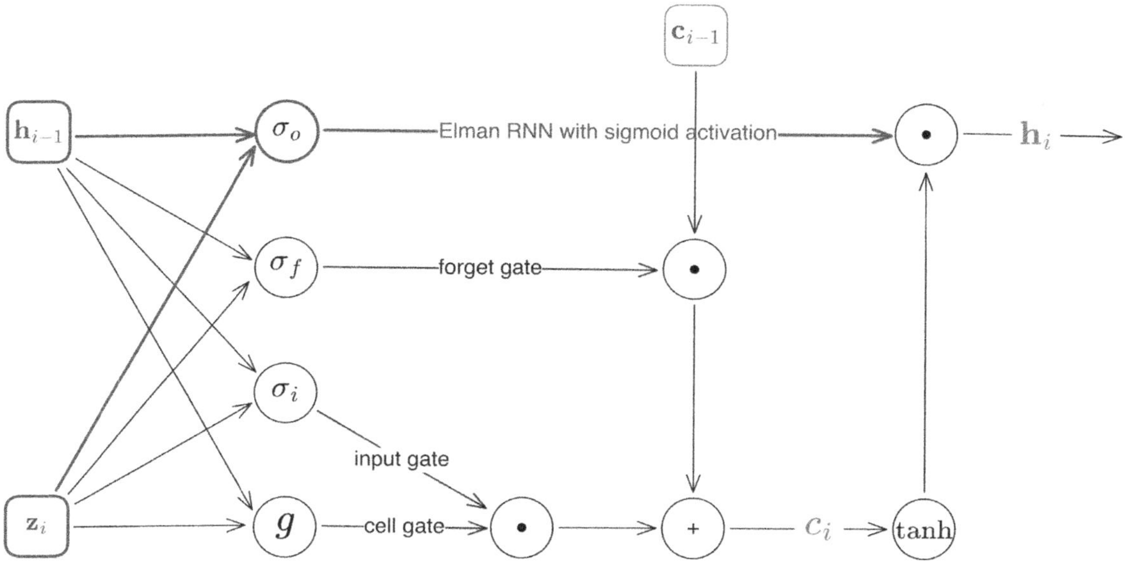

Figure 8.11: Computational graph of LSTM

In the preceding diagram, the $h_{i-1}$and c_{i-1} variables are coming from the previous iteration and the h_i and c_i variables are passed on to the next iteration. The thick blue lines and symbols form the Elman RNN. In this LSTM network, we have the cell state, c, which controls how the information is passed onto the next iteration. We have abstracted some details of the computational graph for better readability and understanding.

In *Figure 8.11*, we sketched the computational graph of an LSTM cell. The σ_o, σ_f, and σ_i functions are sigmoids, including the weights and biases, for example, $\sigma(W_i z_i + W_h h_{i-1} + b)$, which outputs values in the range (0,1). Similarly, the cell gate, g, is $\tanh\left(W_{ig} z_i + W_{hg} h_{i-1} + b_g\right)$, which outputs values in the range $(-1,1)$.

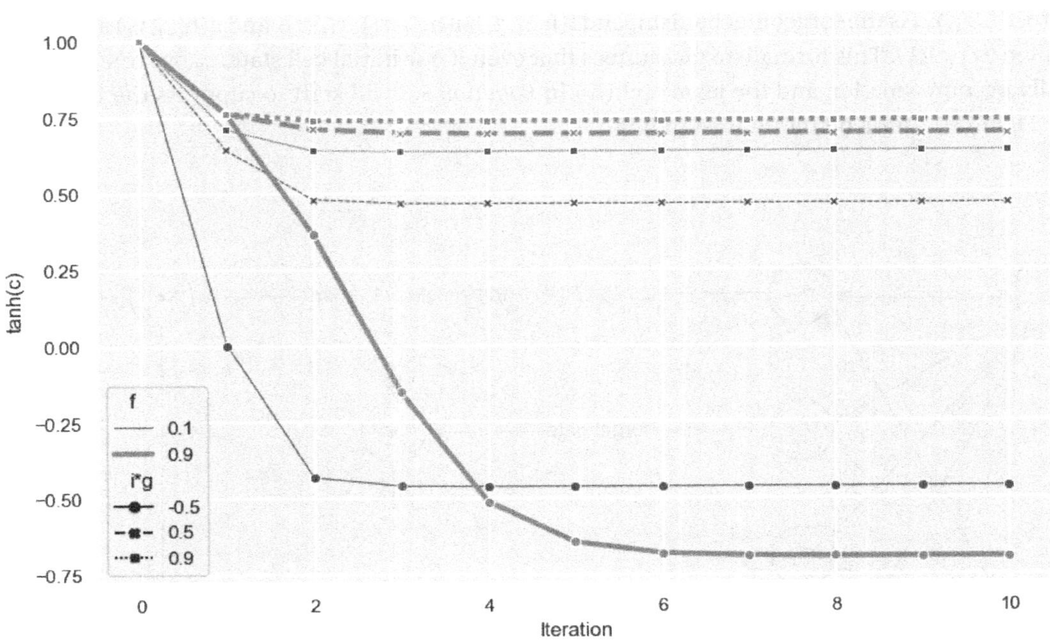

Figure 8.12: LSTM cell state

We compute the term $\boldsymbol{tanh(c)}$assuming the terms $\boldsymbol{f(h_{i-1}, z_i)}$and $\boldsymbol{i(h_{i-1}, z_i)} \cdot \boldsymbol{g(h_{i-1}, z_i)}$remain constant in each iteration. In a real LSTM cell, these terms will vary, but the cell state iterations are similar to what is depicted in this graph because they are constrained to absolute values less than 1.

Introducing GRUs

Once researchers understood the fundamental flaws in simple RNNs, particularly the vanishing or exploding gradient in multiple iterations, they developed various modified models and architectures. The GRU is an alternative to LSTM with fewer parameters. A GRU is simpler than an LSTM cell but with a much simpler idea: instead of applying gates to control the information flow, the GRU simply mixes the state of the previous step and a newly computed state. The following figure illustrates this:

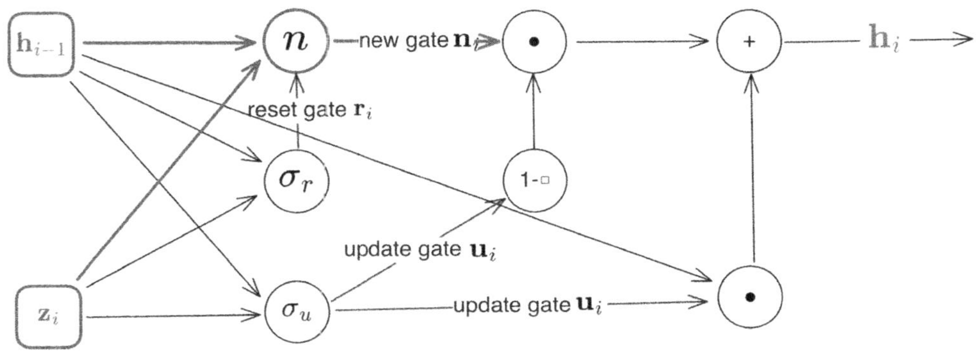

Figure 8.13: Computational graph of a GRU

The new state, $\boldsymbol{n}_i$, is an Elman RNN but weighted by the reset gate, $\boldsymbol{r}_i$. If the reset gate has all elements being $\mathbf{1}$, $\boldsymbol{n}_i$ is an Elman RNN. On top of that, if all elements of the update gate, $\boldsymbol{u}_i$, become 0, the whole GRU is reduced to an Elman RNN.

The computational graph of a GRU is shown in *Figure 8.13*. Compared to the computational graph of the LSTM in *Figure 8.11*, a GRU has fewer gates compared to an LSTM. A GRU has an update gate, $\mathbf{u}_i = \boldsymbol{\sigma}_u(\mathbf{W}_{iu}\mathbf{z}_i + \mathbf{W}_{hu}\mathbf{h}_{i-1} + \mathbf{b}_u)$, that weights the previous hidden state; a newly calculated new gate, $\mathbf{n}_i = \mathbf{tanh}(\mathbf{W}_{in}\mathbf{z}_i + \mathbf{r}_i \odot (\mathbf{W}_{hn}\mathbf{h}_{i-1} + \mathbf{b}_n))$; and a reset gate, $\mathbf{r}_i = \boldsymbol{\sigma}_r(\mathbf{W}_{ir}\mathbf{z}_i + \mathbf{W}_{hr}\mathbf{h}_{i-1} + \mathbf{b}_r)$, that adjusts how the previous state $\mathbf{h}_{i-1}$ is used.

The GRU and LSTM are quite similar in ideas, and there is no conclusion on which one is better. In time-series forecasting, they are mostly on par with each other, but we should perform experiments on both models when it comes to our own forecasting tasks. However, we should be aware that GRU is more compact and, therefore, more computationally efficient when designing our model architecture. You can refer to Yamak, Yujian, and Gadosey 2019; Gao, Wang, and Zhou 2021; Liu, Lin, and Feng 2021; and ArunKumar et al. 2022 in the supplemental reading material at the end of this chapter.

Stacking

Apart from inventing new RNN cell structures, some work has focused on altering network connections of RNNs, including stacking, skip connections, and dilation (Pascanu et al. 2013; Chang et al. 2017). For completeness, we present the basic concepts but omit the mathematical details.

Stacking RNNs in a direction perpendicular to the time axis (the direction of iteration) increases the model's capacity without increasing the number of iterations in a recurrent model. For example, *Figure 8.14* shows a two-layer architecture. Instead of directly outputting the forecasts, we feed the output from the bottom layer into the upper layer. Stacked RNNs usually forecast better than non-stacked RNN models.

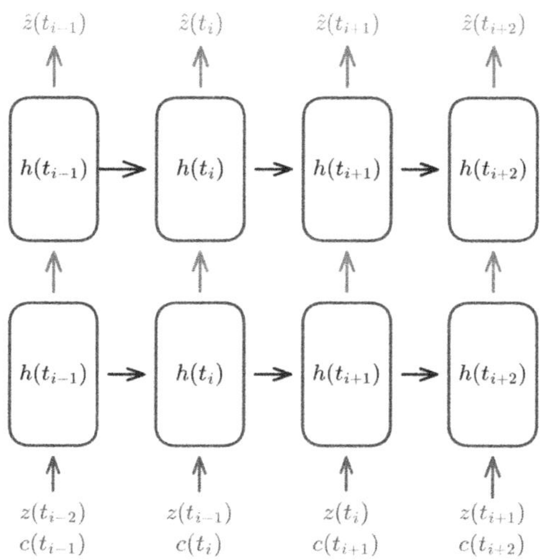

Figure 8.14: RNN stacking. Instead of connecting the RNN units sequentially, we can also stack the RNN units vertically

Forecasting using RNNs

To begin our journey of forecasting using RNNs, we construct a toy problem to show the concepts of building a forecaster using RNNs. We will generate a sinusoid dataset, train an RNN model, and benchmark it against a naive forecaster. In this tutorial, you shall grasp most of the basic concepts of creating a forecaster using PyTorch. We will skip backtesting and hyperparameter tuning in this tutorial.

First of all, we build our dataset using NumPy and pandas. To ensure the model training captures the oscillations in the data, the sinusoid dataset contains many cycles (see *Figure 8.15*):

```
import numpy as np
import pandas as pd
import matplotlib.pyplot as plt

df = pd.DataFrame(
    {
        "t": np.linspace(0, 100, 501),
        "y": np.sin(np.linspace(0, 100, 501))
    }
)

_, ax = plt.subplots(figsize=(5, 3.09))

df.plot(x="t", y="y", ax=ax);
```

Here's the code's output:

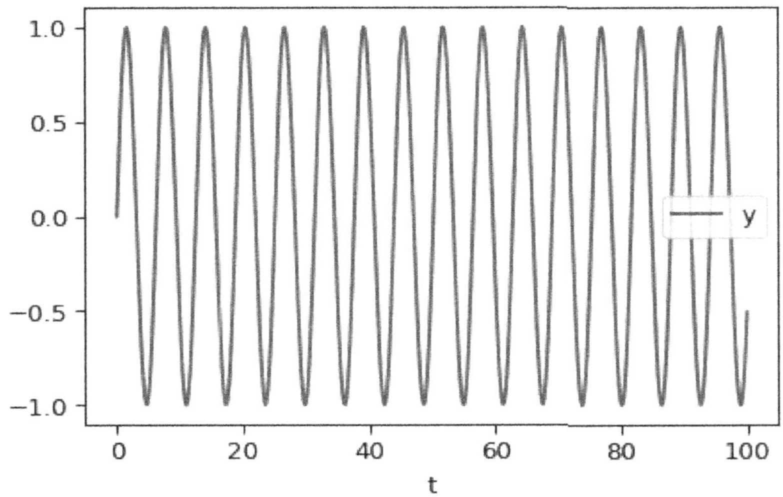

Figure 8.15: Toy dataset generated using a sinusoid

Next, we create an RNN model using PyTorch. We will use the `torch.nn.RNN` model provided by PyTorch. When building our models, we always assume that the tensors for time series have the dimension (batch size, sequence length, number of variables).

First of all, we create a dataclass to hold all the model parameters:

```
import dataclasses
import torch
from torch import nn

@dataclasses.dataclass
class TSRNNParams:
    """A dataclass to be served as our parameters
    for the RNN model.

    :param hidden_size: number of dimensions
        in the hidden state
    :param input_size: input dim
    :param num_layers: number of units stacked
    """

    input_size: int
    hidden_size: int
    num_layers: int = 1
```

With the model parameters ready, we can create an RNN model using PyTorch, which accepts the model parameters as well as the forecast horizon when instantiated. Note that a PyTorch module requires us to implement the `forward` method:

```
class TSRNN(nn.Module):
    """RNN for univaraite time series modeling.

    :param history_length: the length of the input history.
    :param horizon: the number of steps to be forecasted.
    :param rnn_params: the parameters for the RNN network.
    """

    def __init__(
        self, history_length: int,
        horizon: int,
        rnn_params: TSRNNParams
    ):
        super().__init__()
```

```
        self.rnn_params = rnn_params
        self.history_length = history_length
        self.horizon = horizon

        self.regulate_input = nn.Linear(
            self.history_length,
            self.rnn_params.input_size
        )

        self.rnn = nn.RNN(
            input_size=self.rnn_params.input_size,
            hidden_size=self.rnn_params.hidden_size,
            num_layers=self.rnn_params.num_layers,
            batch_first=True,
        )

        self.regulate_output = nn.Linear(
            self.rnn_params.hidden_size,
            self.horizon
        )

    @property
    def rnn_config(self):
        return dataclasses.asdict(self.rnn_params)

    def forward(
            self, x: torch.Tensor
        ) -> torch.Tensor:
        x = self.regulate_input(x)
        x, _ = self.rnn(x)

        return self.regulate_output(x)
```

We define our forecasting problem to take in a history length of `100` and a forecast horizon of `1` for simplicity:

```
history_length_1_step = 100
horizon_1_step = 1
```

To be able to feed the data into our model, we build a class (`DataFrameDataset`) that converts the pandas DataFrame into a dataset for PyTorch. Since we are focusing on the actual model in this chapter, we have hidden it in a Python package (`ts_bolt`), and you need only to focus on the model creation and training.

To make the lightning training code simpler, we will build a LightningDataModule (`DataFrameDataModule`) and a LightningModule (`RNNForecaster`):

1. First of all, we create the forecaster using Lightning by asking it to accept the TSRNN RNN model we created in the previous step. We implement `training_step`, `validation_step`, and `predict_step` in the LightningModule, as well as `configure_optimizers` to configure our SGD optimizer. When we are training, the trainer will pick up the `training_step` method. Similarly, the training will pick up the `validation_step` when we are performing validations. We have removed the docstrings in the class for easier readability in the printed version of the book; however, it is highly recommended to write docstrings when you write your own code:

```
import lightning as L

class RNNForecaster(L.LightningModule):
    """Forecaster based on RNN

    :param rnn: RNN model that takes time
        series input and predicts time series
        of the specified horizon.
    """
    def __init__(self, rnn: nn.Module):
        super().__init__()
        self.rnn = rnn

    def configure_optimizers(self) -> torch.optim.Optimizer:
        optimizer = torch.optim.SGD(self.parameters(), lr=1e-3)
        return optimizer

    def training_step(
            self,
            batch: tuple[torch.Tensor],
            batch_idx: int
        ) -> torch.Tensor:
        x, y = batch
        x = x.squeeze().type(self.dtype)
        y = y.squeeze(-1).type(self.dtype)

        y_hat = self.rnn(x)

        loss = nn.functional.l1_loss(y_hat, y)
        self.log_dict({"train_loss": loss}, prog_bar=True)
        return loss
```

```
    def validation_step(
            self,
            batch: tuple[torch.Tensor],
            batch_idx: int
        ) -> torch.Tensor:
        x, y = batch
        x = x.squeeze().type(self.dtype)
        y = y.squeeze(-1).type(self.dtype)

        y_hat = self.rnn(x)

        loss = nn.functional.l1_loss(y_hat, y)
        self.log_dict({"val_loss": loss}, prog_bar=True)
        return loss

    def predict_step(
            self,
            batch: tuple[torch.Tensor],
            batch_idx: int
        ) -> tuple[torch.Tensor]:
        x, y = batch
        x = x.squeeze().type(self.dtype)
        y = y.squeeze(-1).type(self.dtype)

        y_hat = self.rnn(x)
        return x, y_hat

    def forward(self, x: torch.Tensor) -> tuple[torch.Tensor]:
        x = x.squeeze().type(self.dtype)
        return x, self.rnn(x)
```

2. The data module is built based on the DataFrame. We take 100 steps as input and the next step as the value to be predicted. We slide forward with step size 1 to create our dataset:

```
from ts_bolt.datamodules.pandas import DataFrameDataModule

pdm_1_step = DataFrameDataModule(
    history_length=history_length_1_step,
    horizon=horizon_1_step,
    dataframe=df[["y"]],
)
```

```
With all the ingredients prepared, we instantiate our model:
from torchinfo import summary


ts_rnn_params_1_step = TSRNNParams(
    input_size=100,
    hidden_size=64,
    num_layers=1
)

ts_rnn_1_step = TSRNN(
    history_length=history_length_1_step,
    horizon=horizon_1_step,
    rnn_params=ts_rnn_params_1_step,
)

summary(ts_rnn_1_step)
```

Here's our output:

```
=================================================================
Layer (type:depth-idx)                   Param #
=================================================================
TSRNN                                    --
├─Linear: 1-1                            10,100
├─RNN: 1-2                               10,624
├─Linear: 1-3                            65
=================================================================
Total params: 20,789
Trainable params: 20,789
Non-trainable params: 0
=================================================================
```

We can also observe that our model is quite small as it has a small number of parameters. The whole training will be fast even on our personal laptops.

3. We create the actual encapsulated forecaster using the RNN model we instantiated in the previous step:

```
rnn_forecaster_1_step = RNNForecaster(rnn=ts_rnn_1_step)
```

4. PyTorch Lightning provides an easy-to-use trainer for our model. We create the trainer and fit our model:

```
from lightning.pytorch.callbacks.early_stopping import EarlyStopping

logger_1_step = L.pytorch.loggers.TensorBoardLogger(
    save_dir="lightning_logs",
    name="rnn_ts_1_step"
)

trainer_1_step = L.Trainer(
    precision="64",
    max_epochs=100,
    min_epochs=5,
    callbacks=[
        EarlyStopping(
            monitor="val_loss",
            mode="min",
            min_delta=1e-7,
            patience=3
        )
    ],
    logger=logger_1_step,
)
GPU available: False, used: False
TPU available: False, using: 0 TPU cores
HPU available: False, using: 0 HPUs
trainer_1_step.fit(
    model=rnn_forecaster_1_step,
    datamodule=pdm_1_step
)
```

This code gives the following output:

```
  | Name | Type  | Params | Mode
-----------------------------------------
0 | rnn  | TSRNN | 20.8 K | train
-----------------------------------------
20.8 K    Trainable params
0         Non-trainable params
20.8 K    Total params
0.083     Total estimated model params size (MB)
```

5. During fitting, the trainer takes the training data loader and validation data loader prepared in the `pdm_1_step` data module. Once the model is trained, we can forecast using the `predict` method, which takes the `predict` data loader and performs a forward pass:

```
predictions_1_step = trainer_1_step.predict(
    model=rnn_forecaster_1_step, datamodule=pdm_1_step
)
```

6. To build up intuitions of our forecasts, we also produce forecasts using the last observation, that is, forecast $x[t + 1] \equiv x[t]$:

```
from ts_bolt.naive_forecasters.last_observation import
LastObservationForecaster

trainer_naive_1_step = L.Trainer(precision="64")

lobs_forecaster_1_step = LastObservationForecaster(
    horizon=horizon_1_step)
lobs_1_step_predictions = trainer_naive_1_step.predict(
    model=lobs_forecaster_1_step,
    datamodule=pdm_1_step
)
```

```
GPU available: False, used: False
TPU available: False, using: 0 TPU cores
HPU available: False, using: 0 HPUs
```

7. To evaluate our model performance, we compare the RNN model output and the naive last observation forecaster in a plot. We observe that the RNN model seems to be better than the naive model:

```
from ts_bolt.evaluation.evaluator import Evaluator

evaluator_1_step = Evaluator(step=0)

fig, ax = plt.subplots(figsize=(10, 6.18))

ax.plot(
    evaluator_1_step.y_true(
        dataloader=pdm_1_step.predict_dataloader()
    ),
    "g-",
    label="truth",
)
```

```
ax.plot(evaluator_1_step.y(predictions_1_step), "r--",
label="predictions")

ax.plot(evaluator_1_step.y(lobs_1_step_predictions), "b-.", label="naive
predictions")

plt.legend()
```

The output is this plot:

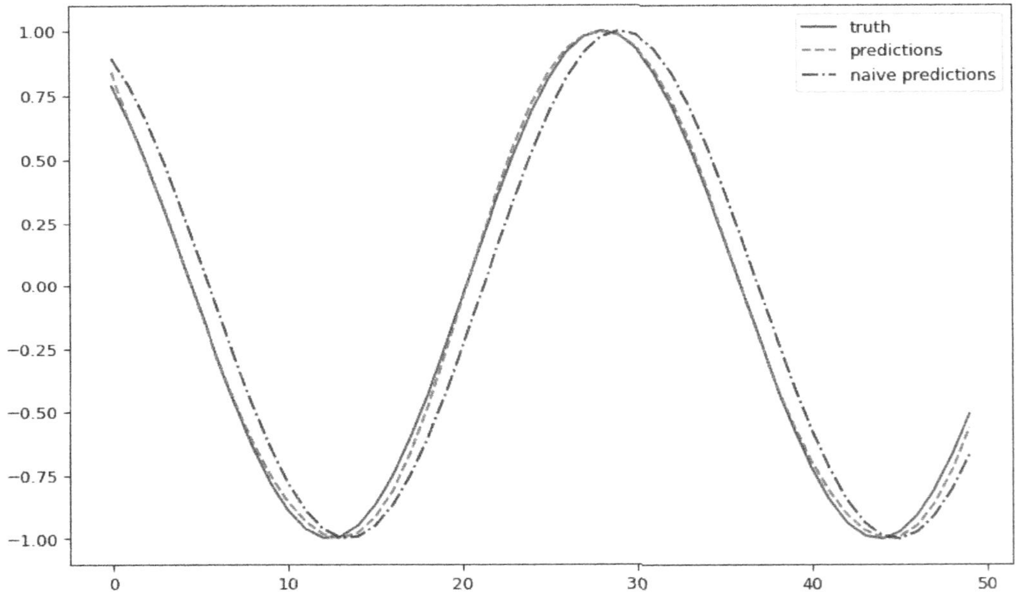

Figure 8.16: Comparing the RNN with the naive forecaster

8. To quantify the model comparison, we compute the metrics for the two models. The RNN model performs better in terms of every metric we computed:

```
pd.merge(
    evaluator_1_step.metrics(
        predictions_1_step, pdm_1_step.predict_dataloader()),
    evaluator_1_step.metrics(
        lobs_1_step_predictions, pdm_1_step.predict_dataloader()),
    how="inner",
    left_index=True, right_index=True,
    suffixes=["_rnn", "_last_obs"]
)
```

Here's our output:

```
values_rnn  values  _last_obs
MeanAbsoluteError  0.022116  0.127596
MeanAbsolutePercentageError  0.045366  0.789616
MeanSquaredError  0.000824  0.019861
SymmetricMeanAbsolutePercentageError  0.043565  0.410083
```

While we have successfully built an RNN model, we had to create a series of utilities, which takes a lot of time. Note that we have not yet performed any backtesting in our RNN example, nor have we tuned the hyperparameters of the model. In a real-world forecasting project, both backtesting and hyperparameter tuning are core to the workflow.

In reality, we usually don't write our barebones model from scratch but use packages. Most deep learning-based forecasting packages provide RNN models. For example, Darts provides `RNNModel`, which packages together training, inference, model saving, and even simple hyperparameter tuning. Similarly, Nixtla also provides the `RNN` model as well as a toolkit to use a pandas DataFrame as input data, which simplifies our workflow a lot. Both Darts and Nixtla provide easy-to-use evaluation utilities.

Summary

In this chapter, we explored how recurrent neural networks (RNNs) can model sequential dependencies in time series forecasting. We began with simple RNNs, noting their limitations in capturing long-term patterns due to vanishing and exploding gradients. We then introduced LSTM and GRU architectures, which use gating mechanisms to preserve relevant information. Stacked RNNs were discussed as a way to increase model capacity. As a practical example, we implemented an RNN forecaster in PyTorch on a sinusoidal dataset, comparing it with a naive baseline. The RNN outperformed the baseline, highlighting its strengths, though real-world use requires backtesting, tuning, and often packaged implementations.

Recommended reading

- Bandara, K., Bergmeir, C., and Smyl, S. 2017. "Forecasting Across Time Series Databases Using Recurrent Neural Networks on Groups of Similar Series: A Clustering Approach." *arXiv* [cs.LG]. `http://arxiv.org/abs/1710.03222`.
- Bengio, Y., Simard, P., and Frasconi, P. 1994. "Learning Long-Term Dependencies with Gradient Descent Is Difficult." *IEEE Transactions on Neural Networks*, 5(2): 157–166. `https://doi.org/10.1109/72.279181`.
- Bianchi, F. M., Maiorino, E., Kampffmeyer, M. C., Rizzi, A., and Jenssen, R. 2017. "An Overview and Comparative Analysis of Recurrent Neural Networks for Short-Term Load Forecasting." *arXiv* [cs.NE]. `http://arxiv.org/abs/1705.04378`.
- Chang, S., Zhang, Y., Han, W., Yu, M., Guo, X., Tan, W., Cui, X., Witbrock, M., Hasegawa-Johnson, M., and Huang, T. S. 2017. "Dilated Recurrent Neural Networks." *NeurIPS 2017*. `https://proceedings.neurips.cc/paper_files/paper/2017/file/32bb90e8976aab5298d5da10fe66f21d-Paper.pdf`.

- Cho, K., van Merrienboer, B., Gulcehre, C., Bahdanau, D., Bougares, F., Schwenk, H., and Bengio, Y. 2014. "Learning Phrase Representations Using RNN Encoder–Decoder for Statistical Machine Translation." In *Proceedings of the 2014 Conference on Empirical Methods in Natural Language Processing (EMNLP)*. Stroudsburg, PA: Association for Computational Linguistics. `https://doi.org/10.3115/v1/d14-1179`.
- Graves, A. 2013. "Generating Sequences with Recurrent Neural Networks." *arXiv* [cs.NE]. `http://arxiv.org/abs/1308.0850`.
- Hochreiter, S., and Schmidhuber, J. 1996. "LSTM Can Solve Hard Long Time Lag Problems." In *Advances in Neural Information Processing Systems*, 473–479. `https://proceedings.neurips.cc/paper_files/paper/1996/file/a4d2f0d23dcc84ce983ff9157f8b7f88-Paper.pdf`.
- Hochreiter, S., and Schmidhuber, J. 1997a. "Long Short-Term Memory." *Neural Computation*, 9(8): 1735–1780. `https://doi.org/10.1162/neco.1997.9.8.1735`.
- Karpathy, A. 2015. "The Unreasonable Effectiveness of Recurrent Neural Networks." `https://karpathy.github.io/2015/05/21/rnn-effectiveness/`.
- Kolen, J. F., and Kremer, S. C. 2009. "Gradient Flow in Recurrent Nets: The Difficulty of Learning Long-Term Dependencies." In *A Field Guide to Dynamical Recurrent Networks*, 237–243. IEEE. `https://doi.org/10.1109/9780470544037.ch14`.
- Lipton, Z. C., Berkowitz, J., and Elkan, C. 2015. "A Critical Review of Recurrent Neural Networks for Sequence Learning." *arXiv* [cs.LG]. `http://arxiv.org/abs/1506.00019`.
- Pascanu, R., Gulcehre, C., Cho, K., and Bengio, Y. 2013. "How to Construct Deep Recurrent Neural Networks." *arXiv* [cs.NE]. `http://arxiv.org/abs/1312.6026`.
- Salinas, D., Flunkert, V., and Gasthaus, J. 2017. "DeepAR: Probabilistic Forecasting with Autoregressive Recurrent Networks." *arXiv* [cs.AI]. `http://arxiv.org/abs/1704.04110`.
- Schmidhuber, J. 2020. "Deep Learning: Our Miraculous Year 1990–1991." *arXiv* [cs.NE]. `http://arxiv.org/abs/2005.05744`.
- Sherstinsky, A. 2020. "Fundamentals of Recurrent Neural Network (RNN) and Long Short-Term Memory (LSTM) Network." *Physica D: Nonlinear Phenomena*, 404: 132306. `https://doi.org/10.1016/j.physd.2019.132306`.
- Zhang, A., Lipton, Z. C., Li, M., and Smola, A. J. 2023. *Dive into Deep Learning*. Cambridge, UK: Cambridge University Press. `https://d2l.ai/chapter_recurrent-neural-networks/bptt.html`.
- Zhang, S., Wu, Y., Che, T., Lin, Z., Memisevic, R., Salakhutdinov, R., and Bengio, Y. 2016. "Architectural Complexity Measures of Recurrent Neural Networks." *NeurIPS 2016*. `https://proceedings.neurips.cc/paper/2016/file/860320be12a1c050cd7731794e231bd3-Paper.pdf`.
- Zhao, J., Huang, F., Lv, J., Duan, Y., Qin, Z., Li, G., and Tian, G. 2020. "Do RNN and LSTM Have Long Memory?" *arXiv* [stat.ML]. `http://arxiv.org/abs/2006.03860`.

9

Transformers

"... the Transformer, a model architecture eschewing recurrence and instead relying entirely on an attention mechanism to draw global dependencies between input and output." From *Attention Is All You Need*

In this new Generative AI era, most of us have heard about the famous transformer model, or the **T** in ChatGPT. In the seminal paper *Attention is All You Need*, the legendary transformer architecture was born (Vaswani et al. 2017). The transformer model was initially invented for language tasks. Since time series data and language are quite similar due to their dependency on sequential positions, models that work on natural language processing are mostly tested on time series forecasting tasks too. In fact, companies such as Zalando already productionized transformers to forecast sales and demand, even before we see any publications about transformers for time series forecasting. We have discussed recurrent neural networks in the previous chapter. There we learned that recurrent structure is useful in time series forecasting; in this chapter, we will show that a different approach called attention is also a good forecaster.

Before we get to the details of how transformer works, beware that transformer has evolved a lot in the past few years and there are a galaxy of variants (Amatriain 2023). You can see the evolution of the transformer models in *Figure 9.1*. It is still a hot topic and there are new ideas to replace or enhance the so-called attention mechanism in a transformer. We will not be able to cover all the recent research; however, you shall be able to follow the research by yourself after reading this chapter.

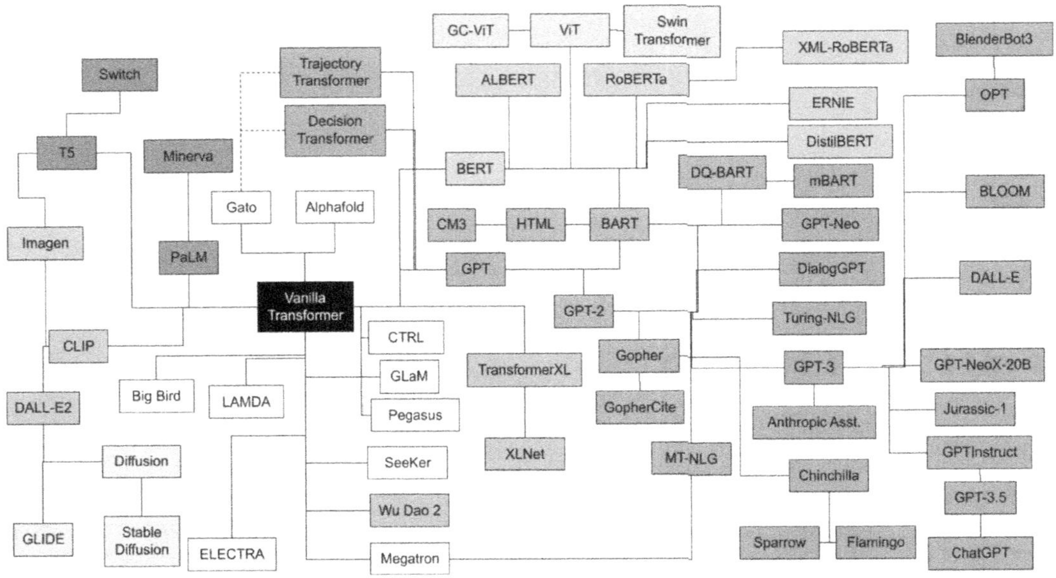

Figure 9.1: Amatriain et al surveyed the history of transformers in 2023

You may recognize a lot of familiar names in this figure.

In this chapter, we will first focus on the details of the vanilla transformer but extend to transformer-based forecasting models. This chapter may seem to have many equations and formulas, but most of them are simply reiterations, with detailed examples to show how the model works. There are already papers and articles that demystify transformers, for example, *The Illustrated Transformer* by Alammar J., available: `http://jalammar.github.io/illustrated-transformer/`. We stick to the time series use case and build up intuitions by computing intriguing examples for each component.

Figure 9.2: The history of transformers in time series forecasting

In the preceding figure, the horizontal axis represents the time of publication while the vertical axis represents the citations in log scale. The lines indicate papers being cited by other papers. The black markers are foundation models for time series forecasting. We included them because they are also transformer-based. However, the details of time series foundation models will be discussed in later chapters. This figure shows that transformers are having a big impact on time series forecasting, as the time series forecasting papers that are influenced by transformers are growing fast.

In particular, we will be covering the following topics:

- The vanilla transformer
- Implementation of transformers
- The M5 competition

The vanilla transformer

The vanilla transformer model is an encoder-decoder structure. In the context of time series forecasting, an encoder is responsible for encoding the historical time series information, and the decoder will make predictions. If our problem setting has future covariates in it, the decoder will also consume them together with the encoding output. Encoder-decoder defines how the past and future variables are taken care of in general, however the time series is mixed up and processed using the so-called **attention mechanism**. In fact, attention is the core of a transformer model. In this section, we will go through the encoder-decoder structure and the principles of attention.

Encoder-decoder

We have briefly discussed the encoder-decoder architecture with RNN in the previous chapter. Needless to say, the encoder-decoder architecture is a handy tool to deal with the inhomogeneities between the input and output. In forecasting tasks, it is easy to design an encoder-decoder model to work with whatever shape of input or output we need. It is also convenient to pass future covariates during inference. The information passed on to the encoder can be dramatically different from the information passed on to and generated by the decoder.

> Refer to color images on the Graphic bundle or the ebook.

As discussed before, a vanilla transformer is an encoder-decoder architecture. In *Figure 9.3*, we illustrate how the encoder-decoder consumes historical information and forecasts based on future covariates together with the encoded historical information.

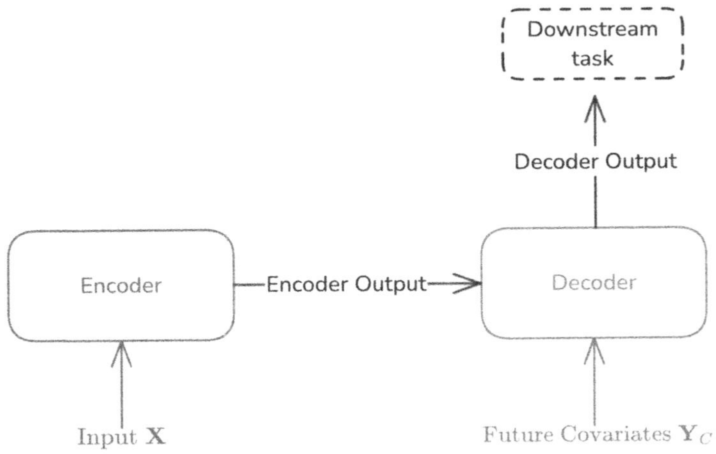

Figure 9.3: Encoder-decoder architecture of a transformer forecasting model

In the preceding figure, the Input $\boldsymbol{X}$ contains historical information and may also include covariates. The output of the decoder can be used as forecasts directly or for other downstream tasks. Note that the illustration is simplified and there are other transformations in between the steps. We also assume that the Input $\boldsymbol{X}$ is already embedded and converted to tensors.

In an ideal world, the encoder-decoder is capturing the correlations of the historical time series and use them to map the future. The **Input X** is first encoded and compressed by attention mechanism which weights the values and selects the time slots that are more useful for forecasting. The selection of the time slots is based on the historical time series itself. The output from the encoder is then used to help selecting the future covariates if any before generating the forecasts. In some **models**, the output from the decoder is directly used as forecasts, however, it is also a common practice to use the output of the decoder as input to another module which actually generates the forecasts.

Attention mechanism

The key to a transformer is the attention mechanism, which looks into the relations among the different time steps of the embeddings (Vaswani et al. 2017; A. Zhang et al. 2023). To understand the attention mechanism, we need to understand three seemingly unrelated terms: query ($\mathbf{Q}$), key ($\mathbf{K}$), and value ($\mathbf{V}$). In the end, the attention mechanism is a classifier that outputs the usefulness of the elements in the value, and the usefulness is represented using a matrix formed by the query and the key. Different models employ different attention mechanisms. In this section, we stick to the vanilla transformer dot-product attention. The original paper discussed text embeddings while we are focusing on time series. For better understanding, we adapted it to make it easier to understand in our time series domain.

$$\text{Attention}(\mathbf{Q}, \mathbf{K}, \mathbf{V}) = \sigma\left(\frac{\mathbf{Q}\mathbf{K}^{\mathrm{T}}}{\sqrt{d_k}}\right)\mathbf{V}, \qquad (9.1)$$

where d_k is the dimension of the key $\mathbf{K}$, and σ = softmax. In a self-attention setting, we can construct the **query**, **key**, and **value** by applying a linear layer to the same input $\mathbf{X}$. It is more explicit when we write down the component form of matrix multiplications.

Conventions

We follow the convention that the first index of $\mathbf{X}$ is the index for the input element. For example, if we have two time steps in our input time series, the element X_{0j} is the representation of the first time step and X_{1j} is that of the second.

For simplicity, we also use Einstein notation for matrix multiplications in this section, i.e., $X_{ik}W_{kj} \equiv \sum_k X_{ik} W_{kj}$ as the index $_k$ appears in both X_{ik} and W_{kj}.

Name	Definition	Component Form	Comment
Query $\mathbf{Q}$	$\mathbf{Q} = \mathbf{X}\mathbf{W}^Q$	$Q_{ij} = X_{ik}W^Q_{kj}$	Note that the weights $\mathbf{W}^Q$ can be used to adjust the size of the query.
Key $\mathbf{K}$	$\mathbf{K} = \mathbf{X}\mathbf{W}^K$	$K_{ij} = X_{ik}W^K_{kj}$	In the vanilla scaled-dot attention, the dimension of the key is the same as the query. This is why $\mathbf{Q}\mathbf{K}^\mathrm{T}$ works.
Value $\mathbf{V}$	$\mathbf{V} = \mathbf{X}\mathbf{W}^V$	$V_{ij} = X_{ik}W^V_{kj}$	

Table 9.1: Mathematical details of Query, Key, and Value

The dot product $\mathbf{Q}\mathbf{K}^\mathrm{T}$ in *Equation 9.1* is the score and it becomes:

$$\mathbf{A} \equiv \mathbf{Q}\mathbf{K}^\mathrm{T} = \mathbf{X}\mathbf{W}^Q{\mathbf{W}^K}^\mathrm{T}\mathbf{X}^\mathrm{T} = X_{im}W^Q_{mk}W^K_{kn}X_{ni}. \qquad (9.2)$$

which is a square matrix that determines how the elements in the value tensor, $A_{ij}V_{jk}$, are mixed. For identity $\mathbf{A}$, we do not mix the values of $\mathbf{V}$.

In Equation 9.1, $\sigma\left(\frac{\mathbf{Q}\mathbf{K}^\mathrm{T}}{\sqrt{d_k}}\right)$ serves as a similarity function that computes how similar $\mathbf{Q}$ and $\mathbf{K}$ are. For example, if the second element in both $\mathbf{Q}$ and $\mathbf{K}$ are similar, $\mathbf{Q}\mathbf{K}^\mathrm{T}$ will create a matrix that has large values on the second row and second column (see Figure 9.4).

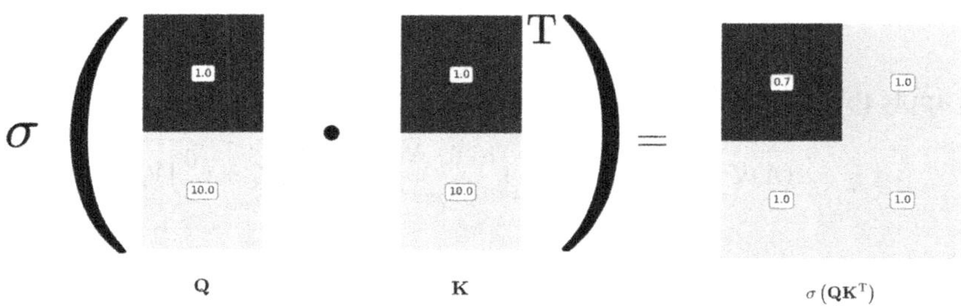

Figure 9.4: Attention and Similarity

The second elements of $\boldsymbol{Q}$ and $\boldsymbol{K}$ are large, which leads to large elements in the second row and second column in $\sigma(\boldsymbol{QK}^T)$.

We are not limited to dot-product and softmax based similarity functions. In general, any function that gives us a sense of similarity between **Q** and **K** will work; for example, we can use a general form.

$$\text{Attention}(\mathbf{Q}, \mathbf{K}, \mathbf{V}) = f(\mathbf{Q}, \mathbf{K})\mathbf{V}.$$

> The dot-product attention is like a linear model that outputs the usefulness of the elements in **V**. After training, **A** should be able to combine the different elements in **V** in a way that works well for the task.

As an example, we show the different effects of **A** on **V**. We assume that $\mathbf{V} \to (1{,}10)$. Since we have fixed dimensions in our matrices, we also ignore $\sqrt{d_k}$ in *Equation 9.1* as it only changes the scales.

The selecting process can be visualized by writing down the component form of $\sigma(\mathbf{A})\mathbf{V}$ explicitly, for example.

$$\sigma(\mathbf{A})\mathbf{V} \to \sigma(A_{ij})V_j \to \begin{pmatrix} \sigma(A_{1j}) \\ \sigma(A_{2j}) \\ \vdots \end{pmatrix} V_j,$$

Where:

$$\begin{pmatrix} \sigma(A_{1j}) \\ \sigma(A_{2j}) \\ \vdots \end{pmatrix}$$

selects where to pay more attention on each value V_j. We calculate a few examples using 2-dimensional vectors. The first example is keeping the original value vector **V** with identity matrix $\sigma(\mathbf{A})$, for example:

$$\begin{pmatrix} \sigma(A_{11}) \\ \sigma(A_{21}) \end{pmatrix} = \begin{pmatrix} 1 \\ 0 \end{pmatrix}$$

for $j = 1$ and,

$$\begin{pmatrix} \sigma(A_{12}) \\ \sigma(A_{22}) \end{pmatrix} = \begin{pmatrix} 0 \\ 1 \end{pmatrix}$$

for $j = 2$. We apply this specific example of $\sigma(\mathbf{A})$ onto value **V**,

$$\sigma(\mathbf{A})\mathbf{V} = \begin{pmatrix} \sigma(A_{11}) \\ \sigma(A_{21}) \end{pmatrix} V_1 + \begin{pmatrix} \sigma(A_{12}) \\ \sigma(A_{22}) \end{pmatrix} V_2 = \begin{pmatrix} 1 \\ 0 \end{pmatrix} V_1 + \begin{pmatrix} 0 \\ 1 \end{pmatrix} V_2.$$

We observe that:

$$\begin{pmatrix} 1 \\ 0 \end{pmatrix}$$

Keeps V_1 on the first element and,

$$\begin{pmatrix} 0 \\ 1 \end{pmatrix}$$

Keeps V_2 on the second element.

In a second example, we have:

$$\begin{pmatrix} \sigma(A_{11}) \\ \sigma(A_{21}) \end{pmatrix} = \begin{pmatrix} 0 \\ 1 \end{pmatrix}$$

for $j = 1$ and,

$$\begin{pmatrix} \sigma(A_{12}) \\ \sigma(A_{22}) \end{pmatrix} = \begin{pmatrix} 1 \\ 0 \end{pmatrix}$$

for $j = 2$, which leads to:

$$\sigma(\mathbf{A})\mathbf{V} = \begin{pmatrix} 0 \\ 1 \end{pmatrix} V_1 + \begin{pmatrix} 1 \\ 0 \end{pmatrix} V_2.$$

In this example, the position of V_1 and V_2 switched. Through the two examples, we build up the intuition that each column in the matrix $\sigma(\mathbf{A})$ will decide how each element V_j of $\mathbf{V}$ is weighted.

The following table shows some special cases to help you understand how it works with numbers.

$\mathbf{A}$	$\sigma(\mathbf{A})$	$\sigma(\mathbf{A})\mathbf{V}$	Note
$\begin{pmatrix} 1 & 0 \\ 0 & 1 \end{pmatrix}$	$\begin{pmatrix} 0.73 & 0.5 \\ 0.5 & 0.73 \end{pmatrix}$	$\begin{pmatrix} 5.7 \\ 7.8 \end{pmatrix}$	The larger element is still larger.
$\begin{pmatrix} 1 & -10 \\ -10 & 1 \end{pmatrix}$	$\begin{pmatrix} 0.73 & 4.5 * 10^{-5} \\ 4.5 * 10^{-5} & 0.73 \end{pmatrix}$	$\begin{pmatrix} 0.73 \\ 7.3 \end{pmatrix}$	Roughly the original ratio between the two elements.
$\begin{pmatrix} -10 & 1 \\ 1 & -10 \end{pmatrix}$	$\begin{pmatrix} 4.5 * 10^{-5} & 0.73 \\ 0.73 & 4.5 * 10^{-5} \end{pmatrix}$	$\begin{pmatrix} 7.3 \\ 0.73 \end{pmatrix}$	Roughly the original ratio between the two elements but switched order.

Table 9.2: Examples of attention

In principle, it is good enough if we can come up with good choices for the elements of $\mathbf{A}$. Keep in mind that we need $\mathbf{A}$ to adapt based on $\mathbf{X}$ because the attention should be different for different input patterns. That being said, the elements of $\mathbf{A}$ should be dynamic, i.e., $\mathbf{A} = f(\mathbf{X})$. The simplest approach is the linear combination we have shown in *Table 9.1*.

Attention in a real transformer model

In the previous examples, we showed how to calculate the attention assuming the elements of $\mathbf{Q}$, $\mathbf{K}$, and $\mathbf{V}$ are scalars, i.e., real numbers. In an actual transformer model, the elements will be tensors. We construct a simple example to illustrate how attention works inside a real transformer model.

Suppose our univariate data input data $\mathbf{X}$ has the size $(B = 1, L = 10)$, where B is the batch size and L is the sequence length. A transformer model will first encode the input data, for example, into $\mathbf{X}_{\text{embedded}}$ with size $(B = 1, L = 10, E = 4)$ where E is the embedding size.

The embedded input $\mathbf{X}_{\text{embedded}}$ is then used to construct the query, key, and value, as shown in *Table 9.1*. The corresponding sizes are

1. $\mathbf{Q}$: $(B = 1, L = 10, E = 4)$
2. $\mathbf{K}$: $(B = 1, L = 10, E = 4)$, and
3. $\mathbf{V}$: $(B = 1, L = 10, E = 4)$

The score $\mathbf{A} \equiv \mathbf{Q}\mathbf{K}^{\mathrm{T}}$ in the attention in *Equation 9.1* is only contracting the embedding dimension, i.e., the size of $\mathbf{A}$ is $(B = 1, L = 10, L = 10)$. That being said, we have calculated the inner product of each embedding dimension in $\mathbf{Q}$ and $\mathbf{K}$. The sizes of the relevant variables are shown in *Figure 9.5*:

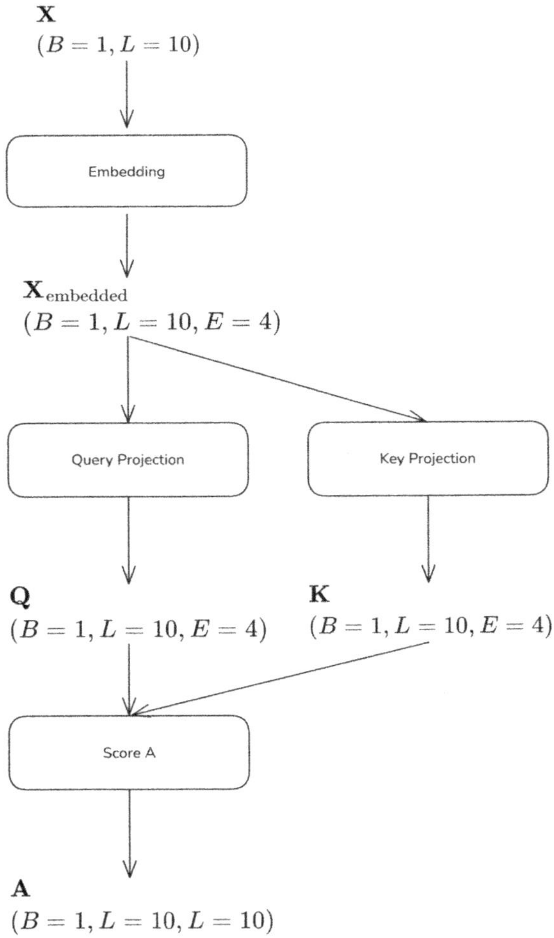

Figure 9.5: The variable sizes in the attention mechanism

There are different ways of using the attention mechanism on time series. In the above example, we choose to use the attention to find the correlations in the temporal dimension. In the iTransformer model (iTransformer: Inverted Transformers Are Effective for Time Series Forecasting available at http://arxiv.org/abs/2310.06625), the author proposed to contract the temporal dimension, i.e., the size of $\mathbf{A}$ is $(B = 1, E = 4, E = 4)$ (Liu et al. 2023).

In practice, we also don't use one single attention in our transformer model. In the original transformer paper, the authors repeated the attention 8 times, each time on a part of the embedded input. This is called **Multi-Head Attention**.

Knowledge of temporal information

The attention mechanism is a brilliant invention to deal with sequential data. Before we jump into writing the code for transformers, there is one extra problem to be solved: the temporal information, aka, position information, of each element in the time series is not yet known by the model. In RNN models, each element can be fed into the model sequentially, but a transformer consumes a given chunk of input time series and spits out the predictions at once. The attention mechanism doesn't know which element in the chunk of input time series comes first. A similar problem exists in language model transformers. The authors of the transformer model came up with an idea for the model to distinguish the different positions of the input sequence using the so-called position encoding.

In position encoding, positional information, or time order information for time series input, is represented by adding different perturbations that depend on the position of each element. There are different positional encoders; the authors of the original transformer paper chose to encode the position using trigonometric functions (Vaswani et al. 2017):

$$
\begin{aligned}
p(\text{position}, 2i) &= \sin(\text{position}/10000^{2i/d_{\text{model}}}) \\
p(\text{position}, 2i+1) &= \cos(\text{position}/10000^{2i/d_{\text{model}}}),
\end{aligned}
$$

where position corresponds to the time step in the input time series, d_{model} is the embedding dimension for each time step. The encoding is the interleaving of cosine and sine. For the 0th time step, the position encoding is always the alternating of 0 and 1. The positional encoding is simply added to the input embedding $\mathbf{X}_{\text{embedded}}$, as shown in Figure 9.6, which augments Figure 9.3.

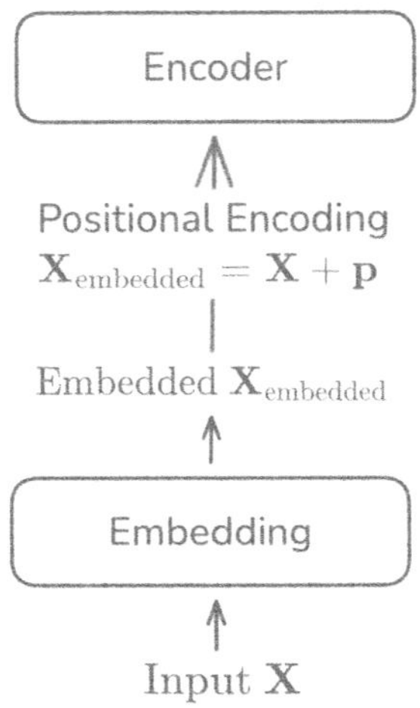

Figure 9.6: Positional encoding is added to the embedded input

In the previous figure, the input data is first embedded into a hidden space, which is in turn summed up with the positional encoding before being fed into the encoder.

The positional encoding is implemented as follows. It is a simple translation of the two equations mentioned in the previous paragraph.

```
import math
import numpy as np
class PositionalEncodingSimple:
    """Positional encoding for our transformer
    written in numpy.
    :param d_model: hidden dimension of the encoder
    :param max_len: maximum length of our positional
        encoder. The encoder can not encode sequence
        length longer than max_len.
    """
    def __init__(self, d_model: int, max_len: int = 100):
        position = np.expand_dims(
            np.arange(max_len), axis=1
        )
        div_term = np.exp(
            np.arange(0, d_model, 2)
            * (-math.log(10000.0) / d_model)
        )
        self.pe = np.zeros((max_len, d_model))
        self.pe[:, 0::2] = np.sin(position * div_term)
        self.pe[:, 1::2] = np.cos(position * div_term)
```

```
    def __call__(self, x: np.ndarray) -> np.ndarray:
        """
        :param x: input to be encoded
            with shape
            `(batch_size, sequence_length, embedding_dim)`
        """
        return self.pe[: x.shape[1]]
```

An intuitive visualization of the positional encoding is to plot out all the elements. In *Figure 9.7* we show that each temporal dimension has its own unique values, which passes the temporal information onto the encoder.

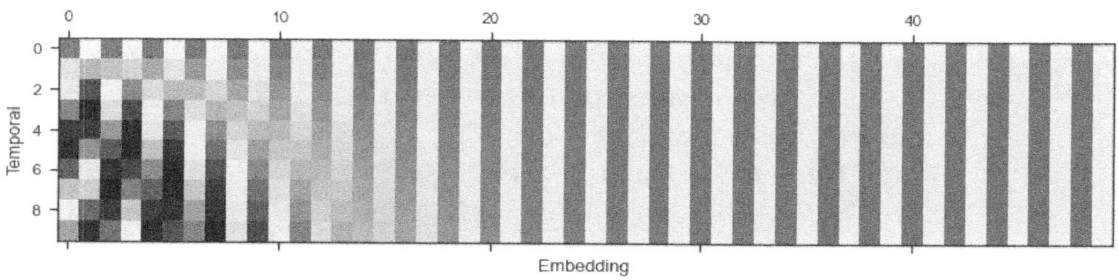

Figure 9.7: Positional encoding written in NumPy

The grayscale/color in *Figure 9.7* indicates the values of the positional encoding. Note that each temporal dimension (each row) is unique. When each temporal dimension is added on top of the embedded input, we annotate the positional information of each time step.

The previous implementation is to help us build up the intuitions. To be able to use positional encoding in a transformer model, we need to implement it as a `torch.nn.Module`. The following is one implementation of the positional encoding using PyTorch which can be plugged into a transformer model. We also included dropout as in the original transformer implementation.

```
from torch import nn
import torch
class PositionalEncoding(nn.Module):
    """Positional encoding to be added to
    input embedding.
    :param d_model: hidden dimension of the encoder
    :param dropout: rate of dropout
    :param max_len: maximum length of our positional
        encoder. The encoder can not encode sequence
        length longer than max_len.
    """
    def __init__(
            self, d_model: int,
```

```
            dropout: float = 0.1,
            max_len: int = 5000,
        ):
        super().__init__()
        self.max_len = max_len
        self.dropout = nn.Dropout(p=dropout)
        position = torch.arange(max_len).unsqueeze(1)
        div_term = torch.exp(
            torch.arange(0, d_model, 2)
            * (-math.log(10000.0) / d_model)
        )
        pe = torch.zeros(max_len, d_model)
        pe[:, 0::2] = torch.sin(position * div_term)
        pe[:, 1::2] = torch.cos(position * div_term)
        self.register_buffer("pe", pe)
    def forward(self, x: torch.Tensor) -> torch.Tensor:
        """
        :param x: input embedded time series,
            shape `[batch_size, seq_len, embedding_dim]`
        """
        history_length = x.size(1)
        x = x + self.pe[: history_length]
        return self.dropout(x)
```

Positional encoding is a broad topic with various approaches. There are different versions of positional encodings. The positional encoding we introduced is static, meaning that it doesn't change based on the data. There are also learned positional encodings during training. Positional encoding is still actively researched. In a recent paper by Zhang et al., *Intriguing Properties of Positional Encoding in Time Series Forecasting*, the authors discovered that using positional encoding based on the similarity of each time step can be beneficial, and they designed new positional encoding to include semantic similarity between each time step (J. Zhang et al. 2024).

In the end, position encoding is an important topic for a transformer model; whether we decide to apply it or not, or use a static version or dynamic version, those are important decisions when developing a transformer model. With this in mind, we can move on to create our transformer model.

Implementation of transformers

Implementation of a transformer is straightforward, as PyTorch already includes most of the modules we need. We will reuse the same dataset from the previous chapter just for the purpose of this tutorial. We assemble a transformer model following the architecture illustrated in *Figure 9.3* and *Figure 9.7*.

To do this, we follow these steps:

1. First, we define a `dataclass` that holds all the hyperparameters of the model. A `dataclass` is very useful when passing the hyperparameters to our model.

```
import dataclasses
@dataclasses.dataclass
class TSTransformerParams:
    """A dataclass that contains all
    the parameters for the transformer model.
    """
    d_model: int = 512
    nhead: int = 8
    num_encoder_layers: int = 6
    dropout: int = 0.1
```

2. We implement a simple transformer for time series, which consists of embedding, encoder, and decoder. We choose the simplest approaches for pedagogical purposes. In the following code block, we use a linear layer as embedding, which *expands* each time series value into a long vector in the embedding space. We skip the details of the attention implementation in an encoder and take `nn.TransformerEncoder` from PyTorch directly. The decoder is also a linear layer.

```
class TSTransformer(nn.Module):
    """Transformer for univariate time series forecasting.
    :param history_length: the length of the input history.
    :param horizon: the number of steps to be forecasted.
    :param transformer_params: all the parameters.
    """
    def __init__(
        self,
        history_length: int,
        horizon: int,
        transformer_params: TSTransformerParams,
    ):
        super().__init__()
        self.transformer_params = transformer_params
        self.history_length = history_length
        self.horizon = horizon
        self.embedding = nn.Linear(
            1, self.transformer_params.d_model
        )
        encoder_layer = nn.TransformerEncoderLayer(
```

```
            d_model=self.transformer_params.d_model,
            nhead=self.transformer_params.nhead,
            batch_first=True,
        )
        self.encoder = nn.TransformerEncoder(
            encoder_layer,
            num_layers=self.transformer_params.num_encoder_layers
        )
        self.reverse_embedding = nn.Linear(
            self.transformer_params.d_model, 1
        )
        self.decoder = nn.Linear(
            self.history_length, self.horizon
        )
    @property
    def transformer_config(self) -> dict:
        """all the param in dict format
        """
        return dataclasses.asdict(
            self.transformer_params
        )
    def forward(self, x: torch.Tensor) -> torch.Tensor:
        """
        :param x: input historical time series,
            shape `[batch_size, seq_len, n_var]`
        """
        x = self.embedding(x)
        encoder_state = self.encoder(x)
        decoder_in = self.reverse_embedding(
            encoder_state
        ).squeeze(-1)
        return self.decoder(decoder_in)
```

3. Similar to our RNN implementations and experiments, we also create a class to hold all the training, validation, and prediction functionalities

```
import lightning as L
class TransformerForecaster(L.LightningModule):
    """Transformer forecasting training, validation,
    and prediction all collected in one class.
    :param transformer: pre-defined transformer model
    """
    def __init__(self, transformer: nn.Module):
        super().__init__()
        self.transformer = transformer
    def configure_optimizers(
            self
        )-> torch.optim.Optimizer:
        optimizer = torch.optim.SGD(
            self.parameters(), lr=1e-3
        )
        return optimizer
    def training_step(
            self, batch: tuple[torch.Tensor],
            batch_idx: int
        ) -> torch.Tensor:
        x, y = batch
        y = y.squeeze(-1).type(self.dtype)
        y_hat = self.transformer(x)
        loss = nn.functional.mse_loss(y_hat, y)
        self.log_dict(
            {"train_loss": loss},
            prog_bar=True
        )
        return loss
    def validation_step(
            self, batch: tuple[torch.Tensor],
```

```
            batch_idx: int
        ) -> torch.Tensor:
        x, y = batch
        y = y.squeeze(-1).type(self.dtype)
        y_hat = self.transformer(x)
        loss = nn.functional.mse_loss(y_hat, y)
        self.log_dict(
            {"val_loss": loss},
            prog_bar=True
        )
        return loss
    def predict_step(
            self, batch: list[torch.Tensor],
            batch_idx: int
        )-> tuple[torch.Tensor]:
        x, y = batch
        y = y.squeeze(-1).type(self.dtype)
        y_hat = self.transformer(x)
        return x, y_hat
    def forward(self, x: torch.Tensor) -> tuple[torch.Tensor]:
        return x, self.transformer(x)
```

4. In this example, we set the history input length to be `100` to be able to capture some trends in the input data and our objective is to forecast one step in the future, i.e., the forecast horizon is set to `1`

```
history_length_1_step = 100
horizon_1_step = 1
```

5. The hyperparameters can be set using the `dataclass` `TSTransformerParams`

```
ts_transformer_params_1_step = TSTransformerParams(
    d_model=192, nhead=6, num_encoder_layers=1
)
ts_transformer_params_1_step
TSTransformerParams(
    d_model=192, nhead=6, num_encoder_layers=1, dropout=0.1
)
```

6. With the hyperparameters defined, we pass them on to our transformer `TSTransformer`. The layers are printed out to read about the details inside the encoder layer.

```
ts_transformer_1_step = TSTransformer(
    history_length=history_length_1_step,
    horizon=horizon_1_step,
    transformer_params=ts_transformer_params_1_step,
)
ts_transformer_1_step
TSTransformer(
  (embedding): Linear(in_features=1, out_features=192, bias=True)
  (encoder): TransformerEncoder(
    (layers): ModuleList(
      (0): TransformerEncoderLayer(
        (self_attn): MultiheadAttention(
          (out_proj): NonDynamicallyQuantizableLinear(
            in_features=192, out_features=192, bias=True)
        )
        (linear1): Linear(in_features=192, out_features=2048,
                          bias=True)
        (dropout): Dropout(p=0.1, inplace=False)
        (linear2): Linear(in_features=2048, out_features=192,
                          bias=True)
        (norm1): LayerNorm((192,), eps=1e-05,
                           elementwise_affine=True)
        (norm2): LayerNorm((192,), eps=1e-05,
                           elementwise_affine=True)
        (dropout1): Dropout(p=0.1, inplace=False)
        (dropout2): Dropout(p=0.1, inplace=False)
      )
    )
  )
  (reverse_embedding): Linear(in_features=192, out_features=1,
                              bias=True)
  (decoder): Linear(in_features=100, out_features=1, bias=True)
)
```

7. The transformer can be inserted into the forecaster `TransformerForecaster` and can be trained exactly the same way as we have shown in the previous chapter

```
transformer_forecaster_1_step = TransformerForecaster(
    transformer=ts_transformer_1_step
)
```

The M5 competition

We have discussed how transformers work as well as the Python code to build your own transformer forecaster. The code provided in the previous section is fully executable and can be readily used in your project. We did not show the training process in the previous section because it will be exactly the same as in *Chapter 8*. Instead, we fulfill our promise and demonstrate how to use PyTorch-based forecasting packages. In this section, we use two models, **VanillaTransformer** and **iTransformer** from `https://nixtlaverse.nixtla.io/neuralforecast` package which is written in Python, based on PyTorch.

In this example, we use the M5 competition dataset. The M5 competition is the fifth M competition focusing on sales forecasting based on the sales of Walmart (more details about the competition can be found here `https://www.sciencedirect.com/science/article/pii/S0169207021001187`) (Makridakis, Spiliotis, and Assimakopoulos 2022). We simplify the dataset by selecting only California stores and aggregate the sales to the state level, thus we only forecast one single time series: the daily sales of all Walmart stores in California. The following is a small sample of the data.

ds	y	
0	2011-01-29	14195
1	2011-01-30	13805
2	2011-01-31	10108
3	2011-02-01	11047
4	2011-02-02	9925
5	2011-02-03	11322
6	2011-02-04	12251
7	2011-02-05	16610
8	2011-02-06	14696
9	2011-02-07	11822

Table 9.3: Sample of the dataset

In the following graphs, the upper panel shows the full time series of sales in all California stores in Walmart. The middle panel shows the details of the year 2015. We observe the cyclic features of the time series. The bottom panel plots the autocorrelation of the time series. As expected, there is a 7-day cycle in sales.

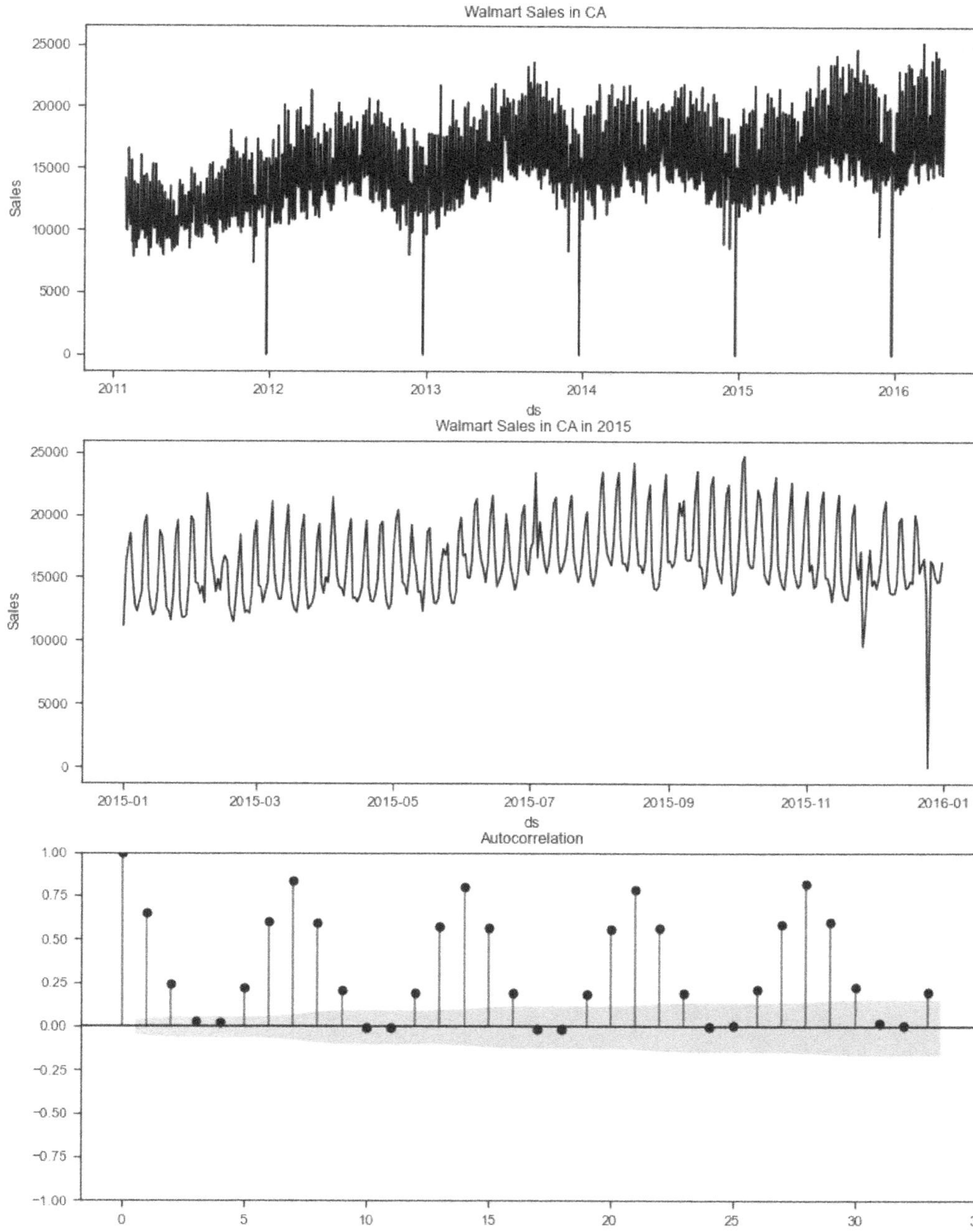

Figure 9.8: Walmart sales data extracted from the M5 dataset

Our objective is to build a transformer model that forecasts the sales of the next 3 days. To do this, we will follow these steps:

1. To be able to justify the performance of our model, we split the data at 2016-01-01 using the code:

```
train_test_split_date = "2016-01-01"
df_train = df[df.ds<=train_test_split_date]
df_test = df[df.ds>train_test_split_date]
horizon = 3
```

2. Before we start building the transformer forecaster, we construct a baseline forecaster. Here we choose an ARIMA model as we have clear cyclic behavior in the data. We stick to the `Nixtlaverse` and use their `statsforecast` package.

```
from statsforecast import StatsForecast
from statsforecast.models import AutoARIMA
sf = StatsForecast(
    models = [AutoARIMA(
        seasonal=False
    )],
    freq = 'd'
)
```

3. We also create a function to help us evaluate the tests. We perform the test with a fixed horizon, but the input time series is sliding into the future with a step size of 1 day.

```
def forecast_test(
    forecaster_pred: callable,
    df_train: pd.DataFrame,
    df_test: pd.DataFrame,
) -> pd.DataFrame:
    """Generate forecasts for tests
    We gradually increase the historical information
    by appending part of the test data and
    predict based on the new historical information.
    :param forecaster_pred: forecasting function/method
    :param df_train: train dataframe
    :param df_test: test dataframe
    """
    dfs_pred = []
    for i in range(len(df_test)):
        logger.debug(f"Prediction Step: {i}")
        df_pred_input_i = pd.concat(
            [df_train, df_test[:i]]
```

```
            ).reset_index(drop=True)
            df_pred_output_i = forecaster_pred(df_pred_input_i)
            df_pred_output_i["step"] = i
            dfs_pred.append(
                df_pred_output_i
            )
        df_y_hat = pd.concat(dfs_pred)
        return df_y_hat
```

4. We obtain all the baseline model forecasts by calling the forecast method of a `StatsForecast` model:

```
df_y_hat_arima = forecast_test(
    lambda x: sf.forecast(df=x, h=horizon),
    df_train=df_train,
    df_test=df_test
)
```

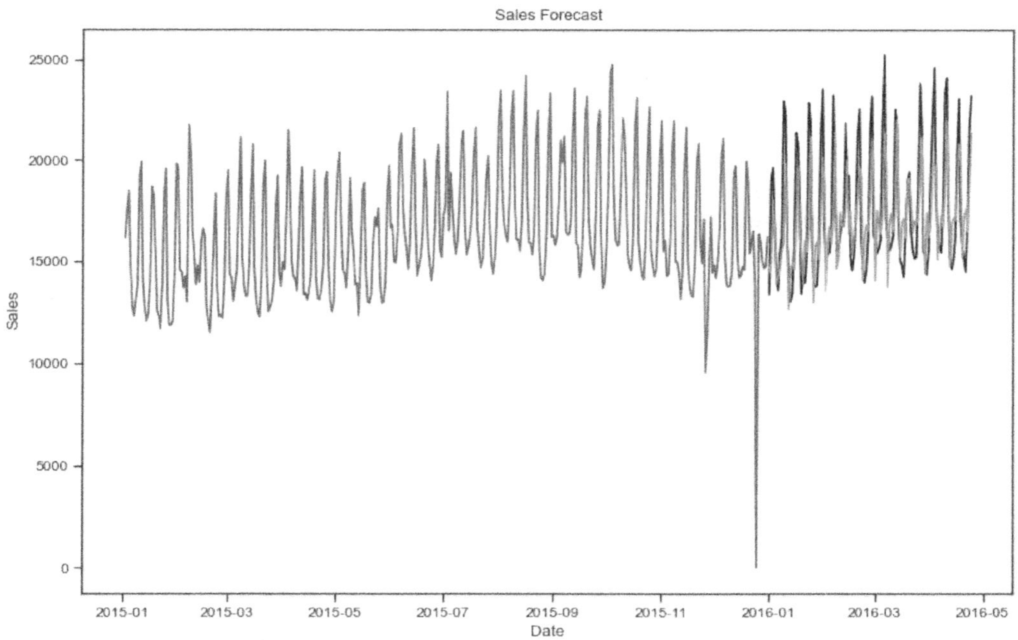

Figure 9.9: Forecast by ARIMA model

The solid lines are the true values, and the dashed line is the forecast, but only taking the 1-step-ahead forecast.

The baseline forecasts from ARIMA are not bad at all. It is expected because sales data is indeed seasonal and ARIMA is good at capturing seasonalities.

With this baseline ready, we will start building our transformer forecaster. It is extremely easy to create a transformer forecaster in Nixtla's `neuralforecast`. In fact, we are creating two transformer models, `VanillaTransformer` and `iTransformer`. `iTransformer` is just like a vanilla transformer, but instead of embedding each time step, it transposes the data and embeds each time series independently (*iTransformer: Inverted Transformers Are Effective for Time Series Forecasting* by Yong Liu et al., available at `https://arxiv.org/abs/2310.06625`) (Liu et al. 2023).

```
from neuralforecast import NeuralForecast
from neuralforecast.models import VanillaTransformer, iTransformer
models = [
    VanillaTransformer(
        hidden_size=128,
        n_head=4,
        learning_rate=0.0001,
        scaler_type="robust",
        max_steps=500,
        batch_size=32,
        windows_batch_size=512,
        random_seed=16,
        input_size=30,
        step_size=3,
        h=horizon,
    ),
    iTransformer(
        input_size=30,
        h=horizon,
        max_steps=50,
        n_series=1,
    )
]
nf = NeuralForecast(models=models, freq="d")
```

Model training is a one-liner: calling the `fit` method. Behind the scenes, what happens is that the `fit` method calls the `Trainer` from `lightning` package, which we have been using to build the forecasters in the previous section as well as the previous chapter. Nixtla's `neuralforecast` is handy because we don't need to manually create the trainer and perform the training.

```
nf.fit(df=df_train)
```

We generate the forecasts for the whole test range similar to what we have calculated for the baseline model.

```
df_y_hat_transformer = forecast_test(
    nf.predict, df_train=df_train, df_test=df_test
)
```

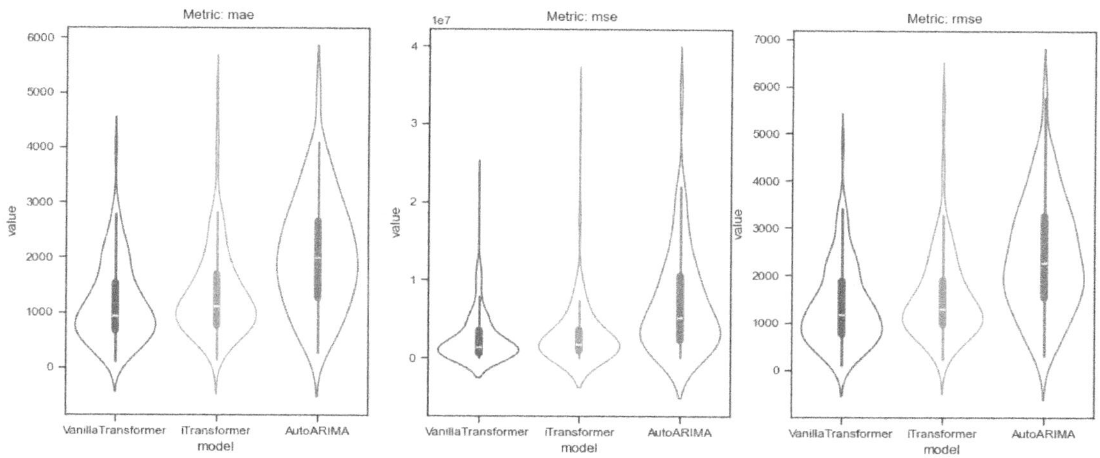

Figure 9.10: Comparing the metrics of transformer models and our ARIMA baseline

Transformers are performing better than our baseline, judging by the distribution of the metrics on the full test dataset.

In *Figure 9.10*, we observe that our transformers are better than the baseline. What is even more interesting is that we only trained our transformer model once and only using the data before 2016, while we have retrained our ARIMA model for each evaluation. In principle, the ARIMA baseline has utilized more data than our transformer models. But it still falls behind the transformers. The reason that transformers are better in this example is that transformers forecast better for more distant future steps:

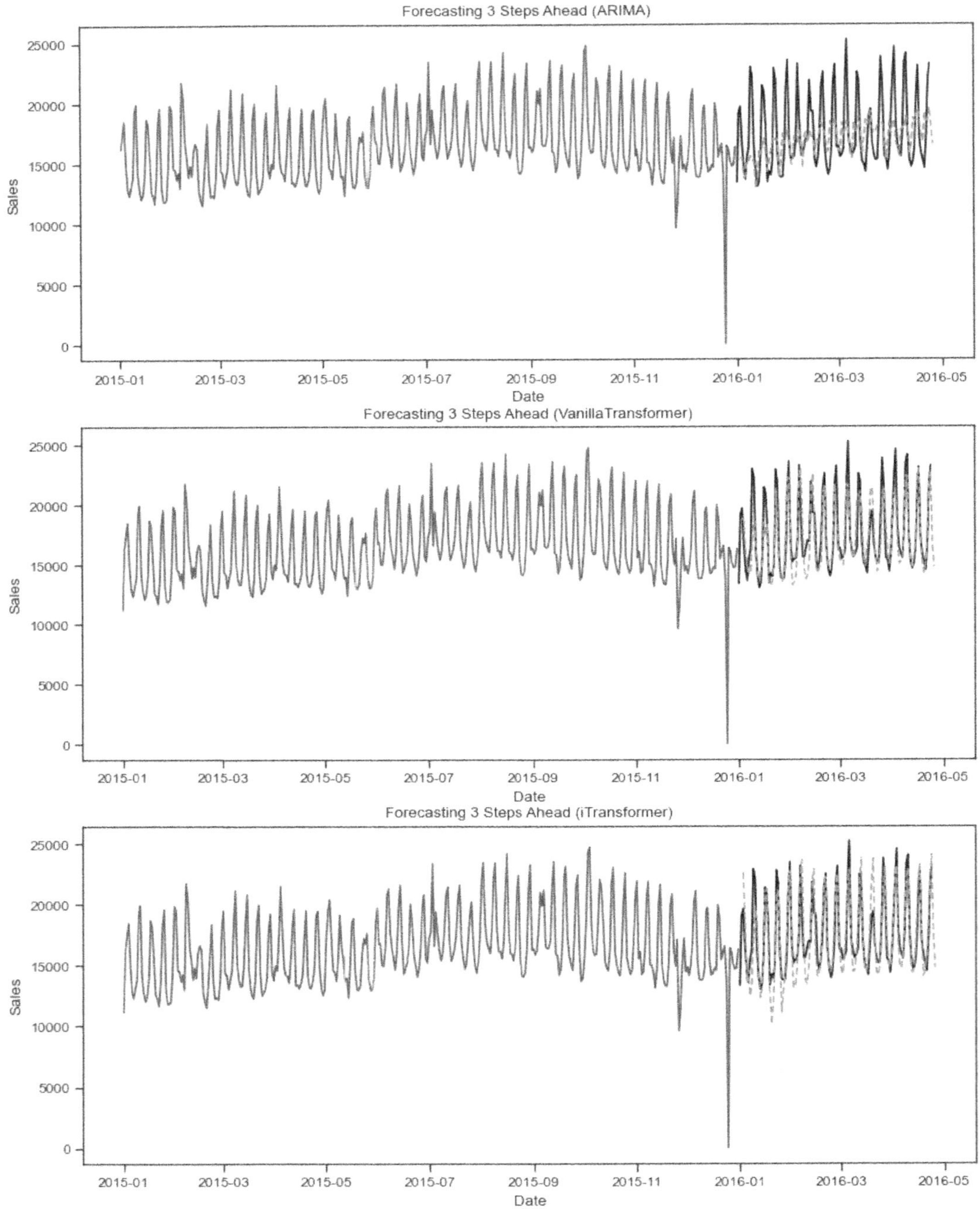

Figure 9.11: Comparing the 3 steps ahead forecasts of all three models

As you can see, ARIMA is visually much worse than the transformers.

We will stop here and won't spend more time calculating the metrics as we have done it so many times in the previous chapters. Using this section, we wanted to show you that using open source packages such as `nixtla` saved us a lot of work in experiments. In fact, this is a key message we want to convey: we need to understand the models thoroughly, but we can also use open source packages to speed up our experimentation without implementing the details of the model.

Summary

In this chapter, we have investigated the theories of transformers, built our own transformer using PyTorch, and applied transformer models to the M5 competition dataset. Transformer is a very active area of research in forecasting, and we get new transformer-based models every now and then. However, you can easily catch up with the new research with a good understanding of the vanilla transformer. For example, we briefly mentioned a model called `iTransformer` whose architecture is almost the same as the vanilla transformer. As long as we have built the big picture of the architecture and understand how the attention mechanism works, `iTransformer` is intuitive to understand. With both recurrent models and transformers explained and experimented, we will discuss more neural networks that can be used as forecasters in the upcoming chapter.

In the next chapter, we will show you even more neural network models. We will glance over basic neural networks such as MLPs as well as advanced models such as **Kolmogorov-Arnold Networks** (**KAN**).

Recommended reading

We discussed many different topics about transformers, but we could not cover all the details. The following list of papers and blog posts provides more details.

- Amatriain, Xavier. 2023. "Transformer Models: An Introduction and Catalog." *arXiv [Cs.CL]*. `https://doi.org/10.48550/ARXIV.2302.07730`.
- Liu, Yong, Tengge Hu, Haoran Zhang, Haixu Wu, Shiyu Wang, Lintao Ma, and Mingsheng Long. 2023. "ITransformer: Inverted Transformers Are Effective for Time Series Forecasting." *arXiv [Cs.LG]*. `http://arxiv.org/abs/2310.06625`.
- Makridakis, Spyros, Evangelos Spiliotis, and Vassilios Assimakopoulos. 2022. "The M5 Competition: Background, Organization, and Implementation." *International Journal of Forecasting* 38 (4): 1325–36. `https://doi.org/10.1016/j.ijforecast.2021.07.007`.
- Vaswani, Ashish, Noam Shazeer, Niki Parmar, Jakob Uszkoreit, Llion Jones, Aidan N Gomez, Lukasz Kaiser, and Illia Polosukhin. 2017. "Attention Is All You Need." 2017. `http://arxiv.org/abs/1706.03762`.
- Zhang, Aston, Zachary C. Lipton, Mu Li, and Alexander J. Smola. 2023. Dive into Deep Learning. Cambridge, England: Cambridge University Press. `https://d2l.ai/chapter_recurrent-neural-networks/bptt.html`.
- Zhang, Jianqi, Jingyao Wang, Wenwen Qiang, Fanjiang Xu, Changwen Zheng, Fuchun Sun, and Hui Xiong. 2024. "Intriguing Properties of Positional Encoding in Time Series Forecasting." *arXiv [Cs.AI]*. http://arxiv.org/abs/2404.10337.

10

Other Neural Structures

"We need diversity of thought in the world to face the new challenges." — Tim Berners-Lee

In recent years there has been a Cambrian explosion of new models created, each claiming to be **State of the Art** (**SOTA**). As you may know by now, there is no single modeling approach that is perfect (`https://openreview.net/pdf?id=wEc1mgAjU-`), and with so many different approaches it is difficult to stay on top of them all.

While it can take a substantial effort to keep afloat of all the models, we can classify most networks into families, which gives us an intuitive starting point for understanding how a network may attempt to model our data. This chapter discusses a number of neural network families, and briefly looks at a selection of models from these families that we cannot cover in full detail. We will apply our selected models on data from the M5 competition (all this code will be available in the accompanying notebook in the GitHub repository).

In this chapter, we will:

- Setting up our M5 dataset
- Exploring some neural network families
- Multilayer perceptrons
- Convolutional neural networks
- Graph neural networks
- Kolmogorov-Arnold networks
- Foundation models

Refer to color images on the Graphic bundle or the ebook.

Setting up our M5 dataset

We use the M5 dataset in this chapter. For more background about this dataset, you can refer to the M5 competition in *Chapter 1* and the exploration of the dataset in *Chapter 2*. In this section, we will load the dataset and perform some basic analysis.

1. **Configuring our dataset:** First, the related imports and configurations. We place the dataset inside a folder called `data` alongside the script/notebook where our code runs.

```
import pandas as pd
import numpy as np
import matplotlib.pyplot as plt
import seaborn as sns
from datetime import datetime
from pathlib import Path
custom_palette = [
    "#000000", "#0072B2", "#D55E00",
    "#009E73","#CC79A7", "#56B4E9","#E69F00"
]
class CFG:
    data_folder = Path.cwd().parent / "data"
    img_dim1 = 12
    img_dim2 = 6
    fontsize = 18
plt.rcParams.update({'figure.figsize': (CFG.img_dim1,CFG.img_dim2)})
```

2. Loading and preprocessing the data: To explore the dataset, we need to load it and convert it into a pandas `dataframe`. We build a function to achieve this.

```
def load_m5_subset(data_folder: Path) -> pd.DataFrame:
    # Construct filepath
    filepath = data_folder / 'M5_t20_ABC.csv'
    # Verify file exists
    if not filepath.exists():
        raise FileNotFoundError(
            f"Data file not found at {filepath}")
    # Read data
    df = pd.read_csv(filepath, index_col=0)
    # Convert date to datetime
    df['date'] = pd.to_datetime(df['date'])
    # Remove last 28 days of each series
    df_sorted = df.sort_values(['item_id', 'date'])
    def remove_tail_values(group):
        return group.iloc[:-28]
```

```
    df = df_sorted.groupby(
        'item_id', group_keys=False
    ).apply(remove_tail_values).reset_index(drop=True)
    # Add time-based features
    df['year'] = df['date'].dt.year
    df['month'] = df['date'].dt.month
    df['dayofweek'] = df['date'].dt.dayofweek
    df['quarter'] = df['date'].dt.quarter
    df['week_of_year'] = df['date'].dt.isocalendar().week
    return df
```

3. **Summarizing the dataset:** We also need to get some basic statistics of the dataset. We use the following function to compute the summary statistics of the M5 competition dataset.

```
def analyze_dataset(df: pd.DataFrame) -> dict:
    # Get ABC distribution by unique products
    abc_dist = df.groupby(
        'item_id'
    )['ABC_class'].first().value_counts().to_dict()
    analysis = {
        'n_series': df['item_id'].nunique(),
        'n_stores': df['state_id'].nunique(),
        'date_range': (df['date'].min(), df['date'].max()),
        'total_sales': df['sold'].sum(),
        'mean_price': df['sell_price'].mean(),
        'abc_distribution': abc_dist,
        'sales_by_class': df.groupby(
            'ABC_class')['sold'].sum().to_dict(),
        'avg_price_by_class': df.groupby(
            'ABC_class')['sell_price'].mean().to_dict()
    }
    return analysis
```

4. Visualizing patterns: And finally, we need to build a function to plot out some examples of the sales data:

```
def plot_sales_analysis(series: pd.DataFrame, figsize=(15, 10)):
    fig, axes = plt.subplots(2, 1, figsize=figsize)
    # Sales over time
    axes[0].plot(
        series['date'], series['sold'],
        color = custom_palette[1], alpha=0.7)
    axes[0].set_title(
```

```
        f"Sales Over Time for {series['item_id'].iloc[0]}")
    axes[0].set_xlabel('Date')
    axes[0].set_ylabel('Units Sold')
    # Sales by day of week
    sns.boxplot(
        data=series, x='dayofweek', y='sold',
        ax=axes[1], color = custom_palette[2]
    )
    axes[1].set_title('Sales Distribution by Day of Week')
    axes[1].set_xlabel('Day of Week')
    axes[1].set_ylabel('Units Sold')
    plt.tight_layout()
    plt.show()
    # Print summary statistics
    print(
        "\nSummary Statistics: "
        f"{series['sold'].describe()}"
    )
```

5. **Modeling a subset of the data:** To be able to investigate multiple models quickly, we create a small subset of time series taken from the M5 competition, from `locn_id (location)` CA. Let's quickly load this data:

```
df = load_m5_subset(CFG.data_folder)
# Analyze dataset
analysis = analyze_dataset(df)
analysis
{'n_series': 60,
 'n_stores': 1,
 'date_range': (Timestamp('2011-01-29 00:00:00'),
  Timestamp('2016-05-22 00:00:00')),
 'total_sales': 1815356,
 'mean_price': 2.719599776747381,
 'abc_distribution': {'C': 20, 'B': 20, 'A': 20},
 'sales_by_class': {'A': 1456556, 'B': 271653, 'C': 87147},
 'avg_price_by_class': {'A': 1.7382524471921692,
  'B': 2.8056069036579085,
  'C': 3.614939979392066}}
```

A quick analysis of our data shows we have 60 unique items, tracked over 5.5 years (January 2011 until June 2016). The items are evenly split across the ABC classification, with 20 products in each group. `'A'` items are high-volume products that make up about 80% of total sales (`1.46 million` out of `1.82 million` units). `'B'` items are medium-volume and account for around

15% (272,000 units), while 'C' items are low-volume and make up the remaining 5% (87,000 units). This kind of difference in sales volume is normal, but in this case, we created the distribution on purpose. In the original data, you'd see something like: A = 15%, B = 25%, and C = 60% of items. We can also see a clear trend between sales volume and pricing — high-volume 'A' items tend to be cheaper (averaging $1.74), while low-volume 'C' items are more expensive (averaging $3.62).

6. **Exploring variation in different time series:** Let's plot out some series to investigate some of the patterns in our dataset:

```
# Select series to plot
series_to_plot = [
    'FOODS_3_090', 'FOODS_3_714',
    'HOUSEHOLD_2_437', 'HOUSEHOLD_1_448'
]
# Create figure with subplots
fig, axes = plt.subplots(2, 2, figsize=(15, 10))
fig.tight_layout(pad=3.0)
# Flatten axes for easier iteration
axes = axes.flatten()
for idx, (series_id, ax) in enumerate(zip(
    series_to_plot, axes)
):
    # Get single series
    series = df[df['item_id'] == series_id].copy()
    series = series.sort_values('date')
    # Get last 36 months
    last_date = series['date'].max()
    start_date = last_date - pd.DateOffset(months=36)
    series = series[series['date'] >= start_date]
    # Plot
    ax.plot(series['date'], series['sold'],
            color=custom_palette[idx],
            alpha=0.7)
    ax.set_title(f"Sales for {series_id}", fontsize=14)
    ax.set_xlabel('Date', fontsize=16)
    ax.set_ylabel('Units Sold', fontsize=16)
    ax.tick_params(axis='x', rotation=45)
plt.subplots_adjust(hspace=0.4, wspace=0.2)
plt.show()
```

The code results in the following plot:

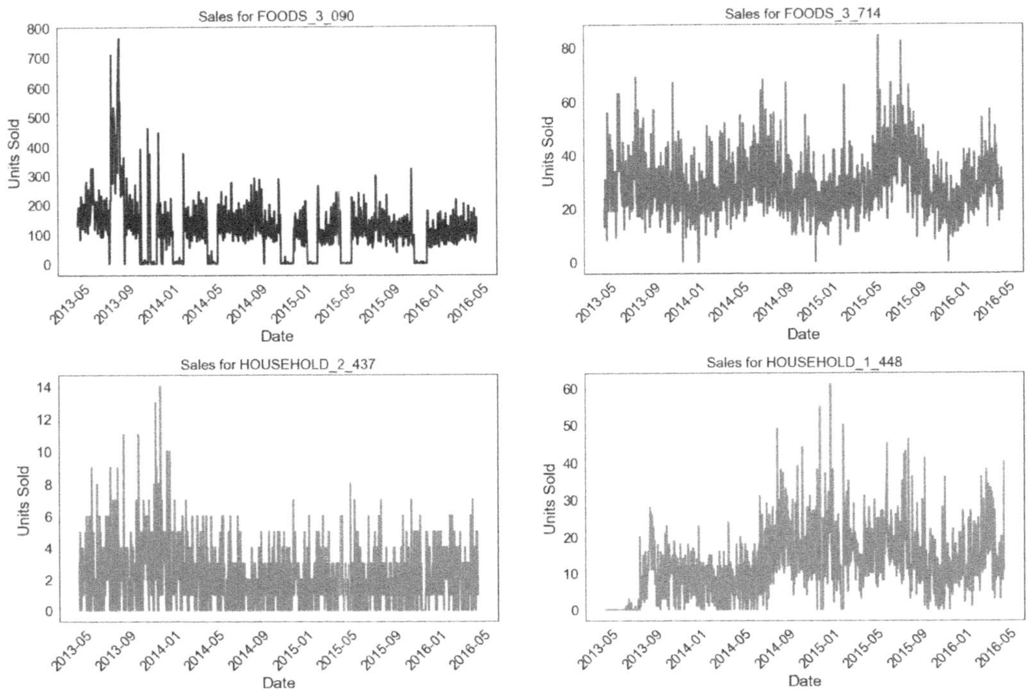

Figure 10.1: Investing patterns using plots

Looking at a few individual time series, we can see some variation. The first item (**FOOD_3_090**), which we have seen earlier, has a high magnitude of sales but has several data drop offs. There are several such instances in the dataset, and it is unlikely that we can model these sets perfectly without significant data wrangling and feature engineering. The rest of the time series mostly show straightforward patterns, ranging from high to low volume selling items.

As our goal here is to discuss and demonstrate modeling with different architectures, we are not going to do too much data wrangling but simply model all items in the subset. We will first need to set up our data quickly for modeling with `neuralforecast`. We will attempt to model the last 90 days of each series to get an idea of how they perform on a typical demand dataset. `Neuralforecast` requires a specific format of the dataset. In our Jupyter notebook for *Chapter 10* on GitHub, you will find the details of how we set up the datasets. Just be aware that this could be done better with more hyperparameter tuning, and different datasets will see different results.

Exploring some neural network families

While there is an ever-increasing number of neural architectural designs, most designs are variations stemming from general structures, which we can refer to as a family. Having some understanding of each neural network family can be useful when encountering new models. These can largely be broken down into: **Multi-Layer Perceptrons** (**MLPs**, which we have already discussed), **Convolutional**

Neural Networks (**CNN**), **Recurrent Neural Networks** (**RNN**), **Gated Neural Networks**, **Transformers**, **Foundation Models** (for example LLMs, which Karpathy suggests are best called **Autoregressive Transformers**), and the latest family at time of writing, **Kolmogorov-Arnold Networks** (**KAN**). There are also hybridized architectures that mix together elements of different networks. In this chapter we will use Nixtla's Neuralforecast, which covers many neural network families and the most popular network designs within them, as you can see in the table.

The full table of models and their capabilities (including univariate/multivariate support and forecast type) can be found in the Nixtla's NeuralForecast documentation. In the following table, we show each model's family and exogenous variable support, as this is particularly relevant when deciding which model to use with your data.

Model	**Family**	**Exogenous Support**
Autoformer	Transformer	F
BiTCN	CNN	F/H/S
DeepAR	RNN	F/S
DeepNPTS	MLP	F/H/S
DilatedRNN	RNN	F/H/S
FEDformer	Transformer	F
GRU	RNN	F/H/S
HINT	Any	F/H / S
Informer	Transformer	F
iTransformer	Transformer	–
KAN	KAN	F/H/S
LSTM	RNN	F/H/S
MLP	MLP	F/H/S
MLPMultivariate	MLP	F/H/S
NBEATS	MLP	–
NBEATSx	MLP	F/H/S
NHITS	MLP	F/H/S
NLinear	MLP	–
PatchTST	Transformer	–
RMoK	KAN	–
RNN	RNN	F/H / S
SOFTS	MLP	–
StemGNN	GNN	–

Model	Family	Exogenous Support
TCN	CNN	F/H/S
TFT	Transformer	F/H/S
TiDE	MLP	F/H/S
TimeMixer	MLP	-
TimeLLM	LLM	-
TimesNet	CNN	F
TimeXer	Transformer	H/S
TSMixer	MLP	-
TSMixerx	MLP	F/H/S
VanillaTransformer	Transformer	F
XLinear	MLP	F/H/S
xLSTM	mLSTM	F/H/S

Table 10.1: Models in neuralForecast and their capabilities; adapted from `https://nixtlaverse.nixtla.io/neuralforecast/docs/capabilities/overview.html`

In this table:

- **Family** refers to the main neural architecture type.
- F refers to **Future**, H to **Historical**, and S to **Static** variable support; – indicates no exogenous support.
- **HINT** is a modular framework that can combine different neural architectures with specialized mixture probability and hierarchical reconciliation strategies.

Multilayer perceptron

MLPs are generally considered a misnomer. As we discussed earlier this is more to do with the origin of neural networks which initially used activation functions like sigmoid and tanh before alternatives became common.

Generally, MLP refers to networks with that **Feed-Forward Networks** (**FFNs**). So, by this standard, CNNs are technically MLPs, although we consider them a separate family. As we used MLPs/FFNs to discuss the fundamentals of *deep learning* in *Chapters 4 to 6*, we will not be discussing them in great detail here.

You usually see MLPs/FFNs used directly, such as the MLP and MLPMultivariate models in `neuralforecast`. However, FFNs can be used in various ways, i.e., restricted outputs such as DeepNPTS, or to combine information from other *blocks* of a network together, i.e., SOFTS.

In *Table 10.1*, among the many models we presented—MLP, MLPMultivariate, DeepNPTS, and SOFTS—are considered true MLP-based models. Let's look at these in depth here:

- **MLP:** This is a straightforward MLP model using `tanh` or ReLU activation functions.
- **MLPMultivariate** : MLPMultivariate is MLP adapted for multivariate modeling.
- **DeepNPTS:** This is a non-parametric (distribution free) approach which uses the **Neural Network** (**NN**) to learn sampling probabilities over the observed inputs. It makes predictions by sampling from the past values, not creating new values, which can be useful.
- **Series-cOre Fused Time Series forecaster** (**SOFTS**): Models multivariate time-series, by attempting to balance signals in independent channels while accounting for cross-channel interactions through a centralized architecture, instead of using an attention mechanism discussed in *Chapter 9*, that is compares every channel with every other channel. SOFTS aggregates information into a global *core* representation, then fuses this to individual channel data. In theory this approach might help in forecasting data in panels like traffic networks, energy grids, or retail sales across multiple stores, where attention-based models might be computationally expensive or prone to over-fitting noisy channel relationships.

While the previous architectures directly use MLPs/FFNs networks, there are several popular methods that are classified under the MLP family, due to these structures using MLPs as the base blocks for underpinning these networks. Several of these networks are hybrids that are specialized for time-series forecasting.

- **Neural Basis Expansion Analysis for Interpretable Time Series** (**NBEATS**)/**NBEATSx:** This is a specialized architecture that uses MLP blocks within a broader architectural framework. It combines backward and forward residual links with specialized blocks for trend and seasonality.
- **Neural Hierarchical Interpolation for Time Series** (**NHITS**): Similar to NBEATS, it's a hybrid architecture that extends NBEATS with interpolation and pooling.
- **Time-series Dense Encoder** (**TiDE**): While using MLP components, it's more accurately described as a transformer-inspired architecture without the attention mechanism. It uses time embeddings and specialized layers.
- **TimeMixer/TSMixer/TSMixerx:** These are actually MLP-Mixer inspired architectures, adapting the transformer-like mixing concept but using MLPs instead of attention. They're closer to transformer variants than traditional MLPs.
- **NLinear:** This is a linear model with neural components, not a traditional MLP.

We will cover two of these in the following sections.

N-BEATS

N-BEATS was perhaps the first pure deep learning approach to surpass statistical modeling approaches, as was famously established on M4 competition data in the original paper Oreshkin et al. (2020). It outperformed the competition winning entry (an ES-RNN) by 3%. N-BEATS achieved this by using neural basis expansion to augment target (y) data, by projecting it with basis functions, which allowed the model to learn non-linear relationships. You may wonder what basis functions are, and why they help. Let's use a classic example - if we were attempting to predict ice cream sales based on temperature, a simple model might appear linear:

```
# Generate example data
temp = np.array([14, 16, 18, 20, 22, 24, 26, 28, 30]).reshape((-1, 1))
sales = np.array([50, 55, 60, 80, 110, 160, 180, 190, 200])
# Linear regression
from sklearn.linear_model import LinearRegression
model = LinearRegression()
model.fit(temp, sales)
plt.scatter(temp, sales, color=custom_palette[0])
plt.plot(temp, model.predict(temp), color=custom_palette[1])
plt.xlabel('Temperature')
plt.ylabel('Sales')
plt.tight_layout(pad=3.0)
plt.show()
```

The result of this code is:

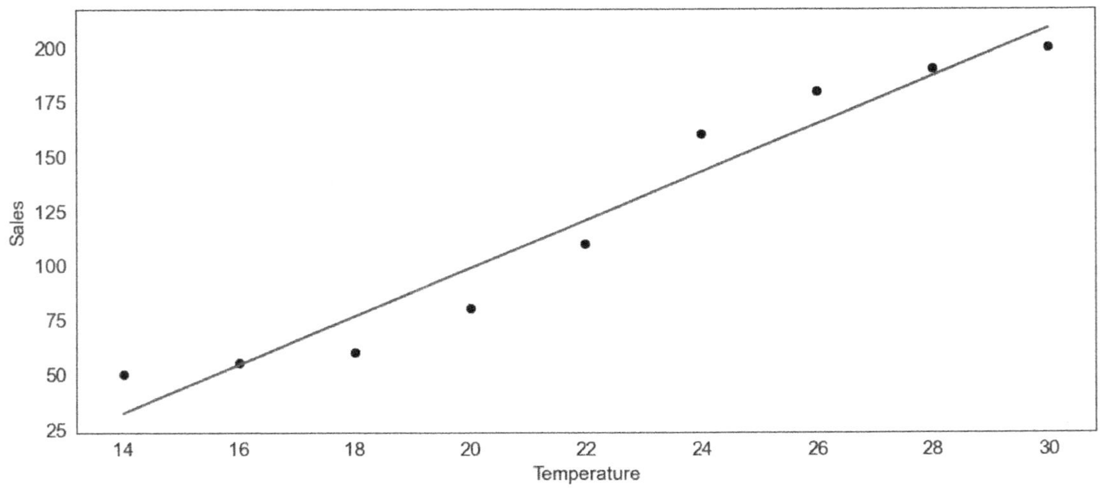

Figure 10.2: Relationship between Sales and Temperature

Despite the linear function performing OK, we can see that the relationship isn't so simple. **Sales** increase slowly at first, then rapidly as **Temperature** rises. Clearly a *linear* model is too simple, a non-linear model would be better. This is where *basis expansion* comes into play, by applying polynomial terms up to the third degree, we can expand our *basis*, allowing our model to learn both quadratic and cubic non-linear relationships.

```
# Polynomial basis expansion
from sklearn.preprocessing import PolynomialFeatures
poly = PolynomialFeatures(degree=3, include_bias=False)
temp_poly = poly.fit_transform(temp)
model = LinearRegression()
model.fit(temp_poly, sales)
plt.scatter(temp, sales, color=custom_palette[0])
plt.plot(temp, model.predict(temp_poly), color=custom_palette[1])
plt.xlabel('Temperature')
plt.ylabel('Sales')
plt.show()
```

The result of this code:

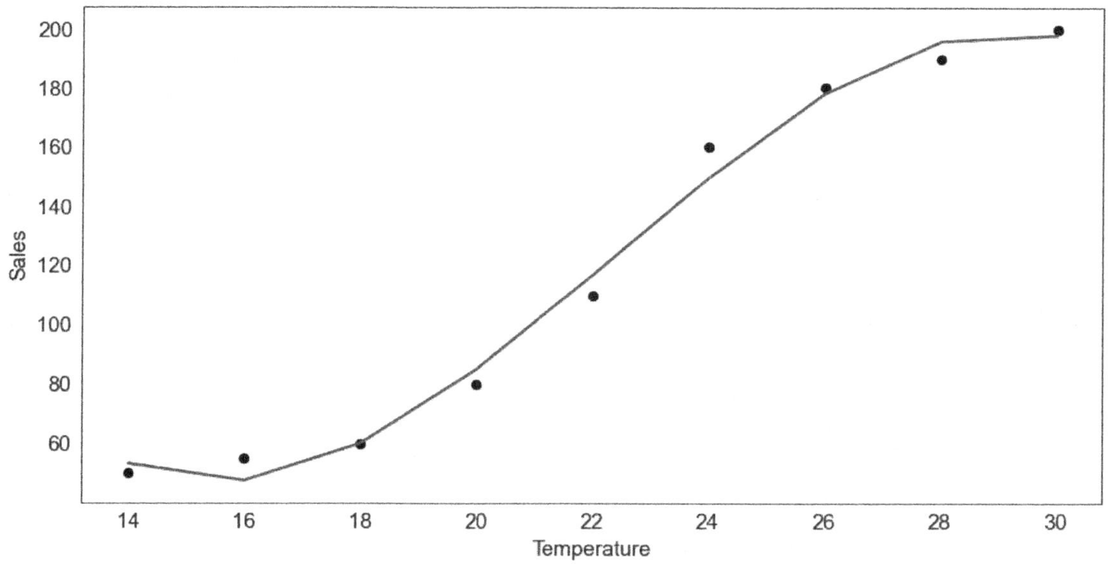

Figure 10.3: Applying basis expansion on the example dataset

There are many different basis expansions, such as logarithms, piecewise-linear splines, binning, natural cubic splines, and squares, but these are restrictive. We can create customized basis expansions. As the precise basis expansion is unknown to us, it is often better to learn it from our data. This is exactly what *neural basis expansion* in N-BEATS attempts to do: automatically select and optimize for the best basis expansion.

The precise architecture of N-BEATS is as follows:

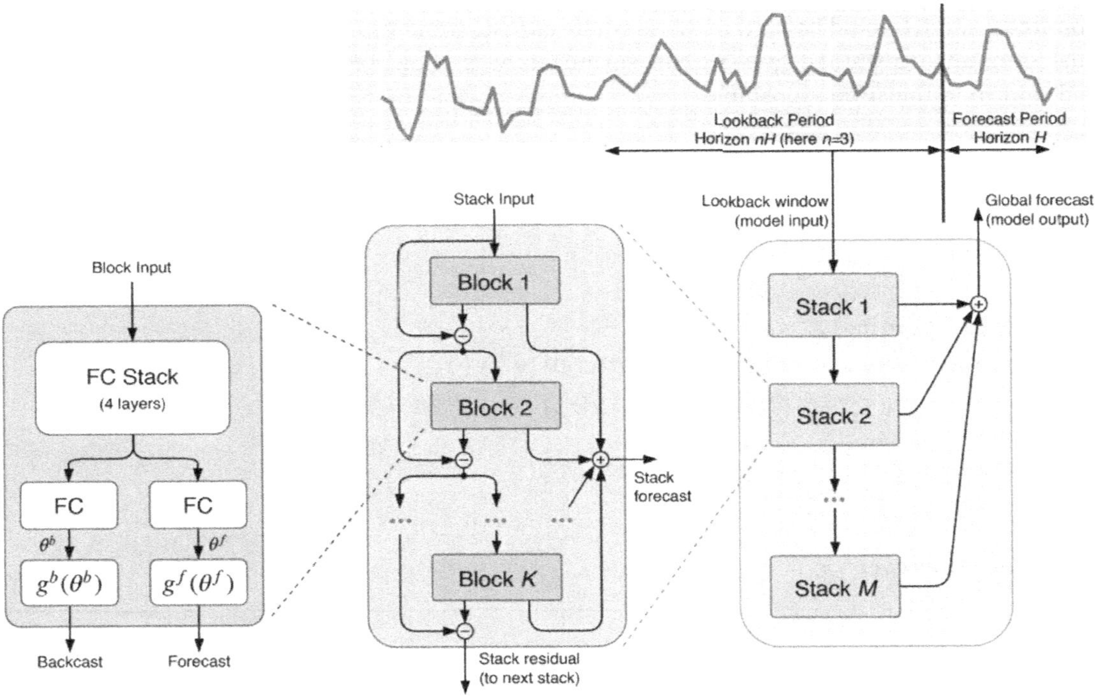

Figure 10.4: The N-BEATS architecture

Figure 10.4 is inspired from B. N. Oreshkin, D. Carpov, N. Chapados, and Y. Bengio, "*N-BEATS: Neural basis expansion analysis for interpretable time series forecasting*," arXiv [cs.LG], May 24, 2019. Available: `http://arxiv.org/abs/1905.10437`. [Accessed: Jan. 18, 2025]

This diagram shows that N-BEATS processes time series data by first splitting it into lookback and forecast periods. The length of the lookback window is important for capturing relevant patterns; originally, the authors of N-BEATS used 2x forecast horizon for generic architecture, and up to 7x for interpretable versions (see Oreshkin et al., 2020). However, the optimal lookback length depends on your data's seasonality structure. We represent this input sequence of historical values as $x = [x_1, \ldots, x_s]$.

The third rectangle shows that N-BEATS employs a hierarchical structure of stacks (second rectangle), each stack specializing in modeling specific time series components like trend or seasonality. While it is useful to think like this, typically the stacks operate on the same input (lookback period), with the outputs of each stack being summed to produce a final forecast. The strength of this approach comes from allowing each stack to learn different aspects of an underlying data generating process.

Each block (detailed in the first rectangle) implements a novel dual-output architecture. The network splits into two branches that simultaneously produce:

- A backcast that attempts to reconstruct the input data,
- A partial forecast of future values, which is mathematically represented as:

$$\text{Block}(x) = \big(b(x), f(x)\big),$$

where $b(x)$ is the backcast and $f(x)$ is the forecast.

Within each block, a fully connected neural network with four layers learns expansion coefficients θ. These are then combined with basis functions $g(t)$ through neural basis expansion. For interpretable trend modeling:

$$g_{\text{trend}}(t) = [1, t, t^2, \ldots, t^p]$$

And for seasonal patterns with period T:

$$g_{\text{seasonal}}(t) = [\sin(2\pi t/T), \cos(2\pi t/T), \ldots, \sin(2\pi ht/T), \cos(2\pi ht/T)]$$

Blocks connect through residual links within each stack, which allows each block to focus on modeling residual patterns not captured by previous blocks. The stack aggregates partial forecasts from its blocks through summation, and the final forecast combines outputs across all stacks.

This residual stacking is what enables N-BEATS to decompose time series, somewhat similarly to traditional statistical approaches like ARIMA. The primary difference is that ARIMA focuses on quality of the residuals and needs manual adjustments based on domain knowledge, while N-BEATS utilizes more powerful basis functions, thus is more expressive. N-BEATS' interpretable variant enforces specific basis functions—polynomial for trend and Fourier for seasonality—enabling clear component decomposition.

N-BEATSx extends this architecture by incorporating exogenous variables X alongside a target variable y. While N-BEATS processes only historical target values through its blocks, N-BEATSx processes a covariate matrix X containing external features (Olivares et al., 2023). For example, we could include variables like price and calendar features, or any other relevant external data. The core architectural elements—stacks, blocks, and residual connections—remain the same, but each block in N-BEATSx learns from both the target and exogenous variables, allowing N-BEATSx to capture relationships between the target and external influences, along with time-varying patterns. Of course, adding exogenous variables is not guaranteed to improve performance.

N-HITS

In 2022, N-HiTS improved upon N-BEATS by addressing two limitations for long forecast horizons: decreasing accuracy and computational cost as forecast horizons increase in length. It maintains the core neural basis expansion concept of N-BEATS, while introducing two changes: multi-rate data sampling and hierarchical interpolation (Challu et al., 2022).

Multi-rate data sampling occurs at the block level through *MaxPool* layers. Each stack's blocks apply MaxPool with a specified kernel size to smooth the input signal. That allows each stack to look at a series at different scales.

```
# Original data
data = np.array([3, 6, 4, 0, 8, 5, 3, 2, 7, 9, 6, 0, 5, 3, 2])
x = np.arange(len(data))
# Function to perform max pooling
def max_pool(data, kernel_size):
    output = []
    for i in range(0, len(data), kernel_size):
        end = min(i + kernel_size, len(data))
        output.append(np.max(data[i:end]))
    return np.array(output)
# Calculate pooled data
pool_3 = max_pool(data, 3)
pool_5 = max_pool(data, 5)
# Create x values for pooled data (centered in their windows)
x_pool_3 = np.arange(1, len(pool_3) * 3, 3)[0:len(pool_3)]
x_pool_5 = np.arange(2, len(pool_5) * 5, 5)[0:len(pool_5)]
# Create the plots
fig, (ax1, ax2) = plt.subplots(2, 1, figsize=(12, 7))
# Top plot - data with alternating kernel=3 bands
ax1.plot(x, data, 'o-', color=custom_palette[0], linewidth=1.5, markersize=6,
zorder=5)
colours_k3 = [custom_palette[1], custom_palette[5]]
for idx, i in enumerate(range(0, len(data), 3)):
    end = min(i + 3, len(data)) - 1
    ax1.axvspan(i - 0.5, end + 0.5, alpha=0.15, color=colours_k3[idx % 2],
                label='Kernel Size 3 windows' if idx == 0 else None)
# Horizontal bars for kernel=5 at the top
bar_y = 9.8
for idx, i in enumerate(range(0, len(data), 5)):
    end = min(i + 5, len(data)) - 1
    ax1.plot([i, end], [bar_y, bar_y], color=custom_palette[6],
             linewidth=4, solid_capstyle='butt',
```

```
                label='Kernel Size 5 windows' if idx == 0 else None)
        ax1.plot([i, i], [bar_y - 0.3, bar_y + 0.3],
                color=custom_palette[6], linewidth=2)
        ax1.plot([end, end], [bar_y - 0.3, bar_y + 0.3],
                color=custom_palette[6], linewidth=2)
    ax1.set_ylim(-0.5, 10.5)
    ax1.set_title('Original Time Series with Pooling Windows')
    ax1.legend(loc='upper right', fontsize=11, framealpha=0.9)
    # Bottom plot - pooled results
    ax2.plot(x_pool_3, pool_3, 'o-', label='Kernel Size 3',
            color=custom_palette[1], linewidth=2)
    ax2.plot(x_pool_5, pool_5, 'o-', label='Kernel Size 5',
            color=custom_palette[6], linewidth=2)
    ax2.legend(loc='upper right', fontsize=11)
    ax2.set_title('Max Pooled Results')
    plt.tight_layout()
    plt.show()
```

The resulting plot is:

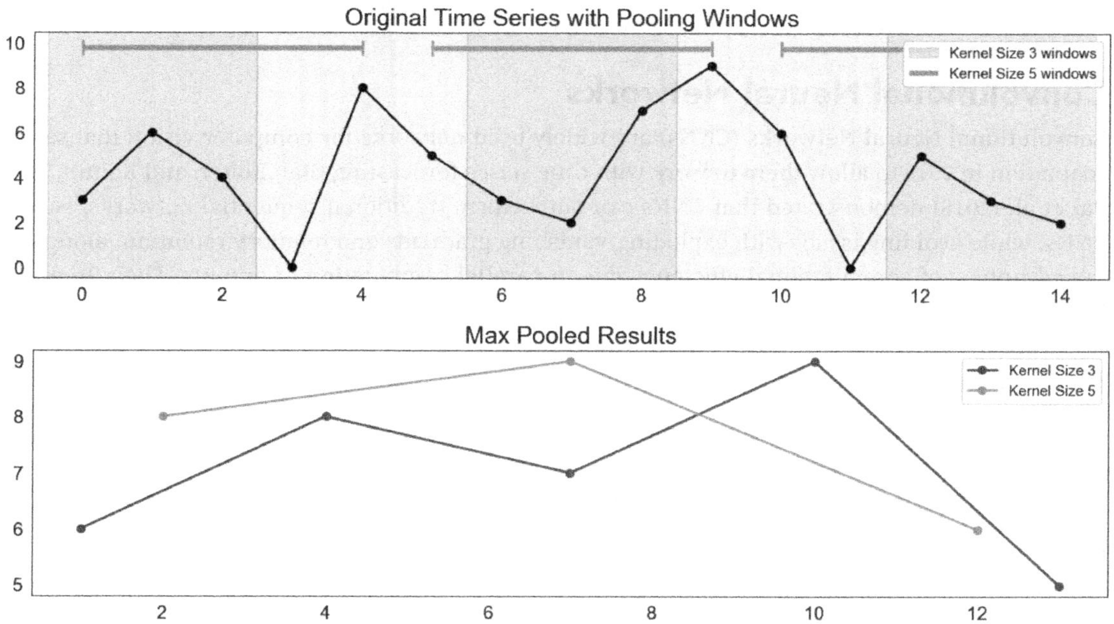

Figure 10.5: Illustration of multi-rate data sampling in N-HiTS using MaxPool operations

Larger kernel sizes create stronger smoothing, emphasizing long-term patterns, while smaller sizes preserve short-term details. N-HiTS uses a top-down approach where initial stacks focus on long-term effects (larger kernels) and later stacks capture short-term patterns (smaller kernels).

Hierarchical interpolation reduces computational load by allowing stacks to forecast at different time scales. This is controlled by an expressiveness ratio that determines how many predictions each stack makes within the forecast horizon. For example, with a ratio of 1/2, a stack predicts every second value of a final forecast.

$$\text{predictions}_{\text{stack}} = \text{horizon}_{\text{l}}\text{ength} * \text{expressiveness\_ratio}$$

The final prediction combines interpolated forecasts from all stacks.

$$\hat{y} = \sum_{s=1}^{S} \text{interpolate}\left(\hat{y}_s, \text{horizon\_length}\right)$$

where $\hat{y}_s$ is the forecast from stacks, and $\text{interpolate}(\hat{y}_s, \text{horizon\_length})$ upsamples the prediction to match the forecast horizon length.

As a result of this, N-HiTS allows earlier stacks to focus on weekly patterns with fewer predictions, while later stacks handle daily or hourly patterns with higher resolution. The expressiveness ratio typically increases exponentially between stacks or aligns with known cyclical patterns in the data. All of this helps N-HiTS models to be more efficient and accurate for predicting long horizons.

Convolutional Neural Networks

Convolutional Neural Networks (**CNNs**) are widely used networks for computer vision that saw an adaptation in 2018 to allow them to work with time series forecasting (Bai, Kolter, and Koltun, 2018). Bai et al. (2018) demonstrated that CNNs can outperform traditional sequential networks, such as RNNs, while avoiding issues with exploding/vanishing gradients and memory retention, along with the advantage of computational efficiency due to parallel computation of outputs. Their **Temporal Convolutional Network** (**TCN**) architecture was a promising start to using CNNs for forecasting. While CNNs are not currently the most popular family for forecasting, architectures like BiTCN, TCN and TimesNet (particularly TCN) have shown promising results.

CNNs processes three-dimensional tensors, with 1D CNNs operating on input shape (`batch_size`, `input_length`, `input_size`) and output shape (`batch_size`, `input_length`, `output_size`). For forecasting, `input_size` represents features at each timestep, `output_size` for predicted variables.

From the input tensor (`batch_size`, `input_length`, `input_size`), kernels operate along the temporal dimension (`input_length`). Kernels are learnable weight matrices that move across the sequence of input data, processing fixed-size windows of data. For a single-channel input, a kernel of size `k` examines `k` consecutive timesteps, element-wise multiplying its weights with these values and summing the results. When processing multi-channel inputs (like `temperature`, `price`, `volume`), the kernel becomes a 2D matrix of shape (`kernel_size × input_channels`), performing calculations across all channels simultaneously to capture cross-feature relationships.

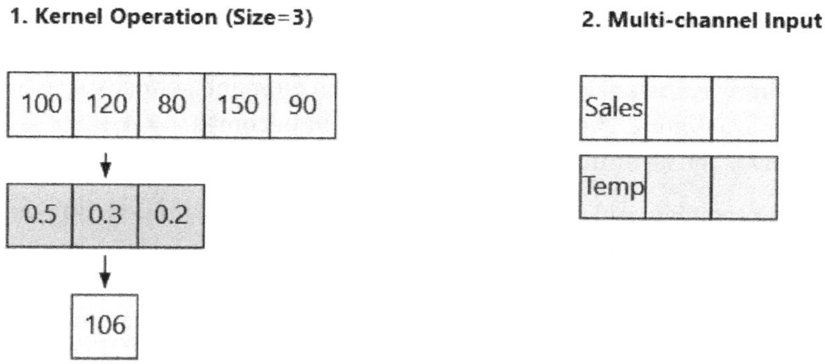

Figure 10.6: CNN Kernel and MultiChannel diagram

Kernels slide across sequences according to stride length, typically we would want `stride=1` to avoid skipped timesteps. At each position, the kernel performs a weighted sum across k timesteps. Each kernel of size k performs a weighted sum across k timesteps. For example, with kernel [0.5,0.3,0.2]and sales data [100,120,80], the operation computes $(0.5 \times 100 + 0.3 \times 120 + 0.2 \times 80) = 106$Multiple kernels operate in parallel, each learning different temporal patterns, with their count determining the layer's `output_channels`. Each layer uses multiple parallel kernels ($kernel\ size \times input\ channels \times output\ channels$) to learn temporal patterns.

Causality in sequence prediction requires that each output depends only on current and previous inputs in the sequence, never future ones. This is achieved through left-side zero padding, where zeros are added only to the start of the sequence. For a kernel size k and dilation rate d, we add $p = (k - 1) \times d$ zeros to the left side. This asymmetric padding ensures the kernel's window never sees future values while maintaining the desired output length.

In multi-channel inputs (when using exogenous variables), kernels extend across all input channels, with dimensions ($kernel\ size \times input\ channels \times output\ channels$), enabling the network to learn cross-feature relationships.

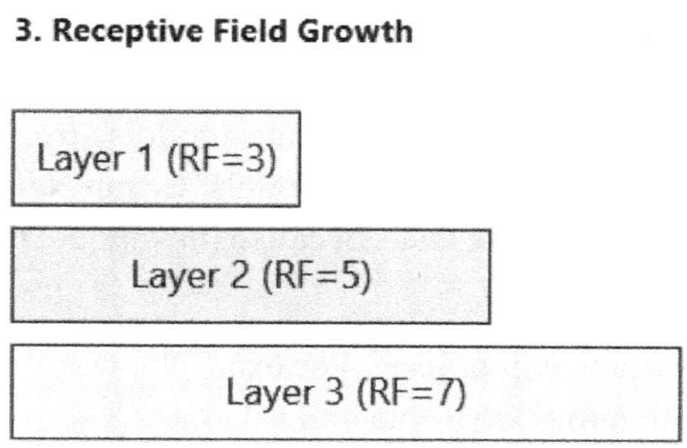

Figure 10.7: CNN receptive field growth diagram

A **receptive field**—span of input positions that influence each output—grows with network depth according to $r = 1 + \Sigma(k-1)$ for n layers with kernel size k. This basic formulation can be enhanced through dilation, where kernels skip $d-1$ inputs between elements, effectively expanding the field without increasing parameters. The dilated receptive field becomes $r = 1 + \Sigma(k-1)d^i$, where d is the dilation rate and i is the layer index.

To prevent holes in a receptive field—gaps where certain inputs fail to influence an output—the kernel size k must exceed dilation base b. Consider a six-timestep sequence with kernel size 2 and dilation base 3: the model would see timesteps $[t, t-3, t-6]$ while completely missing $[t-1, t-2, t-4, t-5]$. These holes create blind spots in the network's view of the input sequence, severely impacting model performance. For a sequence of length l, the minimum required layers for full coverage (without holes) is $n = \lceil log_k(l(k-1)+1) \rceil$.

For forecasting specifically, input and target sequences are created through a sliding windows approach. Given a forecast horizon h, each input sequence generates a target sequence shifted h steps forward. The maximum forecast horizon is limited by `output_length`, with window stride controlling training sample overlap.

CNNs and RNNs handle memory requirements differently across training and inference phases. During training, CNN memory usage scales with sequence length, batch size, and number of parameters (`kernel_size` × `input_channels` × `output_channels` per layer). During inference, RNNs maintain only their hidden state, while CNNs require the complete effective history of inputs, creating a memory-speed tradeoff.

These architectural principles—causality, dense predictions (`stride=1`), and dilated convolutions—form the foundation for advanced models like TCNs. When adapting models between domains with different memory requirements, careful adjustment of kernel size (`k`) and dilation rate (`d`) is crucial for maintaining performance while managing computational resources.

Graph neural networks

In the sales data we analyzed in the beginning of this chapter, forecasting the sales of all the goods requires a model to capture the **inter-series** relationships, for example, people buy more party items during the weekend, and the **intra-series** relationships, for example, people grabbing a pizza might also buy a cola. Many statistical time series forecasting models already attempt to capture such relations, for example, **Vector Autoregressive** (**VAR**) models. **Graph Neural Networks** (**GNN**) is a good candidate for such multivariate forecasting tasks because they model the interactions between series using a more generic concept: graph.

A graph describes relationships between objects. For example, we can use the following to describe the idea that people buy pizza together with cola and salad:

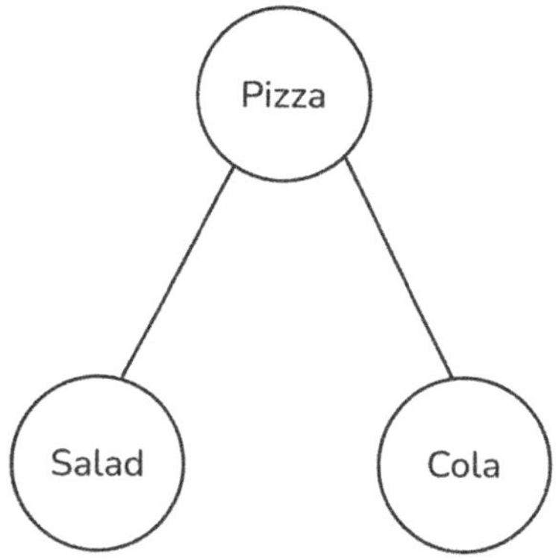

Figure 10.8: Pizza, Cola, and Salad graph

In *Figure 10.8*, the circles **Pizza, Salad** and Cola are the vertices, and the lines connecting them are the so-called edges. The same information can be described using a matrix, where the elements of it are 1 if there is an edge between the corresponding vertices, e.g., Pizza-Salad, and 0 otherwise, e.g., Salad and **Cola.** This matrix is called the adjacency matrix, which looks like this for our pizza-salad-cola example:

$$\mathbf{A} = \begin{pmatrix} 0 & 1 & 1 \\ 1 & 0 & 0 \\ 1 & 0 & 0 \end{pmatrix}. \qquad (1)$$

The first row in *Equation 1* is for **Pizza**, the second for **Salad**, and the third for **Cola.**

GNN models operate on such graph structures. To demonstrate how it works, we discuss one of the most intuitive GNN strategies called message passing. First of all, we have a representation of each vertex which is usually an embedding tensor, **h**.

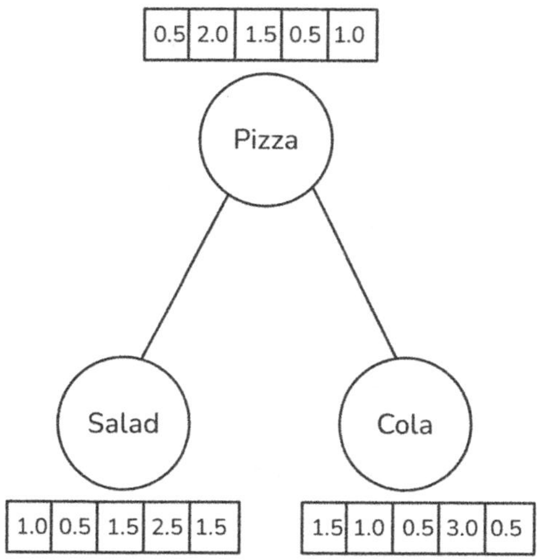

Figure 10.9: Representations of Pizza, Salad, and Cola

The idea is simple: each vertex aggregates all the information from its neighbors and combines it with the information it carries. In the **Pizza-Salad-Cola** example, the **Pizza** vertex will aggregate the information from **Salad** and **Cola** vertices, and then combine it with its own information. For example, we update the representation of **Pizza** as follows:

Aggregate the information from **Salad** and **Cola** vertices:

$$\mathbf{m}_P = \mathbf{h}_{\text{Salad}} + \mathbf{h}_{\text{Cola}} = \begin{pmatrix} 1.0 \\ 0.5 \\ 1.5 \\ 2.5 \\ 1.5 \end{pmatrix} + \begin{pmatrix} 1.5 \\ 1.0 \\ 0.5 \\ 3.0 \\ 0.5 \end{pmatrix}.$$

Combine the aggregated information with the information of **Pizza** itself.

$$\begin{aligned} \mathbf{h}_{\text{Pizza}} &= \sigma\left(\mathbf{W}_{\text{self}} \cdot \mathbf{h}_{\text{Pizza}} + \mathbf{W}_{\text{neighbor}} \cdot \mathbf{m}_P\right) \\ &= \sigma\left(\mathbf{W}_{\text{self}} \cdot \begin{pmatrix} 0.5 \\ 2.0 \\ 1.5 \\ 0.55 \\ 1.0 \end{pmatrix} + \mathbf{W}_{\text{neighbor}} \cdot \left(\begin{pmatrix} 1.0 \\ 0.5 \\ 1.5 \\ 2.5 \\ 1.5 \end{pmatrix} + \begin{pmatrix} 1.0 \\ 0.5 \\ 1.5 \\ 2.5 \\ 1.5 \end{pmatrix}\right)\right), \end{aligned}$$

where σ is an activation function.

Repeat the above process for all vertices in the graph, and for multiple iterations until we reach convergence.

There are many variations of the above algorithm; for example, we can apply graph convolutions to update the representations of the vertices (Kipf and Welling, *Semi-Supervised Classification with Graph Convolutional Networks*, available at `http://arxiv.org/abs/1609.02907`), or use attention mechanisms to aggregate the neighbor information (Veličković et al., *Graph Attention Networks*, available at `http://arxiv.org/abs/1710.10903`). In any case, the key to the GNN message passing algorithm is to capture the information of each vertex and the relationships between them.

A more specific GNN model for forecasting is the **Stem Graph Neural Network** (**StemGNN**) model for multivariate time series forecasting. StemGNN first builds a graph for the correlations between the input multivariate time series using a self-attention mechanism, followed by graph convolutions before spitting out the backcast and forecast. Note that the S in the name StemGNN is for *Spectral*, which refers to the fact that graph convolution is performed in the Fourier domain. As of writing, StemGNN is already included in the neuralforecast library.

Kolmogorov–Arnold networks

KAN is a new family of neural networks that use superposition of base functions to approximate any continuous functions. For example, if we have a continuous function $f(x_{p=1}, x_{p=2})$ of two variables $x_{p=1}$ and $x_{p=2}$, the Kolmogorov-Arnold representation theorem states that there exists a decomposition:

$$
\begin{aligned}
& f(x_{p=1}, x_{p=2}) \\
= & \sum_{q=1}^{5} \Phi_q \left(\phi_{q,p=1}(x_{p=1}) + \phi_{q,p=2}(x_{p=2})\right) \\
= & \sum_{q=1}^{2n+1} \Phi_q \left(\sum_{p=1}^{n} \phi_{q,p}\left(x_p\right) \right),
\end{aligned}
$$

where $n = 2$ as we have two variables. We require that the domain of $\phi_{q,p}$ be $[0,1]$, such as the famous B-spline B, or a shifted Legendre polynomial $\tilde{P}_n(x)$. Note that the Kolmogorov-Arnold representation theorem works for more than two variables, and we only choose two variables for simplicity.

The KAN model is constructed using a recursive structure. First, we obtain the sum $\sum_{p=1}^{n} \phi_{q,p}\left(x_p\right)$ for $q = 1$, as shown in the computational graph in *Figure 10.10*.

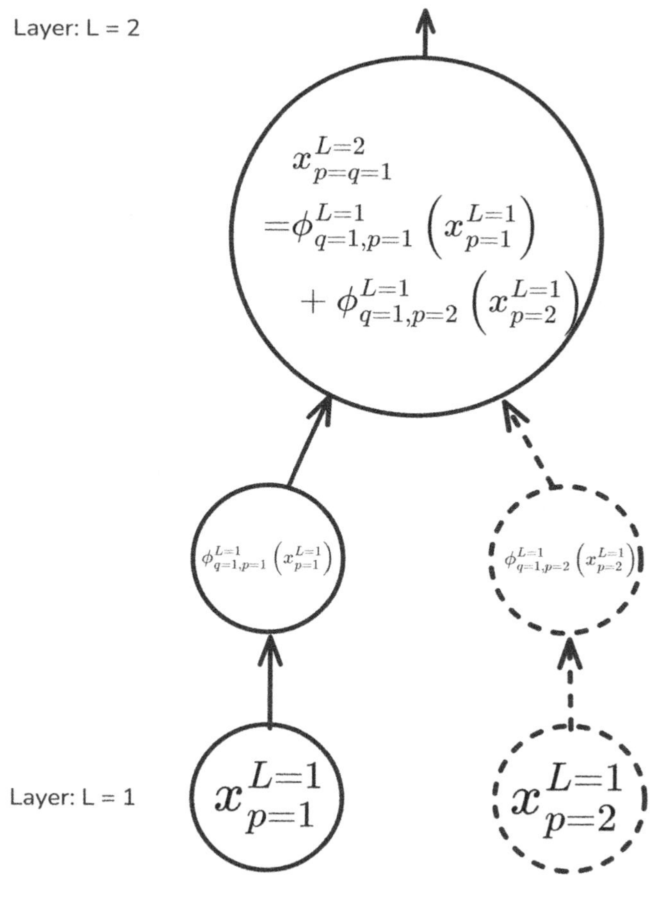

Figure 10.10: Computational graph of the q = 1 component of a KAN model

Mathematically, the $q = 1$ component is written in the top circle *of Figure 10.10*. As *Figure 10.10* is tedious to draw and read, we simplify it into *Figure 10.11*.

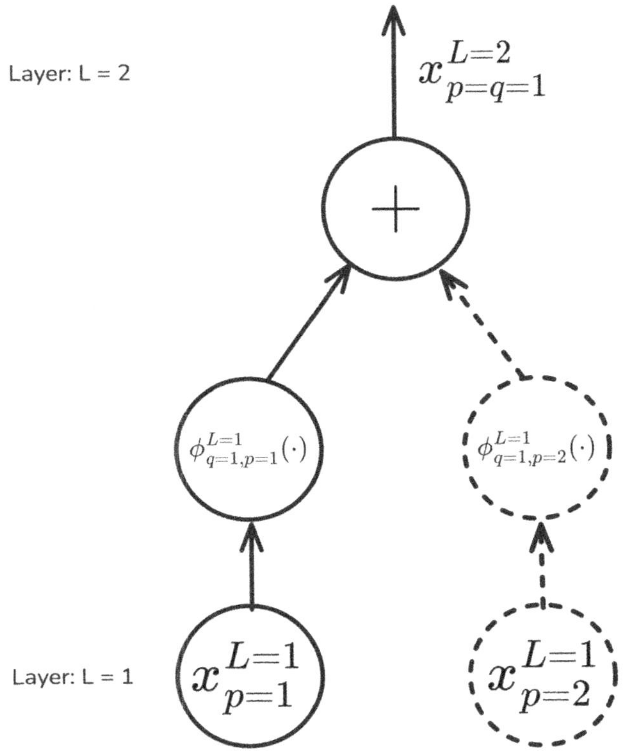

Figure 10.11: Simplified computational graph of the q=1 component of a KAN model

We can use the output $x_{p=1}^{L=2}$ from the first layer $L = 1$ as the input for the next layer $L = 2$. However, before we move on to build a second layer, we need to finish our layer $L = 1$ first. We add a second output from layer 1 in *Figure 10.12*. Note that we just need to repeat five times to get all our outputs for the first layer. Using the five outputs from the first layer, we can build the second layer, also shown in *Figure 10.12*.

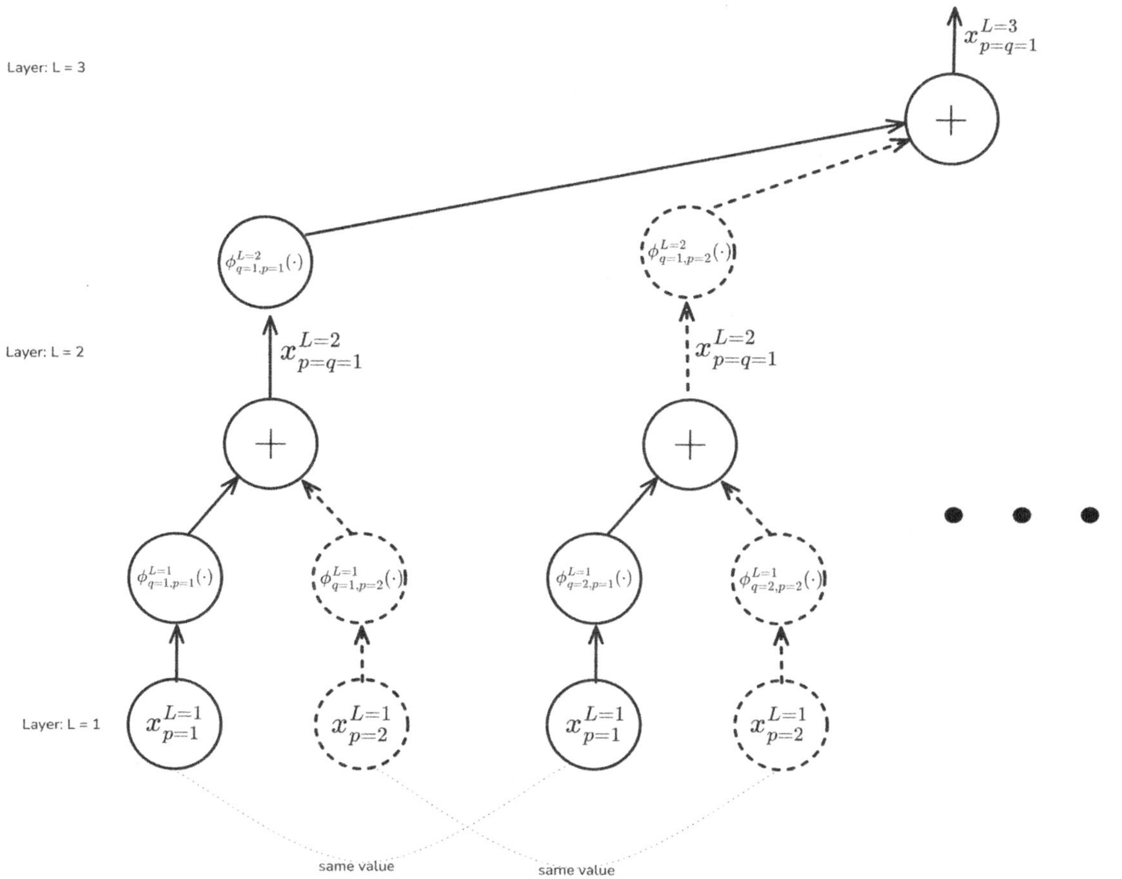

Figure 10.12: Computational graph of the first layer of a KAN model

Training a KAN model is usually slower than an MLP with a comparable amount of parameters. However, KAN can be used to estimate more complicated patterns with the same amount of parameters. Similar to other models mentioned above, KAN is also included in the `neuralforecast` library by Nixtla.

Foundation models

Foundation models such as **LLMs**, are taking the tech world by storm. A pre-trained large language model can be used to perform a variety of language-related zero-shot tasks. Similarly, a pre-trained large time series forecasting model can perform zero-shot forecasting tasks, i.e., forecasting without any model training.

At the time of writing, there are a few time series foundation models. We will briefly discuss three types of them: language models adapted for time series forecasting, models pretrained using time series data, and a special model called TabPFN that is pretrained with tabular data.

LLMs are trained on vast amounts of text data, learning to predict the next word in a sentence. The same concept inspired people working on time series forecasting. In fact, the first approaches were to use LLM as the time series foundation model. However, reusing LLM for time series forecasting requires us to answer questions such as how the time series data is encoded in an LLM. Some methods, such as Time-LLM (M. Jin et al., *Time-LLM: Time series forecasting by reprogramming large language models,* available at: `http://arxiv.org/abs/2310.01728`), use reprogramming, e.g., aligning time series data with text data using a small network so that the time series data is translated into language data.

On the other hand, language models are specialized in language tasks. Considering the fact that time series data is also sequential just like natural language, we can directly pre-train a model from scratch using time series data. Time-GPT was the first model of its kind, pre-trained on time series data, and it is based on the GPT architecture. While some technical details are missing, Time-GPT is a vanilla transformer model, trained on a bunch of diverse datasets with different frequencies and different forecast horizons. Since Time-GPT is not open source, we will focus more on an open-source time series foundation model called Lag-Llama.

To get started with the core concept of the Lag-llama model, we come back to our mental model of a swing. A swing oscillates but the amplitude diminishes. The time series will show this clearly. However, there is a different perspective to visualize and capture the idea. We plot out our swing angle at time t against the swing angle at time $t-1$. The plot will show a pattern that the swing angle is oscillating but becoming smaller, as shown in *Figure 10.13*. This diagram is called a **phase space** diagram, and it captures the dynamics of the time series data.

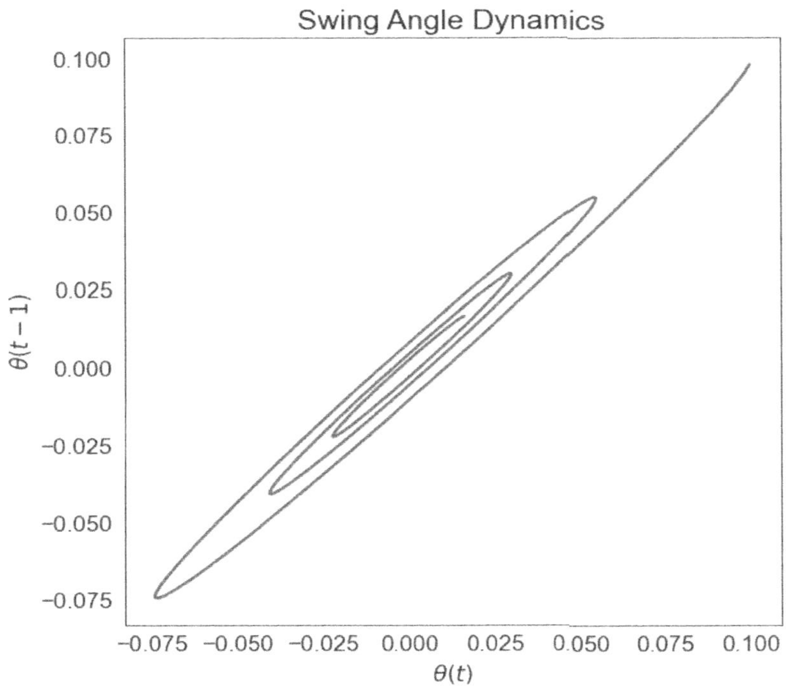

Figure 10.13: Angle of a swing compared to its previous time step

The swing starts from the top right of this so-called phase space diagram. The pattern indicates that the swing angle is oscillating but becoming smaller.

Lag-Llama is based on the idea that a phase space of the time series captures the dynamics of the time series data. The input of the Lag-Llama model is the original time series with different lags. For example, given a time series , 1,2,3,4, we convert it into the following input by lagging it by 1:

$$\text{Input} = \begin{pmatrix} 1 & 2 \\ 2 & 3 \\ 3 & 4 \end{pmatrix}.$$

The *first* column represents the original data, while the *second* column lags by 1, and we can add more columns by lagging more.

Each row is a point in a high-dimensional phase space similar to the one we have in *Figure 10.13*. The core model, as the name indicates, is a **Llama** architecture. Now that we have the data input and model architecture, we can train it on a diverse set of data. After training, Lag-Llama can be used as a zero shot model for time series forecasting.

The third category of foundation models, as we mentioned earlier, is TabPFN. TabPFN is a model that is pretrained on tabular data. Similar to other foundation models, TabPFN is also a transformer-based model. However, TabPFN utilizes an attention mechanism that helps the model to capture both column-wise and row-wise relationships in the tabular data. Thus, it is useful in time series forecasting tasks too. Normally, we have the train-evaluate-test paradigm for each forecasting problem we are solving, where we understand the forecasting problem, build a model, train, and evaluate using a dataset, and finally perform out-of-sample tests using a holdout dataset. TabPFN dropped the paradigm, and it uses the so-called **In-Context-Learning** (**ICL**) for zero shot forecasting. In ICL, the model receives the input data and the task together and makes predictions based on all the input without additional training. Similar to LLMs, TabPFN is a pretrained model. TabPFN is a good candidate for time series forecasting on small datasets. We encourage the reader to check out the original paper by N. Hollmann et al., "Accurate predictions on small data with a tabular foundation model," available at `https://www.nature.com/articles/s41586-024-08328-6`.

Summary

In this chapter, we discussed several neural network models for time series forecasting. We will not dive deep into these models; however, we recommend the reader at least understand the basic concepts of these models. Time series forecasting is a complicated task, and different domains and different datasets require different models. With more intuition about different models, the reader can better choose a model for the task at hand. We also discussed several foundation models, which are pretrained models that can be used for zero-shot forecasting tasks. These models are useful when we want to quickly perform a forecasting task or as baselines for our models if we have no financial constraints.

Recommended reading

- Bai, S., Kolter, J. Z., and Koltun, V. 2018. "An Empirical Evaluation of Generic Convolutional and Recurrent Networks for Sequence Modeling." *arXiv*. `http://arxiv.org/abs/1803.01271`.
- Challu, C., et al. 2022. "N-HiTS: Neural Hierarchical Interpolation for Time Series Forecasting." *arXiv*. `http://arxiv.org/abs/2201.12886`.
- Hollmann, N., et al. 2024. "Accurate predictions on small data with a tabular foundation model." *Nature*. `https://www.nature.com/articles/s41586-024-08328-6`.
- Jin, M., et al. 2023. "Time-LLM: Time series forecasting by reprogramming large language models." *arXiv*. `http://arxiv.org/abs/2310.01728`.
- Kipf, T. N., and Welling, M. 2016. "Semi-Supervised Classification with Graph Convolutional Networks." *arXiv*. `http://arxiv.org/abs/1609.02907`.
- Olivares, K. G., et al. 2023. "Neural Basis Expansion Analysis with Exogenous Variables: Forecasting Electricity Prices with NBEATSx." *International Journal of Forecasting*, 39 (2): 884–900. `https://doi.org/10.1016/j.ijforecast.2022.03.001`.
- Oreshkin, B. N., et al. 2020. "N-BEATS: Neural Basis Expansion Analysis for Interpretable Time Series Forecasting." *arXiv*. `http://arxiv.org/abs/1905.10437`.
- Veličković, P., et al. 2017. "Graph Attention Networks." *arXiv*. `http://arxiv.org/abs/1710.10903`

Join our community on Discord

Join our community's Discord space for discussions with the authors and other readers:

`https://discord.gg/VXpz9YPUTW`

11

Transfer Learning and Global Modelling

"Transfer learning is probably where neural networks have a unique advantage over shallow models." —Andriy Burkov, The Hundred-Page Machine Learning Book

We learned earlier in the book that a key aspect of machine learning models, such as neural networks, is that they learn to map the data we input relative to their outputs. In order to adapt to the data we set these models up with various hyperparameters that allow our models to capture the optimal function for data that is passed through. This process is a means of controlling model complexity to adjust functions that suit the underlying patterns in our data, although we can also conceptualize this process as a complex form of memory compression.

This approach is often applied to series individually in datasets, i.e. one item of clothing, food, energy region, etc., which limits a trained model to a specific dataset and its underlying distribution of data used for training. When seeing new data that differs greatly from its training distribution, whether in scale, features, or relationships, these bespoke models will often make predictions with high error. Training individual bespoke models for each time series we encounter means that they often do not generalize well to a new set of data, despite the fact that we can pass them similar inputs to what a model was trained on to generate predictions. This means we have to instantiate a fresh model and train it from scratch. This is wasteful in many ways because:

- Building a model for each time series is computationally expensive, especially if you have thousands to millions of series to model
- You are likely to have access to multiple data sources, some from different domains
- Many data sources share the same underlying patterns

But can we adjust a model, and particularly its training, to generalize beyond one dataset/series? Transfer learning addresses the limitation of training individual models, enabling neural networks to leverage knowledge gained from one context (source domain) to improve performance in another (target domain). Rather than treating each forecasting task as an isolated learning problem, transfer learning allows a network to adapt and reuse knowledge across different but related domains. We will learn the fundamental principles that underpin the autoregressive transformer (LLM) *AI* revolution, and how the same principles can be useful for time series forecasting.

In this chapter, we will learn:

- What is transfer learning?
- Structuring time series data for transfer learning
- Global modeling
- Building GFMs with PyTorch
- Transfer learning with NeuralForecast

What is transfer learning?

What exactly is **Transfer Learning** (**TL**)? It isn't a single method, strategy or technique, but we could probably narrow it down to a methodological framework for optimizing the predictions of neural networks. In essence, transfer learning can improve forecasting performance by leveraging knowledge from related tasks or domains. Traditional ML methods assume that training and test data originate from the same distribution; under TL we relax this assumption by allowing for knowledge to transfer between similar *domains*.

> Much of the code for this chapter is reduced so that you don't have to look at a wall of code. The full code available in our companion notebooks can be found in the book's GitHub repository.

While TL has gained a lot of attention recently, its mathematical foundations were laid down by Bozinovski and Fulgosi in 1976 (Bozinovski 2020). The field evolved through key developments like Pratt's discriminability-based transfer algorithm (1992; Pratt (n.d.)) and gained formal theoretical grounding by 1998 with the emergence of multi-task learning. However, TL's practical applications became more obvious after the release of ImageNet (2009), which allowed practitioners like us to download and train models with 14 million open source images. **Convolutional Neural Networks** (**CNNs**) like AlexNet (Yoss 2020) could be built with this data for a multitude of computer vision tasks on similar but unrelated datasets. For example, you could build a model, i.e., AlexNet, for classification of medical x-ray images as potentially containing a tumour or not. For many CV tasks, practitioners will begin training models with a huge dataset, or downloading a model someone has kindly *pre-trained* on large-scale datasets like ImageNet, from databases like Hugging Face. The next significant advancement came with transformers in **Natural Language Processing** (**NLP**), which showed that *meaningful* representations could be learned from unlabeled text data, leveraging vast corpora like Wikipedia and Common Crawl. Of course, this is largely what is behind LLMs like ChatGPT.

> Refer to color images on the Graphic bundle or the ebook.

If you haven't used pretrained models before, you may be missing out on a critical aspect of data science; this tutorial on Hugging Face and related material can start you off on a journey down the rabbit hole (`https://huggingface.co/docs/transformers/en/quicktour`). Hugging Face is one of a number of places we can get pretrained models; we can also find them on Kaggle (`https://www.kaggle.com/models`), directly from companies like Meta (`https://www.llama.com/`), and from GitHub (i.e., DeepSeek). Some companies give us access to certain models via APIs rather than the fully parameterized model.

The successes of TL for computer vision and NLP tasks led forecasters to think about how it can be applied to time series. This may seem like a rational step, but it is less obvious to formulate; *what does it mean to learn features across domains over time*? Time-series data is not like images or text; gathering a large and diverse library of consistently sampled time series datasets is difficult (Li and Law 2024). This is one reason why classic statistical methods, like exponential smoothing or ARIMA, have remained popular for time series forecasting; they simply don't require as much data to build a decently performing model. This has changed somewhat over the last few years as access to large bodies of good quality data increased. As data scientists started exploring deep learning for time series, a 2019 paper investigated zero-shot learning for time series by pre-training on a large, diverse time series dataset and then transferring to a target domain (Oord, Li, and Vinyals 2019).

Specifically, we say transfer learning involves two key components as established in seminal work (Pan and Yang 2010): a source, which we want to transfer knowledge from, and a target, which we want to apply that knowledge to. We can formally express this as:

- A source domain (D_S) and its associated learning task (T_S)
- A target domain (D_T) and its associated learning task (T_T)

The goal of transfer learning is simply to improve learning of a target's predictive function $f_T(\cdot)$ in our target domain (D_T) by leveraging what we know from our source domain (D_S) (Pan and Yang 2010). We aim to do this where the domains ($D_S \neq D_T$) or tasks ($T_S \neq T_T$) are not equal to each other, scenarios where traditional machine learning assumptions can break down. This splits TL into two subcategories:

- **Domain Adaptation (DA)**: Where our domains differ ($D_S \neq D_T$), but we're trying to solve the same kind of task. For example, forecasting energy consumption across different buildings; the underlying task is the same (predicting energy use), but the domains (different buildings) have unique characteristics, i.e. transferring a model trained on electricity demand for commercial to residential buildings. Under such scenarios, the fundamental patterns (i.e. daily cycles, seasonal variations) may differ somewhat, but the prediction task is identical (Zhang et al. 2022).
- **Task Adaptation**: When we're dealing with different tasks ($T_S \neq T_T$), even if they might be in the same domain. For example, we might build a model with hourly stock price values and predictions, then adapt it to predicting daily aggregates with prediction intervals. The domain (financial) remains the same but we are transforming the nature of the model's prediction task from point forecasting to uncertainty quantification (see *Chapter 7*).

Clearly, it is possible to apply both domain and task adaptation simultaneously, although this is less common with mixed results in forecasting. Progress has been made with autoregressive transformers (aka LLMs) such as TimeGPT (https://github.com/Nixtla/nixtla), Lag-Llama (`https://github.com/time-series-foundation-models/lag-llama`), and Amazon's Chronos (`https://github.com/amazon-science/chronos-forecasting`), although it could be argued this is simply domain adaptation, due to the number and variation of datasets each model is trained on.

Types of transfer learning

Transfer learning can be utilized under a number of different data settings and scenarios, depending on the availability of labeled data, which defines the type of transfer learning we are working with:

- **Inductive Transfer Learning** (the *Informed* case is called so because we can *induce* the target mapping from labeled examples in both domains. This represents our most common scenario in practice (Zhuang et al. 2020; Pan and Yang 2010). We have labeled data in our source domain, and some labels in our target domain, so we attempt to infer artificial labels or work directly with feature relationships. Imagine you're a retail forecaster attempting to build forecasts for a new store. Your new store only has three months of sales data, but you have years of historical sales from other stores in the retailer's organization. Having new data and historic patterns allows models to learn both general retail patterns and location-specific characteristics.
- **Transductive Transfer Learning** addresses scenarios where we have no data from our target domain (cold start problem). In this situation as a retailer, our store hasn't opened yet, but we have to make forecast predictions. While we have extensive historical data from existing stores, we have no actual sales data from the new location. We must rely on indirect relationships, and must carefully selecting our data to what we think (based on some comparative analysis) might fit best with how the store may perform.
- **Unsupervised Transfer Learning**: is the most challenging scenario, which emerges when we have no labeled data across either domain (source or target). Under our retailer scenario we might be creating a website that sells our products, but we have no idea how this will perform compared to the performance of our stores.

Applications of transfer learning

There are different ways we can employ transfer learning to address these challenges.

Model-based transfer is the approach we'll focus on here. It involves reusing and adapting model parameters from a source domain to improve predictions in a target domain. In our retail forecasting scenario, this might involve taking a model pre-trained across multiple stores to capture universal sales patterns, i.e. a global model. A single model learns to forecast multiple time series. This approach implicitly transfers knowledge across different series through shared parameters and architectural elements, while maintaining the ability to capture series-specific nuances (Li and Law 2024). The effectiveness depends on the particulars of the modeling pipeline, i.e. selection of data for models. Global modeling is an effective and efficient approach to modeling at scale, which has led to its increasing prominence in modern forecasting systems, particularly with the advent of sophisticated neural architectures, although there is some loss of information when adjusting to multiple datasets.

Figure 11.1 illustrates the relationship between source and target domains in transfer learning:

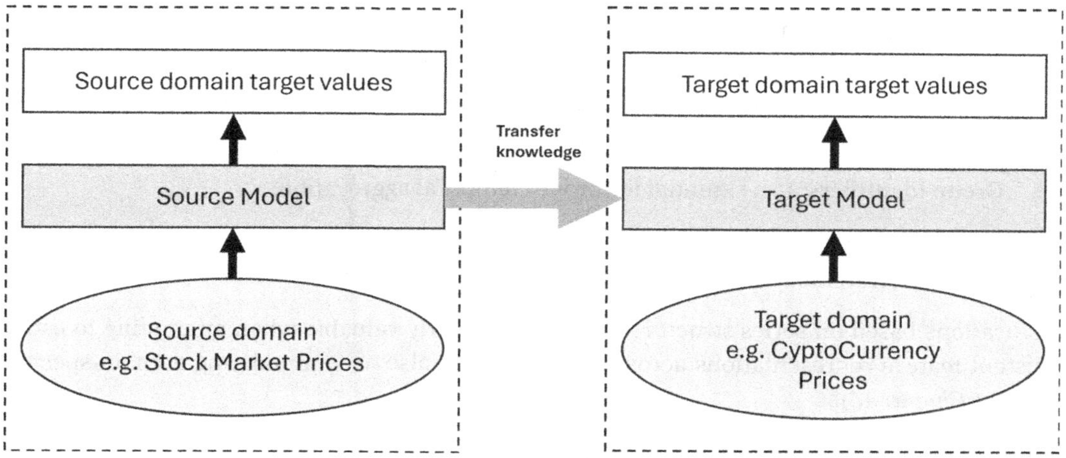

Figure 11.1: Domain Learning Diagram

Feature-based transfer focuses on finding and transforming data from different domains into a common representational feature space that we can use to construct abstract features that capture underlying patterns independent of domain-specific characteristics. The idea is that the features built by this approach transfer across seemingly disparate contexts. We often see feature-based TL applied as a complementary component to model-based transfer rather than as a standalone solution.

Instance-based transfer operates through selective weighting or sampling of source domain data based on relevance to our target domain. It relies on the assumption that some historical instances provide more valuable information for target prediction tasks than others. We consider this a simpler solution, with its effectiveness heavily dependent on the quality of similarity metrics between domains.

Relational-knowledge transfer is not particularly useful for forecasting tasks due to the sequential nature of time series, as this method focuses on complex relational structures, which rarely arise in traditional forecasting tasks.

Structuring time series data for transfer learning

Practical implementation of transfer learning requires some thought about data structure. Unlike traditional ML tasks, where observations maintain independence, time series data demands attention to sequential dependencies, scale variations, and hierarchical relationships. In which case we need to organize and classify our data appropriately. Features in time series transfer learning can be broken down into a taxonomy like so:

- **Static features:** Time-invariant characteristics for series identification:
    - **Categoricals:** Unchanging categorical attributes that define fundamental groupings (e.g., product hierarchies, geographical designations).
    - **Reals:** Fixed numerical characteristics that capture baseline behaviors (e.g., historical consumption patterns, facility capacities).

- **Dynamic features:** These capture the temporal evolution of our series:
    - **Time-varying known:** Features with deterministic future values (e.g., bank holidays, planned promotions).
    - **Time-varying unknown:** Features requiring prediction (e.g., target variables, dependent measurements).
- **Group identifiers:** It is essential for proper temporal aggregation:
    - Unique series identifiers for panel structures.
    - Hierarchical indicators for nested relationships.

Classifications based on series structure can be particularly valuable when attempting to establish consistent feature representations across domains while also accommodating domain-specific nuances (see *Chapter 15*).

Temporal windowing and sequential organization

Of course, the efficacy of transfer learning with time series data depends on our approach to temporal structuring. Consider a time series $X = x_1, \ldots, x_T$ with associated features $F = f_1, \ldots, f_T$. A typical windowing strategy might take the following form, where lookback defines how many past observations the model sees as input and horizon sets how many future steps it predicts:

```
def __init__(self, lookback: int, horizon: int):
    self.lookback = lookback  # Historical context length
    self.horizon = horizon    # Forecast horizon

def create_window(self, series: np.ndarray, t: int) -> tuple:
    history = series[t-self.lookback:t]
    future = series[t:t+self.horizon]
    return history, future
```

This windowing approach facilitates consistent temporal context across source and target domains, the preservation of critical sequential dependencies, and efficient batch processing during model training.

Scale normalization and feature transformation

Another crucial consideration in transfer learning is the handling of varying scales and distributions between domains. Typically we employ normalization strategies that preserve relative patterns while enabling effective transfer.

While we could employ a standard scaler ($x_{normalised} = \frac{x-\mu}{\sigma}$) to transform values to have a mean of 0 and unit variance across a whole dataset, i.e. panel, a better approach is often to apply a group-wise normalisation:

$$x_{normalised} = \frac{x - \mu_g}{\sigma_g}$$

Where μ_g and σ_g represent group-specific statistics i.e. mean and standard deviation of a specified grouping of your data. For instance, when forecasting retail sales across multiple locations, group-wise normalisation by store helps maintain local patterns while enabling cross-store learning.

You don't have to use standard scaling or min-max.

$$x_{normalised} = \frac{x - \min(x)}{\max(x) - \min(x)}$$

Works fine for some datasets, and can also be applied at the group level. Which normalization approach to select, is best tested as it depends on your data characteristics. Even with scaling the magnitudes of difference between series within the grouped data for a GFM can cause issues.

Temporal feature engineering

Beyond basic normalization, we can create features to represent temporal patterns consistently across domains that we expect to see. We have listed three approaches here that are helpful.

- Cyclical encoding of temporal features i.e. days of the week numbered 1 to 7 breaks continuity; instead, we can use sine and cosine waves like so...

```
# For day of week (d = 0-6)
day_sin = sin(2 * π * d/7)
day_cos = cos(2 * π * d/7)
```

- Relative temporal indicators are incredibly useful so things like new products, end of line items, changes in contract terms i.e. count since start of series, count down until change/event/holiday.
- Domain-invariant seasonality representations. Although seasonal patterns may vary across domains, they can still share underlying structures, which we can capture with:
    - Fourier terms for regular seasonal patterns
    - Moving averages to capture evolving seasonality
    - Seasonal decomposition techniques to separate trend, seasonal, and residual components into features

Global modeling

The classical approach to constructing forecasting models focuses on building individual models for each series. But when we are dealing with multiple related series, such as panel datasets of sales across multiple stores, the bespoke approach becomes increasingly inefficient and suboptimal. Global modeling is an alternative paradigm, which forms the backbone of transfer learning. It allows us to build potentially more robust and efficient forecasting models. Fundamentally, **Global Forecasting Models** (GFMs) operate on the premise that related series share some underlying structural patterns and data-generating processes (Joseph and Tackes 2024, Ch. 15).

By allowing a model to capture information from multiple series, we are effectively enabling it to develop a more sophisticated knowledge of the shared information (*cross-learning*), while reducing computational overhead. When multiple series share underlying patterns or generative processes, their individual noise components tend to be independent. By learning jointly across series, GFMs effectively average out independent noise components while reinforcing genuine structural patterns. This averaging effect occurs naturally through the shared parameter space, allowing the model to distinguish between persistent signals and transient fluctuations far more effectively than local models, which can struggle with limited data. You can imagine that this is valuable in sparse and volatile data domains, an extremely common problem we often deal with in practice.

Figure 11.2 illustrates the conceptual difference between local and global modeling approaches.

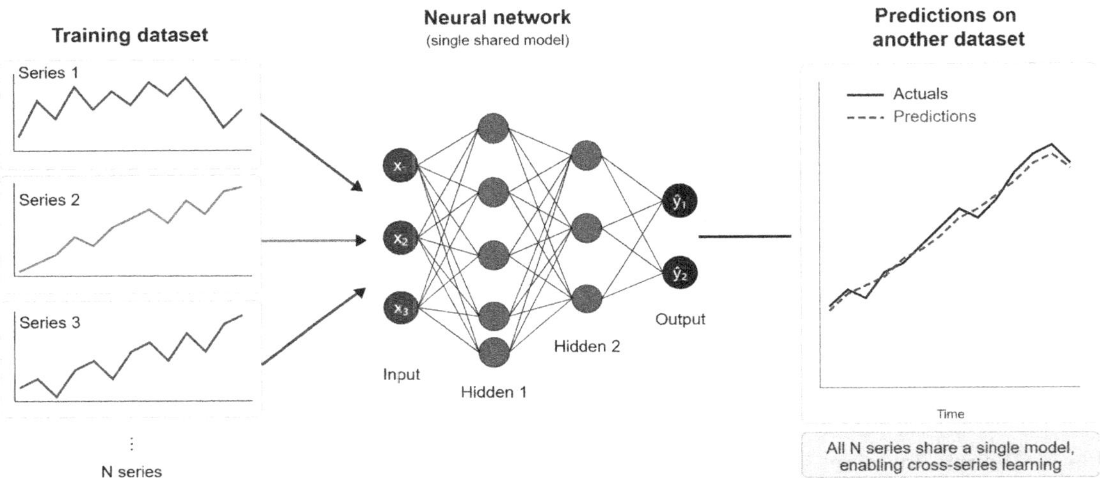

Figure 11.2: GFM conceptual diagram

Within a GFM framework, we consider a collection of K time series, each representing discrete observations of potentially related stochastic processes. Traditional local models estimate K separate mapping functions from feature space X to target space Y, implicitly assuming independence between series. GFMs, conversely, learn a single unified mapping that captures both shared patterns and series-specific characteristics. We can build some intuition for why this helps by looking at learning-theoretic generalisation bounds, which draw on concentration inequalities like *Hoeffding's inequality*. Hoeffding's inequality provides probabilistic bounds on the deviation between empirical and expected means. For a random variable X bounded in $[a, b]$, we have:

$$P(|\bar{X} - E[X]| \geq t) \leq 2e^{-\frac{2nt^2}{(b-a)^2}}$$

Using arguments derived from this inequality (via a union bound over the hypothesis space), we can express how well a model's in-sample performance approximates its expected out-of-sample performance. Here, complexity refers to the hypothesis space dimensionality, which effectively quantifies the range of functions our model class can represent. For a single local model trained on one time series of length T, the error bound takes the form:

$$E_{out}^{(i)} < E_{in}^{(i)} + \sqrt{\frac{\log(|H|) + \log(2/\delta)}{2T}}$$

Where $|H|$ represents hypothesis space complexity and δ is our confidence parameter. When we train K separate local models, we get K independent bounds like this, each relying on only T observations. The aggregate expected error is an average of these individual bounds. A clear limitation of this is that each local model's generalisation depends on T alone. With short series (small T), the bound is loose, meaning the gap between in-sample and out-of-sample performance can be large, and we have limited evidence that our model has learned genuine patterns rather than noise.

By comparison, global modeling yields a tighter bound by pooling data.

$$E_{out}^{Global} < E_{in}^{Global} + \sqrt{\frac{\log(|J|) + \log(2/\delta)}{2TK}}$$

A GFM's effective sample size is TK rather than T, because it pools observations across all K series. Even though the hypothesis space $|J|$ may be larger than any single $|H|$ (the model needs capacity to capture patterns across diverse series), the growth in this effective data typically dominates. We can summarise this as GFM's having a statistical advantage due to variance reduction through data pooling. Shared representations average out series-specific noise while reinforcing genuine cross-series patterns, which is why the resulting bounds on out-of-sample error can be significantly tighter than training K separate models.

This theoretical advantage manifests in practical benefits through multi-task learning, where each series contributes to a shared representational structure while maintaining individual characteristics. The tighter bound for global models reflects the intuition that pooling data across related time series provides more generalizable predictions than isolated modeling approaches. Of course, we can realize these advantages of GFMs in PyTorch! You can see a simple toy example in *Figure 11.3* (see `simple_gfm_example_nf.ipynb`) where we used PyTorch models of LSTM and NHITS (NeuralForecast implementation) with two seemingly unrelated series which share similar structures.

Figure 11.3 shows forecasts from a simple GFM trained jointly on passengers and milk production data.

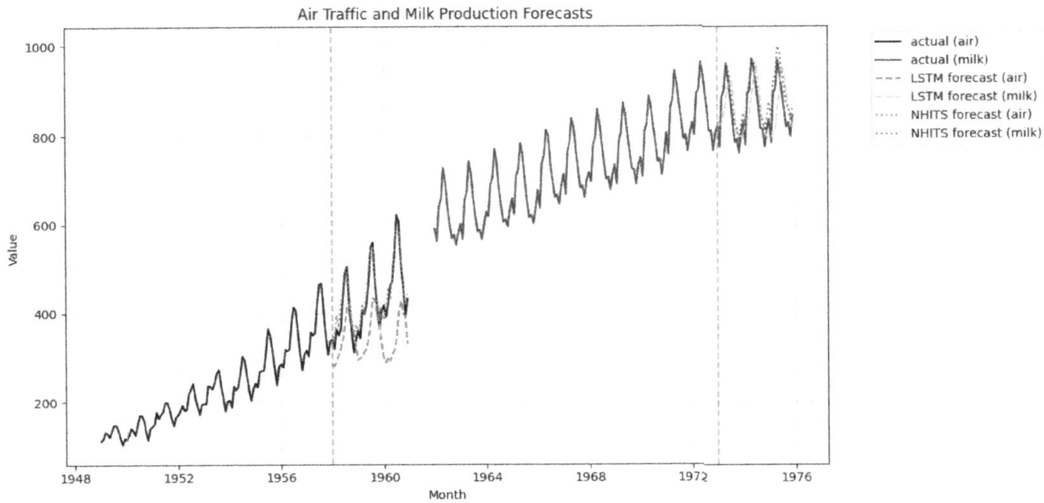

Figure 11.3: Simple GFM model with NeuralForecast

Building GFMs with PyTorch

Let's put transfer learning into practice by building our own GFM with PyTorch to develop our understanding! It seems fairly simple on the surface; instead of training separate models for each series, we'll build one model that learns patterns across multiple series, allowing knowledge transfer between similar series while preserving their unique characteristics. Building a GFM in PyTorch requires a few changes to the general structure of networks designed for single series processing, so that they can efficiently handle multiple time series while maintaining their temporal relationships and individual characteristics. Typically we want the architecture to consist of shared parameters for capturing common patterns, along with mechanisms for handling series-specific information. The main topics we'll cover here are:

- **Series embeddings** which are learnable representations for each time series
- **Feature processing** which are shared layers for processing temporal features
- **Pattern extraction**, the core architecture (LSTM/Transformer) for capturing temporal dependencies
- **Integration mechanism** which is how series-specific and shared information combines
- **Prediction head** which are the final layers for generating forecasts

We'll demonstrate these by building the following PyTorch Lightning LSTM module, and you can see how we have applied these in the companion script in the GitHub repo for the book, which is paired with additional supporting scripts due to the need to separate out code to prevent conflicts in memory for PyTorch.

```
import pytorch_lightning as L
import torch
```

```
import torch.nn as nn
import torch.nn.functional as F

class GlobalForecastingModel(L.LightningModule):
    def __init__(
        self,
        num_series: int,
        input_dim: int = 1,
        hidden_dim: int = 64,
        num_layers: int = 2,
        dropout: float = 0.1,
        learning_rate: float = 1e-3
    ):
        super().__init__()
        self.save_hyperparameters()

        # Project raw features into consistent representation
        self.feature_projection = nn.Linear(input_dim, hidden_dim)

        # Learn unique patterns for each series
        # Embedding dimension controls capacity for series-specific information
        self.series_embeddings = nn.Embedding(num_series, hidden_dim)
        self.embedding_norm = nn.LayerNorm(hidden_dim)

        # Shared temporal pattern extractor
        # All series share these LSTM parameters to learn common dynamics
        self.temporal_encoder = nn.LSTM(
            input_size=hidden_dim,
            hidden_size=hidden_dim,
            num_layers=num_layers,
            dropout=dropout if num_layers > 1 else 0,
            batch_first=True
        )

        # Combine global and series-specific patterns for prediction
        self.decoder = nn.Sequential(
            nn.Linear(hidden_dim * 2, hidden_dim),
            nn.ReLU(),
            nn.Dropout(dropout),
            nn.Linear(hidden_dim, 1)
        )
```

Our LSTM architecture incorporates a few elements that allow the network to capture information across multiple series. Firstly, we have a `feature_projection` layer that transforms input features into consistent internal representations; these are then projected to `self.series_embeddings = nn.Embedding(num_series, hidden_dim)`. `series_embeddings` is a learnable representation space where our network captures series-specific characteristics, while maintaining shared latent features. We use the embedding dimension (`hidden_dim`) to control how much unique information a model stores about each series. Layer normalization (`self.embedding_norm`) is used to help stabilize embeddings during training. This is the most common way to achieve this. The network also has a `temporal_encoder`, implemented as an LSTM here, which extracts the shared *patterns* across all series while maintaining temporal coherence. Most *common* neural architectures like LSTM, GRU, FFN, TCN simply follow the same three modifications to become global models:

- Add series embeddings to create learnable series-specific representations
- Implement an integration mechanism to combine these embeddings with the architecture's temporal features (typically through concatenation, attention, or gating)
- Structure data loading to preserve temporal ordering within each series while enabling efficient batching across series

These modifications fundamentally alter how a model processes information, enabling simultaneous learning of shared patterns while preserving series-specific characteristics. Although not immediately apparent, the computational efficiency gains grow as the scale (K series) does. Reiterating what we discussed previously, parameter complexity grows as $O(P + KE)$ for global models compared to $O(KP)$ for local models, where K represents the number of series, P the base model parameters, and E the embedding dimension.

It's worth pointing out here that models like N-BEATS, TFT and DeepAR are designed for global modeling, albeit with different approaches.

Feature integration mechanisms

There are three primary approaches to combining global patterns with series-specific characteristics, each offering different advantages:

- **Simple concatenation**: This is the most straightforward approach concatenates temporal features with series embeddings. This can be effective for series with similar scales and patterns. While computationally efficient, this approach assumes equal importance of global and series-specific features.

```
def forward(self, x, series_ids):
    # Project and encode features - shared patterns
    projected_features = self.feature_projection(x)
    temporal_features, _ = self.temporal_encoder(projected_features)

    # Get series specific patterns/embeddings
    series_embed = self.series_embeddings(series_ids)
    series_embed = self.embedding_norm(series_embed)
```

```
        # Combine them
        combined = torch.cat([temporal_features[:, -1, :], series_embed],
    dim=1)
        return self.decoder(combined)
```

- **Attention-based integration:** For more complex relationships, attention mechanisms can dynamically weight the importance of temporal features based on series context. This makes it particularly useful for heterogeneous series with varying temporal dependencies. We capture the weights in our attention layer, and apply them to the temporal features.

```
class AttentionIntegration(nn.Module):
    def __init__(self, hidden_dim):
        super().__init__()
        self.query_projection = nn.Linear(hidden_dim, hidden_dim)
        self.key_projection = nn.Linear(hidden_dim, hidden_dim)
        self.value_projection = nn.Linear(hidden_dim, hidden_dim)
        self.scale = np.sqrt(hidden_dim)

    def forward(self, temporal_features, series_embed):
        # Project series embedding to create attention query
        query = self.query_projection(
            series_embed).unsqueeze(1)  # [batch, 1, dim]

        # Project temporal features to create keys and values
        keys = self.key_projection(
            temporal_features)    # [batch, seq_len, dim]
        values = self.value_projection(temporal_features)

        # Compute scaled dot-product attention
        attention_scores = torch.matmul(
            query, keys.transpose(-2, -1)) / self.scale
        attention_weights = F.softmax(attention_scores, dim=-1)

        # Weight temporal features by attention
        attended_features = torch.matmul(
            attention_weights, values).squeeze(1)

        return attended_features
```

This mechanism allows the model to focus on relevant temporal patterns conditioned on series-specific context, particularly valuable when series exhibit varying degrees of temporal dependence.

- **Gated integration**: Gating provides a learnable way to control the flow of information from both global and series-specific sources through learned gates, which is useful when our series vary in their adherence to global patterns. It does this by learning dimension-specific mixing coefficients through an activation function, in this case sigmoid, to control the balance between temporal and series-specific features. Unlike attention, which weights temporal relationships, gating enables granular feature selection at each dimension; when a gate value approaches 1, temporal features dominate, and when it approaches 0, series-specific characteristics prevail.

```
class GatedIntegration(nn.Module):
    def __init__(self, hidden_dim):
        super().__init__()
        self.gate = nn.Linear(hidden_dim * 2, hidden_dim)

    def forward(self, temporal_features, series_embed):
        # Compute adaptive gates
        gate_input = torch.cat(
            [temporal_features[:, -1, :], series_embed], dim=1)
        gates = torch.sigmoid(self.gate(gate_input))

        # Gated combination
        return gates * temporal_features[:, -1, :] + (1 - gates) *
series_embed
```

If we were inclined, we could modify our LSTM-GFM to incorporate any of these integration mechanisms:

```
class GlobalForecastingModel(L.LightningModule):
    def __init__(self, ..., integration_type='attention'):
        super().__init__()
        # Previous initialization code...

        # Select integration mechanism
        if integration_type == 'attention':
            self.integration = AttentionIntegration(hidden_dim)
        elif integration_type == 'gated':
            self.integration = GatedIntegration(hidden_dim)
        else:
            self.integration = None  # Use simple concatenation
```

While libraries like `neuralforecast` have selected these in their code, it can be worth the time to try out different approaches on your own data, as the integration mechanism impacts model performance. In our experience, the following can help:

- Use concatenation when series are similarly structured and equally important
- Choose attention when different temporal patterns matter for different series
- Implement gating when series vary significantly in their adherence to global patterns

In the repository for this book, we ran a brief experiment on 60 series (20 each of high, medium, and low frequency selling items) to compare the different integration methods. This particular composition of data makes it somewhat difficult for a global model (He, Pang, and Si 2019; Zhang et al. 2022) as the patterns are quite different. In *Chapters 14* and *15*, we'll investigate methods that can help improve the performance of these models. You can see that overall all three models beat a seasonal naive with a 7-period window (days in this case).

```
             mse      rmse     mae      mape
Model
concat 126.5379 11.2489 5.6073 122.9569
att    125.3221 11.1947 6.0231 160.7044
gated  147.2588 12.1350 5.8526 111.3186
naive  327.7558 18.1040 9.1612 231.0724
```

Ignoring MAPE, the attention (`att`) and concatenation methods work best here, as they have the best MSE and RMSE. The MAE, however, isn't so good for the attention method. Overall, they appear to be picking up on a mean average with some seasonality. In either case, we would just pick the fastest.

Figure 11.4 shows the predictions from each integration approach.

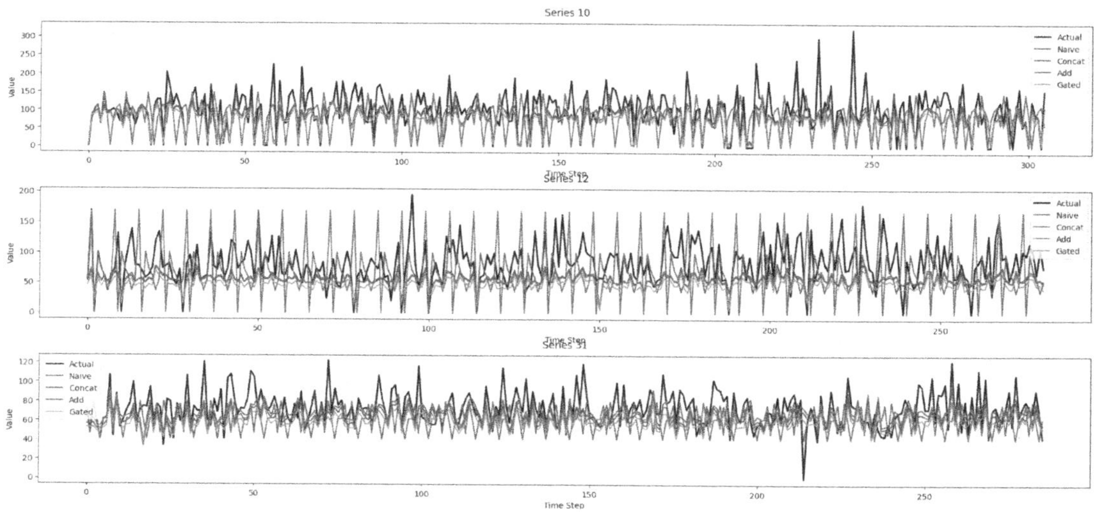

Figure 11.4: Integration performance predictions

Data organization and batching

Designing data structures for GFMs demands balancing computational efficiency with effective modeling performance. We need to handle multiple series while preserving temporal coherence and memory efficiency. Our PyTorch implementation utilizes index management and sliding windows to efficiently access data and preserve temporal sequences.

```
class TimeSeriesDataset(torch.utils.data.Dataset):
    def __init__(self, series_dict, lookback, horizon, min_samples=100):
        self.lookback = lookback
        self.horizon = horizon
        self.samples = []

        # Create mapping from series ID to numerical index
        self.series_to_idx = {
            sid: idx for idx, sid in enumerate(series_dict.keys())}

        # Preprocess and validate each series
        for series_id, data in series_dict.items():
            if len(data) < lookback + horizon:
                warnings.warn(f"Series {series_id} too short, skipping")
                continue

            series_idx = self.series_to_idx[series_id]
            valid_windows = self._generate_windows(data, series_idx)
            self.samples.extend(valid_windows)

        if len(self.samples) < min_samples:
            raise ValueError("Insufficient valid samples for training")

    def _generate_windows(self, data, series_idx):
        windows = []
        for t in range(len(data) - self.lookback - self.horizon + 1):
            windows.append({
                'series_idx': series_idx,
                'temporal_idx': t,
                'input': self._normalize_window(data[t:t+self.lookback]),
                'target': data[t+self.lookback:t+self.lookback+self.horizon]
            })
        return windows

    def _normalize_window(self, window):
```

```
        # Add scale-invariant normalization
        mean = window.mean()
        std = window.std()
        return (window - mean) / (std + 1e-8)
```

Our TimeSeriesDataset class addresses multi-series temporal data organization by using a mapping (`series_to_idx`) to create a consistent indexing scheme across series, and a sliding window approach to capture temporal dependencies. The sliding window generation creates input-target pairs by moving through each series, with lookback defining historical context and horizon determining prediction length. This standard approach provides efficient batch processing through PyTorch's DataLoader, but can be memory-intensive for large datasets. For substantial datasets, we can use *on-the-fly* window generation by computing windows only when requested during training, storing only metadata and window indices rather than materialized windows. Instead of being memory intensive, this is computationally expensive because windows are repeatedly generated rather than retrieved from storage, potentially creating I/O bottlenecks that can impact training throughput, particularly when GPU utilization outpaces CPU-based data generation. So you should keep watch over CPU/worker usage.

Training and balanced sampling

Training a GFM requires only a slight change from single-series training: batching through series. We wrap this into a `configure_training` function that wires together the data loaders, balanced sampling, and trainer configuration.

```
def configure_training(model, train_dataset, val_dataset, **kwargs):
    # Configure data loaders with balanced sampling
    train_sampler = create_balanced_sampler(train_dataset)
    train_loader = DataLoader(
        train_dataset,
        batch_size=kwargs.get('batch_size', 32),
        sampler=train_sampler,
        num_workers=kwargs.get('num_workers', 4)
    )

    val_loader = DataLoader(
        val_dataset,
        batch_size=kwargs.get('batch_size', 32),
        shuffle=False,
        num_workers=kwargs.get('num_workers', 4)
    )

    # Configure trainer with early stopping
    trainer = pl.Trainer(
        max_epochs=kwargs.get('max_epochs', 100),
        accelerator='gpu' if torch.cuda.is_available() else 'cpu',
```

```
            devices=1,
            gradient_clip_val=0.1,
            callbacks=[
                EarlyStopping(
                    monitor='val_loss',
                    patience=kwargs.get('patience', 10),
                    mode='min'
                )
            ]
        )

        return trainer, train_loader, val_loader
```

This implementation is deceptively simple, masking some underlying complexity; each batch has to maintain coherent temporal windows while balancing representation across different series. As with batching in general, there are advantages and disadvantages to the impact of this on computational efficiency and statistical robustness.

The architectures we've explored provide a foundation for knowledge transfer across series, but their effectiveness depends heavily on how we expose the model to different patterns during training. Default sampling approaches can create subtle biases that limit a model's ability to learn transferable patterns, particularly when series vary in length or characteristics. If we consider how sampling works in practice: when generating training batches, we typically take windows from all available series and sample randomly from this pool. An issue here is that a series with more observations contributes more windows to our sampling pool than a short series. Given two series:

```
# Series lengths
series_A = 1000  # ~970 possible windows
series_B = 100   # ~70 possible windows

# Probability of sampling from each in random batch
p_series_A = 970 / (970 + 70)  # ≈ 93%
p_series_B = 70 / (970 + 70)   # ≈ 7%
```

Our model would see patterns from longer series far more frequently, creating a training bias that undermines transfer learning. The solution is weighted sampling that balances representation across series:

```
def create_balanced_sampler(dataset):
    """Create sampler that evenly weights series representation"""
    # Calculate series frequencies
    series_counts = Counter(
        sample['series_idx'] for sample in dataset.samples
    )
```

```
    # Inverse frequency weighting
    weights = torch.zeros(len(dataset))
    for idx, sample in enumerate(dataset.samples):
        weights[idx] = 1.0 / series_counts[sample['series_idx']]

    # Normalize weights
    weights = weights / weights.sum()

    return WeightedRandomSampler(
        weights=weights,
        num_samples=len(dataset),
        replacement=True
    )
```

We apply this in our DataLoader, like so.

```
train_loader = DataLoader(
    train_dataset,
    batch_size=32,
    sampler=create_balanced_sampler(train_dataset),
    num_workers=4
)
```

Under this weighting scheme each series has equal probability of appearing in our training batches, regardless of length. The impact of balanced sampling extends beyond simple fairness; it fundamentally affects what patterns our model learns and how well they transfer, providing more diverse pattern exposure during training, better gradient updates for underrepresented series, improved generalization across varying series lengths, and reduced overfitting to dominant patterns. Our experiments show this balanced approach typically improves transfer performance, though training may take longer as the model sees more varied patterns. The sampling approach can extend beyond series length to balance any relevant characteristic, be it magnitude of values, frequency of events, or other domain-specific features.

Note

The companion notebooks include additional diagnostic utilities (`analyze_sampling_distribution`, `check_batch_bias`) for examining how different series contribute to your training batches and verifying that your sampling strategy is working as intended.

Optimizers, learning rate scheduling, and batching

Learning rate scheduling proves particularly crucial for global models due to the complex loss landscape created by optimizing across multiple series simultaneously. We use an adaptive learning rate schedule through PyTorch's `ReduceLROnPlateau` scheduler, alongside the Adam optimizer (introduced in *Chapter 6*):

```
def configure_optimizers(self):
    optimizer = torch.optim.Adam(
        self.parameters(),
        lr=self.hparams.learning_rate
    )

    scheduler = torch.optim.lr_scheduler.ReduceLROnPlateau(
        optimizer,
        mode='min',
        patience=5,
        factor=0.5
    )

    return {
        'optimizer': optimizer,
        'lr_scheduler': {
            'scheduler': scheduler,
            'monitor': 'val_loss'
        }
    }
```

The scheduler monitors validation loss and reduces the learning rate when performance plateaus. This adaptation proves essential as the optimizer must navigate a loss surface that combines both shared and series-specific patterns. The patience parameter allows the model to explore potentially promising regions before reducing the learning rate, while the factor controls the magnitude of adjustments.

Batch processing fundamentally influences both computational efficiency and model learning dynamics. Our experiments suggest optimal batch sizes typically range between 32-128 series windows, though this varies with dataset characteristics. We default to 32 in our examples as a conservative starting point that works well for moderately sized datasets; larger batches expedite training through increased parallelization, yet risk averaging out important temporal dependencies. For substantial datasets, memory constraints may necessitate streaming approaches to batch generation, where windows are grouped by series to preserve temporal ordering within each series while randomizing series order between epochs. Multiple workers facilitate efficient data loading through parallel processing, particularly valuable when handling extensive panel datasets; however, you must ensure that the data loading process preserves the temporal integrity of each series.

Validation demands particular attention in global modeling contexts. Our approach incorporates both temporal and transfer validation, assessing performance on recent data from known series as well as entirely held-out series.

```
def validation_step(self, batch, batch_idx):
    y_hat = self.forward(batch['input'], batch['series_ids'])
    loss = F.mse_loss(y_hat, batch['target'])

    self.log('val_loss', loss)
    self._log_series_metrics(y_hat, batch)

    return loss
```

Once we have set up our network and optimizers, it is simply a case of structuring our code. The following example is the general structure you should follow for initializing and training a GFM model.

```
# Model initialization
model = GlobalForecastingModel(
    num_series=len(series_dict),
    input_dim=feature_dim,
    hidden_dim=64
)

# Data preparation
train_dataset = TimeSeriesDataset(
    series_dict=train_series,
    lookback=24,
    horizon=12
)

# Training configuration
trainer = pl.Trainer(
    max_epochs=100,
    accelerator='gpu' if torch.cuda.is_available() else 'cpu',
    devices=1,
    gradient_clip_val=0.1
)

# Execute training
trainer.fit(
    model,
    DataLoader(
        train_dataset,
```

```
            batch_size=32,
            sampler=create_balanced_sampler(train_dataset),
            num_workers=4
        )
    )
```

PyTorch Lightning API change

The `gpus=1` parameter in PyTorch Lightning's Trainer is deprecated. Current versions use `accelerator='gpu', devices=1` (or `accelerator='cpu'` for CPU training). The code examples throughout this chapter use the updated API, but if you are following older tutorials or documentation you may encounter the legacy syntax.

The modifications for converting standard architectures to GFMs are fairly straightforward to implement while maintaining the core adaptability of the base architecture and enabling *cross-series learning*. Most modern deep learning frameworks support these adaptations with minimal code changes. The complexity lies not in the modifications themselves, but in selecting integration mechanisms and embedding dimensions based on the characteristics of your time series ensemble. For example, concatenation suits homogeneous series, while attention mechanisms may better serve heterogeneous patterns.

Transfer learning with NeuralForecast

Libraries like Darts and Nixtla's NeuralForecast are far more developed than we can build by ourselves, and both allow us to easily run global models. So it makes sense to make use of these libraries. In the companion notebooks and scripts for this section, we have set up and run several experiments, adapted from Darts' transfer learning tutorial (`https://unit8co.github.io/darts/examples/14-transfer-learning.html`). This is a great structure for testing and comparing transfer learning, and a great opportunity to develop your learning by applying the same code to your own data!

In the sections that follow we will explain the basics of Nixtla's modeling set-up, build models to predict international airline departures using individual statistical models, GFMs with decision-tree models, GFMs with neural networks, and pretrained GFMs with neural networks using data from another domain (retail). We will then compare the performance of all the models.

While global modeling allows us to train a model with many time series, the trained model can be used for meta-learning (*transfer learning*), which allows our models to adapt to new tasks, such as forecasting on new time-series data at inference without any additional training.

Data preparation

Our example here takes the international report departures dataset (`https://data.transportation.gov/api/views/xgub-n9bw/rows.csv`), which has several hundred series. After selecting for minimum size and resampling (`freq = 'ME'`; this is data aggregated by month, with month end as the date) to reduce gaps, we are left with 439 time series that we split into a training (`air_train`) and test set (air_test). Having a long enough series is important for models to simply have enough information to work with; not having enough is particularly problematic for neural networks. It is important that there are no gaps in time as the predictions will fail, so make sure you check your data if you use a different dataset. The time series do not have to cover the same starting and end dates, just have the same forecast horizon (fh = 18) with rows to match in the test set. As you know from previous chapters, we set up dataframes for Nixtla libraries so that we have three columns:

- unique_id which identifies each time series
- ds that contains timestamps
- y containing target values of the time series we want to predict

You can specify custom column names if you wish, but it's easier to simply use the standard naming convention so you don't have to provide the arguments to the function.

After wrangling and splitting the data you are left with something like the following univariate data. If you have exogenous variables like price, events etc., they should be included here too.

```
air_train = pd.concat(air_train, axis=0)
air_test = pd.concat(air_test, axis=0)

air_test.head(n=3)

            ds     y  unique_id
222 2008-07-31  4437      19386
223 2008-08-31  4319      19386
224 2008-09-30  3939      19386
```

It's worth noting at this point that we have not scaled the data; Nixtla libraries do this for us when we specify the scaler directly, which you will see later.

Statistical baselines

Once we have our data set up we instantiate our models in a list, or provide an individual model. Models are then passed to the appropriate library orchestrator, i.e. Theta to StatsForecast, XGBoost to MLForecast, and NHITS to NeuralForecast. It is important to pass the frequency of the data being modeled as this affects what the model expects and what it will forecast. We can also pass the number of workers/processors to be used if using CPU, although we highly recommend using GPU for neural networks. So long as GPUs are active and n_jobs is not specified, they will be used automatically. Once the instantiation is set up we pass training data (df=air_train) to the .fit method to train our models. You can see this in the first part of our modeling, where we set up several statistical models (SeasonalNaive, AutoARIMA, AutoETS, Theta) with StatsForecast.

```
from statsforecast.models import SeasonalNaive, AutoARIMA, AutoETS, Theta
from statsforecast import StatsForecast

fh = 18
```

```
models = [
    SeasonalNaive(season_length=12),
    AutoARIMA(),
    AutoETS(),
    Theta()
]

sf = StatsForecast(
    models=models,
    freq='ME',
    n_jobs=-1
)

sf.fit(df=air_train)
```

This code creates a single model for each series in the dataset, for each model type. We then take the fitted object and create forecasts by simply passing the forecast horizon length (`h=fh`) to the .forecast method. This creates a dataframe containing the predictions, with the predictions for each model in a separate column, which we can join to the actuals from the test set (`air_test`) to create a dataframe for evaluating model performance (`eval_df`).

```
fcst = sf.forecast(h=fh)
fcst = fcst.reset_index()

eval_df = fcst.merge(
    air_test[['unique_id', 'ds', 'y']],
    on=['unique_id', 'ds'],
    how='left'
)
```

Evaluation is simply a case of running the evaluate function on your forecast results. This function computes the performance of our model(s) using a set of specified metrics from `utilsforecast.losses`, such as **Mean Absolute Error** (**MAE**), **Root Mean Squared Error** (**RMSE**), etc. We pass these metrics, model names (matching column names), standard columns (`id_col`, `time_col`, `target_col`), and the training data (`air_train`) which is used to create metrics like MASE that require a baseline (e.g., a naive forecast) for scaling.

```
from functools import partial
from utilsforecast.evaluation import evaluate
from utilsforecast.losses import mae, mse, rmse, mase, mape, smape

series_metrics = evaluate(
    eval_df,
```

```
        models=[model_name], # Change to match column name
        metrics=metrics,
        train_df=air_train,
        id_col='unique_id',
        time_col='ds',
        target_col='y'
    )
```

This results in a long-format table of error metrics for each series and each error metric, that we can use for our comparisons.

```
   unique_id metric   value          model
0      19386    mae 673.778  SeasonalNaive
1      19392    mae  45.500  SeasonalNaive
2      19393    mae 357.778  SeasonalNaive
```

At this point, we have built individual models for each statistical modeling algorithm. We take the median results for each metric and algorithm and put them in a table for our comparison. However, it is wise to look at and statistically compare the distribution of errors across the different metrics (generally RMSE, MAE and some relative metric).

You can see below that AutoARIMA tends to perform better on most metrics, although it is beaten by the Seasonal Naive on both MAE and MASE (<1 is better than naive). We could make the argument that some feature engineering could enhance the AutoARIMA, but overall these models are not impressing greatly beyond a naive. They are, however, all extremely quick to build.

```
        model  time    mae  mape  mase      mse   rmse  smape
SeasonalNaive 0.089 32.389 0.415 0.986 1652.067 40.646  0.147
    AutoARIMA 0.173 32.719 0.408 1.035 1552.600 39.403  0.142
      AutoETS 0.092 34.445 0.449 1.122 1969.052 44.374  0.151
        Theta 0.901 31.873 0.420 1.083 1707.977 41.328  0.146
```

Decision-tree global models

Now let's extend our approach to include Global Forecasting Models (GFMs), which as stated earlier are trained on multiple series. We use decision-tree-based algorithms such as LightGBM, XGBoost, and Random Forest. These are highly effective with tabular data because they can capture complex, non-parametric, and non-linear relationships in our data, which makes them well-suited for time series forecasting tasks. We highly recommend reading more about decision trees in Manu Joseph's *Modern Time Series Forecasting with Python, 2nd edition*.

The MLForecast code deviates slightly from the StatsForecast approach to better support the ML models we are using: LinearRegression, XGBoost, LightGBM and Random Forest, although we only show two here. First we directly import the model, i.e. `lightgbm as lgb`, and `LinearRegression` from `sklearn`. We then embed these within the MLForecast orchestrator. Here we utilize MLForecast's feature engineering to add time-specific features that support model builds: lags, moving averages, and date features.

For decision tree based models like LightGBM specifically, we introduce additional lag transformations such as `RollingMean(12)`, `ExponentiallyWeightedMean(alpha=0.3)`, and `ExpandingMean()`, which enrich the feature set and help the model capture trends and patterns more effectively. All models use date-based features like month and year, extracted from the timestamp column (ds), to account for seasonality and time-related patterns.

The code follows the same pattern of model training and forecasting, although it does each separately. The `.fit` method trains each model on the training dataset (`air_train`), using the same column mappings for `unique_id`, `ds`, and `y`. After training, we apply the `.predict` method to generate forecasts for the specified horizon (`h=fh`). Although it's not shown here, we also captured the training time for each model, as this is an important factor to consider when building models in production; this allows us to compare the computational efficiency of different algorithms. We'll look at this at the end.

Notice that we have not specified any hyperparameters. While it is common to compare models before tuning, it is worth noting that this will affect results.

```
from mlforecast import MLForecast
from mlforecast.lag_transforms import ExpandingMean, ExponentiallyWeightedMean,
RollingMean
import lightgbm as lgb
from sklearn.linear_model import LinearRegression
from utilsforecast.feature_engineering import time_features


# Train Linear Regression
linear_model = MLForecast(
    models={'linear': LinearRegression()},
    freq='ME',
    lags=range(1,12),
    date_features=['month', 'year']
)

linear_model.fit(
    df=air_train,
    id_col='unique_id',
    time_col='ds',
    target_col='y',
    static_features=[]
)

linear_preds = linear_model.predict(h=fh)

# Train LightGBM
```

```
lgb_model = MLForecast(
    models=[lgb.LGBMRegressor(verbosity=-1,random_state=352, objective='mae')],
    freq='ME',
    lags=range(1,12),
    lag_transforms={
        1: [RollingMean(12), ExponentiallyWeightedMean(alpha=0.3)],
        12: [ExpandingMean()]
    },
    date_features=['month', 'year']
)

lgb_model.fit(air_train, id_col='unique_id', time_col='ds', target_col='y',
static_features=[])

lgb_preds = lgb_model.predict(h=fh)
```

So how do our ML global models with default hyperparameters do? Well, not great on MASE, which is over 1 for all of the ML models. But on MAE and RMSE we see some parity with AutoARIMA and the Seasonal Naive. All the decision tree models do well on RMSE, with Random Forest doing the best.

model	time	mae	mape	mase	mse	rmse	smape
SeasonalNaive	0.089	32.389	0.415	0.986	1652.067	40.646	0.147
AutoARIMA	0.173	32.719	0.408	1.035	1552.600	39.403	0.142
LinearReg	0.001	36.149	0.625	1.589	2041.514	45.183	0.186
LGBMRegressor	0.005	34.614	0.424	1.034	1601.328	40.015	0.138
XGBRegressor	0.005	32.339	0.464	1.220	1659.345	40.721	0.149
RandomForestRegressor	0.076	32.155	0.470	1.116	1489.510	38.594	0.142

Neural network GFMs

For our last direct GFMs we import NBEATS and NHITS PyTorch architectures directly from NeuralForecast. This is quite different from MLForecast; first we set up the model configuration, and you'll notice the forecast horizon is explicitly stated within the configuration object. The hyperparameters are not tuned but chosen to balance performance and efficiency.

Common settings across models in NeuralForecast are learning rates, validation steps, and batch sizes. `learning_rate=1e-3` and `val_check_steps=100` balance stable learning with early overfitting detection. Batch sizes (`1024` for NBEATS, `64` for NHITS) optimize memory usage and gradient stability. `input_size` is also a commonly shared attribute, although we have set different lengths in this example; typically twice the length of the forecast horizon is a good starting point.

For NBEATS specifically, `n_blocks=[1]*20` creates `20` blocks to learn hierarchical patterns, three MLP stacks per block (`mlp_units=[[256, 256]] * 3`) with `identity`, `trend`, and `seasonality` stack types decompose the time series, and a lookback window of `input_size=fh * 2` (twice the forecast horizon) ensures sufficient historical context. For NHITS, `n_pool_kernel_size=[2, 2, 1]` and `n_freq_downsample=[4, 2, 1]` capture patterns at different resolutions, with all `stack_types='identity'` and larger MLPs (`mlp_units=[[512, 512]] * 3`) providing greater capacity to capture patterns.

We recommend using a separate environment for MLForecast and NeuralForecast, as there can be library conflicts between them.

You will notice that we applied lag features, rolling features, and EWMA with our ML models but not with the DL models. This is because architectures like NBEATS and NHITS learn these representations internally from the raw data. However, the NBEATSx variant can also incorporate exogenous variables such as Fourier terms (cosine and sine waves to represent time-variant features like weekdays and months), promotional events, advertising data, etc. Generally this works well for moderate to large data sizes, but one should always consider hyperparameter tuning with blocked backtesting (see *Chapter 6*). We'd encourage you to experiment with these settings; tweaking them can lead to even better results and a deeper understanding of how the model works!

```
import torch
from neuralforecast import NeuralForecast
from neuralforecast.models import NBEATS, NHITS

# NBEATS
nbeats_config = NBEATS(
    h=18,
    input_size= fh * 2,  # sum=36
    n_blocks=[1]*20,
    mlp_units=[[256, 256]]  *3,
    stack_types=['identity', 'trend', 'seasonality'],
    learning_rate=1e-3,
    batch_size=1024,
    random_seed=243,
    max_steps=10,
    val_check_steps=100,
    scaler_type='standard'
)

# Initialize NeuralForecast with NBEATS
nbeats_model = NeuralForecast(
    models=[nbeats_config],
```

```
    freq='ME'
)

# Train NBEATS
nbeats_model.fit(df=air_train)
```

Another critical difference is that when calling .predict on the fitted model we pass a future dataframe (`futr_df=air_test`); we can do this with other libraries in Nixtla's ecosystem. This guides predictions and makes sure we are getting back appropriate predictions for each series, although it can also be useful for holding additional information such as exogenous variables (i.e. sale price) and static variables (i.e. business hierarchy level).

```
# Generate NBEATS forecasts
nbeats_preds = nbeats_model.predict(futr_df=air_test)
```

Looking at the median performance we can see that the DL models are better across the board of metrics (except on sMAPE), with substantially improved MAE, MASE and RMSE. It is fairly clear that NHITS is the superior of these untuned models for this particular dataset, if we don't weight sMAPE metrics highly (we don't), although this comes at a speed cost.

model	time	mae	mape	mase	mse	rmse	smape
SeasonalNaive	0.089	32.389	0.415	0.986	1652.067	40.646	0.147
AutoARIMA	0.173	32.719	0.408	1.035	1552.600	39.403	0.142
LGBMRegressor	0.005	34.614	0.424	1.034	1601.328	40.015	0.138
XGBRegressor	0.005	32.339	0.464	1.220	1659.345	40.721	0.149
RandomForestRegressor	0.076	32.155	0.470	1.116	1489.510	38.594	0.142
NBEATS	0.014	28.412	0.385	0.952	1217.697	34.895	0.151
NHITS	0.263	26.799	0.389	0.903	1170.369	34.211	0.149

Pretraining with external data

The final transfer learning idea we investigate, for now, is *pretraining*. This takes us from training a model with all-time series in a dataset to training with another dataset entirely. This brings us back around to the start of the chapter where we discussed using information from different domains to support modeling; of course the domain can be related or unrelated. Typically we want to pick up on similar patterns, so it can help to use data recorded or resampled to the same frequency. In our example we use the M4 monthly dataset (`https://github.com/Mcompetitions/M4-methods/tree/master/Dataset`), which is retail data. This is a much larger dataset (`n = 48,000`, averaging `200` time steps) than the international report departures dataset, with several thousand time series to gain information from. Large samples like this can take a long time to train as each epoch would be huge, so it's useful to limit the number of training samples used per series; we did this in the previous DL GFMs via `windows_batch_size`, but this is something to be aware of when pretraining and tuning.

If you are using your own data, it will help you to analyze and observe the series in your pretraining domain, to compare them to your target domain (we'll discuss this further in our chapter on time series clustering, *Chapter 15*).

The set-up is identical to the previous neural networks, except we pass in the M4 training data (`m4_train`) to `.fit`. After pretraining, we simply pass the training and test sets of our target domain to the trained model's `.predict` under `df` and `futr_df`.

```
# Set-up
nbeats_model = NeuralForecast(
    models=[nbeats_config],
    freq='ME'
)

# Pretrain with M4 data
nbeats_model.fit(df=m4_train)

# Generate forecasts for air_test
nbeats_preds = nbeats_model.predict(df=air_train, futr_df=air_test)
```

Our previous GFMs took advantage of the shared patterns across each series within our target domain to improve performance. Logically, if we use more data we would hope to improve on this, provided our pretraining data is informative, even if the model has never seen the training data for our test set (`air_test`). This is exactly what happens in this case; we see reductions in error (except APE metrics) for both models. This is a small boost for NBEATS, but huge for NHITS.

Why did NHITS benefit more from pretraining than NBEATS? The key difference is `max_steps`: NBEATS was configured with `max_steps=10`, while NHITS used `max_steps=1000` (which is default in NeuralForecast). With only `10` steps, NBEATS barely scratched the surface of the M4 dataset; with `1000` steps, NHITS had far more opportunity to extract the temporal abstractions from the larger pretraining corpus. The NHITS configuration was otherwise identical to the GFM trained on `air_train` alone; the model simply took more advantage of the additional training steps with the larger dataset. Both pretrained models demonstrate excellent performance that could undoubtedly be improved further with hyperparameter tuning.

model	time	mae	mape	mase	mse	rmse	smape
SeasonalNaive	0.089	32.389	0.415	0.986	1652.067	40.646	0.147
AutoARIMA	0.173	32.719	0.408	1.035	1552.600	39.403	0.142
LGBMRegressor	0.005	34.614	0.424	1.034	1601.328	40.015	0.138
XGBRegressor	0.005	32.339	0.464	1.220	1659.345	40.721	0.149
RandomForestRegressor	0.076	32.155	0.470	1.116	1489.510	38.594	0.142
NBEATS	0.014	28.412	0.385	0.952	1217.697	34.895	0.151
NHITS	0.263	26.799	0.389	0.903	1170.369	34.211	0.149
NBEATS_m4_pretrained	0.210	27.125	0.484	1.049	1145.380	33.843	0.181
NHITS_m4_pretrained	0.717	19.576	0.439	0.910	751.001	27.404	0.155

There are different approaches we could take to improve pretraining performance. For example, we could pretrain on daily data such as the M5 dataset first, then fine-tune on our monthly training data before making predictions. However, we have found that the opposite works better: training on monthly and then fine-tuning on daily. This seems to be because the smoother abstract patterns (e.g., trends, seasonality) learned from monthly data generalize better across domains. If model performance worsens on the same domain, it is likely because of domain mismatch, but fine-tuning does tend to narrow the gap between mismatched domains. Ultimately, pretraining benefits depend on shared temporal abstractions rather than identical frequencies; of course, this also depends on a model's capacity for capturing the complexities within large datasets.

> **Tips for improving GFMs and pretraining**
>
> We should mention some techniques that could improve performance. You will notice that we applied lag features, rolling features, EWMA, and time series meta-features with our ML models but not DL models. However, these can help inform DL models such as NBEATSx too; we could also apply other exogenous variables, i.e. promotional events, advertising, etc. This applies to pretraining as well as GFMs, if the variables align.
>
> - Hyperparameter tuning with Bayesian optimization (*Chapter 6*) is always recommended.
>
> Partitioning wherein you select and group data that is similar in terms of structure or natural grouping, i.e., potatoes and carrots are both vegetables. We will revisit this in *Chapter 15*.
>
> Encoding; not one-hot or ordinal, but frequency and target-mean encoding. We will discuss this further in *Chapter 16*.
>
> Hyperparameter tuning and partitioning your data appropriately can significantly enhance performance and are often the difference between a mediocre and excellent GFM.

Summary

Transfer learning is at the heart of what makes neural architectures shine in almost every domain. It allows networks to leverage patterns learned from one series to improve predictions on others, or to do this at the level of the whole dataset, a capability traditional statistical models lack. While bespoke models excel on individual series, they often struggle with new data distributions, scale variations, and unseen domains. Training separate models for thousands of series is computationally expensive and ignores shared temporal structures. Transfer learning addresses this by treating forecasting as a unified learning problem, capturing domain-invariant patterns through shared parameters and balancing series-specific nuances with global dynamics.

What you have seen mirrors techniques behind modern LLMs, enabling models to adapt to new tasks with minimal retraining. Although it seems that our understanding of this is still in its early stages for time-series forecasting, as can be seen with how LLMs perform poorly so far in this area.

In this chapter we covered:

- The mathematics of global modeling and why it reduces error bounds
- PyTorch techniques for handling multiple series (embeddings, integration mechanisms)
- Implementing domain adaptation with NeuralForecast for tasks like airline passenger forecasting
- Pretraining strategies (e.g., using M4 data to boost airline passenger forecasts)
- Strategies that can improve GFMs and pretraining, including partitioning, encoding, and hyperparameter tuning

Using PyTorch and Nixtla's implementations, we built global models that share learned patterns across series while preserving unique characteristics. Experiments showed neural architectures like NHITS outperformed statistical and tree-based models, with pretraining on external datasets (e.g., M4) further reducing errors. Key steps include structuring data with temporal windows, balancing batch sampling, and selecting integration mechanisms (concatenation, attention, or gating).

Your turn to experiment and learn: start with NeuralForecast's NBEATS or TFT implementations, pretrain on public datasets (M4, Wikipedia traffic), then fine-tune on your domain. The upfront computational cost can pay dividends in scalable, adaptable forecasts. In the next chapter, we will explore synthetic data generation, which can augment cross-learning and transfer learning datasets, particularly useful when your target domain has limited data.

Recommended reading

- Bandara, K., et al. 2020. "Improving the Accuracy of Global Forecasting Models using Time Series Data Augmentation." *arXiv*. `https://arxiv.org/abs/2008.02663`.
- Bozinovski, S. 2020. "Reminder of the First Paper on Transfer Learning in Neural Networks, 1976." *Informatica*, 44 (3). `https://doi.org/10.31449/inf.v44i3.2828`.
- Ehrig, C., et al. 2024. "The Impact of Data Set Similarity and Diversity on Transfer Learning Success in Time Series Forecasting." *arXiv*. `https://arxiv.org/abs/2404.06198`.
- Godahewa, R., et al. 2021. "Monash Time Series Forecasting Archive." *arXiv*. `https://arxiv.org/abs/2105.06643`.
- He, Q., Pang, P. C., and Si, Y. 2019. "Transfer Learning for Financial Time Series Forecasting." *PRICAI 2019: Trends in Artificial Intelligence*, 11671: 24–36. `https://doi.org/10.1007/978-3-030-29911-8_3`.
- Hewamalage, H., Bergmeir, C., and Bandara, K. 2021. "Recurrent Neural Networks for Time Series Forecasting: Current Status and Future Directions." *arXiv*. `https://arxiv.org/abs/1909.00590`.
- Hewamalage, H., Bergmeir, C., and Bandara, K. 2022. "Global Models for Time Series Forecasting: A Simulation Study." *arXiv*. `https://arxiv.org/abs/2012.12485`.
- Joseph, M., and Tackes, J. 2024. *Modern Time Series Forecasting with Python*. Second edition. Packt Publishing Ltd.

- Li, W., and Law, K. L. E. 2024. "Deep Learning Models for Time Series Forecasting: A Review." *IEEE Access*, 12: 92306–27. `https://doi.org/10.1109/ACCESS.2024.3422528`.
- Montero-Manso, P., and Hyndman, R. J. 2021. "Principles and Algorithms for Forecasting Groups of Time Series: Locality and Globality." *arXiv*. `https://arxiv.org/abs/2008.00444`.
- Oord, A. V. D., Li, Y., and Vinyals, O. 2019. "Representation Learning with Contrastive Predictive Coding." *arXiv*. `https://doi.org/10.48550/arXiv.1807.03748`.
- Pan, S. J., and Yang, Q. 2010. "A Survey on Transfer Learning." *IEEE Transactions on Knowledge and Data Engineering*, 22 (10): 1345–59. `https://doi.org/10.1109/TKDE.2009.191`.
- Pratt, L. Y. n.d. "Discriminability-Based Transfer Between Neural Networks."
- Weber, M., et al. 2021. "Transfer Learning With Time Series Data: A Systematic Mapping Study." *IEEE*. `https://ieeexplore.ieee.org/document/9646532`.
- Yoss, A. 2020. "Transfer Learning Using Pre-Trained AlexNet Model and Fashion-MNIST." Toward Data Science. `https://towardsdatascience.com/transfer-learning-using-pre-trained-alexnet-model-and-fashion-mnist-43898c2966fb`.
- Zhang, X., et al. 2022. "Self-Supervised Contrastive Pre-Training For Time Series via Time-Frequency Consistency." *arXiv*. `https://doi.org/10.48550/arXiv.2206.08496`.
- Zhuang, F., et al. 2020. "A Comprehensive Survey on Transfer Learning." *arXiv*. `https://doi.org/10.48550/arXiv.1911.02685`.

12

Synthetic Time Series Data

In the previous chapters, we discussed several models to deal with time series data. In this chapter, we take a step back and rethink time series forecasting from a high level and introduce a brand new and important topic: synthetic data for time series. We will introduce a new model called TimeVAE to generate new synthetic time series data.

To build a deep learning-based time series forecaster, we have four core components that work closely with each other. As shown in the simplified illustration in *Figure 12.1*, the first component is a time series dataset. Based on the dataset and the predetermined task, we come up with a proper forecasting model. The model is then trained with an optimization algorithm of choice using the time series dataset. Finally, we require a set of evaluation methods to justify the performance of our model.

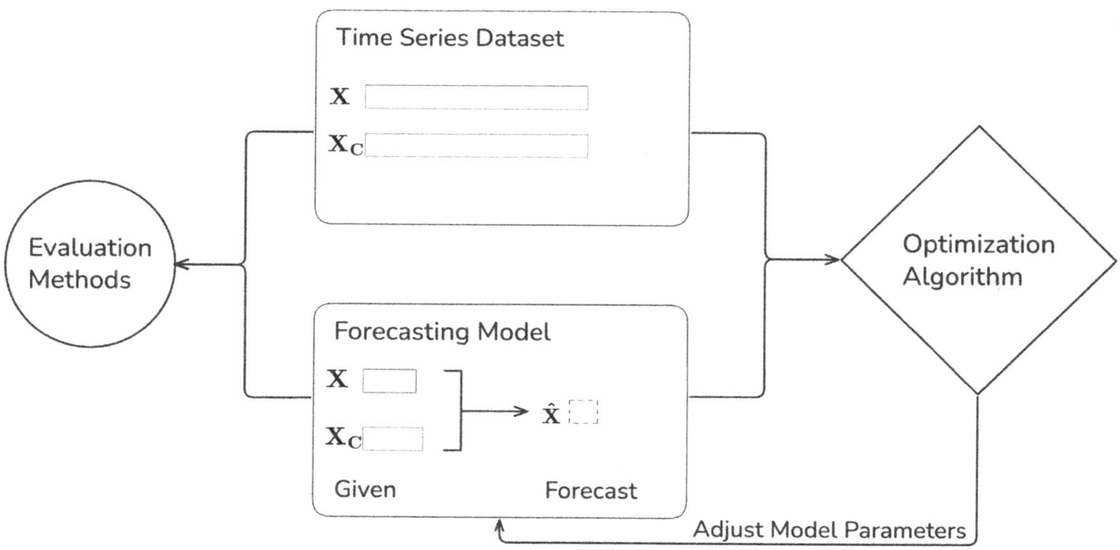

Figure 12.1: Four core components of a time series forecasting system: Dataset, Model, Optimization Algorithm, and Evaluation methods

To develop a high-performing time series forecasting model, it is crucial not only to select an appropriate model but also to ensure the dataset is of high quality. However, real-world datasets often have missing data points due to incidents or the high cost of data collection. Real-world data may also have restricted access due to sensitive information in the data. A commonly used strategy to address these issues is to generate synthetic time series data, which can augment the original dataset and improve its quality or avoid leakage of sensitive information.

In this chapter, we will guide you through several approaches for creating synthetic time series datasets. We will start with classical methods that use abstract mathematical models to generate data. Then, we will explore deep learning techniques for synthesizing time series data.

In the chapter:

- The data generating process
- Data-driven synthetic time series data
- Generating time series data using variational autoencoder
- Implementing TimeVAE
- Beyond VAE

The data-generating process

The data generating process for time series can be rather complex and highly diverse. Before diving into machine learning models, it is essential to develop intuitions of some fundamental data generating processes that underlie time series data. In this section, we provide an example by generating the number of visitors in a kiosk store, focusing on daily cycles.

Simulating visitor numbers can start simply with a Poisson distribution, which models the occurrence of events—in our case, number of visitors—within a set time frame. There is only one single parameter in a Poisson distribution, i.e., the expected number of visitors within 5 minutes. We use the `numpy` library to generate random visitor counts. To make the simulation more realistic, we also introduce a daily pattern. In the following discussions, we will assume that the number of visitors follows a sine wave pattern, where the number of expected visitors peaks at noon.

Guided by the above intuitions, we define a class called `KioskVisitors` to generate visitor counts. This class takes a `daily_profile` argument, incorporating the aforementioned daily pattern into the simulations.

```
import numpy as np
import pandas as pd
from numbers import Number


class KioskVisitors:
    """generate number of visitors for a kiosk store

    :param daily_profile: expectations of visitors
```

```
        within a fixed time window during a day
    """

    def __init__(self, daily_profile: np.ndarray):
        self.daily_profile = daily_profile
        self.daily_segments = len(daily_profile)

    def __call__(
        self,
        n_days: int,
        seed: Number=42,
    ) -> pd.DataFrame:
        """generate number of visitors for n_days

        :param n_days: number of days to generate visitors
        """
        rng = np.random.default_rng(seed)
        visitors = np.concatenate(
            [
                rng.poisson(self.daily_profile)
                for _ in range(n_days)]
        )

        df = pd.DataFrame(
            {
                "visitors": visitors,
                "time": np.arange(len(visitors)),
                "expectation": np.tile(
                    self.daily_profile,
                    n_days
                ),
            }
        )

        return df
```

To represent the assumed `sine` wave pattern in visitor numbers, we develop the `profile_sin` function to produce a daily profile that simulates the fluctuations in visitor counts throughout a typical day.

```
def profile_sin(
    t: np.ndarray,
    lambda_min: float,
```

```
    lambda_max: float,
) -> np.ndarray:
    """generate a sin wave profile for
    the expected number of visitors
    within a few min for each hour during a day

    :param t: time in minutes
    :param lambda_min: minimum number of visitors
    :param lambda_max: maximum number of visitors
    """
    amplitude = lambda_max - lambda_min
    t_rescaled = (t - t.min()) / t.max() * np.pi

    return amplitude * np.sin(t_rescaled) + lambda_min
```

In *Figure 12.2*, we showcase the time series dataset for 3 days of simulated kiosk visitors. We use solid lines to illustrate the actual visitor numbers generated by our model, allowing for direct comparison with the expected visitor counts represented by dashed lines.

Refer to color images on the Graphic bundle or the ebook.

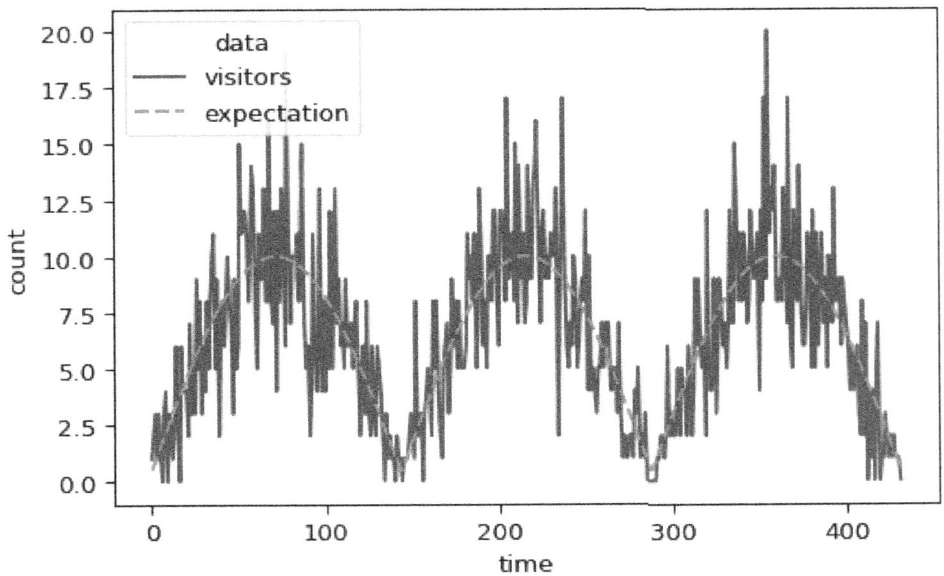

Figure 12.2: Simulation of a kiosk store in a busy area

We have generated a synthetic time series dataset, and we can generate a bunch with different Poisson distribution parameters and daily fluctuation profiles. There are an enormous number of mathematical models that fit each domain in order to generate synthetic time series data. For example, the *Wiener* process in economics can be used to model stock prices. While such mathematical modeling methods provide bases to generate time series data, there are two challenges:

- The **goodness** of the generated dataset heavily relies on our understanding of the real-world problem and our ability to select an appropriate mathematical model.
- The generated data may not fully capture the complexities and nuances found in real-world time series, even with a well-chosen model.

What we are doing is essentially modeling data based on the **Data Generating Process** (**DGP**) behind the scenes using intuition. Then we use the simple mathematical model to generate new data points. While physicists and economists may find this approach appealing for its simplicity, data science practitioners often gravitate toward data-driven approaches. In the next section, we will explore a data-driven approach to remove the manual mathematical modeling step.

Data-driven synthetic time series

Before we delve into deep learning models, let's explore two classes of simpler, data-driven approaches:

- **Randomizing the existing time series data**: For example, we can take the original data, add noise to each data point to alter the amplitudes. We can also randomize in time dimension by chunking the series and rearranging the chunks. Randomization can be applied in the time domain or in a frequency domain using methods like Fourier transformation.
- **Decomposing and mixing existing time series data**: For example, we take the average of multiple time series data or mixing them using an appropriate algorithm. In the paper "*A global averaging method for dynamic time warping, with applications to clustering*" by F. Petitjean et al., the authors proposed a method called **Dynamic Time Warping Barycenter Averaging** which is a weighted time series averaging methods based on the distance defined by **Dynamic Time Warping** (**DTW**).

These methods have proven effective for certain use cases. In a paper by Um et al., available at `http://arxiv.org/abs/1706.00527`, the authors augmented their sensor data using different randomization methods and observed improvements in classification performance. However, these naive methods often struggle to reproduce the intricate patterns and dependencies found in complex real-world time series. Deep learning models offer a more powerful alternative for generating synthetic time series that can capture these complexities.

Generating time series data using variational autoencoders

The field of generative models has exploded in this era of GenAI, with researchers constantly pushing the boundaries of what is possible. In fact, the idea of using deep learning models to generate new data has a long history. In 1994, Christopher Bishop proposed a model called the **Mixture Density Network** (**MDN**). The MDN model is a neural network that generates the means and variances of a Gaussian mixture model. New data can be generated by sampling from the Gaussian mixture model. Since then, generative models have advanced significantly; two prominent architectures are **Variational Autoencoders** (**VAEs**) and **Generative Adversarial Networks** (**GANs**). This section will delve into VAEs and their application to generating synthetic time series data. Although newer models like diffusional models have gained traction recently, VAEs remain a foundational and inspiring architecture, providing valuable insights into the principles behind more state of the art architectures.

As the name suggests, VAEs are built upon two key concepts: the autoencoder structure and the *variational* component. We've already encountered encoder-decoder architectures in previous chapters. Autoencoders are encoder-decoder architectures, but the decoder is set to reconstruct the input to the encoder. Take a look at *Figure 12.3*, it presents a conceptual illustration of autoencoders. The *variational* aspect refers to a specific way of formulating the loss function, incorporating principles of variation of functional. We will discuss more details as we go through the implementation. The implementation follows the paper *TimeVAE: A Variational Auto-encoder for multivariate time series generation* by Desai A et al., which is available at `http://arxiv.org/abs/2111.08095`.

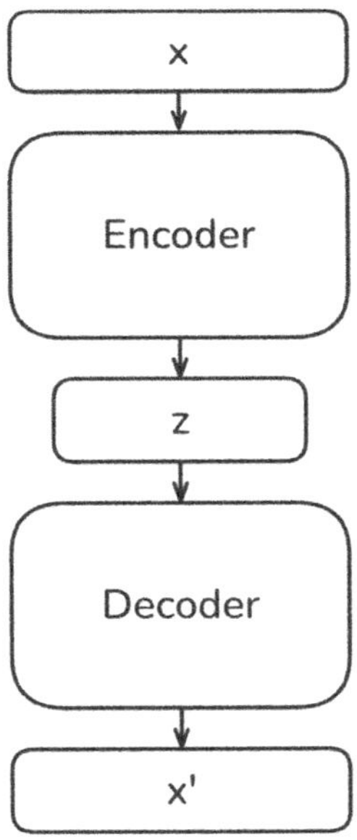

Figure 12.3: Variational Autoencoder consists of an encoder and a decoder

The input x is fed into the encoder and the encoder generates the latent space embedding z. z is then given to the decoder to reconstruct the input, and the output is denoted as x'.

Implementing TimeVAE

This section provides a practical implementation of the TimeVAE model using PyTorch. To streamline readability, we've made some simplifications in our code. For a more detailed implementation, please refer to the official repository `https://github.com/abudesai/timeVAE` or the repository `https://github.com/wangyz1999/timeVAE-pytorch`.

1. As is usual, we start with preparing the time series dataset. As the TimeVAE model is quite complex, we will use a simple synthetic dataset to demonstrate the model. The synthetic dataset is generated using the following code:

```
import math
import pandas as pd
import numpy as np

def generate_sine_wave(
    num_periods: int,
    num_samples_per_period: int,
    amplitude: float = 1.,
    period: float = 20,
) -> pd.DataFrame:
    """generate a sine wave time series

    :param num_periods: number of periods
    :param num_samples_per_period: number of
        samples per period
    :param amplitude: amplitude of the sine wave
    :param period: period of the sine wave
    """
    time_step = period / num_samples_per_period
    steps = []
    time_series = []
    for i in range(
        num_periods * num_samples_per_period
    ):
        t = i * time_step
        angle = (
            amplitude
            * math.cos(2 * math.pi * t / period)
            + 2
        )
        steps.append(t)
        time_series.append(angle)
```

```
    return pd.DataFrame(
        {"t": steps, "theta": time_series}
    )

df = generate_sine_wave(300, 30)
df.head(100).plot(
    x="t", y="theta", style="k.-",
    title="Sin Data Samples for TimeVAE Training",
)
```

The output of the code block is shown in the following figure. It is the sine wave data we will use for our TimeVAE model.

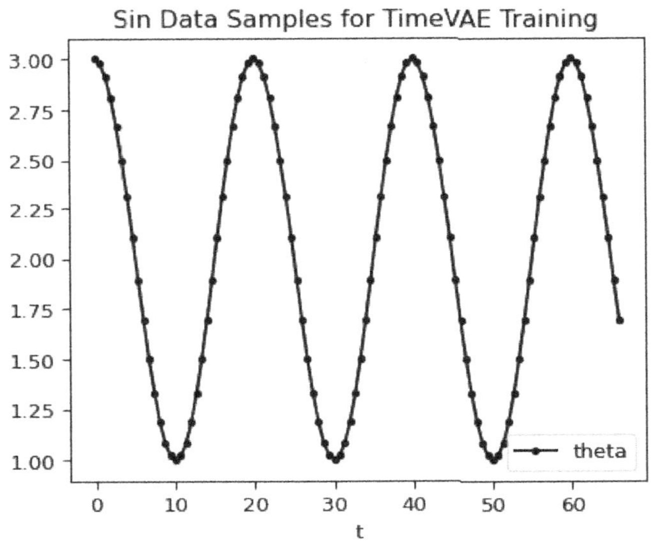

Figure 12.4: Artificial Training Data. We created this sine wave as our training data

The generated training data shown in *Figure 12.4* is **sinusoidal**. The data consists of 300 periods of sine wave, with 30 samples in each period. We will use this data to train the so-called TimeVAE model. Before our discussion about the model itself, we need to create the datasets and dataloaders.

2. Next, we'll create a native PyTorch Dataset, which defines how our data is structured and accessed during training:

```
from typing import Tuple
from torch.utils.data import Dataset
```

```
class TimeVAEDataset(Dataset):
    """A dataset from a pandas dataframe.

    For a given pandas dataframe, this generates a pytorch
    compatible dataset by sliding in time dimension.

    :param dataframe: input dataframe with all the features.
    :param window_size: length of time series slicing chunks
    """

    def __init__(
        self,
        dataframe: pd.DataFrame,
        window_size: int,
    ):
        super().__init__()
        self.dataframe = dataframe
        self.window_size = window_size
        self.dataframe_rows = len(self.dataframe)
        self.length = self.dataframe_rows - self.window_size + 1

    def moving_slicing(self, idx: int) -> np.ndarray:
        return self.dataframe[
            idx : self.window_size + idx
        ].values

    def __getitem__(
        self,
        idx: int
    ) -> np.ndarray | list[np.ndarray]:
        if isinstance(idx, slice):
            if (idx.start < 0) or (idx.stop >= self.length):
                raise IndexError(
                    f"Slice out of range: {idx}"
                )
            step = idx.step if idx.step is not None else 1
            return [
                self.moving_slicing(i)
                for i in
                range(idx.start, idx.stop, step)
            ]
```

```
        else:
            if idx >= self.length:
                raise IndexError("End of dataset")
            return self.moving_slicing(idx)

    def __len__(self) -> int:
        return self.length
```

Using the custom TimeVAEDataset, we can easily create all the data loaders required for the model training, validation, and testing. We use `LightningDataModule` from Lightning to simplify our pipelines. As a convention, we set the default `test_fraction` to `0.3` (meaning 30% of the data is reserved for testing) and `val_fraction` to `0.1` (meaning 10% for validation).

3. We then code the related imports for our data module:

```
from functools import cached_property
import torch
from torch.utils.data import DataLoader
import lightning as L
```

4. First, we define the initialization logic and cached dataset properties for the data module:

```
class TimeVAEDataModule(L.LightningDataModule):
    """
    Lightning DataModule for Time Series VAE.

    This data module takes a pandas dataframe and generates
    the corresponding dataloaders for training, validation and
    testing.

    :param window_size: length of time series slicing chunks
    :param dataframe: input dataframe with all the features.
    :param test_fraction: fraction of the dataframe to be used
        for testing
    :param val_fraction: fraction of the dataframe to be used
        for validation
    :param batch_size: batch size for the dataloaders
    :param num_workers: number of workers for the dataloaders
    """

    def __init__(
        self,
        window_size: int,
        dataframe: pd.DataFrame,
```

```
        test_fraction: float = 0.3,
        val_fraction: float = 0.1,
        batch_size: int = 32,
        num_workers: int = 0,
    ):
        super().__init__()
        self.window_size = window_size
        self.batch_size = batch_size
        self.dataframe = dataframe
        self.test_fraction = test_fraction
        self.val_fraction = val_fraction
        self.num_workers = num_workers

        (
            self.train_dataset,
            self.val_dataset
        ) = self.split_train_val(
            self.train_val_dataset
        )

    @cached_property
    def df_length(self):
        return len(self.dataframe)

    @cached_property
    def df_test_length(self):
        return int(self.df_length * self.test_fraction)

    @cached_property
    def df_train_val_length(self):
        return self.df_length - self.df_test_length

    @cached_property
    def train_val_dataframe(self):
        return self.dataframe.iloc[: self.df_train_val_length]

    @cached_property
    def test_dataframe(self):
        return self.dataframe.iloc[self.df_train_val_length :]

    @cached_property
```

```
    def train_val_dataset(self):
        return TimeVAEDataset(
            dataframe=self.train_val_dataframe,
            window_size=self.window_size,
        )

    @cached_property
    def test_dataset(self):
        return TimeVAEDataset(
            dataframe=self.test_dataframe,
            window_size=self.window_size,
        )
```

5. Next, we add the methods that split the data and return the dataloaders for training, validation, testing, and prediction:

```
    def split_train_val(self, dataset: Dataset):
        return torch.utils.data.random_split(
            dataset, [1 - self.val_fraction, self.val_fraction]
        )

    def train_dataloader(self):
        return DataLoader(
            dataset=self.train_dataset,
            batch_size=self.batch_size,
            shuffle=True,
            num_workers=self.num_workers,
            persistent_workers=(
                True if self.num_workers > 0
                else False
            ),
        )

    def test_dataloader(self):
        return DataLoader(
            dataset=self.test_dataset,
            batch_size=self.batch_size,
            shuffle=False,
            num_workers=self.num_workers,
        )

    def val_dataloader(self):
```

```
        return DataLoader(
            dataset=self.val_dataset,
            batch_size=self.batch_size,
            shuffle=False,
            num_workers=self.num_workers,
            persistent_workers=(
                True if self.num_workers > 0
                else False
            ),
        )

    def predict_dataloader(self):
        return DataLoader(
            dataset=self.test_dataset,
            batch_size=len(self.test_dataset),
            shuffle=False
        )
```

6. Finally, we can instantiate our data module. Here we set `window _size` to 24 as this window size can capture some patterns. We can of course choose a much larger one to capture the patterns, but it will cost a lot of computing power:

```
window_size = 24

time_vae_dm = TimeVAEDataModule(
    window_size=window_size,
    dataframe=df[["theta"]],
    batch_size=32
)
```

Implementing the model

Our TimeVAE model will require some parameter configurations, i.e., deciding the hyperparameters. For example, the encoder requires us to specify the input and output sizes of each layer (`hidden_layer_sizes`), which determine the complexity of each layer in the encoder. We also need to define the input size (`sequence_length`) representing the number of time steps in the model input samples, and the latent space size (`latent_size`) which controls the dimension of the compressed representation in latent space. In our example, we ignore the hyperparameter tuning step to focus on the model itself.

```
import dataclasses
from typing import List


@dataclasses.dataclass
```

```
class VAEParams:
    """Parameters for VAEEncoder and VAEDecoder

    :param hidden_layer_sizes: list of hidden layer sizes
    :param latent_size: latent space dimension
    :param sequence_length: input sequence length
    :param n_features: number of features
    """

    hidden_layer_sizes: List[int]
    latent_size: int
    sequence_length: int
    n_features: int = 1

    @cached_property
    def data_size(self) -> int:
        """The overall dimension
        of the input data when flattened.
        """
        return self.sequence_length * self.n_features

    def asdict(self) -> dict:
        """Convert the parameters to a dictionary
        to make it easier to log the parameters
        during training.
        """
        return dataclasses.asdict(self)
```

The TimeVAE model consists of two key components: an encoder and a decoder. The encoder, illustrated in *Figure 12.5*, uses one-dimensional convolutional layers and activation functions to generate a distribution, which is then used to sample data points from the latent space.

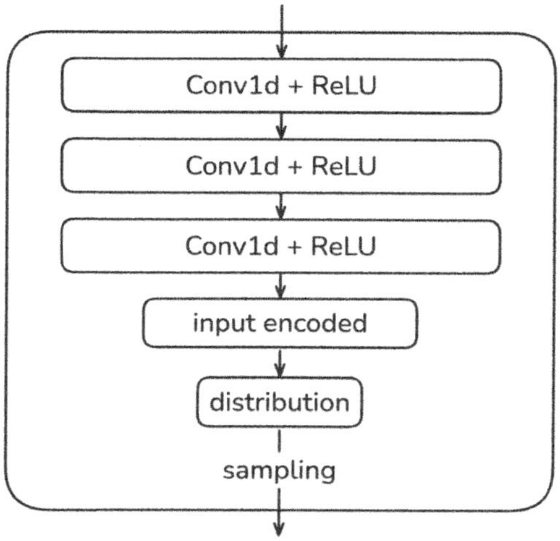

Figure 12.5: The TimeVAE encoder architecture

The input time series is fed into the encoder from the top. The input data goes through a one-dimensional convolution layer then a ReLU layer. This is repeated three times. The result is then used to generate the parameters of a distribution. The final output of the encoder is a sample from the distribution.

Let's import the packages. We will use `torchinfo.summary` to print out the details of the model, so `torchinfo` is also imported.

```
from torch import nn
from torchinfo import summary
```

The encoder outputs the latent space representations `z` by applying several convolutional networks. First, we define the encoder initialization and the forward pass:

```
class VAEEncoder(nn.Module):
    """Encoder of TimeVAE.

    :param params: parameters for the encoder
    """
    def __init__(self, params: VAEParams):
        super().__init__()

        self.params = params
        self.hparams = params.asdict()

        encode_layer_sizes = [
            self.params.n_features
        ] + self.params.hidden_layer_sizes
        self.layers_used_to_encode = [
```

```
            self._conv_block(size_in, size_out)
            for size_in, size_out in zip(
                encode_layer_sizes[:-1],
                encode_layer_sizes[1:]
            )
        ] + [nn.Flatten()]
        self.encode = nn.Sequential(
            *self.layers_used_to_encode
        )
        encoded_size = (
            self.cal_conv1d_output_dim() *
            self.params.hidden_layer_sizes[-1]
        )
        self.z_mean_layer = nn.Linear(
            encoded_size,
            self.params.latent_size
        )
        self.z_log_var_layer = nn.Linear(
            encoded_size,
            self.params.latent_size
        )

    def forward(
        self, x: torch.Tensor
    ) -> Tuple[torch.Tensor, torch.Tensor, torch.Tensor]:
        batch_size, _, _ = x.size()
        x = x.transpose(1, 2)
        x = self.encode(x)

        z_mean = self.z_mean_layer(x).view(
            batch_size,
            self.params.n_features,
            self.params.latent_size
        )
        z_log_var = self.z_log_var_layer(x).view(
            batch_size,
            self.params.n_features,
            self.params.latent_size
        )
        epsilon = torch.randn(
            batch_size,
```

```
            self.params.n_features,
            self.params.latent_size
        ).type_as(x)
        z = z_mean + torch.exp(0.5 * z_log_var) * epsilon

        return z_mean, z_log_var, z
```

Next, we add the helper methods used by the encoder. These methods define the convolutional block and calculate the output size after the convolutional layers:

```
    def _conv_block(
        self, size_in: int, size_out: int
    ) -> nn.Module:
        """1d convolution + ReLU block
        """
        return nn.Sequential(
            *[
                nn.Conv1d(
                    size_in,
                    size_out,
                    kernel_size=3,
                    stride=2,
                    padding=1
                ),
                nn.ReLU(),
            ]
        )

    def cal_conv1d_output_dim(self) -> int:
        """the output dimension of
        all the Conv1d layers
        """
        output_size = (
            self.params.sequence_length *
            self.params.n_features
        )

        for l in self.layers_used_to_encode:
            if l._get_name() == "Conv1d":
                output_size = self._conv1d_output_dim(
                    l, output_size
                )
```

```
            elif l._get_name() == "Sequential":
                for l2 in l:
                    if l2._get_name() == "Conv1d":
                        output_size = self._conv1d_output_dim(
                            l2, output_size
                        )

        return output_size

    def _conv1d_output_dim(
        self,
        layer: nn.Module,
        input_size: int,
    ) -> int:
        """calculate
        the output size of a Conv1d layer
        """
        return (
            (
                input_size
                + 2 * layer.padding[0]
                - layer.kernel_size[0]
            )
            // layer.stride[0]
        ) + 1
```

The encoder is instantiated with hidden layers with sizes `[200, 100, 50]` to compress the information:

```
encoder = VAEEncoder(
    VAEParams(
        hidden_layer_sizes=[200, 100, 50],
        latent_size=8,
        sequence_length=window_size,
    )
)
```

To get a review of what we just built, we use `torchinfo` to print out the layers of the model. We can clearly observe that the first sequential block is the core encoding network.

```
summary(encoder)
=================================================================
Layer (type:depth-idx)                   Param #
=================================================================
```

```
VAEEncoder                                  --
├─Sequential: 1-1                           --
│    └─Sequential: 2-1                      --
│    │    └─Conv1d: 3-1                     800
│    │    └─ReLU: 3-2                       --
│    └─Sequential: 2-2                      --
│    │    └─Conv1d: 3-3                     60,100
│    │    └─ReLU: 3-4                       --
│    └─Sequential: 2-3                      --
│    │    └─Conv1d: 3-5                     15,050
│    │    └─ReLU: 3-6                       --
│    └─Flatten: 2-4                         --
├─Linear: 1-2                               1,208
├─Linear: 1-3                               1,208
=================================================================
Total params: 78,366
Trainable params: 78,366
Non-trainable params: 0
=================================================================
```

The decoder is illustrated in *Figure 12.6*. The decoder takes the latent space and reconstructs the input data using Multi-Layer Perceptron (**MLP**) and **Softplus** activation. The reason we choose **Softplus** is to avoid significant discontinuities in the outputs.

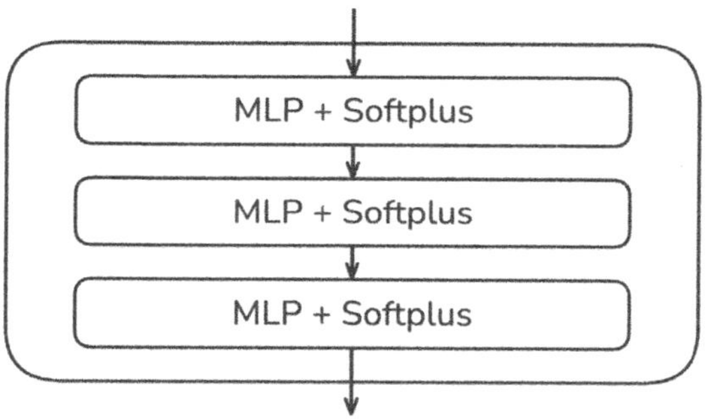

Figure 12.6: The TimeVAE decoder architecture

The latent space sample is fed to the decoder from the top. We attempt to reconstruct the original input using three blocks of MLP followed by Softplus activation.

The decoder implementation is simpler compared to the encoder as we only need to stack a few MLP blocks.

```
class VAEDecoder(nn.Module):
    """Decoder of TimeVAE
```

```
    :param params: parameters for the decoder
    """

    def __init__(self, params: VAEParams):
        super().__init__()

        self.params = params
        self.hparams = params.asdict()

        decode_layer_sizes = (
            [self.params.latent_size]
            + self.params.hidden_layer_sizes
            + [self.params.data_size]
        )

        self.decode = nn.Sequential(
            *[
                self._linear_block(size_in, size_out)
                for size_in, size_out in zip(
                    decode_layer_sizes[:-1],
                    decode_layer_sizes[1:]
                )
            ]
        )

    def forward(self, z: torch.Tensor) -> torch.Tensor:
        output = self.decode(z)
        return output.view(
            -1,
            self.params.sequence_length,
            self.params.n_features
        )

    def _linear_block(
        self, size_in: int, size_out: int
    ) -> nn.Module:
        """create linear block based on
        the specified sizes
        """
        return nn.Sequential(
            *[
```

```
                nn.Linear(size_in, size_out),
                nn.Softplus()
            ]
        )
```

The decoder is instantiated as shown here:

```
decoder = VAEDecoder(
    VAEParams(
        hidden_layer_sizes=[30, 50, 100],
        latent_size=8,
        sequence_length=window_size
    )
)
```

We can compare the decoder to the encoder, and we find that it is a much smaller network. However, keep in mind that the encoder is the part that will generate the representations for us in the end.

```
summary(decoder)
=================================================================
Layer (type:depth-idx)                   Param #
=================================================================
VAEDecoder                               --
├─Sequential: 1-1                        --
│    └─Sequential: 2-1                   --
│    │    └─Linear: 3-1                  270
│    │    └─Softplus: 3-2                --
│    └─Sequential: 2-2                   --
│    │    └─Linear: 3-3                  1,550
│    │    └─Softplus: 3-4                --
│    └─Sequential: 2-3                   --
│    │    └─Linear: 3-5                  5,100
│    │    └─Softplus: 3-6                --
│    └─Sequential: 2-4                   --
│    │    └─Linear: 3-7                  2,424
│    │    └─Softplus: 3-8                --
=================================================================
Total params: 9,344
Trainable params: 9,344
Non-trainable params: 0
=================================================================
```

We create the final TimeVAE model by integrating the encoder and decoder, as illustrated in *Figure 12.3*:

```
class VAE(nn.Module):
    """VAE model with encoder and decoder

    :param encoder: encoder module
    :param decoder: decoder module
    """
    def __init__(
        self,
        encoder: nn.Module,
        decoder: nn.Module
    ):
        super().__init__()
        self.encoder = encoder
        self.decoder = decoder
        self.hparams = {
            **{
                f"encoder_{k}": v
                for k, v
                in self.encoder.hparams.items()
            },
            **{
                f"decoder_{k}": v
                for k, v
                in self.decoder.hparams.items()
            },
        }

    def forward(
        self, x: torch.Tensor
    ) -> tuple[
        torch.Tensor,
        torch.Tensor,
        torch.Tensor
    ]:
        z_mean, z_log_var, z = self.encoder(x)
        x_reconstructed = self.decoder(z)
        return x_reconstructed, z_mean, z_log_var
```

Next, we define the training, validation, and test steps using a `LightningModule` called `VAEModel`. The key here is the loss function of VAE. While a detailed derivation of the loss function is beyond the scope of this chapter, it is beneficial to understand the indications of each component in the loss function. The specific loss function we employ is given as follows:

$$-\mathcal{L}(q) = -\mathbb{E}_q \ln p(x \mid z) + \mathrm{KL}\big(q(z \mid x) \parallel p(z)\big),$$

where $q(z \mid x)$ and $p(x \mid z)$ are representing the encoder and the decoder, respectively. The first term is the reconstruction loss, which is the negative log likelihood of the input data given the latent space. This part is simplified to $-(x - x_{\text{reconstructed}})^2 + \text{const}$ under our Gaussian distribution assumption. If the reconstructed value $x_{\text{reconstructed}}$ diverges too much from the original input x, our model won't be able to faithfully generate data that is similar to our training data. The second term is the Kullback-Leibler (KL) divergence between the encoder and the prior distribution. The KL divergence is a measure of how much one probability distribution differs from a second, expected probability distribution. Thus it is useful to ensure that the latent space produced by the encoder is close to our pre-determined Gaussian distribution. Once the model is trained, we will sample from the normal distribution and feed it into the decoder for it to generate samples. Only if the KL divergence is small enough, we can be confident that the normally distributed sample for the decoder is meaningful. Overall, both components in the loss are better to be small.

For a deep dive into the loss function, including the **Evidence Lower Bound** (**ELBO**), the KL divergence, and the reparameterization trick, we recommend the paper *Autoencoders* by Bank, Koenigstein, and Giryes, available at `http://arxiv.org/abs/2003.05991`.

We now wrap the VAE model in a Lightning module. This lets us define the training, validation, and testing logic in one place.

First, we import the learning rate scheduler:

```
from torch.optim.lr_scheduler import CosineAnnealingLR
```

Next, we define the `VAEModel` class and initialize the key training parameters:

```
class VAEModel(L.LightningModule):
    """VAE model using
    VAEEncoder, VAEDecoder, and VAE

    :param model: VAE model
    :param reconstruction_weight: weight for
        the reconstruction loss
    :param learning_rate: learning rate for
        the optimizer
    :param scheduler_max_epochs: maximum epochs
        for the scheduler
    """
```

```
    def __init__(
        self,
        model: VAE,
        reconstruction_weight: float = 1.0,
        learning_rate: float = 1e-3,
        scheduler_max_epochs: int = 10000,
    ):
        super().__init__()
        self.model = model
        self.reconstruction_weight = reconstruction_weight
        self.learning_rate = learning_rate
        self.scheduler_max_epochs = scheduler_max_epochs

        self.hparams.update(
            {
                **model.hparams,
                **{
                    "reconstruction_weight": reconstruction_weight,
                    "learning_rate": learning_rate,
                    "scheduler_max_epochs": scheduler_max_epochs,
                },
            }
        )
        self.save_hyperparameters(self.hparams)
```

The `forward()` method simply passes the input through the underlying VAE model:

```
    def forward(
        self,
        x: torch.Tensor
    ) -> tuple[
        torch.Tensor,
        torch.Tensor,
        torch.Tensor
    ]:
        return self.model(x)
```

We then define the training step. During training, the model reconstructs the batch, calculates the total loss, and logs the reconstruction and KL-divergence losses:

```
    def training_step(
        self,
        batch: torch.Tensor,
```

```
            batch_idx: int
    ) -> torch.Tensor:
        batch_reconst, z_mean, z_log_var = self.model(batch)
        loss_total, loss_reconst, loss_kl = self.loss(
            x=batch,
            x_reconstructed=batch_reconst,
            z_mean=z_mean,
            z_log_var=z_log_var,
        )

        self.log_dict(
            {
                "train_loss_total": loss_total,
                "train_loss_reconstruction": loss_reconst,
                "train_loss_kl": loss_kl,
            }
        )

        return loss_total
```

The validation and test steps follow the same structure, but log their metrics separately:

```
    def validation_step(
        self,
        batch: torch.Tensor,
        batch_idx: int
    ) -> torch.Tensor:
        batch_reconst, z_mean, z_log_var = self.model(batch)
        loss_total, loss_reconst, loss_kl = self.loss(
            x=batch,
            x_reconstructed=batch_reconst,
            z_mean=z_mean,
            z_log_var=z_log_var,
        )
        self.log_dict(
            {
                "val_loss_total": loss_total,
                "val_loss_reconstruction": loss_reconst,
                "val_loss_kl": loss_kl,
            }
        )
```

```
        return loss_total

    def test_step(
        self,
        batch: torch.Tensor,
        batch_idx: int
    ) -> torch.Tensor:
        batch_reconst, z_mean, z_log_var = self.model(batch)
        loss_total, loss_reconst, loss_kl = self.loss(
            x=batch,
            x_reconstructed=batch_reconst,
            z_mean=z_mean,
            z_log_var=z_log_var,
        )
        self.log_dict(
            {
                "test_loss_total": loss_total,
                "test_loss_reconstruction": loss_reconst,
                "test_loss_kl": loss_kl,
            }
        )
        return loss_total
```

Next, we define the overall VAE loss. It combines the reconstruction loss with the KL-divergence loss:

```
    def loss(
        self,
        x: torch.Tensor,
        x_reconstructed: torch.Tensor,
        z_log_var: torch.Tensor,
        z_mean: torch.Tensor,
    ) -> Tuple[
        torch.Tensor,
        torch.Tensor,
        torch.Tensor
    ]:
        """loss function for VAE
        """
        loss_reconstruction = self.reconstruction_loss(
            x, x_reconstructed
        )
        loss_kl = -0.5 * torch.sum(
```

```
            1 + z_log_var - z_mean**2 - z_log_var.exp()
        )
        loss_total = (
            self.reconstruction_weight
            * loss_reconstruction
            + loss_kl
        )

        return (
            loss_total / x.size(0),
            loss_reconstruction / x.size(0),
            loss_kl / x.size(0),
        )
```

The reconstruction loss measures how closely the reconstructed sequence matches the original input sequence. It also compares the mean of the original and reconstructed sequences:

```
    def reconstruction_loss(
        self,
        x: torch.Tensor,
        x_reconstructed: torch.Tensor
    ) -> torch.Tensor:
        r"""Reconstruction loss for VAE.

        $$
        \sum_{i=1}^{N} (x_i - x_{reconstructed_i})^2
        + \sum_{i=1}^{N}
        (\mu_i - \mu_{reconstructed_i})^2
        $$
        """
        loss = torch.sum(
            (x - x_reconstructed) ** 2
        ) + torch.sum(
            (
                torch.mean(x, dim=1)
                - torch.mean(
                    x_reconstructed, dim=1
                )
            ) ** 2
        )

        return loss
```

Finally, we configure the optimizer and learning rate scheduler:

```
    def configure_optimizers(self) -> dict:
        optimizer = torch.optim.SGD(
            self.parameters(),
            lr=self.learning_rate
        )
        scheduler = CosineAnnealingLR(
            optimizer, T_max=self.scheduler_max_epochs
        )

        return {
            "optimizer": optimizer,
            "lr_scheduler": {
                "scheduler": scheduler,
                "monitor": "train_loss",
                "interval": "step",
                "frequency": 1,
            },
        }
```

It's time to train the model!

Training the model

Let's put everything together and train the model. We set the maximum number of epochs to `2000` and use the `EarlyStopping` callback to prevent overfitting.

```
max_epochs = 2000

vae = VAE(
    encoder=encoder,
    decoder=decoder,
)

vae_model = VAEModel(
    vae,
    reconstruction_weight=3,
    scheduler_max_epochs=(
        max_epochs
        * len(time_vae_dm.train_dataloader())
    )
)
```

The model is trained using the Trainer class from PyTorch Lightning. We also use the `TensorBoardLogger` to log the training process and metrics. The following code block shows how to set up the model and the training process.

```
from lightning.pytorch.callbacks.early_stopping import EarlyStopping


trainer = L.Trainer(
    precision="64",
    max_epochs=max_epochs,
    min_epochs=5,
    callbacks=[
        EarlyStopping(
            monitor="val_loss_total",
            mode="min",
            min_delta=1e-10,
            patience=10
        )
    ],
    logger=L.pytorch.loggers.TensorBoardLogger(
        save_dir="lightning_logs",
        name="time_vae_naive"
    ),
)
```

Training the model involves executing the following lines of code. The training takes some time (about 1 hour on a RTX 4090 GPU) but it is worth running it once and checking out the metrics in tensorboard.

```
trainer.fit(model=vae_model, datamodule=time_vae_dm)
```

Generating synthetic data

Once trained, our model can generate synthetic data by sampling from a normal distribution and feeding these samples into the decoder. The following code generates 100 samples of synthetic data using this process.

```
sampling_z = torch.randn(
    2,
    vae_model.model.encoder.params.latent_size
).type_as(
    vae_model.model.encoder.z_mean_layer.weight
)
sampling_x = vae_model.model.decoder(sampling_z)
```

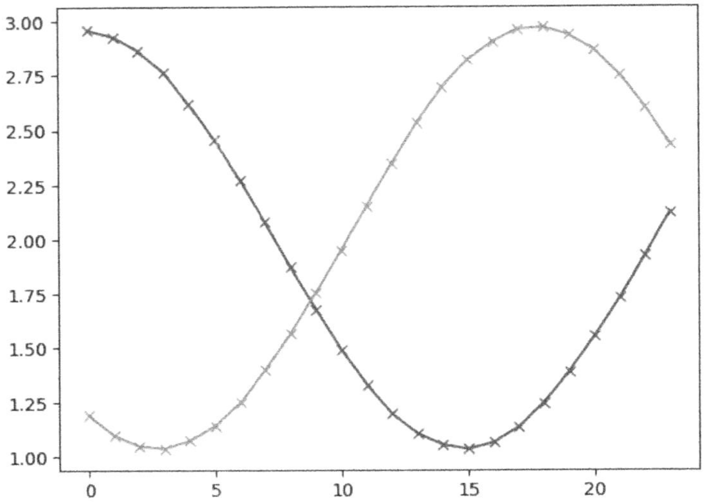

Figure 12.7: Two samples generated by the trained TimeVAE model; both visually close to the input data

To compare the synthetic samples generated by the model and the original data, we use t-SNE to visualize the samples, as shown in *Figure 12.8*. We can see that the patterns of those samples generated by the model are very close to those of the original data.

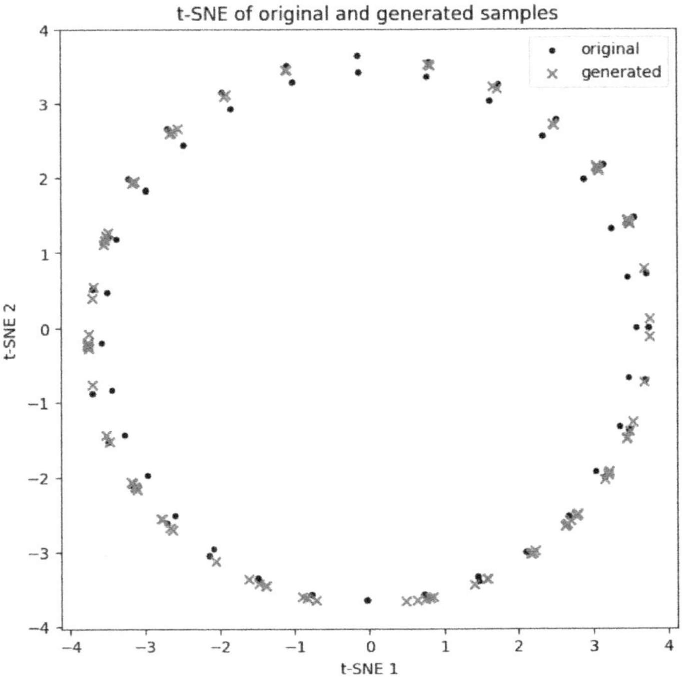

Figure 12.8: Original samples vs generated samples after t-SNE embedding

In this figure we use crosses to represent the generated data and dots for original samples.

The quality of the generated data can also be evaluated by assessing its impact on downstream tasks. For instance, in a time series forecasting project, we can train the same model on the original data, then on the combined data of the original and the generated data. If the model achieves superior performance on the combined dataset, it suggests that the generated data provides valuable information for enhancing the forecasting task. A note of caution: t-SNE is a qualitative evaluation method and should be used in conjunction with other evaluation metrics to ensure the quality of the generated data. We provide a reference on evaluation metrics for synthetic time series data in the reading list.

Generating sine wave data is not yet that amazing. TimeVAE can generate data with way more features; see the original paper. Here we also provide in *Figure 12.9* another example of training and generating data using the Kiosk visitor dataset we generated before. The pattern of generated Kiosk visitors is also very close to the simulated input data using Poisson distributions.

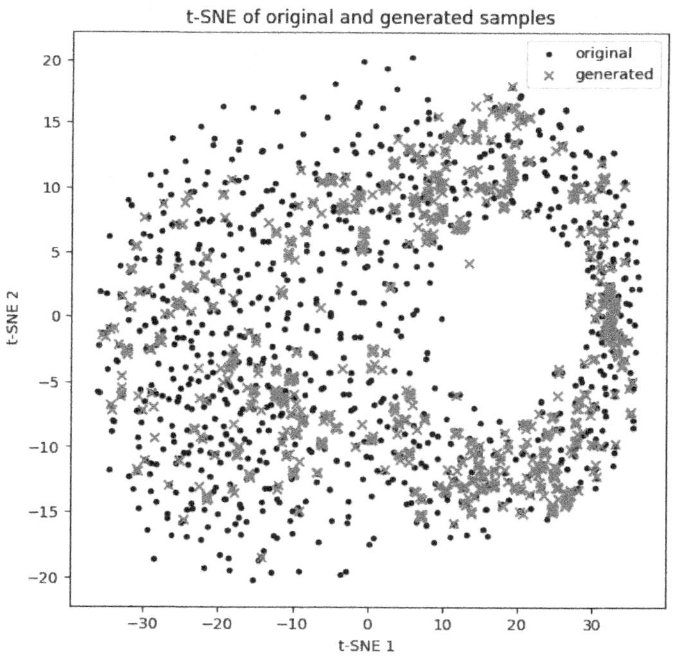

Figure 12.9: Samples of Kiosk visitors simulated using our mathematical model in the beginning of this chapter vs generated samples using TimeVAE after t-SNE embedding

Same as the previous figure, we use crosses to represent the generated data and dots for original samples.

Beyond VAE

The VAE model is a good first deep learning model for generating synthetic time series data. Meanwhile, models such as **Generative Adversarial Network** (**GAN**) are also used to generate time series data. For example, Yoon, Jarrett, and Schaar developed a GAN-based time series generative model in their work *Time-series generative adversarial networks*. Other models such as Neural ODE, which can generate synthetic time series using a mechanism like a first order differential equation, are particularly desirable for irregularly sampled time series data and are useful to interpolate the data point. The field of time series generative modeling remains active and rapidly evolving, with researchers continually exploring new architectures and techniques to enhance the quality and versatility of generated data.

Summary

In this chapter, we introduced the concept of synthetic time series data and discussed why it is useful when real-world data is limited, incomplete, expensive to collect, or sensitive.

We first explored the data generating process and used a kiosk visitor example to show how mathematical assumptions, such as Poisson distributions and daily sine-wave patterns, can be used to simulate time series data.

We then looked at data-driven approaches to synthetic data generation, including adding noise, rearranging time chunks, and combining time series using methods such as Dynamic Time Warping Barycenter Averaging.

Next, we introduced Variational Autoencoders, or VAEs, and implemented a simplified TimeVAE model using PyTorch and Lightning. We built the dataset, dataloaders, encoder, decoder, loss function, and training workflow, then used the trained model to generate new synthetic time series samples.

Finally, we discussed how generated data can be evaluated visually, with methods such as t-SNE, and practically, by testing whether it improves downstream forecasting performance. We closed by looking beyond VAEs to other generative models, such as GANs and Neural ODEs.

Recommended reading

- Bank, D., Koenigstein, N., and Giryes, R. 2020. "Autoencoders." *arXiv*. `http://arxiv.org/abs/2003.05991`.
- Bishop, C. 1994. "Mixture density networks." `https://publications.aston.ac.uk/id/eprint/373/`.
- Desai, A., Freeman, C., Wang, Z., and Beaver, I. 2021. "TimeVAE: A Variational Auto-encoder for multivariate time series generation." *arXiv*. `http://arxiv.org/abs/2111.08095`.
- Petitjean, F., et al. n.d. "A global averaging method for dynamic time warping, with applications to clustering."
- Rubanova, Y., Chen, R. T. Q., and Duvenaud, D. 2019. "Latent ODEs for irregularly-sampled time series." *Advances in Neural Information Processing Systems*. `https://papers.nips.cc/paper_files/paper/2019/hash/42a6845a557bef704ad8ac9cb4461d43-Abstract.html`.

- Stenger, M., et al. 2024. "Evaluation is key: a survey on evaluation measures for synthetic time series." *Journal of Big Data*, 11 (1). `https://doi.org/10.1186/s40537-024-00924-7`.
- Um, et al. 2017. (Topic: Data Augmentation for Sensor Data). `http://arxiv.org/abs/1706.00527`.
- Yoon, J., Jarrett, D., and Schaar, M. 2019. "Time-series generative adversarial networks." *Advances in Neural Information Processing Systems*, 32: 5509–5519. `https://papers.nips.cc/paper/2019/hash/c9efe5f26cd17ba6216bbe2a7d26d490-Abstract.html`.

Join our community on discord

Join our community's Discord space for discussions with the authors and other readers.

`https://discord.gg/VXpz9YPUTW`

13

Diffusion Models

Up until this chapter, we have discussed models that are very bland: those are very abstract models that are mostly inspired by statistical learning. In this chapter, we will introduce a completely new scheme of models called diffusion models. We will go through the basics of diffusion process and also build a diffusion model for forecasting in this chapter.

We will take an in-depth look at:

- Introduction to probabilistic forecasting
- Diffusion as a real-world phenomenon
- Naive diffusion model applied on time series data
- Denoising diffusion model

Introduction to probabilistic forecasting

We have discussed model architectures that are deterministic: The prediction is fixed forecast for a certain given input of the historical values, i.e., the prediction is a point value. On the other hand, in *Chapter 12*, we also explained the model called TimeVAE for synthetic data generation. As you may have noticed, TimeVAE, different from other model architectures we have discussed, is probabilistic: each prediction is generated by sampling from the model, thus the prediction is made of a sample of different result. In this chapter, we will discuss a new probabilistic model architecture called diffusion model that utilizes generative models as forecasters.

Diffusion model is a unique approach. In TimeVAE, we use a latent space to capture the information contained in the original input. The encoder compresses the data into tensors in the latent space, while the decoder uses the compressed data to generate data. In diffusion models, instead of directly compressing the input data using one single encoder, we add noise to the original data using the so-called diffusion process through multiple steps. We then use a reverse process to remove the noise, generating our newly generated time series. In this chapter, you will also learn different ways of incorporating brand new concepts into time series forecasting problems. To understand this process better, let's see where the word **diffusion** comes from in the first place.

Diffusion as a real-world phenomenon

Diffusion is in fact a quite complicated phenomenon. On the other hand, understanding this phenomenon will help us a lot when building the diffusion models. In this section, we will introduce diffusion and use code to simulate the results to help us understand what diffusion is.

When a drop of ink is added to a glass of water, the ink will spread out, and we will observe that original location of the ink is darker and the edge of the color blob is lighter. This physics behind this phenomenon is the diffusion process. Let's put our statistician hat on and examine the diffusion process again, we realize that information is gradually lost as the ink spreads out. However, if we have a method to reverse the process, we can recover the original ink drop. This is the idea behind the diffusion model. The simplest mathematical model for the process is called the **Brownian motion**. We know that the ink drop consists of multiple small and opaque ink particles such as carbon. As shown in *Figure 13.1*, the idea is that water molecules will exert random forces on the ink particles, causing them to move randomly in the water. The ink particles will eventually spread out in the water due to the stochastic forces.

Refer to color images on the Graphic bundle or the ebook.

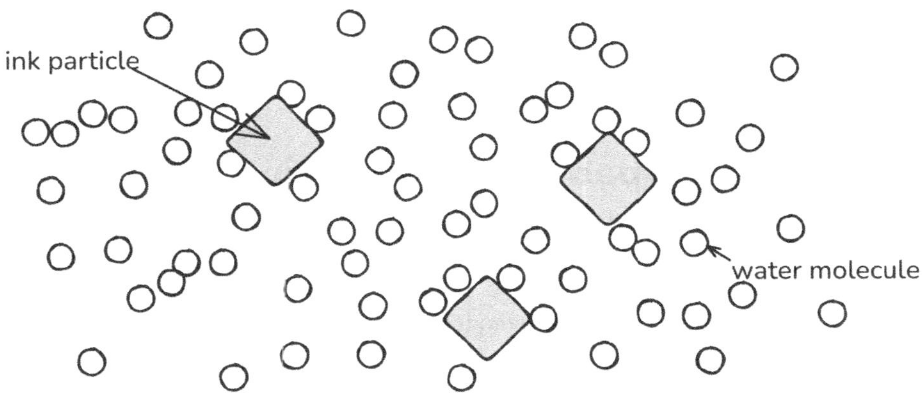

Figure 13.1: Ink particles diffuse in water

In *Figure 13.1* the circles represent the water molecules, and the diamonds are the ink particles. The water molecules hit the ink particles all the time but the net forces on the ink particle are different if we examine it at a specific time. Thus, the ink particles will move stochastically.

To make it easier to illustrate the idea, imagine we have a thin tube of water. We add a drop of ink to particular range of positions in the tube, say a range [0,1] on our one dimensional coordinate, as shown in *Figure 13.2*.

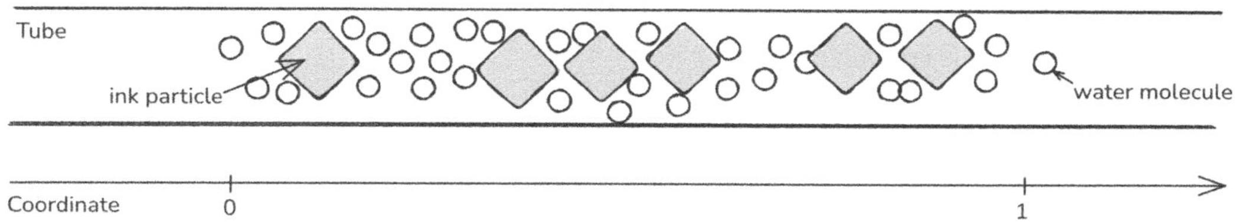

Figure 13.2: Ink particles in a 1D tube and can only move in 1D

The initial positions of the particles, $\{x_i\}$, are almost uniformly distributed between 0 and 1, i.e., $x_i \in [0,1]$.

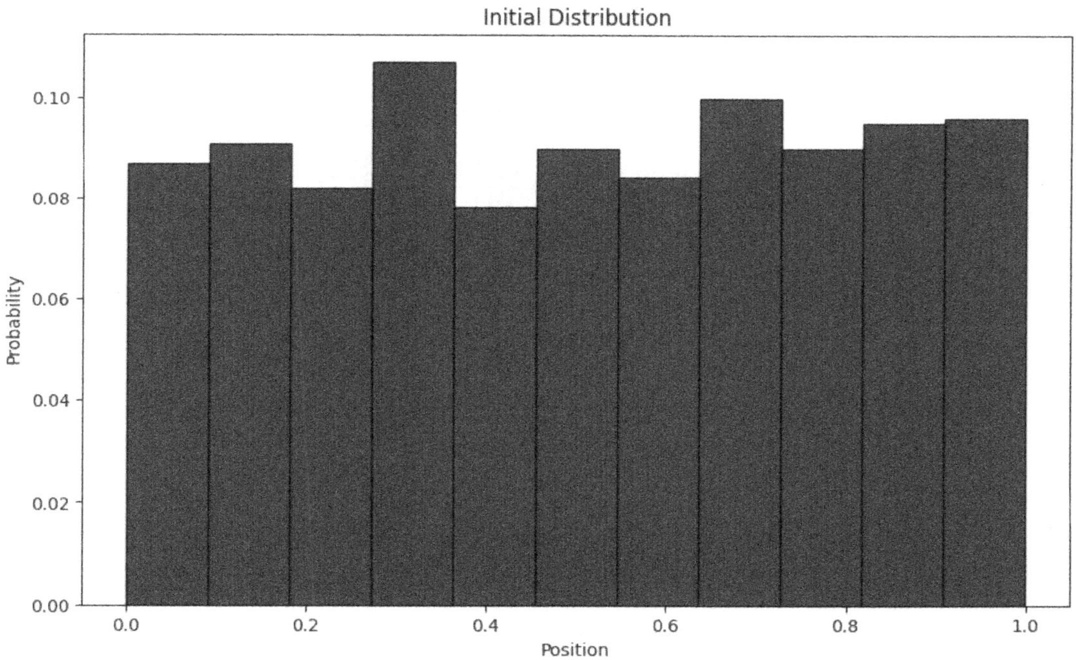

Figure 13.3: Initial distribution of the ink

When the diffusion process starts, the ink particles will move randomly in the tube due to the stochastic forces exerted on the ink particles by the motions of the water molecules. Phenomenologically, the ith ink particle can be described by the Brownian motion,

$$x_i(t+1) = \sqrt{\alpha(t)}x_i(t) + \sqrt{1-\alpha(t)}\epsilon_i(t),$$

where $\epsilon_i(t)$ is a random variable that follows the normal distribution $\mathcal{N}(0,1)$, and $\alpha(t)$ describes the tendency of the particle to stay at its current position, or inertia, physically speaking. When $\alpha(t) = 1$, the particle will remain stationary at its current position, and the stochastic microscopic forces will not have any effect on the ink particle. Usually, $\alpha(t)$ is smaller than 1 as the ink particle will be influenced by the stochastic forces and will gain some velocity to move away from its current position.

In the extreme of $\alpha(t) = 0$, each step is independent and the particle position is completely random without any dependencies on history.

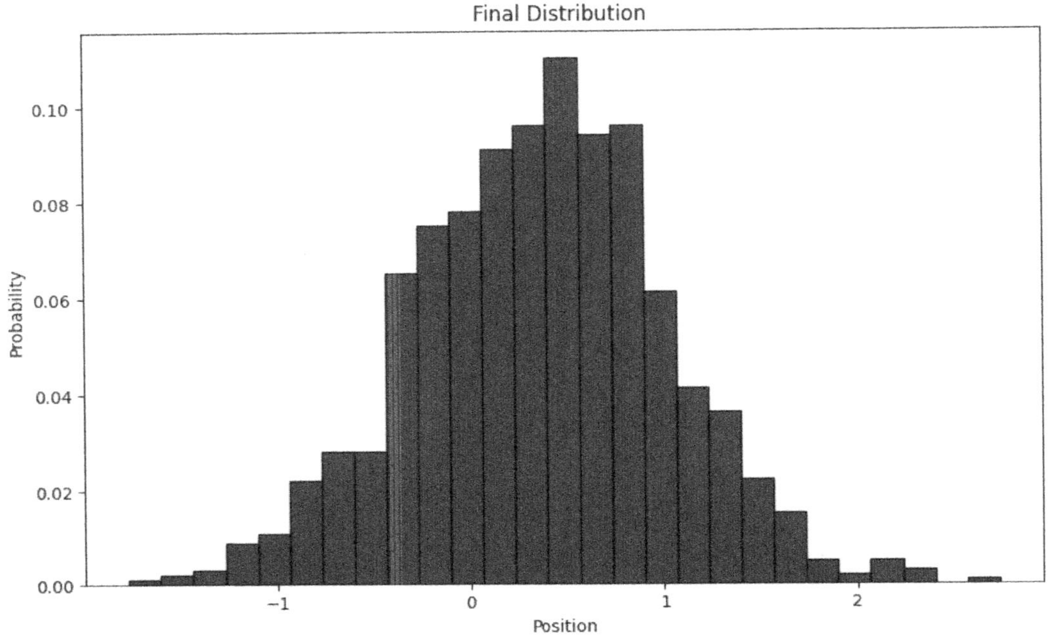

Figure 13.4: The position distribution of the ink particles approaches Gaussian as the diffusion process goes on

In our physical world, we can debate the process is not reversible for an isolated tube of ink and water, as we have never seen the ink blob to spontaneously gather back to the original drop in our daily life. However, mathematically speaking, the above mentioned process we looked at is easily reversed if we know the exact Gaussian noise $\epsilon_i(t)$ used in each step. One could also argue that we can never know the exact noise used in the process in practice. That's the neat part; we don't have to: The goal of reversing the process is to estimate the probability distribution. In fact, we can write down the probabilistic transitions for the reverse process for the previously mentioned Brownian motion.

To build the diffusion model for time series forecasting, let's complete the Python code to simulate the full diffusion process and observe what actually happens.

Simulating the diffusion process

If it is hard to reduce diffusion process to simple ideas and interpretations, we can always calculate some results to understand it. In fact, there is a famous saying about quantum physics by Richard Feynman: "Shut up and calculate". In our case, we can simulate the process to feel the magic. To make it easier to get the intuitions, here we simulate the diffusion process in a one-dimensional setting. This section will be a bit math heavy, but it will be important to understand how a diffusion model works.

First thing first, let's import some utilities.

```
import dataclasses
from functools import cached_property

import numpy as np
import torch
```

A diffusion process can be described by a set of parameters. The most important one is the $\alpha(t)$ we discussed in the previous section. Mathematically, we will see that the expression $1 - \alpha(t)$ repeatedly appears in the equations later. For simplicity, we will use $\beta(t) = 1 - \alpha(t)$. $\beta(t)$ describes the degree of noise added to the original data. The larger $\beta(t)$ is, the more noise we add to the original data. In the following class `DiffusionProcessParams`, we will define and calculate the $\alpha(t)$ and $\beta(t)$ parameters for each diffusion step.

```
@dataclasses.dataclass
class DiffusionProcessParams:
    """Parameter that defines a diffusion process.

    :param steps: Number of steps in the diffusion process.
    :param beta: Beta parameter for the diffusion process,
        representing the degree of noise added at each step.
        A higher beta value corresponds to more noise being
        added, which accelerates the diffusion process.
    """
    steps: int
    beta: float

    @cached_property
    def alpha(self) -> float:
        r"""$\alpha = 1 - \beta$"""
        return 1.0 - self.beta

    @cached_property
    def beta_by_step(self) -> np.ndarray:
        """the beta parameter for each step
        in the diffusion process.
        """
        return np.array([self.beta] * self.steps)

    @cached_property
```

```
    def alpha_by_step(self) -> np.ndarray:
        """the alpha parameter for each step
        in the diffusion process."""
        return np.array([self.alpha] * self.steps)
```

With the parameters defined, we can now define the diffusion process. We will calculate the forward process using `_forward_process_by_step` and the reverse process using `_inverse_process_by_step`. We use protected methods whose name starts with an underscore (_) in our Python class `DiffusionProcess` to indicate that this is only a demonstration of the process, and we will not use them directly in a real-world diffusion model.

The first step is initialization and setting up the alpha parameter:

```
class DiffusionProcess:
    """
    Simulate the diffusion process step by step.

    :param params: DiffusionParams that defines
        how the diffusion process works
    :param noise: noise tensor,
        shape is (batch_size, params.steps)
    """

    def __init__(
        self,
        params: DiffusionProcessParams,
        noise: torch.Tensor,
        dtype: torch.dtype = torch.float32,
    ):
        self.params = params
        self.noise = noise
        self.dtype = dtype

    @cached_property
    def alpha_by_step(self) -> torch.Tensor:
        """The alpha parameter for each step
        in the diffusion process.
        """
        return torch.tensor(
            self.params.alpha_by_step,
            dtype=self.dtype
        )
```

Then we perform the forward and inverse diffusion steps:

```
    def _forward_process_by_step(
        self,
        state: torch.Tensor,
        step: int
    ) -> torch.Tensor:
        r"""This method simulates the diffusion
        Process one time step forward.

        We assume that we know
        the noise at step $t$, the process is

        $$
        x(t) = \sqrt{\alpha(t)}x(t-1)
        + \sqrt{1 - \alpha(t)}\epsilon(t)
        $$

        :param state: The state at step $t-1$.
        :param step: The current step $t$.
        :return: The state at step $t$.
        """
        return (
            torch.sqrt(
                self.alpha_by_step[step]
            ) * state
            + torch.sqrt(
                1 - self.alpha_by_step[step]
            ) * self.noise[:, step]
        )

    def _inverse_process_by_step(
            self,
            state: torch.Tensor,
            step: int
        ) -> torch.Tensor:
        r"""This method simulate the diffusion
        Process one step back in time.

        We assume that we know
        the noise at step $t$, reversing a
```

```
        diffusion process is described by

        $$
        x(t-1) = \frac{1}{\sqrt{\alpha(t)}}
        (x(t) - \sqrt{1 - \alpha(t)}\epsilon(t))
        $$
        """
        return (
            state
            - torch.sqrt(
                1 - self.alpha_by_step[step]
            ) * self.noise[:, step]
        ) / torch.sqrt(
            self.alpha_by_step[step]
        )
```

We have the last missing piece of the puzzle: the noise. We generate the noise using a Gaussian distribution.

```
def gaussian_noise(
        n_var: int,
        length: int
    ) -> torch.Tensor:
    """Generate a Gaussian noise tensor.

    :param n_var: Number of variables.
    :param length: Length of the tensor.
    """
    return torch.normal(
        mean=0,
        std=1,
        size=(n_var, length)
    )
```

We create an example diffusion process of 1000 particles using $\beta = 0.005$ and for 100 steps. To observe the statistical effect of the diffusion process, we will generate a batch of 1000 initial data points and observe their trajectory and distribution during the diffusion process.

```
diffusion_process_params = DiffusionProcessParams(
    steps=100,
    beta=0.005,
)
diffusion_batch_size = 1000
```

```
noise = gaussian_noise(
    diffusion_batch_size,
    diffusion_process_params.steps
)

diffusion_process = DiffusionProcess(
    diffusion_process_params,
    noise=noise
)
```

We can now iteratively apply the `diffusion_process._forward_process_by_step` method to generate the diffusion process. In *Figure 13.5*, we plot out the histogram of the positions of all 1000 particles every ten steps.

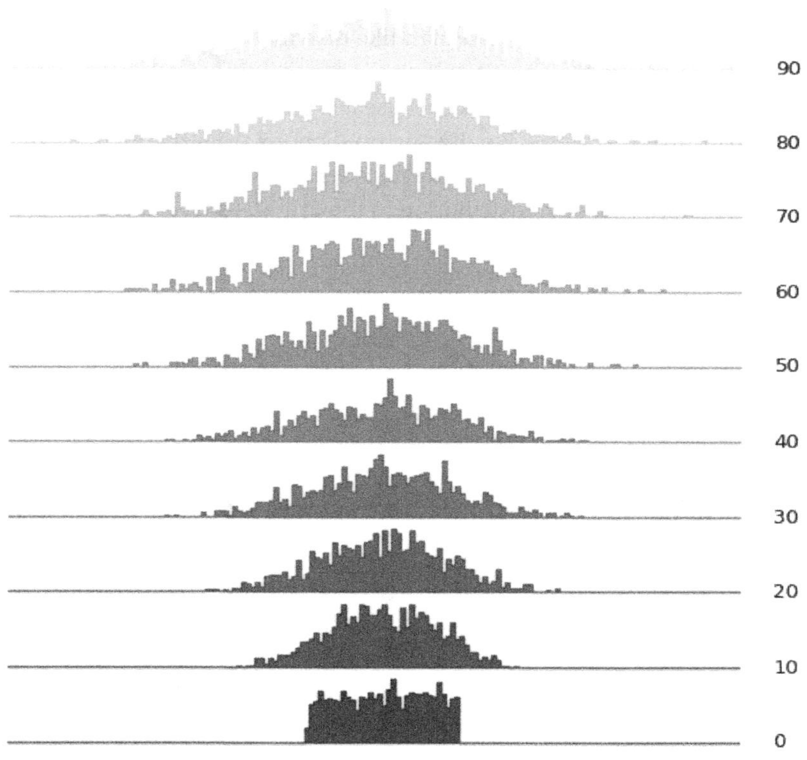

Figure 13.5: Histograms of particle positions at selected steps

In *Figure 13.3*, the numbers on the right vertical axis indicate the corresponding step of the histogram. In the beginning, the distribution is a uniform random distribution, however, as time goes on (reading from bottom to top), the distribution of the particles gradually approaches Gaussian distribution. At the same time, the particles are more dispersed, which also echoes the name diffusion process.

Similarly, we can also apply the reverse process and observe how the particles are reversed to their original positions. However, the reason that we can even reverse it is that we know the exact noise used in the forward process. In the following sections, we will apply the idea to time series forecasting.

Naive diffusion model applied on time series data

We have discussed how diffusion works in a physical world. At first, it seems to be a bit abstract to make the connection between a physical diffusion process and forecasting time series data. One way of establishing the connection is to think of a forecasting problem as generating new data points. If we can generate new data points that follows a desired distribution of the time series data at future time steps, we are then forecasting.

We will first build a naive diffusion model that is explicitly based on the above forward and reverse process. The naive model is not a practical model, but it will help us get our hands on the diffusion process behind a practical diffusion model. For simplicity, we choose a encoder-decoder architecture.

We will encode the time series data using a `DiffusionEncoder` which applies the diffusion process to the data and *diffuses* the data into a Gaussian like distribution, i.e., reading the distribution from bottom to top similar to *Figure 13.5*. An RNN is applied to the diffused data and predicts the next time steps in the latent space. We choose an RNN because it is naturally recursive and can be easily mapped to our step-by-step diffusion process. Finally, the predictions from the RNN are decoded back to the original space using a `DiffusionDecoder`. The decoder applies the reverse diffusion process to the predicted data and restores the data distribution in the original space. An illustration of the model is shown in *Figure 13.6*.

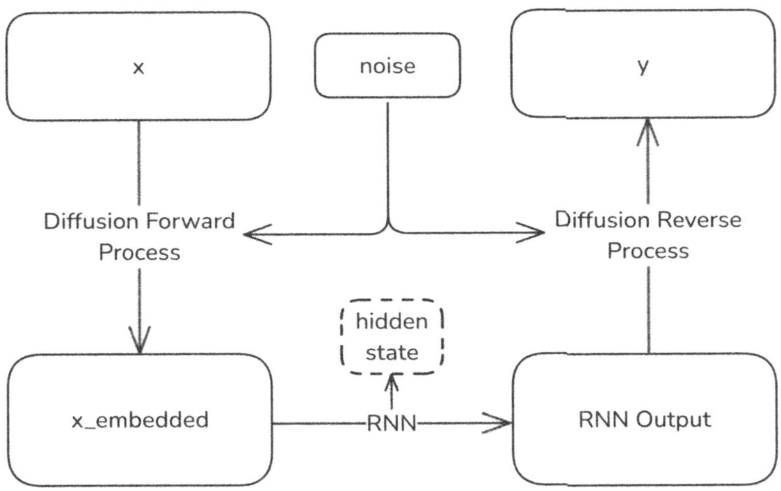

Figure 13.6: Illustration of a Naive Diffusion Model

We will begin, by defining the encoder and decoder in the code. We have written the code in a straightforward way to help you understand the math behind it. In practice, we will use more compact and efficient formalism to simplify the computation.

The encoder is an adaptation of the `_forward_process_by_step` method of the `DiffusionProcess` class, as the encoder is responsible to diffusion our data.

```
import torch.nn as nn


class DiffusionEncoder(nn.Module):
    """Encode the time series
    in the latent space.

    :param params: parameters that
        define a diffusion process.
    :param noise: predefined noise
        for the diffusion process.
    """
    def __init__(
        self,
        params: DiffusionProcessParams,
        noise: torch.Tensor,
    ):
        super().__init__()
        self.params = params
        self.noise = noise

    @staticmethod
    def _forward_process_by_step(
        state: torch.Tensor,
        alpha_by_step: torch.Tensor,
        noise: torch.Tensor,
        step: int,
    ) -> torch.Tensor:
        r"""Assuming that we know the noise
        at step $t$,

        $$
        x(t) = \sqrt{\alpha(t)}x(t-1)
        + \sqrt{1 - \alpha(t)}\epsilon(t)
        $$
        """
        batch_size = state.shape[0]
        return torch.sqrt(
            alpha_by_step[step]
```

```
        ) * state + (
            torch.sqrt(
                1 - alpha_by_step[step]
            ) * noise[:batch_size, step]
        ).reshape(batch_size, 1)

    def forward(
        self, x: torch.Tensor
    ) -> torch.Tensor:
        """Encoding the latent space
        into a distribution.

        :param x: input data, shape
            (batch_size, history_length, n_features)
        """
        alpha_by_step = torch.tensor(
            self.params.alpha_by_step
        ).to(x)
        self.noise = self.noise.to(x)

        diffusion_steps_step_by_step = [x]

        for i in range(0, self.params.steps):
            i_state = self._forward_process_by_step(
                diffusion_steps_step_by_step[-1],
                alpha_by_step=alpha_by_step,
                noise=self.noise,
                step=i,
            )
            diffusion_steps_step_by_step.append(
                i_state
            )

        return diffusion_steps_step_by_step[-1]
```

Similarly, the decoder is an adaptation of the `_inverse_process_by_step` method of the `DiffusionProcess` class, as the decoder is responsible to restore the diffused data, thus a reverse process.

```
class DiffusionDecoder(nn.Module):
    """Decode the latent space into a distribution."""

    def __init__(
```

```
        self,
        params: DiffusionProcessParams,
        noise: torch.Tensor,
    ):
        super().__init__()
        self.params = params
        self.noise = noise

    @staticmethod
    def _inverse_process_by_step(
        state: torch.Tensor,
        alpha_by_step: torch.Tensor,
        noise: torch.Tensor,
        step: int
    ) -> torch.Tensor:
        r"""Assuming that we know the
        noise at step $t$,

        $$
        x(t-1) = \frac{1}{\sqrt{\alpha(t)}}
        (x(t) - \sqrt{1 - \alpha(t)}\epsilon(t))
        $$
        """
        batch_size = state.shape[0]
        return (
            state - (
                torch.sqrt(
                    1 - alpha_by_step[step]
                ) * noise[:batch_size, step]
            ).reshape(
                batch_size, 1
            )
        ) / torch.sqrt(alpha_by_step[step])

    def forward(self, x: torch.Tensor) -> torch.Tensor:
        """Encoding the latent space into a distribution.

        :param x: input data, shape
            (batch_size, history_length, n_features)
        """
        alpha_by_step = torch.tensor(
```

```
            self.params.alpha_by_step
        ).to(x)
        self.noise = self.noise.to(x)

        diffusion_steps_reverse = [x]

        for i in range(
            self.params.steps - 1, -1, -1
        ):
            i_state = self._inverse_process_by_step(
                state=diffusion_steps_reverse[-1],
                alpha_by_step=alpha_by_step,
                noise=self.noise,
                step=i,
            )
            diffusion_steps_reverse.append(i_state)

        return diffusion_steps_reverse[-1]
```

Once the data is encoded, we then apply a RNN to perform the forecasting task. In this example, we choose GRU as our RNN architecture. We define the parameters for the RNN using the `LatentRNNParams` class.

```
@dataclasses.dataclass
class LatentRNNParams:
    """Parameters for Diffusion process.

    :param history_length: input sequence length
    :param latent_size: latent space dimension
    :param num_layers: number of RNN layers
    :param n_features: number of features in data
    :param initial_state: initial state of the RNN
    """
    history_length: int
    latent_size: int = 100
    num_layers: int = 2
    n_features: int = 1
    initial_state: torch.Tensor = None

    @cached_property
    def data_size(self) -> int:
        """The dimension of the input data
```

```
        when flattened.
        """
        return self.sequence_length * self.n_features

    def asdict(self) -> dict:
        return dataclasses.asdict(self)
```

We use a minimal GRU-based RNN model to predict the next steps in latent space. As we are using the dataclass `LatentRNNParams` dataclass to specify the parameters of model, we pass the dataclass to our RNN model `LatentRNN`. We have omitted the params in the docstrings as they are already clear by the variable names.

```
class LatentRNN(nn.Module):
    """Forecasting the next step in latent space."""

    def __init__(self, params: LatentRNNParams):
        super().__init__()

        self.params = params
        self.hparams = params.asdict()

        self.rnn = nn.GRU(
            input_size=self.params.history_length,
            hidden_size=self.params.latent_size,
            num_layers=self.params.num_layers,
            batch_first=True,
        )

    def forward(
        self, x: torch.Tensor
    ) -> tuple[torch.Tensor, torch.Tensor, torch.Tensor]:
        """
        :param x: input data, shape
            (batch_size, history_length * n_features)
        """
        outputs, _ = self.rnn(
            x, self.params.initial_state
        )

        return outputs
```

With the encoder `DiffusionEncoder`, decoder `DiffusionDecoder`, and RNN model `LatentRNN` defined, we can now assemble the forecasting model. The model will take the input time series data, encode it into the latent space using the encoder, apply the RNN to predict the next steps in the latent space, and finally decode the predictions back to the original space using the decoder.

```
class NaiveDiffusionModel(nn.Module):
    """A naive diffusion model that explicitly calculates
    the diffusion process.
    """
    def __init__(
        self,
        rnn: LatentRNN,
        diffusion_decoder: DiffusionDecoder,
        diffusion_encoder: DiffusionEncoder,
        horizon: int = 1,
    ):
        super().__init__()
        self.rnn = rnn
        self.diffusion_decoder = diffusion_decoder
        self.diffusion_encoder = diffusion_encoder
        self.horizon = horizon
        self.scale = nn.Linear(
            in_features=self.rnn.params.latent_size,
            out_features=self.horizon,
        )

    def forward(self, x: torch.Tensor) -> torch.Tensor:
        x_latent = self.diffusion_encoder(x.squeeze(-1))
        y_latent = self.rnn(x_latent)
        y_hat = self.diffusion_decoder(y_latent)

        y_hat = self.scale(y_hat)

        return y_hat
```

As usual, we will use Lightning to wrap the model for model training and evaluation. The following `NaiveDiffusionForecaster` class is a Lightning module that encapsulates our diffusion model. The structure of the code is similar to the previous chapters, where we define the optimizer, training step, validation step, and prediction step. By default, we choose a **Mean Squared Error** (**MSE**) loss function. We will not tune the model as it is meant to be a demonstration of the diffusion process applied to time series forecasting.

```
import lightning.pytorch as pl
from lightning import LightningModule
```

```
class NaiveDiffusionForecaster(LightningModule):
    """A assembled lightning module
    for the naive diffusion model.
    """
    def __init__(
        self,
        model: NaiveDiffusionModel,
        loss: nn.Module = nn.MSELoss(),
    ):
        super().__init__()
        self.model = model
        self.loss = loss

    def configure_optimizers(
        self
    ) -> torch.optim.Optimizer:
        optimizer = torch.optim.SGD(
            self.parameters(),
            lr=1e-3
        )
        return optimizer

    def training_step(
        self,
        batch: tuple[torch.Tensor],
        batch_idx: int
    ) -> torch.Tensor:
        x, y = batch
        x = x.type(self.dtype)
        y = y.type(self.dtype)
        batch_size = x.shape[0]

        y_hat = self.model(
            x
        )[:batch_size, :].reshape_as(y)

        loss = self.loss(y_hat, y).mean()
        self.log_dict(
            {"train_loss": loss},
            prog_bar=True
```

```
        )

        return loss

    def validation_step(
        self,
        batch: tuple[torch.Tensor],
        batch_idx: int
    ) -> torch.Tensor:
        x, y = batch
        x = x.type(self.dtype)
        y = y.type(self.dtype)
        batch_size = x.shape[0]

        y_hat = self.model(
            x
        )[:batch_size, :].reshape_as(y)

        loss = self.loss(y_hat, y).mean()
        self.log_dict(
            {"val_loss": loss},
            prog_bar=True
        )
        return loss

    def predict_step(
        self,
        batch: tuple[torch.Tensor],
        batch_idx: int
    ) -> tuple[
        torch.Tensor,
        torch.Tensor
    ]:
        x, y = batch
        x = x.type(self.dtype)
        y = y.type(self.dtype)
        batch_size = x.shape[0]

        y_hat = self.model(
            x
        )[:batch_size, :].reshape_as(y)
```

```
        return x, y_hat

    def forward(
        self, x: torch.Tensor
    ) -> torch.Tensor:
        x = x.to(self.model.rnn.rnn.weight_ih_l0)
        return self.model(x)
```

To see the model in action, we will `forecast` our classical sine wave data as we had in the previous chapters. We will use a history length of 100 and a horizon of `1` step. The training batch size is set to `64`. As we have done in the previous section, we will also generate the Gaussian noise for the diffusion process.

```
history_length_1_step = 100
horizon_1_step = 1
training_batch_size = 64

training_noise = gaussian_noise(
    training_batch_size,
    diffusion_process_params.steps
)
```

The encoder, decoder, and the RNN model are then instantiated using the parameters we defined earlier. The `DiffusionDecoder` and `DiffusionEncoder` will use the same diffusion process parameters and the predefined noise.

```
diffusion_decoder = DiffusionDecoder(
    diffusion_process_params,
    training_noise
)
diffusion_encoder = DiffusionEncoder(
    diffusion_process_params,
    training_noise
)

latent_rnn_params = LatentRNNParams(
    history_length=history_length_1_step,
    latent_size=diffusion_process_params.steps,
)

latent_rnn = LatentRNN(latent_rnn_params)
```

As the last step of the model assembly, we instantiate the lightning module `NaiveDiffusionForecaster` and prepare we are all set for training.

```
naive_diffusion_model = NaiveDiffusionModel(
    rnn=latent_rnn,
    diffusion_decoder=diffusion_decoder,
    diffusion_encoder=diffusion_encoder,
)
naive_diffusion_forecaster = NaiveDiffusionForecaster(
    model=naive_diffusion_model.float(),
)
```

The training and evaluation will be exactly the same as we have done in the previous chapters thus we will not repeat the training and evaluation code here. We choose a naive forecaster that simply predicts the last value in the history as our baseline. As an intuition, plot out the predictions of the diffusion model on the sine wave data as compared to the baseline.

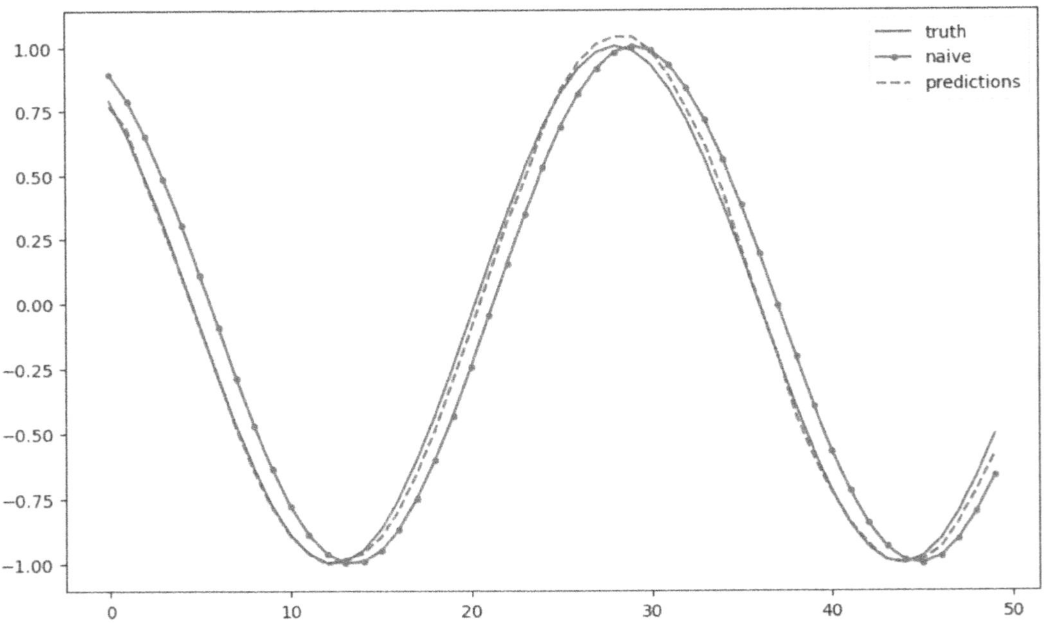

Figure 13.7: The predictions of the diffusion model on the sine wave data. The solid line is the ground truth, the dashed line is the prediction of the diffusion model, and the line with dot marker is the prediction of the naive forecaster baseline

Figure 13.7 shows that the diffusion model is able to capture the trend of the sine wave data, while some of the predictions are not perfect, it is much better than the baseline. The following table shows the metrics on the test data for our diffusion model and the baseline. The diffusion model outperforms the baseline in all metrics.

Metric	Diffusion	Baseline
MAE	0.026833	0.127596
MAPE	0.085078	0.789616
MSE	0.001127	0.019861
sMAPE	0.078725	0.410083

Table 13.1: Metrics on the test data for our diffusion model and the baseline.

Though the model works as expected, we have serious issues with the concepts in the model itself.

- We have used a fixed noise for the diffusion process. Not only it reduces the complexity of the hyperparameters, it also makes the model less flexible. We should learn the noise from the data.
- We are not using the full potential of the diffusion process. The diffusion model we built is only using the diffusion process to encode the data and decode the predictions. As we will observe later, a diffusion process, however, is much more powerful.

To resolve the issues, we will introduce the famous denoising diffusion models in the next section.

Denoising diffusion model

We take a step back and re-think about our time series forecasting problem. Time series forecasting can be formulated as transition probability estimation problems. To understand the concept, we establish a simple stage where we only have 5 time steps, t_1, t_2, t_3, t_4, t_5. We want to estimate the distribution of the time steps t_4, t_5, denoted as $q(x_4, x_5|x_1, x_2, x_3)$, using the given historical values t_1, t_2, t_3, the chain rule of probability tells us

$$q(x_4, x_5|x_1, x_2, x_3) = q(x_4|x_1, x_2, x_3)q(x_5|x_1, x_2, x_3, x_4).$$

Following the research by K. Rasul et. al. (*Autoregressive Denoising Diffusion Models for Multivariate Probabilistic Time Series Forecasting available*), we simplify the calculations, we assume that we can approximate the distributions using the following:

$$\begin{aligned} q(x_4|x_1, x_2, x_3) &\sim p(x_4|h_3) \\ q(x_5|x_1, x_2, x_3, x_4) &\sim p(x_5|h_4), \end{aligned}$$

where h_3 and h_4 model the temporal dynamics of the time series data. For example, h_3 can be an RNN hidden state that carries the information of the historical values t_1, t_2, t_3. With these assumptions, we can formulate a time series forecasting problem as a sequence of conditional probability estimations, i.e.,

$$q(x_4, x_5|x_1, x_2, x_3) \sim p(x_4|h_3)p(x_5|h_4).$$

More generally, we can formulate the forecasting problem as the following:

$$\begin{aligned} &q\left(x_{t_0}, x_{t_0+1}, \dots, x_{t_0+H-1}|x_1, x_2, \dots, x_{t_0-1}\right) \\ \sim\ & p\left(x_{t_0}|h_{t_0-1}\right)p\left(x_{t_0+1}|h_{t_0}\right)\cdots p\left(x_{t_0+H-1}|h_{t_0+H-2}\right), \end{aligned}$$

where H is the forecasting horizon. To understand the relationship between this formulation and a diffusion model, we use an example of horizon $H = 1$. The above equation can be rewritten as

$$q(x_{t_0} | x_1, x_2, \dots, x_{t_0-1}) \sim p(x_{t_0} | h_{t_0-1}). \qquad (13.1)$$

We establish a diffusion process of N steps that starts from the initial value x_{t_0} and ends at the value $x_{t_0}^N$. For consistency of notation, we denote the initial value as $x_{t_0}^0 = x_{t_0}$. Our task is to find out $p(x_{t_0} | h_{t_0-1})$. The first row in *Figure 13.8* demonstrates the mapping between *Equation 13.1* and a diffusion process. Each diffusion step uses the RNN hidden state from the original time series data to estimate the reverse diffusion process. That being said, the results of forward diffusion process, $x_{t_0}^N$ and $x_{t_0+1}^N$ are approximately a Gaussian distribution, as shown in *Figure 13.5*.

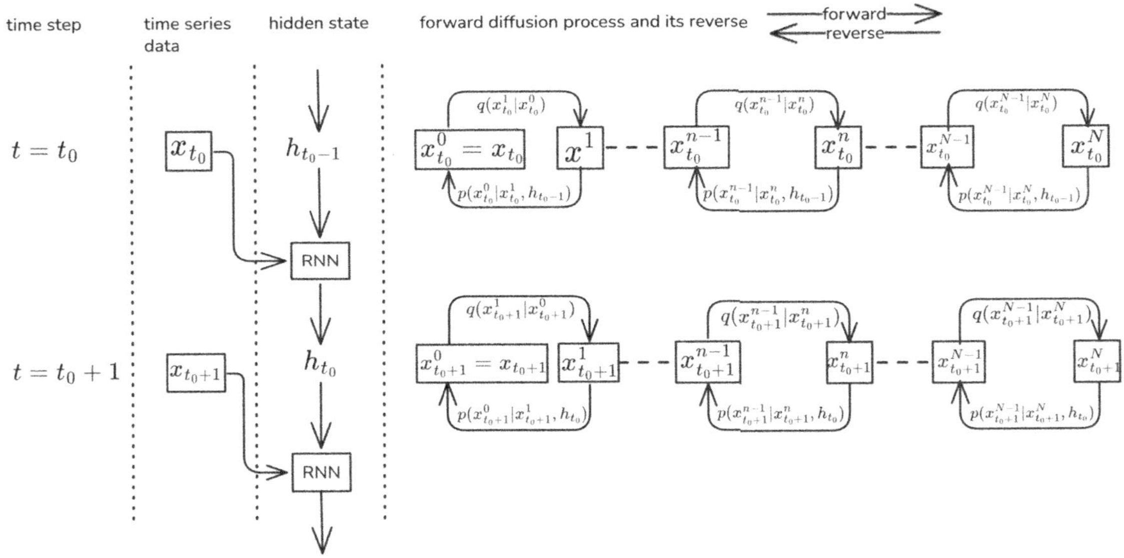

Figure 13.8: Illustration of the autoregressive denoising diffusion model.

From top to bottom, we are showing the time steps to be forecasted. We use the original time series data to build up the hidden states to be used in the denoising process. The column on the right shows the diffusion process and the reverse. Illustration created based on K. Rasul, C. Seward, I. Schuster, and R. Vollgraf, "*Autoregressive Denoising Diffusion Models for Multivariate Probabilistic Time Series Forecasting*".

Once the model is trained, we will obtain the hidden states such as h_{t_0-1}, forecasts are then obtained by generating a sample from a Gaussian distribution $x_{t_0}^N$, and going through the reverse diffusion process based on the transition probability $p(x_{t_0}^{n-1} | x_{t_0}^n, h_{t_0-1})$ until we have reached $x_{t_0}^0$. Figure 13.9 shows inference for a single forecasting step.

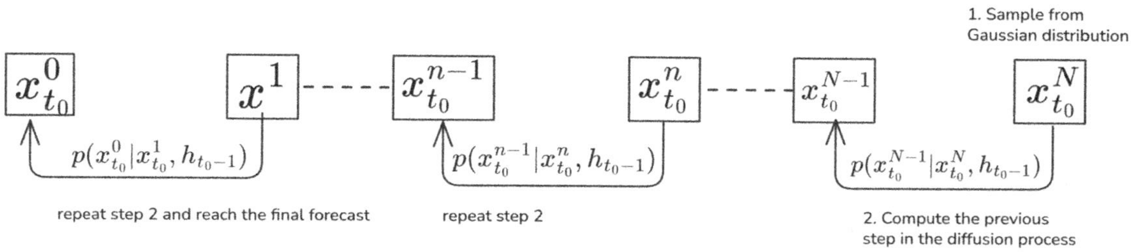

Figure 13.9: Inference process of a trained autoregressive denoising diffusion model. Only a single forecasting step is shown here

The implementation of the autoregressive denoising diffusion model or, TimeGrad dubbed by the original authors, is included in the `pytorch-ts` python package. We will also skip the implementation details here and we provide a very detailed demo of forecasting a few benchmark datasets using the model. You can find the demo in the examples of `pytorch-ts` repository, see `https://github.com/zalandoresearch/pytorch-ts/tree/master/examples`. The model outperforms a bunch of state-of-the-art time series forecasting models on the benchmark datasets.

Comparing the naïve diffusion model and the denoising diffusion model we realize that the denoising diffusion model is more tailored to the time series forecasting problem. Instead of bluntly forcing a model architecture onto the forecasting problems like the naïve diffusion model, we explicitly model it using a probabilistic view and map it to the diffusion process. The first naive diffusion model barely works on a sine wave dataset while the autoregressive denoising diffusion model became the SoTA for multivariate probabilistic forecasting. It is a reminder to all of us that time series forecasting is special and we should always dig deeper to understand how a new model structure works in the context of time series forecasting.

The *Recommended reading* section provides more references on integrating diffusion models into time series forecasting.

Summary

In this chapter, we have studied the diffusion process and applied it to time series forecasting problem using two different approaches. We learned the lesson that adapting models to the time dependency nature of time series data helps when developing new models. In the next chapters, we will discuss a different type of time series task: time series classification.

Recommended reading

Different from the previous chapters, we touched on several new topics in this chapter: diffusion models, probabilistic time series forecasting, and denoising diffusion models. Some of the topics discussed in this chapter may be new to some readers. The following references offer additional resources for those interested in exploring these concepts further.

- C. F. Higham, D. J. Higham, and P. Grindrod, "*Diffusion models for generative artificial intelligence: An introduction for applied mathematicians*," arXiv [cs.LG], Dec. 21, 2023. Available: `http://arxiv.org/abs/2312.14977`. [Accessed: Jan. 30, 2025]
- L. Yan, L. Xinjiang, W. Yaqing, and D. Dejing, "*Generative time series forecasting with diffusion, denoise, and disentanglement*," arXiv [cs.LG], Jan. 08, 2023. Available: `http://arxiv.org/abs/2301.03028`
- K. Rasul, C. Seward, I. Schuster, and R. Vollgraf, "*Autoregressive Denoising Diffusion Models for Multivariate Probabilistic Time Series Forecasting*," arXiv [cs.LG], Jan. 28, 2021. Available: `http://arxiv.org/abs/2101.12072`
- L. Lin, Z. Li, R. Li, X. Li, and J. Gao, "*Diffusion models for time series applications: A survey*," arXiv [cs.LG], Apr. 30, 2023. Available: `http://arxiv.org/abs/2305.00624`
- M. Caspar and Y. C. Lydia, "*The rise of diffusion models in time-series forecasting*," arXiv [cs.LG], Jan. 05, 2024. Available: `http://arxiv.org/abs/2401.03006`

Get this book's PDF version and more

Scan the QR code (or go to `packtpub.com/unlock`). Search for this book by name, confirm the edition, and then follow the steps on the page.

UNLOCK NOW

Note: Keep your invoice handy. Purchases made directly from Packt don't require an invoice.

14

Time Series Classification

"The key is to focus on things that are under your control and make a difference. Don't get distracted by things you can't change." – Jim Simons

Throughout this book we have emphasized that the nature and quality of time series data affect our ability to meaningfully analyze and forecast. The complexity in time series data comes not only from individual sequential/temporal nature, but from the diverse shapes and structures bound within these sequences. We often ask analytical questions of large and varied datasets that we may also wish to build forecasts for, but to do these tasks, it can be useful to group or classify data together.

Consider a manufacturing environment where thousands of sensors monitor equipment performance. The critical task here isn't simply to collect data, but to classify it into meaningful categories, such as normal operations versus potential failure states, efficient versus suboptimal performance patterns, or different types of maintenance needs. Similarly, retail analytics require insights into the volume and frequency of demand, seasonal patterns, price elasticity, and how a product may perform at a new location (store). While in forecasting we might want to find hyper-parameters for forecasting huge panel datasets, without running cross-validation for months on every time-series or group of series in order to improve global forecasting models. Generally, these questions are about identifying and partitioning our data, but how do we answer them? The next few chapters deal with these questions.

If we have labeled our data effectively, we may employ classification techniques; for time series, this depends on understanding the fundamental structures within time series data. From the earliest chapters, we already know that we can extract structural properties from time-series data, such as seasonality, trend, and autocorrelation. But with huge data sets, we can't analyze series individually. Various **Time-Series Classification (TSC)** techniques exist that employ the idea of investigating time series structures. From feature-based methods that transform temporal patterns into comparable metrics, to distance-based approaches that measure similarity between sequences, and recent deep learning techniques that automatically learn relevant patterns, each approach offers advantages.

This chapter will focus on explaining time series structures and how they may be captured and used by us to analyze and classify time series data. In later chapters, we'll advance these ideas into applications for clustering and anomaly detection.

- Types of time-series comparisons
- Distance-based measures
- Applied distances
- Elastic ensemble
- Proximity forest
- Feature-based algorithms
- Shapelets
- Dictionary-based classification
- Convolution-based classification
- Deep learning approaches

Types of time-series comparisons

There are many algorithms that we could use for TSC that have been compared in reproducible comparative studies (i.e. Middlehurst et al. 2023; Ruiz et al. 2021). Generally, algorithms can be grouped by how they extract and learn from temporal patterns in the time series.

Distance-based classifiers use distance metrics like **Euclidean distance** and **Dynamic Time Warping** (**DTW**) with **K-Nearest Neighbors** (**KNN**). For multivariate data, DTW can be independent (calculating distance on each channel separately) or dependent (using all channels in pointwise distance calculation). Feature-based algorithms transform series into features using descriptive statistics. ROCKET (Dempster et al. 2020) generates random convolutions with a ridge classifier, while MultiROCKET (Tan et al. 2022) adds additional pooling operations and extracts features from first-order differences. Deep learning approaches include InceptionTime (Ismail Fawaz et al. 2020), Residual Networks (ResNet), and **Fully Convolutional Neural Networks** (**FCN**). These can be implemented as single models or ensembles. Shapelet-based approaches classify based on phase-independent subseries found in training data, evaluating them with supervised measures like information gain. Dictionary-based algorithms use a bag-of-words approach, classifying based on the frequency of approximated subseries patterns with some degree of supervised selection.

We will cover these in the following section.

Distance-based measures

The simplest way to compare time-series data is to measure the similarity of one temporal sequence of data with another, but how do we do that? Well you could just measure **dissimilarity** which is how far apart one series is from another across a time series.

Norms

Oftentimes data scientists use p-norm distance metrics such as Euclidean, Manhattan, Hamming, and Minkowski to measure the distance between data observations. Commonly, we use these with distance-based algorithms for classification like KNN, in order to find the closest distance between an observation and all observations in a training dataset. Class is determined by the proximity of the closest training observation if $k = 1$ in the KNN instantiation. Another example that uses distance metrics is **Support Vector Machine** (**SVM**), which also uses distance metrics to create a hyperplane to separate observations into classes. Test observations are then calculated with respect to the hyperplane(s), and class is assigned relative to distance from the hyperplanes. The problem with using p-norms is that values are defined by two series with matching lengths, which is not always the case. Secondly, p-norms compare values between series at the same time points, as you can see in *Figure 14.1*, but as we know, most values within a series are often correlated.

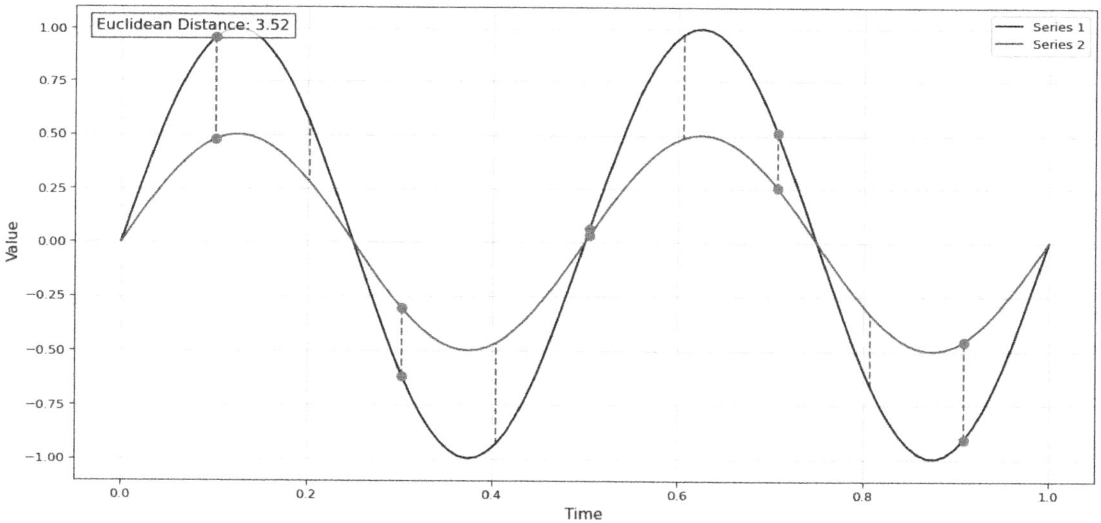

Figure 14.1: Euclidean Distance: Point-to-Point Alignment

This issue is exacerbated when we have sensory data where a pattern may appear slightly later in one device than another because of natural variation or sensor lag. P-norm distances would register these as entirely different patterns despite representing the same underlying phenomenon.

Dynamic time warping

DTW addresses both limitations of p-norm distances by allowing for elastic comparison between time series. DTW works by first calculating a square pointwise distance matrix M between all points in the two series, where $M_{i,j}(a,b) = \left(a_i - b_j\right)^2$, then using dynamic programming to find the minimum-cost path through this matrix:

1. Create a grid where each cell (i, j) represents squared differences between point a_i in the first series and point b_j in the compared series.

2. Starting from the bottom left corner (first points of each series), find the cumulative minimum distance path to the top right (last points of compared series).)
3. At each step, DTW can move horizontally (using the next point in series a but reusing the current point in series b), vertically (using the next point in series b but reusing the current point in series a), or diagonally (advancing both series).)

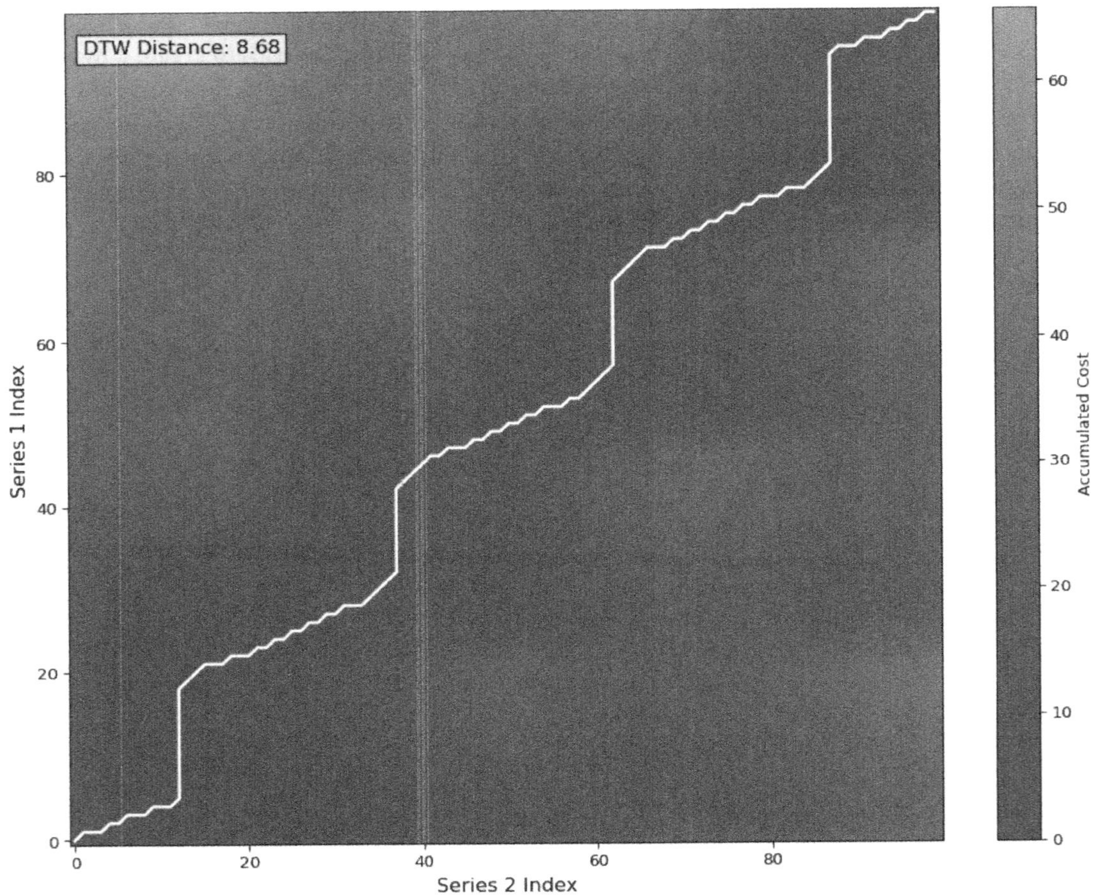

Figure 14.2: DTW accumulated cost matrix

This allows DTW to compare slightly shifted or distorted time series that would be considered dissimilar using standard metrics. The algorithm advances one square at a time through the distance matrix, minimizing the total cost of traversal while accommodating temporal variations. Some implementations include a bounding matrix to constrain the warping path, improving computational efficiency.

Refer to color images on the Graphic bundle or the ebook.

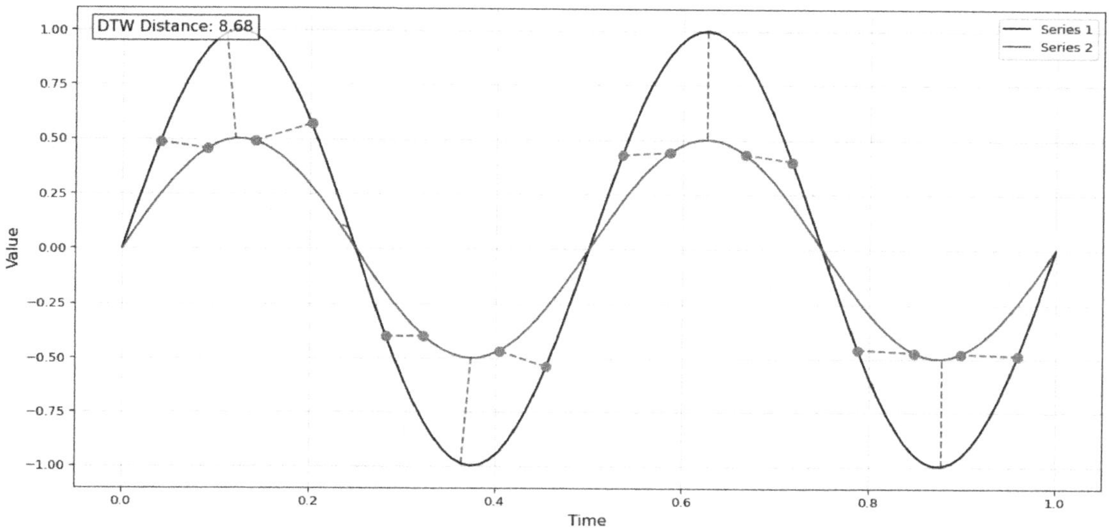

Figure 14.3: DTW: Optimal Path Alignment

While DTW provides excellent flexibility, this comes at a computational cost of $O(n^2)$ for series of length n. For a time series with 1,000 points, Euclidean distance makes 1,000 comparisons, while DTW requires 1,000,000 operations.

- Long time series (e.g., high-frequency sensor data)
- Large datasets where many pairwise comparisons are needed
- Real-time classification requirements

In practice, we rarely need to consider alignments where points at the beginning of one series match points at the end of another. This insight leads to a practical optimization: constraining the warping path.

The most common constraint is Sakoe-Chiba's bounding window, which restricts the warping path to a band along the diagonal of the distance matrix. The width of this band, expressed as a percentage of the series length, controls how much warping is allowed. This constraint not only improves computational efficiency but can also prevent pathological alignments and improve classification accuracy.

Typically 20% is used as a window constraint setting with DTW (often called **constrained DTW** or **cDTW**). Which means we only consider alignments where points can be warped up to 20% of the total series length. Setting the window to 0% reduces DTW to squared Euclidean distance, while 100% allows full warping across the entire series.

```
{'Euclidean': 3.517811819867572,
 'DTW': 8.683920276652852,
 'DTW (window=5%)': 8.73474444914017,
 'DTW (window=10%)': 8.683920276652852,
 'WDTW': 0.0690182068941828}
```

Even with window constraints, DTW remains computationally intensive. A 20% Sakoe-Chiba band reduces the number of calculations to approximately 200,000 operations in our 1,000-point example, but this is still substantially more expensive than p-norm distances.

In addition, the window size parameter requires careful tuning through cross-validation, adding more computational overhead; and DTW's flexibility can create mathematically optimal but semantically meaningless alignments (like matching a brief spike to a prolonged elevation); and its non-metric properties (not satisfying the triangle inequality) prevent the use of many acceleration techniques and indexing structures, further limiting options for improving computational efficiency in large-scale applications.

Retrieving alignment paths

Beyond just calculating distances, DTW provides valuable information about how two series align. The warping path reveals which points in one series correspond to which points in the other, giving insights into temporal differences between patterns.

For classification tasks, examining these alignments can help understand why certain instances are similar or different. The path is returned as a sequence of index pairs $[(i_0, j_0)][(i_0, j_0), (i_1, j_1), ...(i_n, j_n)]$] where each pair (i, j) indicates that point i in the first series aligns with point j in the second.

When developing classification systems, these alignment paths can be useful for:

- Identifying which segments of time series contribute most to class differences.
- Developing preprocessing steps that normalize temporal variations.
- Creating visualizations to explain model decisions to stakeholders.

Most DTW implementations provide a way to retrieve this alignment information alongside the distance value.

Applied distances

As we mentioned earlier, distance calculations are often combined with classification algorithms, such as SVM and KNN. KNN classifies a new time series by finding its k closest neighbors in the training set according to our chosen measure of distance (i.e., Euclid, DTW), then taking a majority vote of their classes.

In this, and the coming chapters, we'll be using the `aeon` library, an offshoot of the forecasting library `sktime`, which focuses on time series classification, clustering, and anomaly detection.

For our experiments, we'll be using the `DodgersLoopDay` dataset; a time series dataset capturing traffic flow in Los Angeles, close to Dodgers Stadium. Activity at the stadium introduces systematic variations in traffic volume across the week, making it a challenging set. Each series is a single 24-hour day split into 288 five-minute intervals. The task is to classify which day of the week (Sunday through Saturday) a series belongs to.

The dataset includes 78 training series and 80 test series, with some missing values denoted by NaN (which we have imputed).

```
   series_id  class source  time  value  value_filled
0          0      1  train     1   12.0          12.0
1          0      1  train     2   18.0          18.0
2          0      1  train     3   11.0          11.0
```

You can see the data is fairly simple, with: `series_id`, `class` (day of the week), `source` (train or test), `time` as sequence of 1:288, `value`, `value_filled`, imputed with catboost (see notebook)

We prepare the data for use with aeon by splitting it back into its training and test sets then split the data in these sets into `y` (class) and `X` (value filled, ordered 1 to 288). Time is implied by the order rather than explicitly used.

To conduct our KNN comparison with Euclidean distance, we import `KNeighborsTimeSeriesClassifier` from `aeon's` distance classifiers, and specify the number of `n_neighbors (k = 1)` and the distance metric we wish to use. Then we `.fit` to the training data and .predict to build the accuracy metrics. Without getting into classification metrics too far, we have simply selected accuracy score here, which is a ratio of correct predictions relative to the set. To illustrate the range of scores, we have a the following figure:

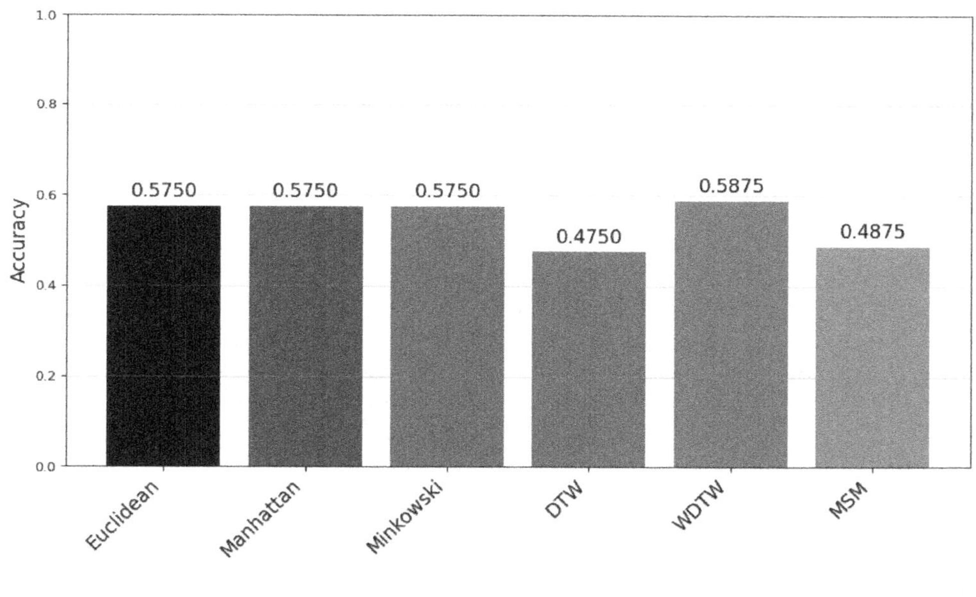

Figure 14.4: Comparing accuracy of KNN classification with distance metrics

With $k = 1$, our simple p-norms give the same result (57.5% accuracy), although performance changes if we use greater k-neighbors, it doesn't improve performance. We could weight neighbor classes (`weights="distance"`) within the KNN function with multiple neighbors (i.e. `k = 3`), but this also does not help. DTW struggles at 47.5%, but WDTW comes out on top at 58.75%, suggesting that adding a penalty for large warpings helps. MSM lands in between at 48.75%. So, WDTW offers a slight edge, but no method is clearly dominant here. Can we do better with distance based classifiers?

Try to run this on your own dataset and experiment with the distance metrics and number of neighbors!

Elastic ensemble

We could ensemble multiple elastic distance metrics together, a method known as **Elastic Ensemble** (**EE**). It combines multiple 1-NN classifiers, each employing distinct elastic distance measures (e.g., DTW, DDTW, WDTW, ERP, MSM, etc.). EE weights each classifier based on its cross-validation accuracy, with weights calculated as $w_i = \frac{\text{acc}_i}{\sum_{j=1}^{M} \text{acc}_j}$, where acc_i is the accuracy of the i-th classifier.

For prediction, each classifier votes for a class, and the final prediction combines these votes using the weights:

$$\hat{y} = \text{argmax}_c \sum_{i=1}^{M} w_i \cdot I(C_i(x) = c)$$

where $I(Ci(x) = c)$ equals 1 when classifier i predicts class c, otherwise 0. This weighting allows the ensemble to adapt by emphasizing the most effective distance measures.

Parameter selection for each distance measure (e.g., warping window size for DTW) is systematically performed via cross-validation.

```
from aeon.classification.distance_based import ElasticEnsemble
ee = ElasticEnsemble(
    distance_measures=["euclidean","wdtw","ddtw","msm", "twe"],
    proportion_of_param_options=0.1,
    proportion_train_in_param_finding=0.3,
    proportion_train_for_test=0.1,
)
ee.fit(X_train, y_train)
ee_preds = ee.predict(X_test)
results["ElasticEnsemble"] = accuracy_score(y_test, ee_preds)
```

EE performed slightly below WDTW here (57.5%), but despite its accuracy, it is computationally expensive, taking a considerable amount of time (41.7 sec) on a relatively small dataset. Its use of many elastic distance metrics and a cross-validation procedure repeated across all measures results in training times that can be orders of magnitude longer than simpler classifiers. During inference, EE also needs the full training set for nearest-neighbor queries, creating significant memory overhead for large-scale applications. Overall, this makes EE less practical in scenarios where time or resource constraints dominate.

Proximity forest

Another approach to using distance-based metrics is Proximity Forests (PF), which combine the strengths of ensemble learning with elastic distance measures. PF builds an ensemble of randomized decision trees where the splitting criteria are based on similarity to exemplars (e) (representative time series) rather than conventional thresholds. At each node, the algorithm:

1. Randomly selects a distance measure δ from a pool of elastic measures (DTW, LCSS, MSM, etc.)
2. Randomly selects one exemplar per class present at the node to form a set E
3. Routes each time series x to the branch whose exemplar is nearest

$$x \to \underset{e \in E}{\operatorname{argmin}}\, \delta(x, e)$$

This creates multi-way splits at each node, with one branch per class. Among multiple candidate splits, the algorithm selects the one maximizing Gini impurity reduction, balancing randomization with discriminative power. When classifying a new time series, each tree independently routes it from root to leaf, assigning the leaf's class label. The final prediction comes from majority voting across all trees.

$$\hat{y} = \underset{c}{\operatorname{argmax}} \sum_{t=1}^{T} I\left(h_t(x) = c\right)$$

Where, $h_t(x)$ is tree t's prediction and I is the indicator function.

A comparative paper across multiple time series classification datasets by Middlehurst et al. (2024) found the PF to be the best performing algorithm overall. Because it leverages a range of distance measures across the ensemble, it can capture complex patterns without explicit feature engineering. It also scales far better than methods like Elastic Ensemble, with average-case learning complexity of $O(n\log(n) \cdot l^2)$ per tree (where n is the number of series and l is series length).

Despite this, its computational costs can still be significant as distance calculations scale quadratically with series length. The model's randomized nature introduces variability, sometimes requiring larger ensembles for stable performance. Also, unlike classic decision tree algorithms, the distance-based splits reduce interpretability, making it difficult to identify the temporal characteristics that drove its classification.

In addition, as with most tree-based algorithms, hyperparameter tuning (i.e., ensemble size, number of candidate splits, and tree depth) significantly impacts performance. Despite these limitations, Proximity Forest represents one of the most effective approaches for TSC tasks, especially when temporal patterns are crucial to accurate prediction.

Let's take a look at how it performs with our data.

```
Training time: 52.07 seconds
```

Performance here is not amazing, especially given that it took 41 secs to run, with untuned parameters selected. Let's try to improve performance by HP tuning with `optuna`. We'll only use the simple train-test the data comes as, with 20 trials (possibly overkill).

While we have used a train-test split here for demonstration purposes, you should definitely use at least a train-validation-test split in practice. Otherwise, you are just training for the test set.

We'll look at the accuracy in a moment, but it is worth stating that the optimization process with a simple train-test split took *FOREVER* to run! Training with the tuned HPs we found with optuna took me from Cenotaph to the start of Umbra on Ghost's Skeletal album. I enjoyed those 8 minutes, but in most production environments, with larger data sets, this would become intolerably long.

Comparing distance methods

Back to the task at hand, let's compare the accuracy of our models so far. Below, it is immediately obvious that the trained PF outperforms all other distance approaches (`accuracy = 0.675`). This is not surprising considering it outperforms most algorithms in the timeseries classification datasets used in the last bake-off redux, a comparison of TS classification algorithms.

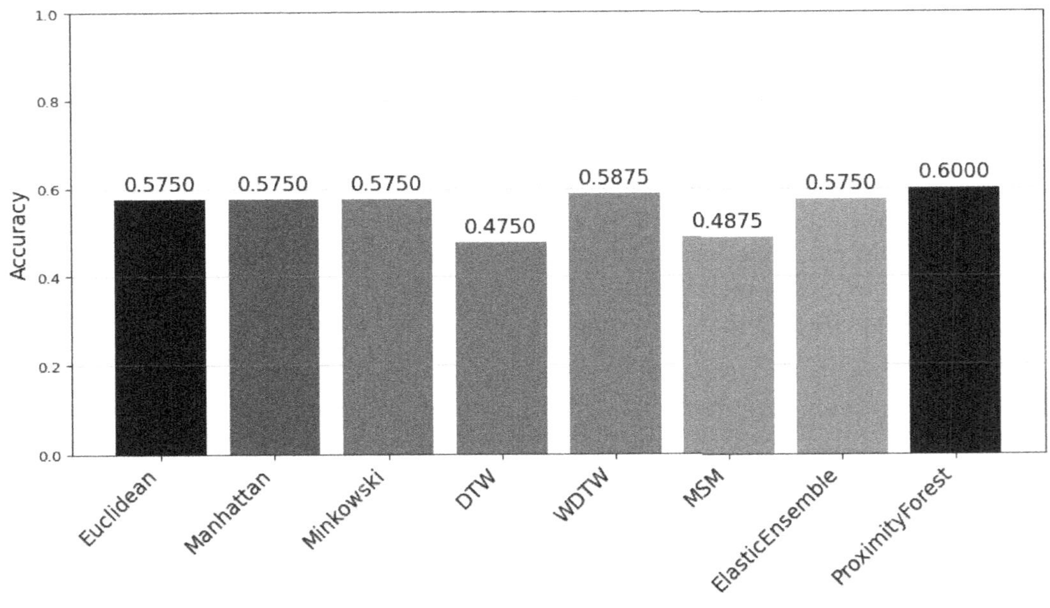

Figure 14.5: Comparing accuracy of all distance classifiers

Although the other advanced methods that utilize elastic distances do not greatly outperform simpler approaches, they have an increased computational complexity. This cost only increases with the size of our data. As a consequence, it can be impractical to use distance-based methods like these.

Feature-based algorithms

Another approach we could take is feature based classification, popular because of its simplicity, and speed. On the surface it's a fairly simple approach, you take your time series data and break it down into summary features that characterize the data.

1. Collect data
2. Wrangle data
3. Build features - with a library
4. Reduce and/or remove excessive features - optional
5. Denoise - optional
6. Pass data to classifier
7. Make classification predictions

Featurization can be done manually, but is best handled by libraries like `hctsa`, `tsfresh`, catch22, and `tsfeatures`, written in faster languages. These libraries can generate thousands of statistical summaries, but this does not obviate the need for domain knowledge, and one should check the documentation. Many of these feature *transforming* libraries assume stationarity or well-behaved signals. Poor planning of preprocessing will lead to misleading results.

You should start by understanding your data's scale and structure.

- Are time series sufficiently long to capture relevant dynamics?
- Is the sampling rate appropriate? Oversampled data can appear trivially autocorrelated, leading some features to be dominated by local continuity rather than structure of interest. In such cases, downsampling or decimation—based on spectral content or autocorrelation structure—can improve the discriminative quality of derived features.

Visual checks can be more informative than summary stats when deciding whether to detrend or denoise. Trends (e.g. sensor drift or low-frequency seasonality) should be removed unless explicitly part of the signal you wish to model. Similarly, transient artefacts or high-frequency noise should be handled with filtering, smoothing, or decomposition methods (e.g. wavelets, EMD, or phase-space reconstruction). If you are doing this at scale, you could generate statistics and subsample, and then apply transformations based on set rules i.e. length, volatility etc.

Standardization across series is often critical, particularly if measurement units vary. Without this, features capturing location or scale (e.g. standard deviation, spectral power) become uninterpretable across observations.

You shouldn't skip EDA. Before applying large-scale featurization, test pipelines using reduced sets like `catch22` or `tsfeatures`. These libraries allow for faster iteration and debugging of assumptions, along with giving a clearer view of what feature classes (e.g. entropy, stationarity, periodicity) are informative for the classification task at hand.

Features built-in

Featurization libraries vary in the number of characteristics they extract from sequential data, with some producing thousands of features, which we could not hope to list. Although, we can categorize most features into a few groups:

- **Distribution features:** Distribution features ignore time ordering, focusing on statistical properties. Features like mean, variance, skewness, and quantiles capture the central tendency, spread, and shape of distributions. For example, heart rate variability might be characterized by the **Standard Deviation Of Intervals** (**SDNN**), while financial time series could be distinguished by their kurtosis (extreme values). Simple, but surprisingly effective at discriminating between time series.

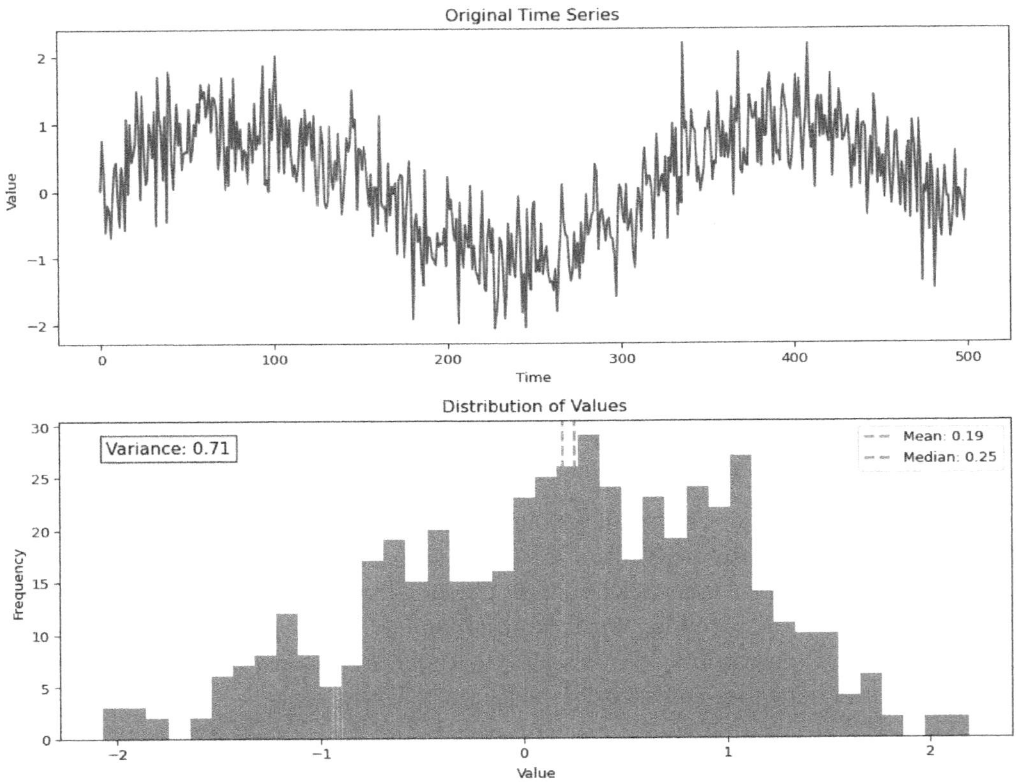

Figure 14.6: Visualizing distribution features

- **Stationarity measures:** Stationarity measures attempt to detect whether the statistical properties of a series change over the sequence time. Stationary is when a series mean, variance, and autocorrelation structure do not change over time. *StatAv* is a metric we use to measure how much the mean varies over time by splitting a sequence into windows, and comparing the variation. Higher values indicate non-stationarity (mean changes significantly across windows), as it does in the following figure (**StatAv** = 0.92).

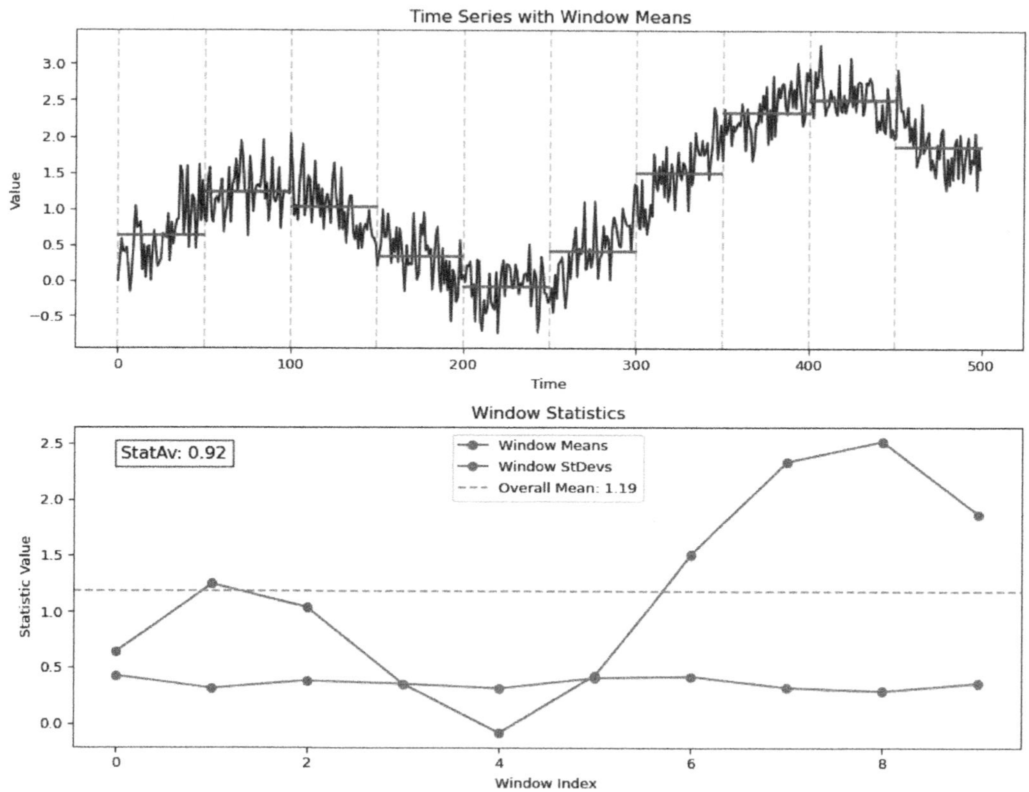

Figure 14.7: Visualizing stationarity measures

- **Autocorrelation features:** As we saw earlier time series are often correlates with lagged versions of themselves, which can reveal temporal dependencies and cyclic patterns in our data. Features like the lag-1 autocorrelation, first zero-crossing, or 1/e crossing point of the autocorrelation function quantify *memory* in the system. High autocorrelation at a specific lag indicates that a series tends to repeat patterns at that interval. For example, weather data often reveals seasonal periodicities in temperature and rainfall over time through peaks in autocorrelation function. Autocorrelations are useful for distinguishing between random, trending, and periodic signals.

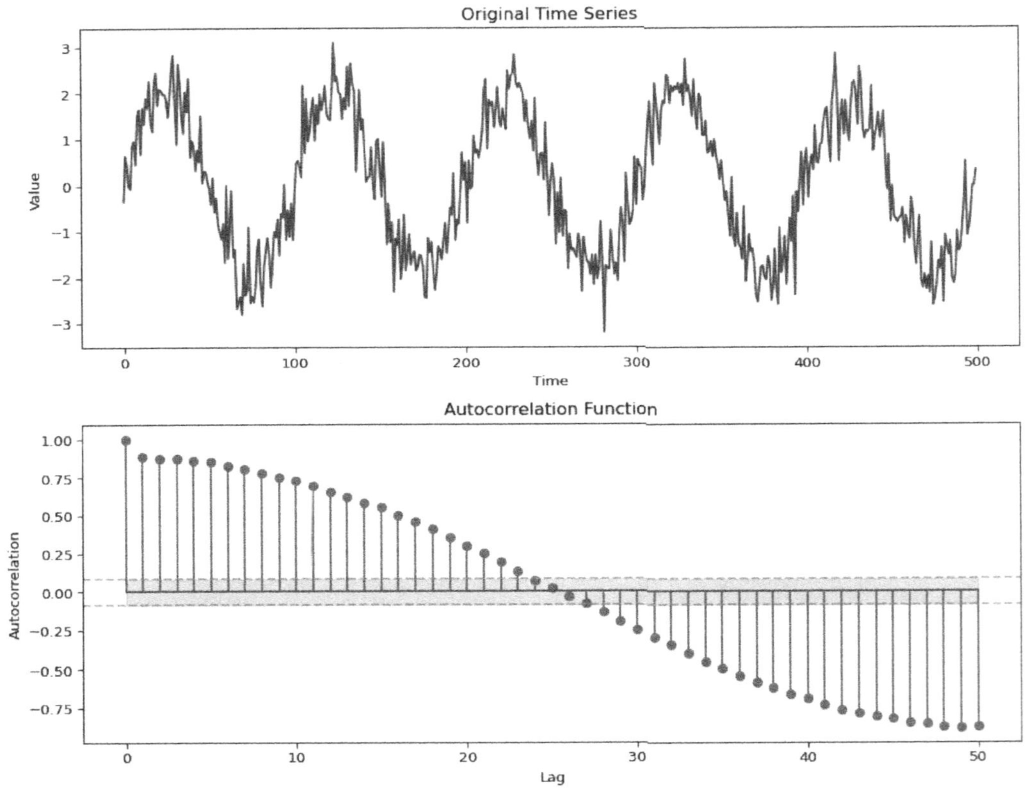

Figure 14.8: Visualizing autocorrelation features

- Incremental difference features: Unlike autocorrelation which captures memory effects across time scales, incremental difference features examine step-by-step changes between consecutive points in a series. They are used to find volatility patterns and local dynamics that global statistics might miss. In the `catch22` library there are two incremental difference features:
    - **High fluctuation** measures the proportion of differences exceeding 4% of standard deviation, distinguishing between relatively stable series and those that *jump around* frequently. This is somewhat of a proxy for volatility.

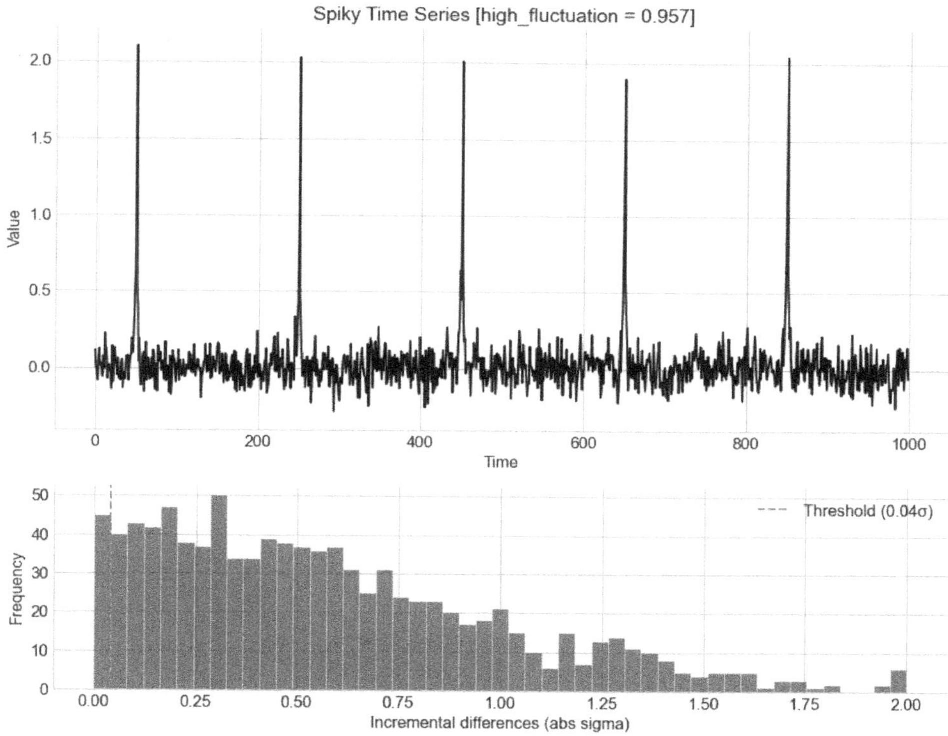

Figure 14.9: High fluctuation feature

- **Whitening timescale** compares autocorrelation decay in the original series versus its differences, and quantifies how differencing affects temporal dependencies

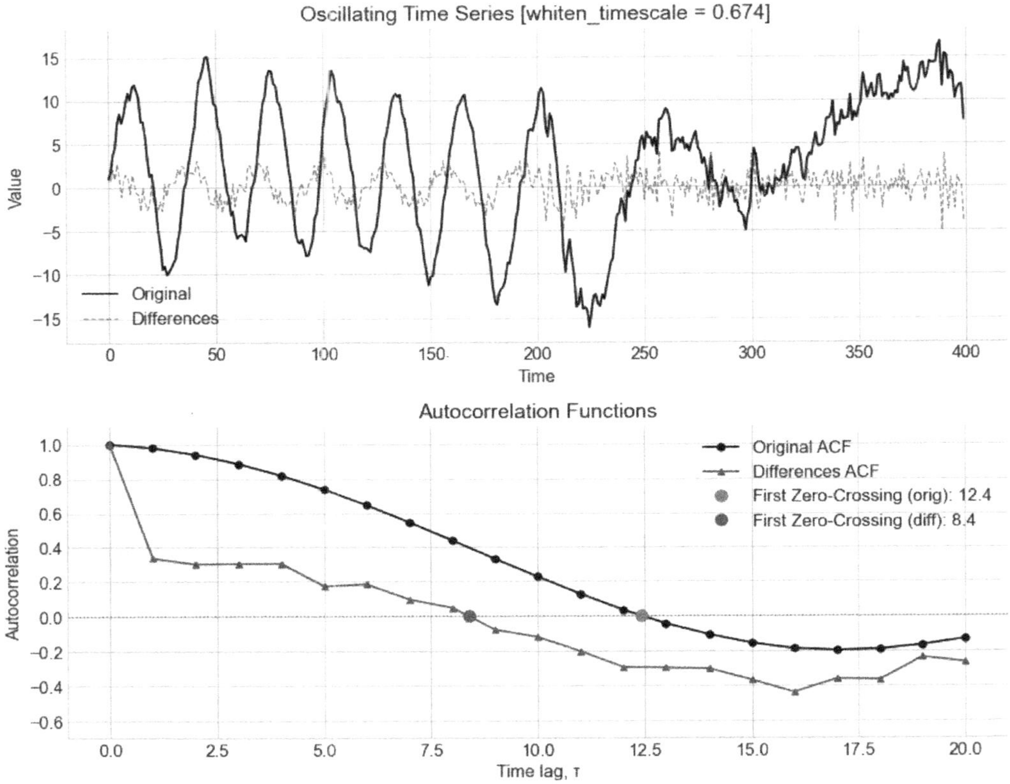

Figure 14.10: Whitening timescale feature

- **Forecasting Features** quantify how predictable a time series is using simple calculations. These features measure our ability to anticipate future values based on past observations, giving us insight into a series' inherent randomness or determinism.
    - **Forecast error** evaluates the standard deviation of residuals using 3 previous values to predict the next point. Forecast errors should have a `stdev <1` if the MA3 is useful. However, forecast errors can be substantial and exceed the variance of the original series.

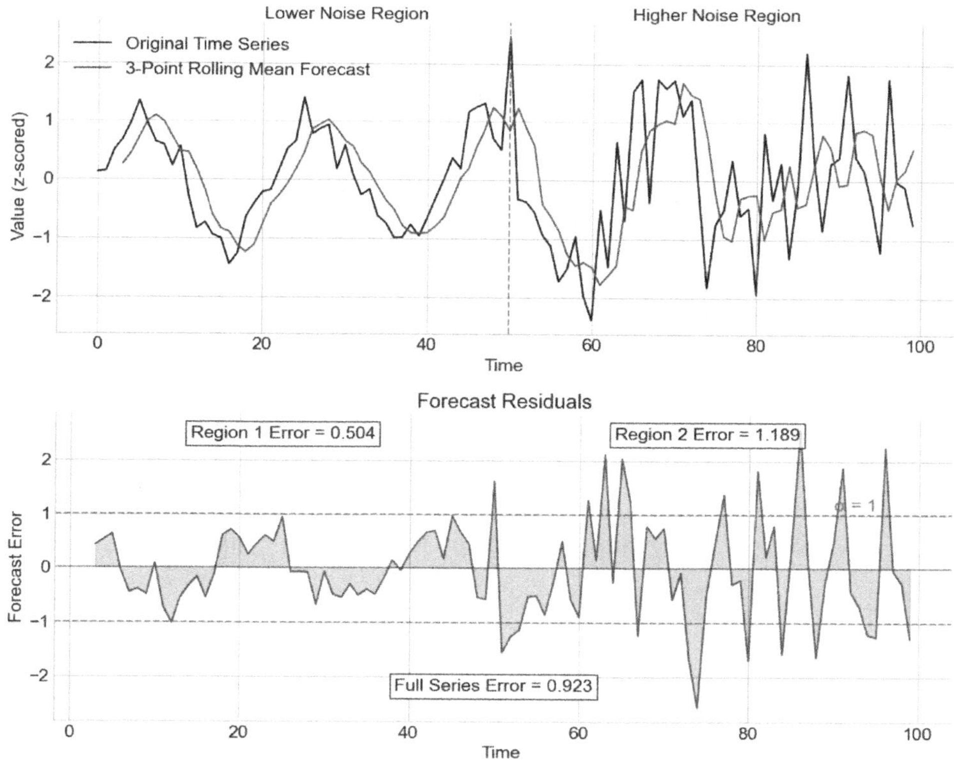

Figure 14.11: Forecast error feature

- **Entropy features**: Entropy measures quantify the predictability, regularity, or complexity of a time series. In particular, *Sample Entropy* measures the likelihood that similar patterns of observations will remain similar when a series/pattern's length is increased by one. We do this because it provides insight into:
    - **Complexity**: Higher entropy indicates more complex, less predictable behavior
    - **Regularity**: Lower entropy suggests more regular, predictable patterns
    - **Randomness**: Completely random signals have high entropy
    - **Deterministic chaos**: Chaotic systems show moderate entropy values

Entropy measures are particularly valuable when the complexity of signals provides important information about the underlying system, i.e., stock market data, neuroscience (EEG signal). They are often used because entropy measures can help detect subtle changes in activity.

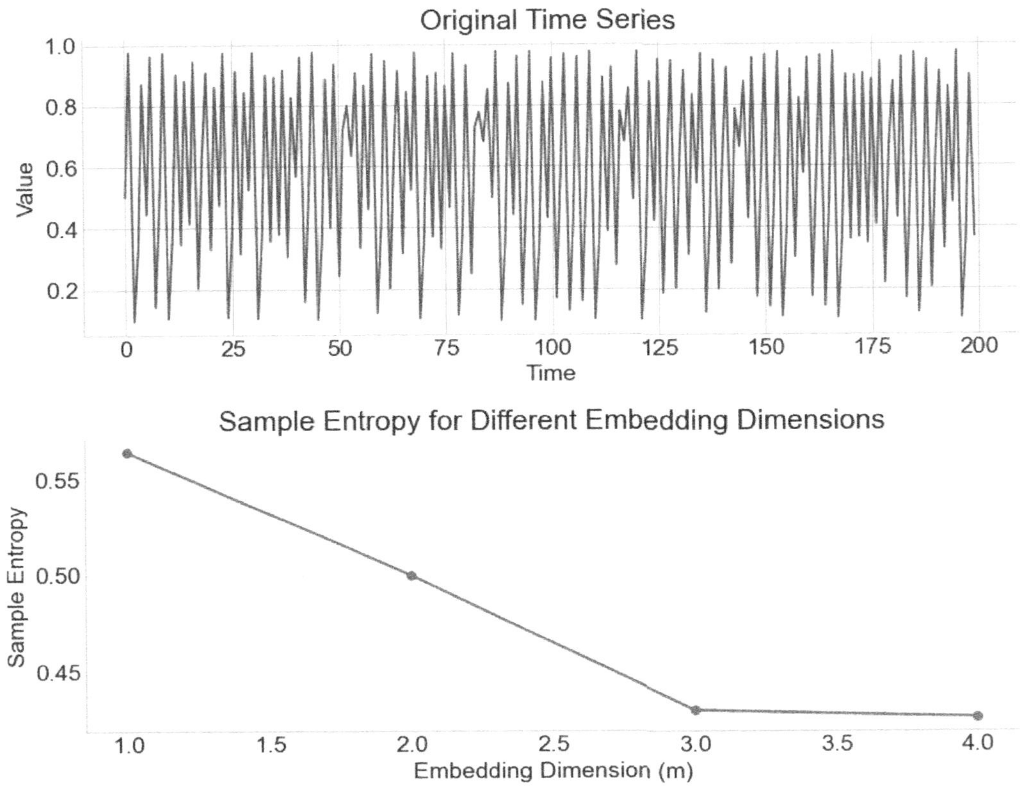

Figure 14.12: Visualising Entropy features

- **Fourier Transform**: Features Fourier transformations convert series into frequency domains, unveiling periodic components as a collection of sinusoidal functions. Think of it like this: the time domain shows how signals change over time; the frequency domain displays which frequencies exist in our data and their strengths. Fourier transformation informs us about:
    - **Dominant frequencies:** Periodic components that are important to a series structure
    - **Power distribution**: How energy is distributed across different frequencies
    - **Harmonic structures:** Related frequency components that form complex patterns

These features help us analyze signals with periodic components, such as audio, electrical signals, or any natural process with cyclical behavior.

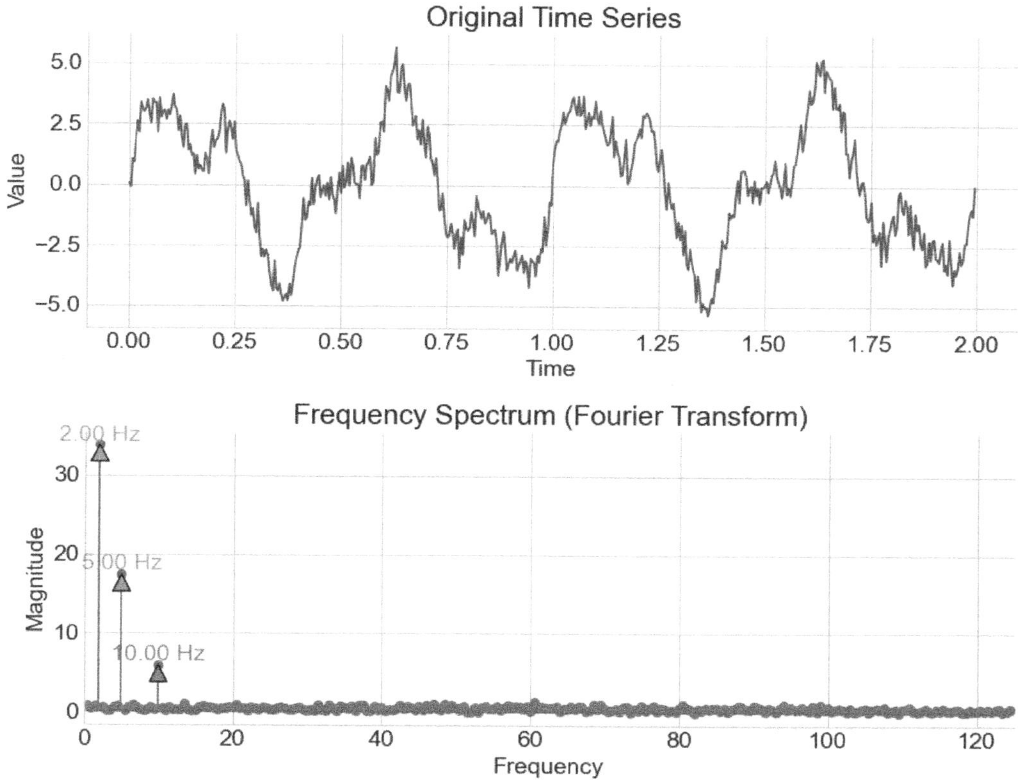

Figure 14.13: Visualizing Fourier features

- **Detrended Fluctuation Analysis (DFA)** unveils long-range correlations and scaling properties in a series. The main output is the scaling exponent α, which characterizes:
    - **Correlation structure:** How fluctuations at different timescales relate to each other
    - **Persistence or anti-persistence:** Whether trends tend to continue ($\alpha > 0.5$) or reverse ($\alpha < 0.5$)
    - **Fractal nature:** If present, how self-similarity manifests across different scales

  Different types of processes have characteristic scaling exponents:
    - White noise: $\alpha \approx 0.5$ (uncorrelated)
    - Pink/1/f noise: α ≈ 1.0 (long-range correlations)
    - Brownian motion: $\alpha \approx 1.5$ (integrated white noise)

DFA works well when we are dealing with complex systems where correlations exist across multiple timescales, such as financial markets, climate data, and natural phenomena with fractal properties.

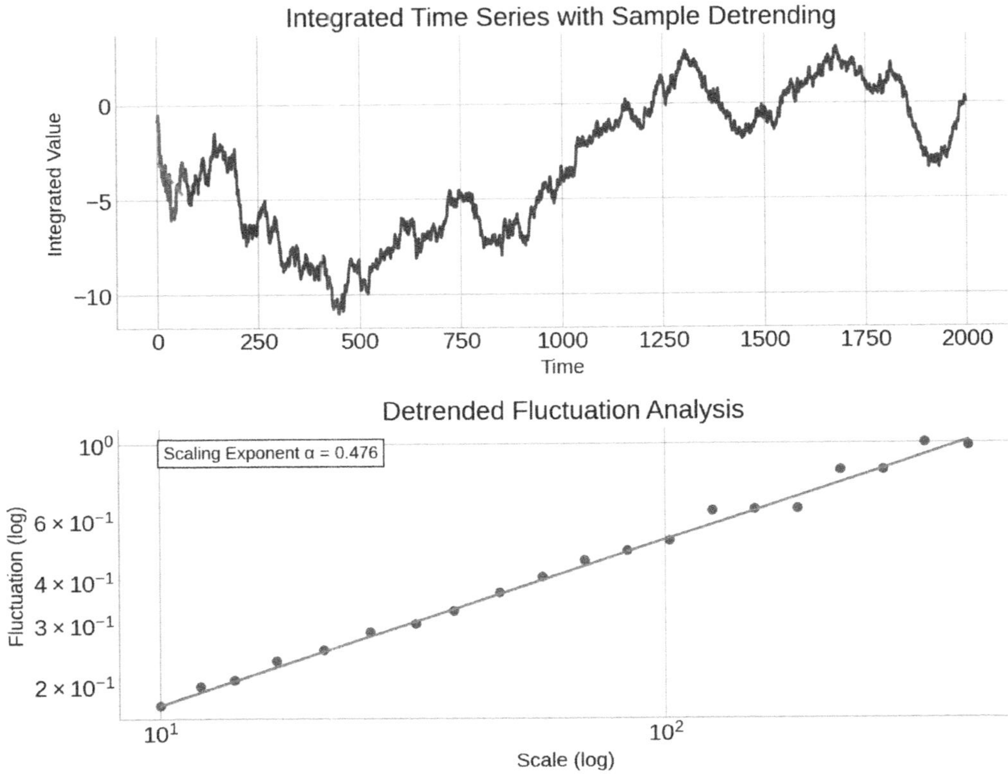

Figure 14.14: Visualising detrended fluctuation analysis

- **Complexity Features:** Extreme event features are used to look at the timings and distribution of outliers. They are supposed to capture unusual events and their distribution throughout your data.
  - **Positive outlier timing** tracks values significantly above the mean that appear in the series
  - **Negative outlier timing** examines when values significantly below the mean occur

We can transform a series data into z-scores and evaluate where extreme values cluster within a series: toward the beginning of the series (output near −1), appear uniformly throughout (output near 0), or concentrate toward the end (output near 1).

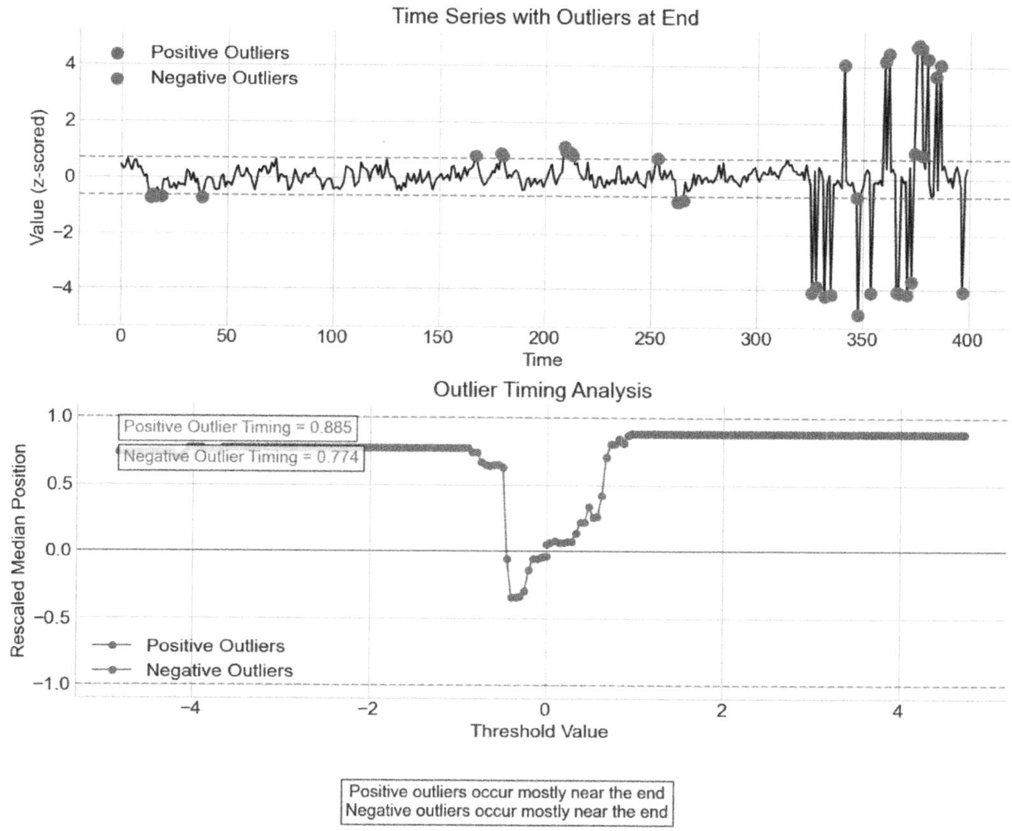

Figure 14.15: Outlier clustering

- **Shapelet-Based Features** Shapelet-based features use algorithms to identify *discriminative subsequence patterns* within a series, allowing them to capture **local** structural patterns. A shapelet is a subsequence of a time series that is highly predictive of class differences. Time series data often exhibits characteristic shapes that are indicative of the class of the time series, making shapelets particularly valuable for classification tasks.

The usefulness of a shapelet can be defined mathematically by measuring the distance between it and a time series:

$$d(s,x) = min_k d(s, x_{k,...,k+l})$$

Here d is a distance value, typically Euclidean, s is the shapelet of length l, and x is the series.

We identify shapelets by:

- Extracting candidate subsequences of varying lengths from training data
- Calculating distances between candidates and all series in a dataset
- Ranking series by distance
- Evaluating how well the ranking separates classes (using information gain)
- Selecting subsequences with maximum discriminative power

Unlike full-sequence distance measures such as DTW, shapelet transforms only measure the distance for subsequences of the data rather than the entire time series. This focused approach can capture localized patterns critical for classification. Some of the highest-performing time series classifiers are ensembles that utilize data transformed through shapelet transforms.

In the following figure we artificially created a shapelet, and then compared it to sections of a series, to demonstrate the idea.

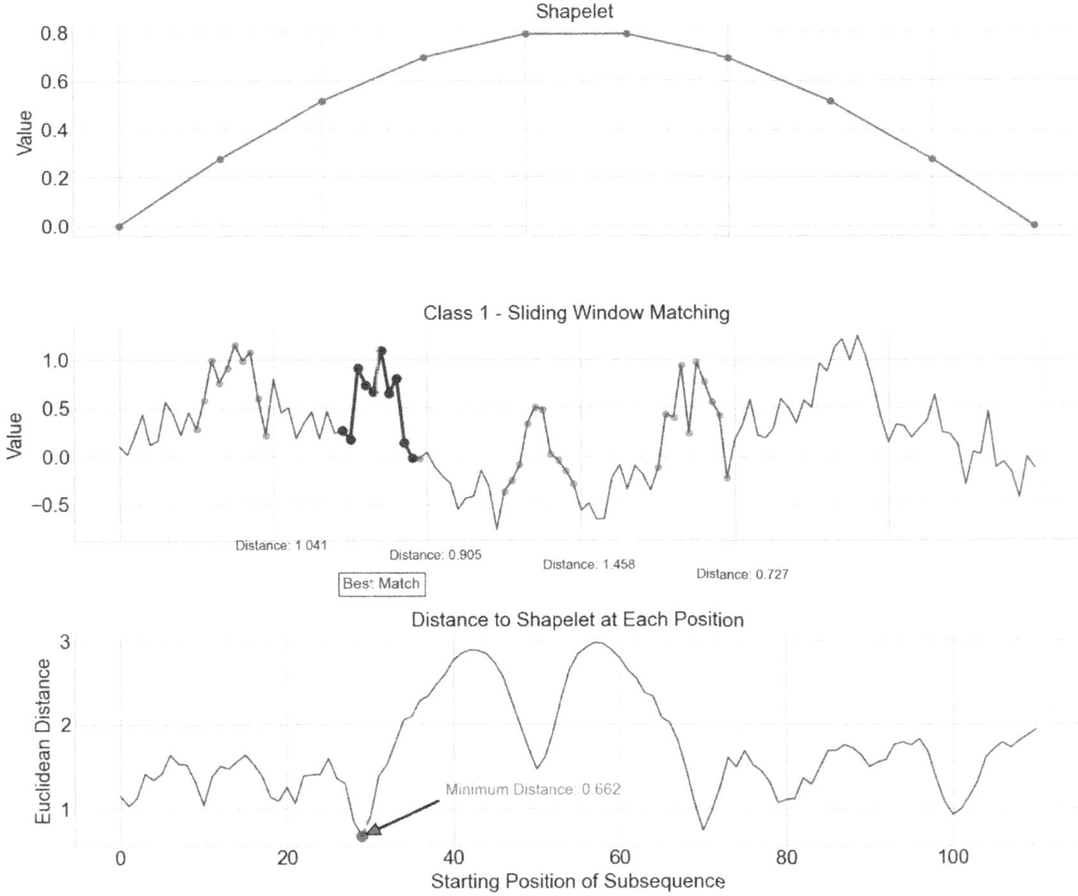

Figure 14.16: Visualizing Shapelets features

Interpreting features

When interpreting time series features, prioritize parsimony. Complex features often echo simpler properties. For instance, lag-1 autocorrelation might separate classes as well as more elaborate features like Lyapunov spectra or recurrence-based measures. When both high- and low-complexity features encode similar structure, the simpler ones are preferable—offering interpretability without sacrificing discriminatory power.

Oddly enough, we often find that temporal structure adds little beyond static distributional differences. Class separation may depend on scale (e.g. variance), skew, and even series length. It's poor practice to attribute predictive performance to intricate descriptors, like fractal dimension or graph-theoretic metrics, if performance is replicated by basic statistical summaries. This is important to remember because feature libraries often build thousands of candidate features, but a small subset of these usually carry most of the signal. Because feature redundancy is common, inspecting correlation matrices of top-performing features or grouping by methodological class can highlight this. Reducing to low-complexity, interpretable sets not only clarifies results but also mitigates overfitting, particularly in low-sample regimes.

Building features

How does featurization perform? Let's use `catch22` (a reduced feature set of 22 features based on a dimensionality reduction of features built by Fulcher's hctsa). We'll also look at Time Series Feature Extraction based on Scalable Hypothesis Tests (TSFresh, around 800 features) with a rotation forest classifier, which we can directly use with `aeon`. You should also note that `catch22` can deal with standardization directly, and has features to look at trend so we don't denoise.

Then convert the series into an array, which we pass to a transformer library to run the calculations, output values for each calculation into a single array of values for each time series (`id`). Here, we have converted this array to a data table with `from aeon.transformations.collection.feature_based import Catch22`.

```
from aeon.transformations.collection.feature_based import Catch22

feature_transformer = Catch22(features = "all",n_jobs=10, replace_nans = True)
series_transformed = feature_transformer.fit_transform(X_train)
feature_names = [
    "DN_HistogramMode_5", "DN_HistogramMode_10",
    "CO_f1ecac", "CO_FirstMin_ac",
    "SB_BinaryStats_mean_longstretch1",
    "DN_OutlierInclude_p_001_mdrmd", "DN_OutlierInclude_n_001_mdrmd",
    "FC_LocalSimple_mean1_tauresrat",
    "CO_trev_1_num",
    "CO_HistogramAMI_even_2_5",
    "IN_AutoMutualInfoStats_40_gaussian_fmmi",
    "MD_hrv_classic_pnn40",
```

```
        "SB_BinaryStats_diff_longstretch0",
        "SB_TransitionMatrix_3ac_sumdiagcov",
        "PD_PeriodicityWang_th0_01",
        "CO_Embed2_Dist_tau_d_expfit_meandiff",
        "IN_AutoMutualInfoStats_40_gaussian_std",
        "FC_LocalSimple_mean3_stderr",
        "CO_HistogramAMI_even_10_5",
        "CO_Embed2_Dist_tau_d_expfit_lambda",
        "MD_hrv_classic_pnn20",
        "SB_BinaryStats_mean_longstretch0"
]

# Create the dataframe with feature values
featurised_df = pd.DataFrame(series_transformed, columns=feature_names)

# Add series_id and labels to the dataframe
featurised_df.insert(0, 'series_id', range(len(series_transformed)))
featurised_df.insert(1, 'label', y_train)
```

The output:

series_id	label	DN_ HistogramMode_5	DN_ HistogramMode_10	SB_ BinaryStats_ mean_ longstretch1	DN_ OutlierInclude_p_001_ mdrmd	DN_ OutlierInclude_n_001_ mdrmd	IN_ AutoMutualInfoStats_40_ gaussian_fmmi
0	1	22.5	20.25	0.429999	88.641115	0.895470	0.074669
1	1	19.0	20.80	0.356294	-14.707317	0.905923	0.059933
2	1	4.2	23.10	0.468194	28.341463	0.923345	0.066859
3	1	6.1	9.15	0.524741	61.982578	0.912892	0.066292
4	1	18.5	16.65	0.403837	-21.724739	0.902439	0.070248

Table 14.1: The features table

Capturing the features is extremely useful, as we will find in the next chapter on clustering. There are a few factors to consider that we have found to improve performance that are not part of the aeon pipeline here:

- Pruning features with ridge regression
- Denoise the data with an autoencoder (we'll do this in chapter 15 on clustering)
- Compare different algorithms for classification i.e. Catboost, Xgboost, Logistic regression
- Tuning the feature transformer and classifier

Continuing with the aeon classification pipeline, we change some HPs, based on tuning some of Catch22's and `randomforests` parameters with `optuna`. Because Catch22 in Aeon is written in `numba` its fair quick so CV is quick.

```
from aeon.classification.feature_based import Catch22Classifier
from sklearn.ensemble import RandomForestClassifier

c22 = Catch22Classifier(
    estimator=RandomForestClassifier(
        n_estimators=291,
        max_depth=5,
        min_samples_split=12,
        min_samples_leaf=5,
        max_features='sqrt',
        bootstrap=True,
        class_weight='balanced',
        random_state=3542
    ),
    features='all',
    catch24=False,
    outlier_norm=True,
    replace_nans=True,
    n_jobs=10,
    random_state=3542
)
c22.fit(X_train, y_train)
c22_preds = c22.predict(X_test)
results["catch22class"] = accuracy_score(y_test, c22_preds)
```

This took seconds to run with `n_jobs = 10`. Now let's run the TSfresh pipeline, which uses a decision tree for classification directly. We did not tune TSfresh as it is notoriously slow. You can select `default_fc_parameters` to be *minimal*, *efficient*, or *comprehensive*; even *efficient* isn't quick.

```
# from aeon.classification.feature_based import FreshPRINCEClassifier
# fp = FreshPRINCEClassifier(default_fc_parameters="efficient", n_jobs=10, n_
estimators = 200, verbose = -1, random_state = 866)
# fp.fit(X_train, y_train)
# fp_preds = c22cls.predict(X_test)
# results["freshprince"] = accuracy_score(y_test, c22_preds)
```

Taking a look at the results (filtered to the best four), we can see that the catch22 pipeline with tuned parameters is the second most accurate option.

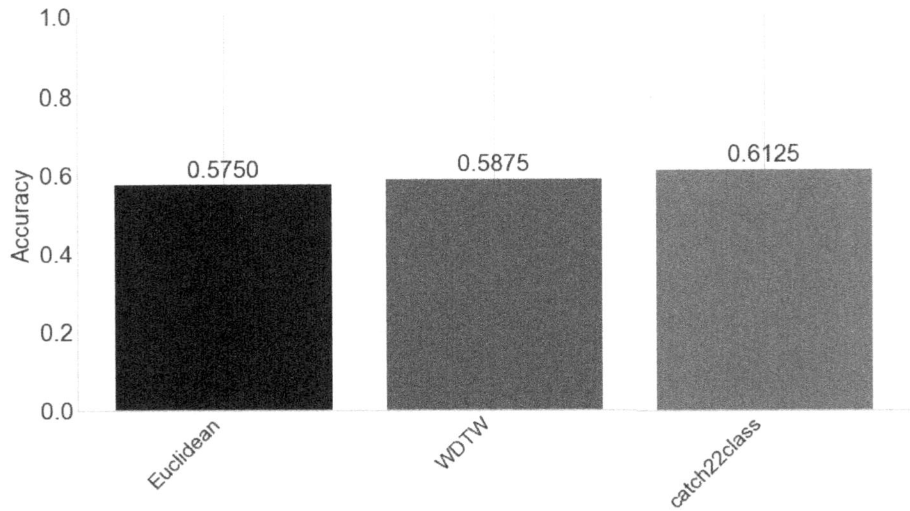

Figure 14.17: Comparing accuracy of feature classifiers

TS feature transformations are useful because the transformed data can be stored for both classification and clustering, while getting decent accuracy compared to other methods. They do require a bit of work on data wrangling, feature building, and selection, along with tuning. It helps to utilize domain knowledge when building features. While not the best performer here, it can work incredibly well with *big data*, and is often the pragmatic solution, even if it is not the optimal.

Shapelets

Shapelets can be used more directly for classification than applied in feature extraction. Shapelet-based classifiers handle the entire pattern recognition process internally. The process of shapelet transformers is as follows:

1. Extracts candidate shapelet patterns from the training data,
2. Transforms series into feature space of shapelet distances,
3. Applies a standard classifier (i.e., Random Forest),
4. Tests on training data

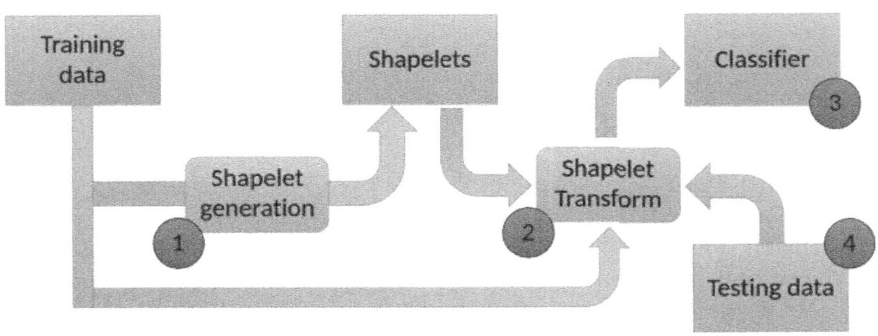

Figure 14.18: Shapelet Classification

We'll use Aeon's ShapeletTransformClassifier for this.

Shapelet algorithms use z-normalization to create shapelets that are scale-invariant, focusing on shape rather than magnitude. This is important for detecting similar patterns across different scales in datasets. Approaches to extraction vary by algorithm:

- Exhaustive evaluates all possible subsequences.
- Heuristic samples a limited number of candidates to reduce computation time (i.e. `RandomShapeletTransform`).
- Self-similarity is also important—algorithms avoid selecting overlapping subsequences to ensure diversity in the pattern set.

Aeon provides some more specialized variants, such as `RDSTClassifier` (**Random Dilated Shapelet Transform**). This uses dilation to capture patterns at different time scales and extracts features from each shapelet, not just the minimum distance. We can also use `SASTClassifier` (**Scalable and Accurate Subsequence Transform**), which is supposed to reduce computational cost by selecting a subset of *reference* series. While `LearningShapeletClassifier` doesn't extract shapelets from the data at all but generates shapelets using optimization techniques, omitted from code below due to requirements (`Tensorflow`).

```
from aeon.classification.shapelet_based import (
    ShapeletTransformClassifier,
    RDSTClassifier,
    SASTClassifier,
    LearningShapeletClassifier
)

# Shared parameters
RANDOM_STATE = 3542

# Common RandomForest parameters
rf_params = {
    'n_estimators': 300,
    'max_depth': 5,
    'min_samples_split': 10,
    'class_weight': 'balanced',
    'random_state': RANDOM_STATE
}

# Define classifiers
classifiers = {
    "ShapeletTransform": ShapeletTransformClassifier(
        estimator=RandomForestClassifier(**rf_params),
        max_shapelets=10,
```

```
        n_shapelet_samples=500,
        random_state=RANDOM_STATE
    ),
    "RDST": RDSTClassifier(
        # For RDST, we pass RF as the estimator instead of direct parameters
        estimator=RandomForestClassifier(**rf_params),
        random_state=RANDOM_STATE
    ),
    "SAST": SASTClassifier(
        classifier=RandomForestClassifier(**rf_params),
    )
}

# Fit classifiers and evaluate
for name, clf in classifiers.items():
    clf.fit(X_train, y_train)
    preds = clf.predict(X_test)
    results[name] = accuracy_score(y_test, preds)
```

So how did these perform on our traffic dataset? Both the simple shapelet (**accuracy = 0.45**) and SAST (**accuracy = 0.525**) transformations performed poorly compared to Euclidean distances (**accuracy = 0.575**). By comparison, RDST (**accuracy = 0.6625**) outperforms catch22's lightweight featurization method and comes close to our tuned proximity forest, even though we cycled the same HPs from the random forest classifier we used with catch22.

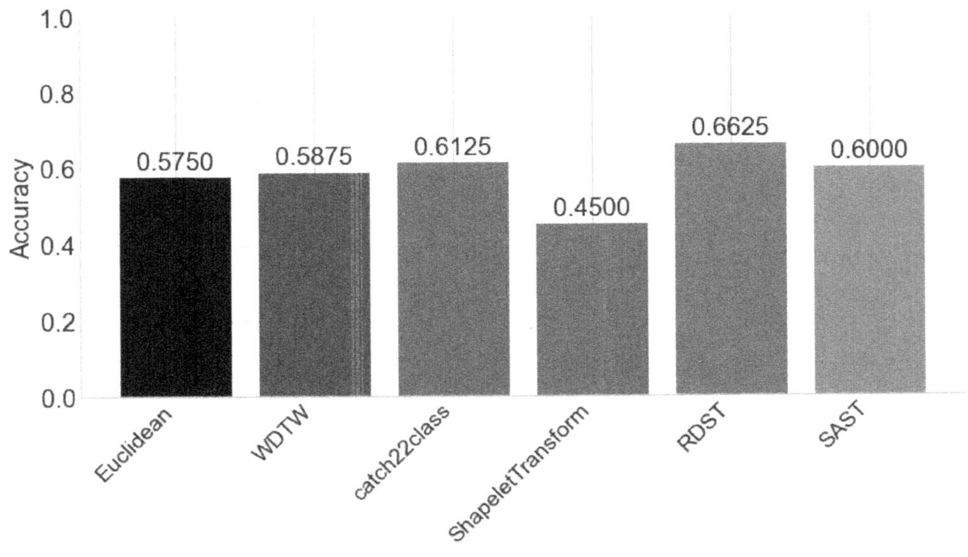

Figure 14.19: Comparing accuracy of shapelet classifiers

Dictionary-based classification

Dictionary-based approaches are somewhat similar to shapelet algorithms, in that they slide windows over a series to find phase-independent subsequences. The difference comes in that they use patterns that appear within a series like words in *text analysis*. In text classification we use a **Bag-of-Words (BoW)** to count word occurrences; similarly, time series dictionaries track frequency of 'pattern' occurrence.

The classic approach to this comes from **Symbolic Aggregate approXimation** (**SAX**), which discretizes a series' segments into a symbol/word-based representation. A **Bag-of-Patterns** (**BoP**) algorithm extends this by sliding a window across each series, converting windowed segments into symbolic words, then using frequencies to build a histogram of pattern occurrences. This feature representation captures recurring motifs while being invariant to their position within the series.

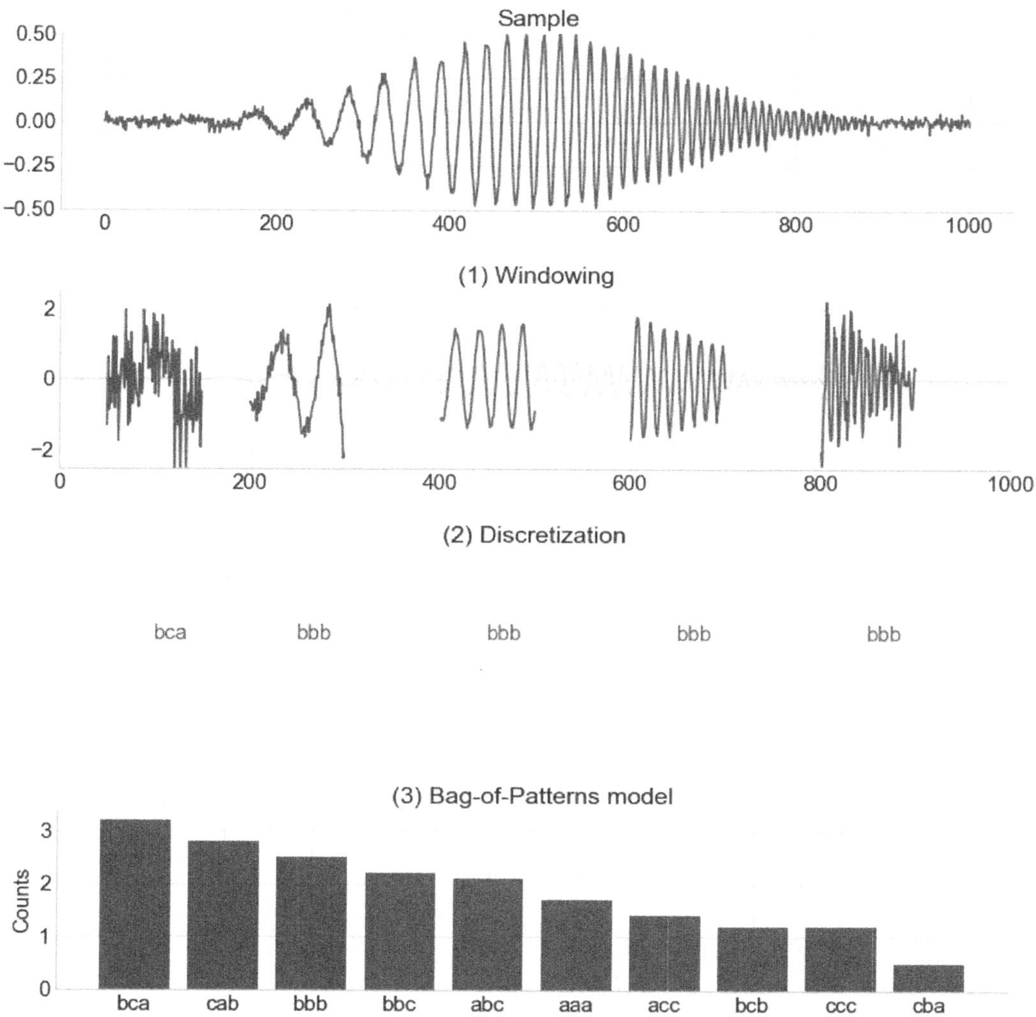

Figure 14.20: Diagram of dictionary approach

In aeon the dictionary methods mostly use **Symbolic Fourier Approximation** (**SFA**) for discretization. Which normalizes each window, then applies Fourier transform to reduce dimensionality, before discretizing the coefficients into symbols from a fixed alphabet. This requires some tuning as the main parameters (window size, reduced series length, alphabet size) significantly affect performance.

Let's try a few of these with our data:

- **Bag of SFA Symbols** (**BOSS**) creates histograms of SFA words
- **Contractable BOSS** (**cBOSS**) for efficiency improvement.
- **WEASEL** (**Word ExtrAction for TSC**) applies bigrams to SFA, and selects features with Anova and Information Gain Binning.
- **Temporal Dictionary Ensemble** (**TDE**) utilizes components of cBOSS and WEASEL, using *Spatial Pyramids*. Spatial pyramids split series into contiguous segments and construct a dictionary.

For more detail, see `https://www.aeon-toolkit.org/en/latest/examples/classification/dictionary_based.html`.

The code to try out methods is as follows:

```
# Dictionary-based classifier implementations
# from aeon.classification.dictionary_based import BOSSEnsemble, WEASEL_V2,
TemporalDictionaryEnsemble

# # BOSS Ensemble
# boss = BOSSEnsemble(
#     max_ensemble_size=5,
#     min_window = 60,
#     random_state=3542,
#     n_jobs=10
# )

# # WEASEL V2
# weasel = WEASEL_V2(
#     random_state=3542,
#     n_jobs=10
# )

# # Temporal Dictionary Ensemble
# tde = TemporalDictionaryEnsemble(
#     n_parameter_samples=250,
#     max_ensemble_size=50,
```

```
#       randomly_selected_params=True,
#       random_state=3542,
#       n_jobs=10
# )

# # Train and evaluate classifiers
# boss.fit(X_train, y_train)
# boss_preds = boss.predict(X_test)
# results["BOSS"] = accuracy_score(y_test, boss_preds)

# weasel.fit(X_train, y_train)
# weasel_preds = weasel.predict(X_test)
# results["WEASEL_V2"] = accuracy_score(y_test, weasel_preds)

# tde.fit(X_train, y_train)
# tde_preds = tde.predict(X_test)
# results["TDE"] = accuracy_score(y_test, tde_preds)

results["BOSS"] = 0.475
results["WEASEL_V2"] = 0.65
results["TDE"] = 0.55
```

`WEASEL_V2` performs just below the proximity forest and `RDST`. But `BOSS` and `TDE` perform poorly, both below `Euclidean` distance. However, one should note that while `WEASEL_V2` is below `RDST`, it runs much faster and has greater scalability.

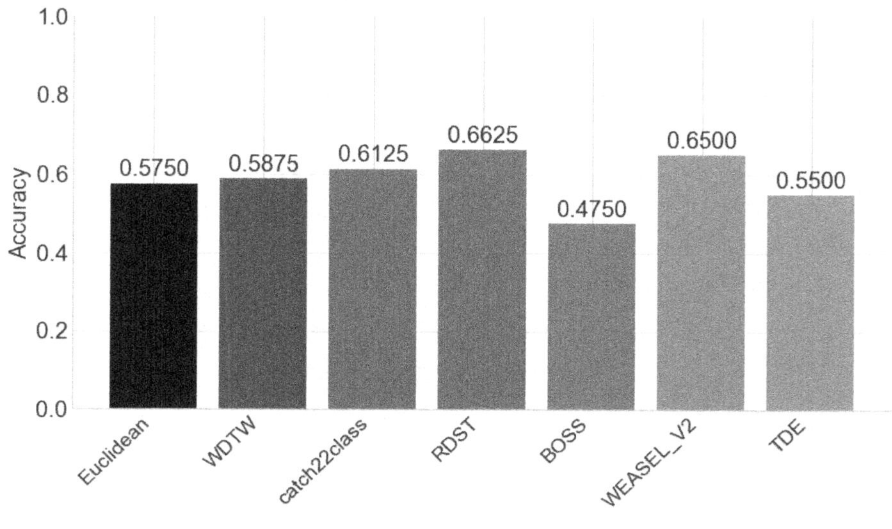

Figure 14.21: Comparing accuracy of dictionary classifiers

Convolution-based classification

Convolution-based classification is similar to shapelets, in that they search for distinctive patterns, but they use a structure called a *kernel* which is a small, weighted pattern detector. As kernels/convolutions slide across a series a dot product is calculated for each step. Creating a new series, which we may refer to as a feature or activation map. For a time series T and a kernel ω, the first value in the feature map is the dot product, like the 3-element kernel (often between 7-11 points):

$$T_{1:3} \cdot \omega = T_1\omega_1 + T_2\omega_2 + T_3\omega_3$$

From this feature map, we use ROCKET (Random Convolutional Kernel Transform) to extract: maximum value (max pooling) and proportion of positive values (PPV). The random part of ROCKET refers to randomization of kernel parameters: length, weights, bias, dilation, and padding.

For example, if a feature map has 10 values and 4 are positive, the PPV would be $\frac{4}{10}$.

These features generated by thousands (unlike shapelets) of random kernels create a transformed dataset that's fed into a classifier, commonly Ridge Regression. If we use this, it's best to use cross-validation to find the optimal L_2-regularization parameter α. Other classifiers can certainly be used, such as logistic regression for large datasets.

ROCKET, like dictionary methods, creates profiles of pattern distributions. It is also efficient because it can capture patterns without overly expensive pattern discovery, via randomized kernels to sample the pattern space.

MiniROCKET simplifies and speeds up the process with fewer, fixed-parameter kernels. Generally, kernel approaches work well when series have recurring patterns and/or complex structures.

```
# ROCKET classifier implementation
from aeon.classification.convolution_based import RocketClassifier,
MiniRocketClassifier
from sklearn.linear_model import RidgeClassifierCV

# Standard ROCKET
rocket = RocketClassifier(
    n_kernels=10000,          # Num random convolutional kernels
    estimator=RidgeClassifierCV(
        alphas=np.logspace(-3, 3, 11)
    ),
    class_weight=None,
    n_jobs=10,
    random_state=3542
)

# MiniROCKET variant - more parameter efficient
mini_rocket = MiniRocketClassifier(
```

```
        n_kernels = 10000,
        max_dilations_per_kernel = 32,
        estimator=RidgeClassifierCV(
            alphas=np.logspace(-3, 3, 11)
        ),
        class_weight=None,
        random_state=3542
)

# Train and evaluate ROCKET
rocket.fit(X_train, y_train)
rocket_preds = rocket.predict(X_test)
results["ROCKET"] = accuracy_score(y_test, rocket_preds)

mini_rocket.fit(X_train, y_train)
mini_rocket_preds = mini_rocket.predict(X_test)
results["MiniROCKET"] = accuracy_score(y_test, mini_rocket_preds)
```

The output:

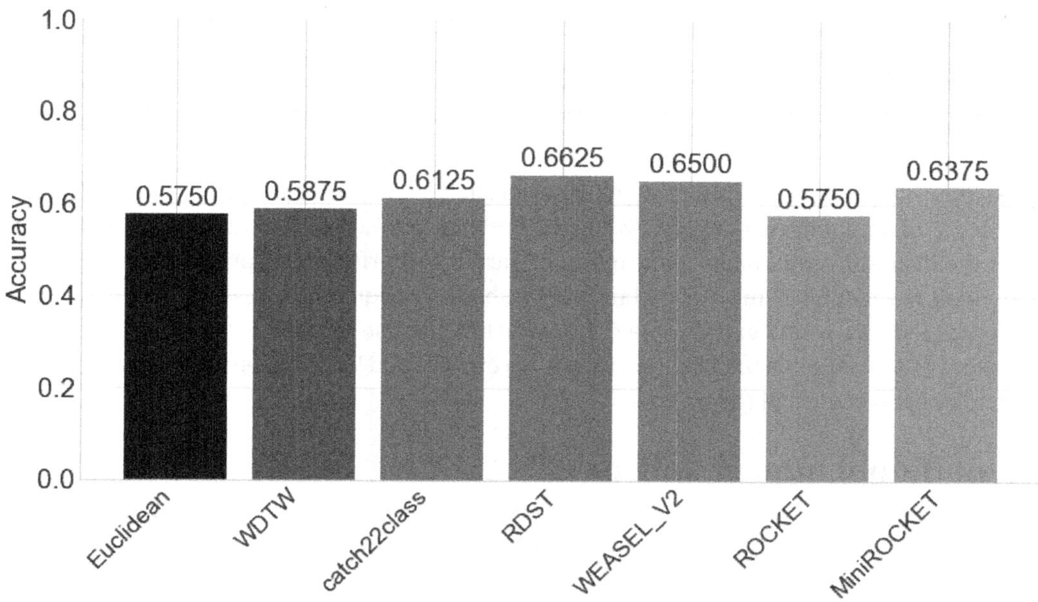

Figure 14.22: Comparing accuracy of kernel classifiers

Performance for kernel approaches are not a magic bullet. There are a few other algorithms worth investigating in this branch of classifiers that we did not have space for here: Arsenal, MultiRocket, and MultiRocketHydra.

Hybrid approaches

While we've reviewed various classification methods individually, the most accurate methods often combine multiple transformation types in a hybrid approach. Shapelet methods, dictionary approaches, and convolution-based methods like ROCKET have individual strengths, which ensemble methods like HIVE-COTE combine to get greater classification accuracy. The latest version, HIVE-COTE v2.0 (*HC2*), includes an ensemble of ROCKET classifiers called Arsenal along with shapelet approaches (*STC*), dictionary methods (*TDE*), and interval-based representations (*DrCIF*). This ensemble captures different aspects of sequential patterns—from specific discriminative subsequences (shapelets) and statistical pattern distributions (ROCKET) to dictionary-based transformations. HC2 is particularly strong as it works with multivariate data, though it is not optimal for small datasets. The main concern with hybrid approaches is the huge computational overhead, with some HIVE-COTE requiring vast amounts of time to train on moderate-sized datasets. Another issue is that ensemble methods sacrifice interpretability in the quest for accuracy as the results are combined and not stored, which is not always desirable. We have not added them here due to time constraints.

Deep learning approaches

It is fair to say deep learning models have struggled to outperform traditional TSC methods until more recently when `InceptionTime` was created. There has since been an explosion of papers on TSC with deep learning, but they often evaluate on multivariate data (easier for DL), and they generally compare against other DL methods (bad practice when the best method varies). As Middlehurst makes brutally clear in the bake-off redux paper, new methods rarely outperform InceptionTime, and its further development H-InceptionTime. Sound familiar? It should, the quality of research in deep learning is often poorer than we would expect. The DL approaches are also incredibly slow at this point in time, often making DL unsuitable for TSC; regardless, we explain and list the methods below as research is leading to improvements all the time.

Convolutional neural networks

There are four DL networks we'll discuss for time TSC, but essentially they are variants of Fukushima's CNNs, applied to one-dimensional sequential data. Your standard CNN architecture consists of a filter ω of length l which slides over a series T, computing a convolution at each position t:

$$C_t = f\big(\omega * T_{t:(t+l)} + b\big) \quad \forall t \in [1 \ldots n - l + 1]$$

Where b is a bias parameter and f is an activation function, typically ReLU. This operation creates feature maps that capture patterns at different temporal scales. CNNs have some advantages worth noting:

- **Parameter efficiency:** Filter weights are shared across all convolutions, which reduces parameters compared to fully-connected networks
- **Translation invariance:** The same pattern can be detected regardless of where it occurs in a sequence
- **Hierarchical feature learning:** Earlier pooling layers capture basic patterns while deeper layers learn more complex representations

Despite this, standard CNNs aren't particularly good classifiers.

ResNet

ResNet introduced residual connections that allow information to flow directly from earlier to later layers, to help address vanishing gradient problems. When used with series, ResNet uses three residual blocks, each containing three convolutional layers followed by batch normalization and ReLU activation. The shortcut connections are what improve gradient flow during training, allowing information to propagate from shallow to deeper layers, and allow the convolutional layer to extract features. These are then passed to **Global Average Pooling** (**GAP**) and a softmax layer (fully connected), with n neurons equal to the k classes. To visualize a ResNet, look at this image:

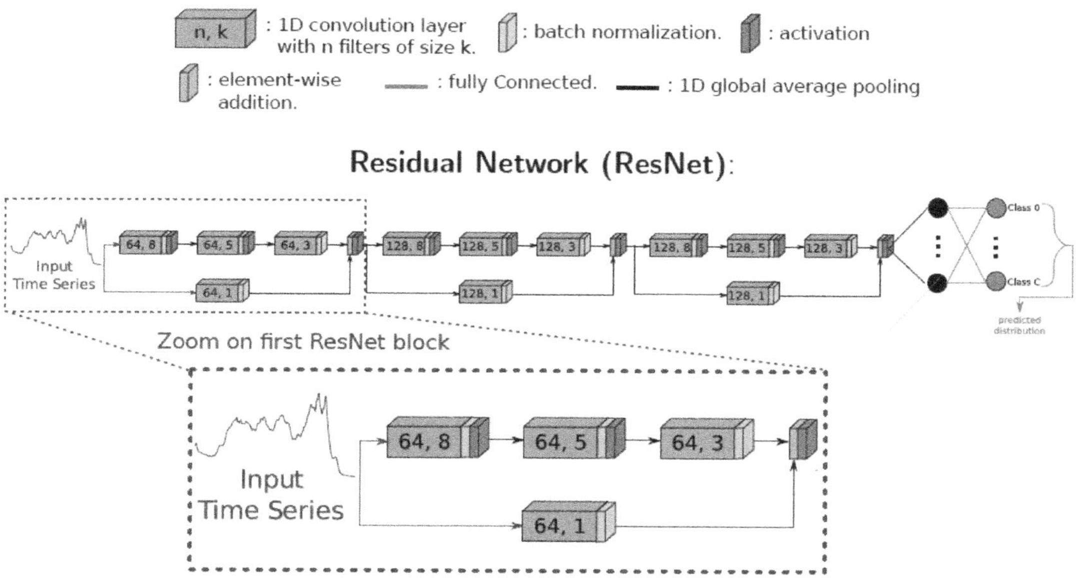

Figure 14.23: ResNet diagram

To apply these concepts in practice, we initialize and train a `ResNetClassifier` using the aeon library with the following configuration

```
from aeon.classification.deep_learning import ResNetClassifier

# Some settings
n_epochs = 200          # Default is typically 1500 - reduced for time
batch_size = 64
random_state = 3542
n_jobs = 10
verbose = False

# ResNetClassifier
# Note. ResNet benefits from multiple residual blocks
resnet = ResNetClassifier(
    n_residual_blocks=3,
    n_conv_per_residual_block=3,
    n_filters=[128, 64, 64],
    kernel_size=[8, 5, 3],
    activation='relu',
    n_epochs=n_epochs,
    batch_size=batch_size,
    random_state=random_state,
    verbose=verbose
)

resnet.fit(X_train, y_train)
resnet_preds = resnet.predict(X_test)
results["ResNet"] = accuracy_score(y_test, resnet_preds)
```

InceptionTime

As we mentioned previously, the next major leap forward came in the form of InceptionTime, which generally outperforms ResNet. It ensembles five neural networks with identical architectures but each using different initial weights. All five networks contain two residual blocks with three inception modules each, bottleneck layers that reduce dimensionality before processing, multiplexing convolution applying multiple filters with different kernel sizes in parallel to capture patterns at various time scales and GAP followed by a softmax layer

It differs from ResNet in that it ensembles the models of the network, makes use of a bottleneck (for multivariate series), multiplexing convolution, and uses only two residual blocks.

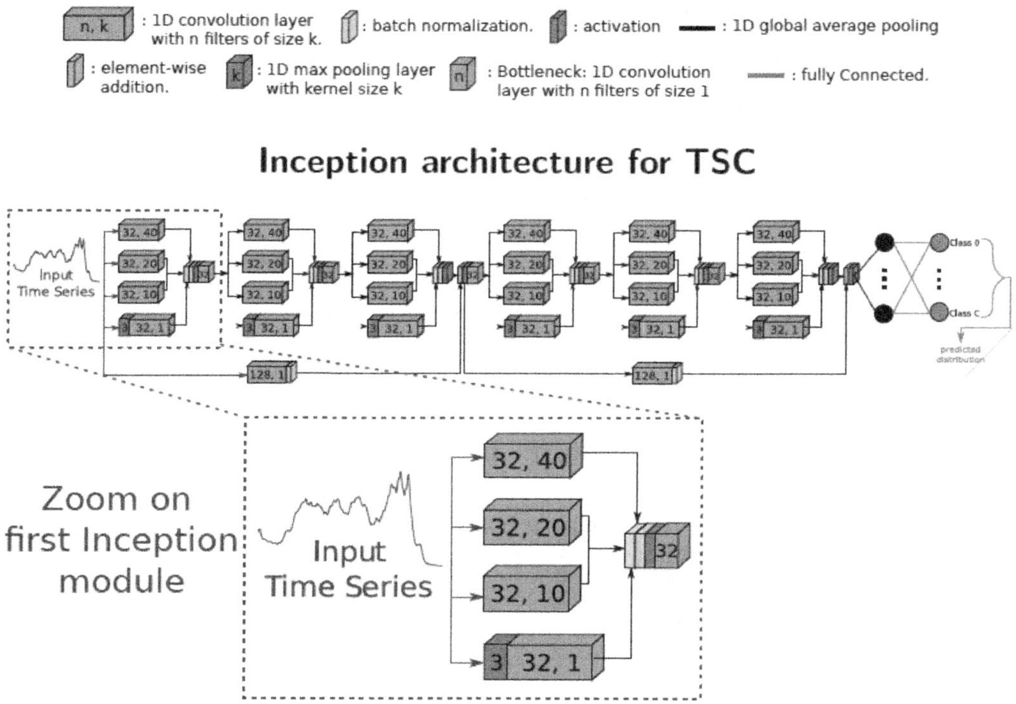

Figure 14.24: InceptionTime diagram

To implement this state-of-the-art classifier, we initialize the `InceptionTimeClassifier` from the aeon library as follows

```
from aeon.classification.deep_learning import InceptionTimeClassifier

inception_time = InceptionTimeClassifier(
    n_classifiers=3,
    n_epochs=n_epochs,
    batch_size=batch_size,
    random_state=random_state,
    verbose=verbose
)
inception_time.fit(X_train, y_train)
inception_time_preds = inception_time.predict(X_test)
results["InceptionTime"] = accuracy_score(y_test, inception_time_preds)
```

H-InceptionTime

H-InceptionTime (**Hybrid InceptionTime**) incorporates *hand-crafted* filters in parallel with the first module of the next to detect three patterns:

- Increasing trends,
- Decreasing trends, and
- Peaks

The use of these filters offers a slight improvement over the classical InceptionTime.

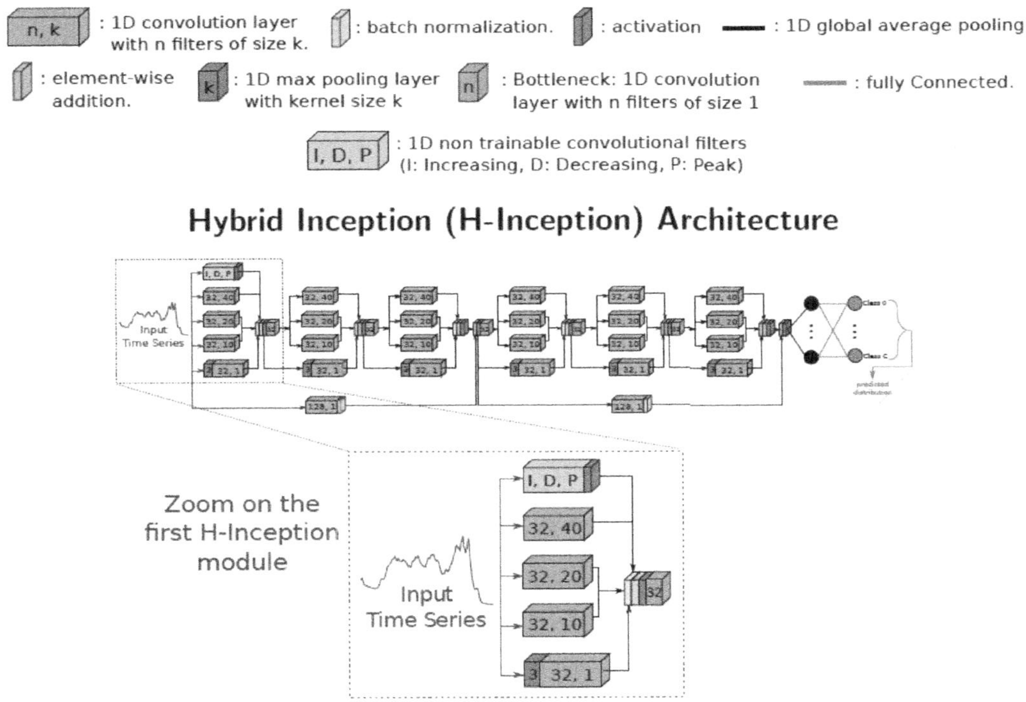

Figure 14.25: H-InceptionTime diagram

To implement this hybrid approach, we initialize and train the `IndividualInceptionClassifier` from the aeon library as follows:

```
from aeon.classification.deep_learning import IndividualInceptionClassifier

h_inception = IndividualInceptionClassifier(
    n_epochs=n_epochs,
    batch_size=batch_size,
    random_state=random_state,
    verbose=verbose
)
```

```
h_inception.fit(X_train, y_train)
h_inception_preds = h_inception.predict(X_test)
results["H-Inception"] = accuracy_score(y_test, h_inception_preds)
```

LiteTime

Unfortunately, the previous three models can be quite slow to train because of the number of parameters in the networks (~500,000). Applying **DepthWise Separable Convolutions** to InceptionTime dramatically decreases parameters. It also utilizes boosting techniques; multiplexing convolution and dilated convolution, which help it to maintain accuracy similar to InceptionTime with reduced computational requirements.

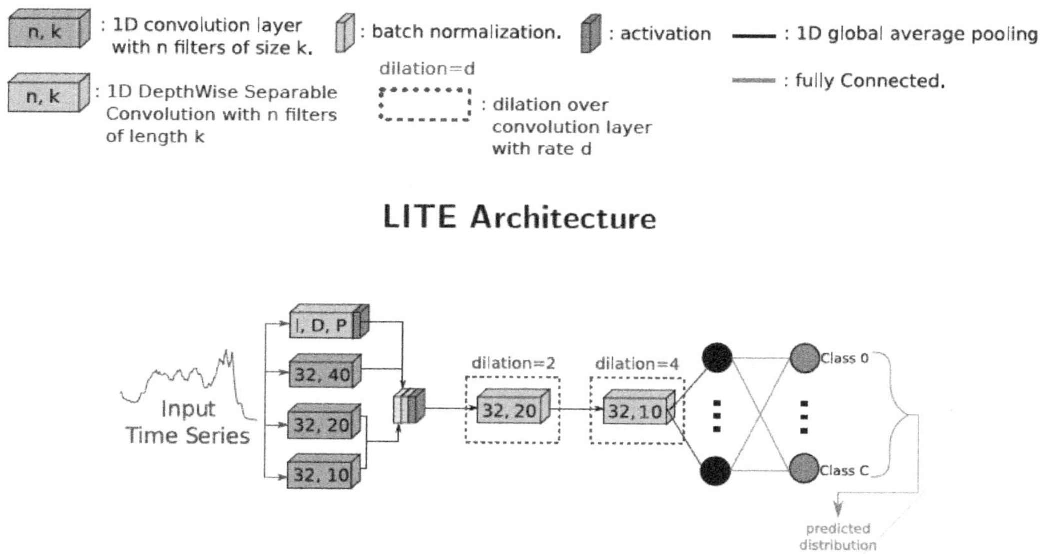

Figure 14.26: LiteTime diagram

To implement this efficient architecture, we initialize and train the `LITETimeClassifier` from the aeon library as follows:

```
from aeon.classification.deep_learning import LITETimeClassifier

lite_time = LITETimeClassifier(
    n_classifiers=3,
    n_epochs=n_epochs,
    batch_size=batch_size,
    random_state=random_state,
    verbose=verbose
)
```

```
lite_time.fit(X_train, y_train)
lite_time_preds = lite_time.predict(X_test)
results["LITETime"] = accuracy_score(y_test, lite_time_preds)
```

As is typical with DL methods elsewhere, you have a few factors to consider when using DL methods:

- **Data requirements:** DL suits larger datasets better. If you have a smaller dataset, shapelets and/or dictionary approaches be better.
- **Computational resources:** Even optimized DL approaches like LiteTime are slower than HIVE-COTE on data with the same hardware, with comparable accuracy.
- **Interpretability:** Current DL models sacrifice explainability compared to approaches like shapelets, and featurized methods.
- **Performance hierarchy:** Empirical studies consistently show: $CNN \approx 1NN - DTW < ResNet < InceptionTime \approx LiteTime < H - InceptionTime$, with $H - InceptionTime$ currently being SOTA in deep learning for TSC.

Despite this DL methods also offer advantages such as pretraining via *transfer learning*. We can also use *data augmentation* techniques, *adversarial training* to improve robustness, and there is the fact that neural architectures automatically discover optimal model structures. Given this it is likely DL for TSC will see further development in time.

Summary

The variety of approaches available to us for TSC gives us options when we need to compare time series. The simplest approaches are distance-based, with specializations like weighted-dynamic time warping; these are usually utilized as baseline comparisons rather than solutions. We can, of course, ensemble (*elastic ensemble*) these together for better performance; unfortunately, it does not scale well to large datasets. One of the best approaches for TSC is *proximity forest*, which, when tuned, performed the best for our specific data set, which is consistent with research. It also scales better than the elastic ensemble, but is still relatively slow and computationally expensive to run. The approach that seems to scale best, depending on the featurization library (which perhaps is the reason for its popularity), is time series featurization. We spent a significant amount of time, because featurization can be adapted via the features we build to tailor it to our area of research. It is incredibly useful for time series clustering too. In our testing, the accuracy of featurization was decent, but not as good as the *proximity forest*, best shapelet, and *convolution*-based approaches. The problem with the other methods is often they handle transformations internally and require rebuilding; they also often fail to scale well despite their superior performance. However, we encourage you to investigate this fast-developing area and use a library like aeon for finding a method before adapting your own for production. The next chapter will expand on many of the concepts here for time series clustering, where you will see another great use of TSC toward the end.

Recommended reading

- Fulcher, B. D. 2018. *"Feature-based time-series analysis."* arXiv. `https://arxiv.org/abs/1709.08055`.
- Ismail Fawaz, H., et al. 2020. *"InceptionTime: Finding AlexNet for Time Series Classification."* Data Mining and Knowledge Discovery. `https://doi.org/10.1007/s10618-020-00710-y`.
- Lubba, C. H., et al. 2019. *"catch22: Canonical Time-series Characteristics."* Data Mining and Knowledge Discovery. `https://doi.org/10.1007/s10618-019-00647-x`.
- Lucas, B., et al. 2019. *"Proximity Forest: An effective and scalable distance-based classifier for time series."* Data Mining and Knowledge Discovery. `https://doi.org/10.1007/s10618-019-00626-2`.
- Middlehurst, M., et al. 2024. *"The Time Series Classification Bake Off: An Evaluation of Recently Proposed Algorithms."* arXiv. `https://arxiv.org/abs/2309.11216`.

Join our community on discord

Join our community's Discord space for discussions with the authors and other readers.

`https://discord.gg/VXpz9YPUTW`

15

Time Series Clustering

"If intelligence were a cake, unsupervised learning would be the cake, supervised learning the icing, and reinforcement learning the cherry on top. We know how to make the icing and the cherry, but we don't yet know how to bake the cake." – Yann LeCun

In the previous chapter, we focused on classification methodologies, where time series data labelled with predefined categories is used to train models through supervised learning (implemented via Aeon). A critical aspect of this approach involved providing algorithms with direct or **transformed** time series data relative to known labels. Here we build on the idea of modelling transformed time series to discover shared patterns that bind temporal data together, in an unsupervised approach without predefined labels.

Time Series Clustering (TSCL) is a challenging *unsupervised* learning problem. As LeCun observes regarding unsupervised learning generally, we don't really know how to do it effectively. This is because there is an inherent uncertainty in the complexity of discovering latent structures in data (temporal or not) without explicit (labelled) guidance. Regardless, we can evaluate unsupervised learning methods, for tasks such as TSCL, by comparing the performance of unsupervised approaches on collecting labelled data together (taking the labels away) across multiple datasets. This is similar to the time series classification bake-off (see Middlehurst, Schäfer, and Bagnall (2024)); in fact we can use the same data. Despite our lack of certainty in method, several clustering approaches have demonstrated reasonable efficacy for forecasting: distance-based, feature-based, and deep learning methods, which all leverage ideas we established in *Chapter 14*.

As with classification, our challenge in TSCL is in defining *similarity* for temporal data, through comparing features or series directly. Unlike *static* data where similarity metrics can be straightforward, time series can be similar in various ways: they might have similar overall trends but different seasonal patterns, identical shapes but different magnitudes, or comparable patterns that occur at different time offsets. This is not simple, as many approaches are based on reducing data to summarized or generalized features. Whatever our approach, we need to capture the temporal dependencies and structural characteristics as best we can that make time series data unique.

Our goal is simply to group similar data; in this case specifically to help us improve model forecast accuracy. This is not a simple task, as we are often trying to understand patterns across thousands to millions of times series. Rather than modelling each series individually, clustering can find series with similar patterns, practitioners, like us, to develop forecasting strategies that are tailored for each group, for example grouping data can be used as embeddings (see *Chapter 16*) for transfer learning, or creating hyperparameter sets that work for each group.

When working with data in real life, we should always conduct exploratory research on the features we generate. We need to be careful to check how data transformations change our data, and how we interpret any modeling built on these transformations.

It is not uncommon for practitioners to create hundreds of time series features for large datasets, and then reduce them with a dimensionality reduction algorithm like PCA or t-SNE. The problem with doing something like this is, what does 20 time series features reduced to 1 actually mean? It may help to get a sense of clusters in visualization, but we shouldn't be overly restricted by it. And we certainly should be careful about building models with features we haven't investigated.

Beyond grouping similar patterns of data, clustering has a few more advantages for us:

- It allows us to target our forecasting approach by grouping series with similar characteristics. For example, with a **Global Forecasting Model** (**GFM**) we could use cluster labels to train a GFM on homogeneous rather than heterogeneous data.
- Cluster labels may facilitate EDA in large datasets, helping us identify relationships, patterns or economic curiosities.
- Clustering can also serve as an effective preprocessing step that can improve subsequent analysis performance by reducing data complexity and mitigating noise effects.

If we consider practical applications, such as a retail chain trying to understand demand patterns across thousands of products and locations, they could use clustering to discover products with similar seasonal behaviors, allowing more targeted forecasting strategies for groups of data. Similarly, in energy management, clustering consumption patterns can help identify distinct customer segments, facilitating load balancing or personalized pricing strategies.

TSCL algorithms broadly fall into four categories, following the taxonomy of Paparrizos, Yang, and Li (2024). Distance-based methods, such as those using **Dynamic Time Warping** (**DTW**) or other shape-aware measures, compute pairwise distances and then apply traditional clustering algorithms like k-means or hierarchical clustering. Distribution-based methods fit probabilistic or statistical models and cluster in parameter or density space. Feature-based approaches transform time series into fixed-length vectors of descriptive statistics, enabling the use of standard clustering techniques. And there are also representation-learning methods, which use neural networks to learn latent *embeddings* that capture temporal patterns, though these models are black boxes with less interpretability than the aforementioned.

This chapter will focus primarily on feature-based clustering, while briefly overviewing the other approaches. As noted by Bandara et al. (2021), feature-based approaches excel where interpretability and resilience to missing and noisy data are required, which is common in production environments. Rather than attempting to capture similarity through expensive point-by-point comparisons, feature methods extract global characteristics like trend, seasonality, autocorrelation structure, and statistical distributions amongst others. Another common necessity is speed and scalability. As we saw in *Chapter 14*, many of the underlying methods to create and compare time series representations (e.g. shapelets) simply don't scale well.

We'll pay some attention to preprocessing techniques that commonly affect clustering performance (as a function of forecasting accuracy), including denoising methods and dimensionality reduction strategies. This emphasis on featurization allows us to focus on experimentation, albeit simplified here for brevity.

In this chapter we will cover:

- What is clustering?
- The TSCL pipeline
- Clustering approaches
- Deep-learning approaches to TSCL
- Applying featurization
- Evaluating clustering results

What is clustering?

Don't fix what is not broken, as the saying goes. Before discussing time series clustering, let's briefly overview clustering in general with the Iris dataset. Feel free to skip to the next section if you already have a good understanding.

The Iris datasets holds measurements from three species of Iris: *Setosa*, *Virginica*, and *Versicolor*. It contains 150 observations of flowers with four features (measurements) for each, of the length and width of each flowers sepals and petals. Sepals are leaf-like structures that are usually green and protect the flower bud during its development, while petals are the vibrant, softer structures that draw the attention of pollinators such as insects and birds. We can load this data with `sklearn`.

```
from sklearn import datasets

iris = datasets.load_iris()
df = pd.DataFrame(iris.data, columns=iris.feature_names)
df['species'] = [iris.target_names[i] for i in iris.target]

display_html_table(df, n_rows=3, cols=6)
```

The output is:

	sepal length (cm)	sepal width (cm)	petal length (cm)	petal width (cm)	species
0	5.1	3.5	1.4	0.2	setosa
1	4.9	3.0	1.4	0.2	setosa
2	4.7	3.2	1.3	0.2	setosa

Table 15.1: First three rows of the Iris dataset with the four measurement columns and species label

Let's quickly look at the data with a pairplot (distribution and scatter plots), to see the relationships in the data that make Iris easy to model.

Refer to color images on the Graphic bundle or the ebook.

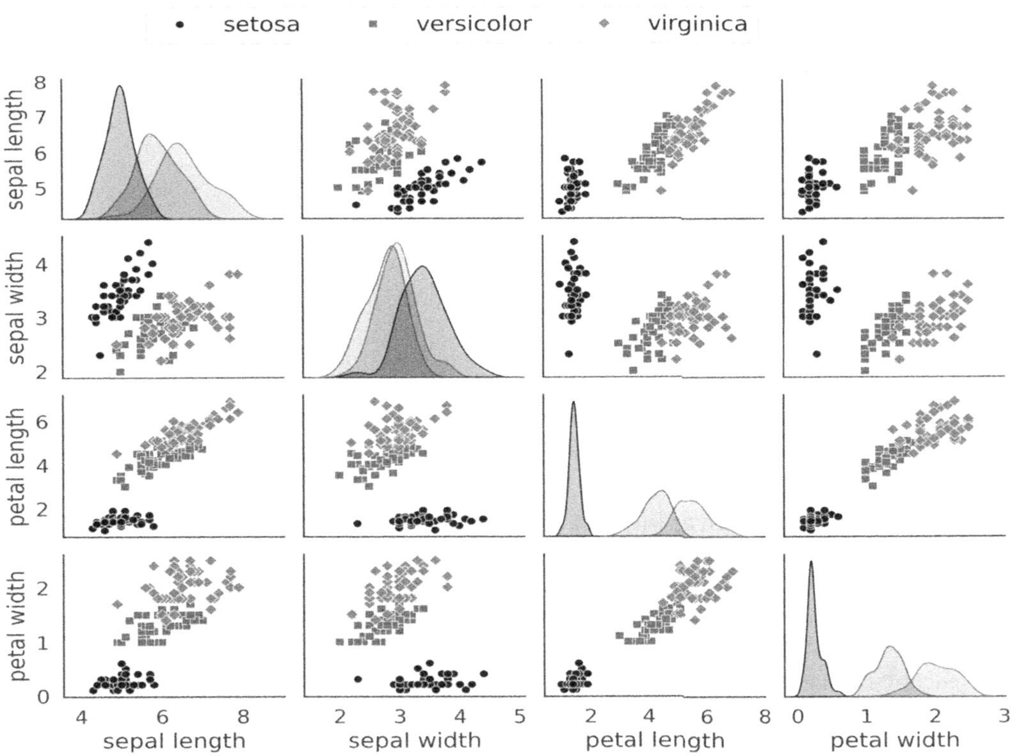

Figure 15.1: Pairplot of the four Iris measurements colored by species, showing setosa as linearly separable and some overlap between versicolor and virginica

You can see across the four variables that the different species are separable. Iris setosa in particular can be easily identified since it is completely linearly separated, as seen in the scatter plots. The difficulty comes in distinguishing the overlap between versicolor and virginica, especially with the sepal measurements.

If we create boxplots, we can see that petal length and width provide slightly better separation.

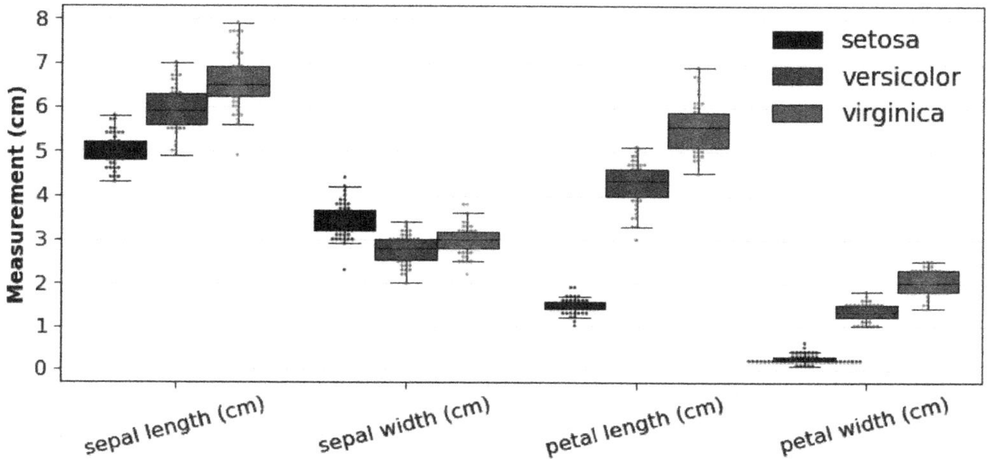

Figure 15.2: Boxplots of the four Iris measurements by species, showing that petal length and width separate the three species more cleanly than the sepal measurements

Clearly the magnitudes are slightly different for each metric, so we would hope that when we use all four features together they'd create a signal that is unique to each group. There is a statistical method for checking if data is suitable for clustering called the *Hopkins statistic* (see note), it measures the degree to which clusters exist in our data. It does this by comparing the distances of randomly generated points within the feature space to the distances of actual data points to their nearest neighbors. A Hopkins statistic close to 1 indicates high clustering tendency, while values around 0.5 suggest randomness or uniform distribution. Values between 0.7 and 0.99 indicate the dataset has a high tendency to form meaningful clusters.

> Hopkins statistic is defined as $H = \frac{\sum_{i=1}^{m} u_i}{\sum_{i=1}^{m} u_i + \sum_{i=1}^{m} w_i}$, where u_i are the nearest-neighbour distances from m uniformly sampled points to the real data, and w_i are nearest-neighbour distances from m randomly selected real points to their nearest real neighbours.
>
> Our implementation uses raw distances without the dimensionality power d, which is a simplified form sometimes called **non-dimensional** or **Lawson-Jurs** variant. Typically, you would want to use a dimensional form.

Hopkins statistic tests the following hypotheses:

- H_0: Dataset is uniformly distributed (no meaningful clusters).
- H_1: Dataset is not uniformly distributed (contains meaningful clusters).

We reject H_0 when the Hopkins statistic is significantly above 0.5 (typically $H > 0.7$), indicating clusterable structure in the data.

Here we have implemented a version of this with `sklearn`.

```
from sklearn.neighbors import NearestNeighbors

def hopkins(X, frac=0.1, random_state=None):
    rng = np.random.default_rng(random_state)
    n, d = X.shape
    m = int(frac * n)
    nbrs = NearestNeighbors(n_neighbors=2).fit(X)

    real_idx = rng.choice(n, m, replace=False)
    real = X[real_idx]

    mins, maxs = X.min(axis=0), X.max(axis=0)
    uniform = rng.uniform(mins, maxs, size=(m, d))

    u_dist = nbrs.kneighbors(uniform, return_distance=True)[0][:, 1]
    w_dist = nbrs.kneighbors(real,    return_distance=True)[0][:, 1]

    return u_dist.sum() / (u_dist.sum() + w_dist.sum())


X = df[['sepal length (cm)', 'sepal width (cm)', 'petal length (cm)', 'petal
width (cm)']].values

h2 = hopkins(X, frac=0.1, random_state=43526)
print(f"Hopkins (fast): {h2:.4f}")
Hopkins (fast): 0.890
```

With a result of 0.89 we clearly have data with a high tendency for clustering. Obviously we already know this, as we already know how many clusters to use, but if we didn't we'd have to guess. There are methods for this such as elbow plots, which visualize how a clustering quality metric changes as we increase the number of clusters (k) and look for the point where adding more clusters yields diminishing returns. To save us time we can use the yellowbrick library to build the plots (see notebook). We'll build nine sets of clusters ($k = 2 - 10$) with k-means clustering (we'll discuss clustering algorithms later), and measure the results with two elbow methods.

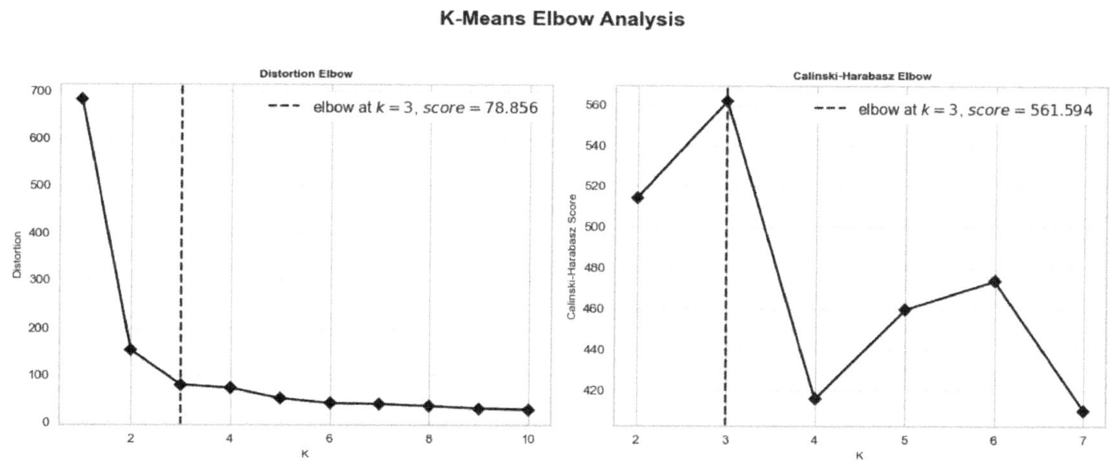

Figure 15.3: K-means elbow analysis on the Iris dataset showing distortion and Calinski-Harabasz scores across values of k, both indicating k=3

Unsurprisingly, both plots select $k = 3$. So let's build our *k*=3 clusters and see how well we do. We'll use sklearn's kmeans for the clustering with the elkan algorithm and k-mean++ initalisation to create our cluster groups. Then we create a 2-dimensional representation of the data by reducing the dimensions down with t-SNE (TSNE), using the `barnes_hut` method, so we can plot the groups on a 2D space as a scatterplot.

```
from sklearn.cluster import KMeans
from sklearn.manifold import TSNE

kmeans = KMeans(
  n_clusters=3,
  init='k-means++',
  n_init='auto',
  max_iter=300,
  algorithm='elkan',
  random_state=5363)
y_kmeans = kmeans.fit_predict(X)

df['cluster'] = y_kmeans

tsne = TSNE(
        n_components=2,
        perplexity=50,
        n_iter=1200,
        random_state=4126,
        method='barnes_hut',
```

```
        init="pca",
        n_jobs=-1
    )

X_tsne = tsne.fit_transform(X)
```

Now let's plot these clusters out.

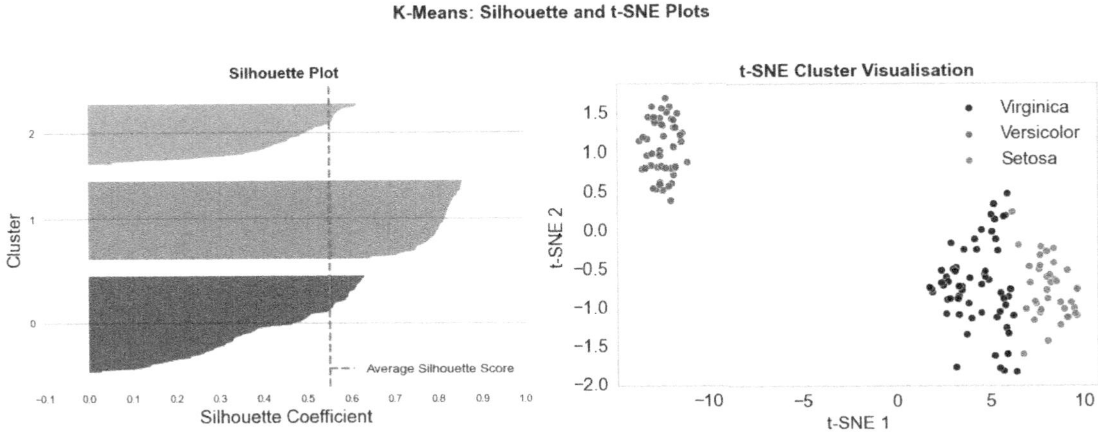

Figure 15.4: Silhouette plot and t-SNE projection of the k=3 k-means clusters on the Iris dataset, with cluster labels mapped to species names

The clusters look reasonably good, but how good are they? Well, let's check them out by comparing the true labels to the cluster labels (arbitrary values, which change each run unless the seed is controlled). Because Iris is easy to cluster, we can use the Hungarian algorithm, which solves the **assignment problem** by finding the minimum cost to assign n workers (cluster labels) to n tasks (species).

Given a cost matrix C where C_{ij} represents the cost of assigning worker i to task j, we want to find a permutation π that minimises:

$$\sum_{i=1}^{n} C_{i,\pi(i)}$$

We then create a benefit matrix from this confusion matrix and convert it to a cost matrix.

$$\text{Cost}_{ij} = \max(\text{Confusion Matrix}) - \text{Confusion Matrix}_{ij}$$

We can apply this with `sklearn` and `scipy` using a crosstab with a linear sum assignment.

```
print(df.groupby(['species', 'cluster']).size().unstack(fill_value=0))

from sklearn.metrics import accuracy_score
from scipy.optimize import linear_sum_assignment
```

```
confusion_matrix = pd.crosstab(df['species'], df['cluster'])
print("\nConfusion Matrix:")
print(confusion_matrix)

cost_matrix = confusion_matrix.max().max() - confusion_matrix
row_indices, col_indices = linear_sum_assignment(cost_matrix)

total_correct = confusion_matrix.values[row_indices, col_indices].sum()
accuracy = total_correct / len(df)
print(f"\nClustering Accuracy: {accuracy:.3f}")

# output:
cluster      0   1   2
species
setosa       0  50   0
versicolor  48   0   2
virginica   14   0  36

Confusion Matrix:
cluster      0   1   2
species
setosa       0  50   0
versicolor  48   0   2
virginica   14   0  36

Clustering Accuracy: 0.89
```

According to this, a clustering accuracy of 89.3% is great! Or is it? Well, you can definitely do better, as we haven't covered preprocessing, denoising, dimensionality reduction and/or different clustering methods yet. But the lesson here is what the Iris dataset shows us about how well clustering, cluster evaluation metrics, and simple methods work with neat spherical clouds in 4-dimensional feature space. In higher-dimensional (i.e. featurized time series) space, we often require additional preprocessing and careful metric interpretation.

Preprocessing for time series clustering

For time series clustering how we preprocess data often matters as much as our clustering approach. This is especially true for time series data that include trends, seasonality, differing scales, noise, varying lengths, and so on. These can mislead clustering if left unhandled. Conversely, we need to be careful in transforming our data before we run clustering, as our choices can embed assumptions in our data and resulting clusters.

For example, **scaling** each series to a common range or distribution is usually a first step. If one series' values range in the thousands and another in the tens, k-means (and many distance metrics) will treat them as far apart even if their general **patterns** over time are similar. By applying something like z-score standardization (so each series has mean 0 and variance 1) or min-max normalization (scaling values between 0 and 1), we ensure that **magnitude** differences don't overpower shape similarities. This way, a series selling 1000 units a day could potentially cluster with one selling 10 units a day if their trends and seasonal patterns align. Equally, you should question if this is a good thing or not!

We also consider **detrending and deseasonalizing** time series, though some representation systems can handle these internally. Clustering might group series just because they all share an upward trend, even if everything else differs. If a long-term trend or strong seasonal pattern isn't what we want defining our clusters, we can remove it (for instance, subtracting a moving average trend or using seasonal decomposition to isolate residuals). The idea is to highlight the features we *do* care about, perhaps short-term fluctuations or overall shape, by removing structural components that could dominate the distance calculations.

Another common preprocessing step is **noise reduction.** Time series data can be noisy (random spikes or drops). Smoothing techniques like moving averages or more sophisticated filters (e.g. a Savitzky-Golay filter) can help suppress these random fluctuations. Reducing noise may help us, but it can also erase meaningful patterns. A lightly smoothed series might cluster more reliably by focusing on the major pattern instead of tiny jitters.

Length differences and their alignment are another key factor for TSCL. If series are of unequal lengths, distance measures might perform poorly, or clustering may focus on length rather than pattern. We might truncate longer series or pad shorter ones (with zeros or by repeating the last value) to create a common length. In this chapter we focus on feature-based approaches where length can be a feature of our data. Typically we want to ensure we're not clustering purely based on one series being longer or starting later than another.

In short, good preprocessing can improve clustering. But there's no perfect approach; the *right* preprocessing depends on our domain knowledge and what we define as similarity. Experimenting with different preprocessing combinations is always worthwhile for unsupervised learning.

Defining time series clustering

At this point it's worth pointing out that **Time Series Clustering** (**TSCL**) can be branched into whole-series, subsequence, and point-based clustering (see *Figure 15.5*). Whole-series clustering aims to group entire series based on information about each series, or group of series (multivariate). This is our focus here, with our intent to improve modeling performance rather than finding specific patterns or points.

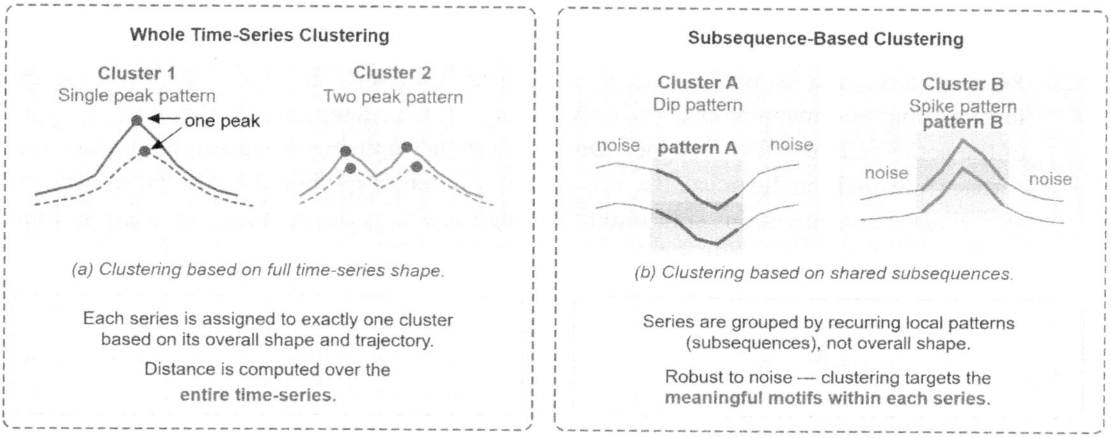

Figure 15.5: Whole time series clustering: subsequence-based vs. entire-series-based

While these categories differ in what they attempt to group, they generally share a common methodological structure. In practice, most TSCL approaches can be viewed as a sequence of transformations: first converting raw series into a more informative representation, then measuring similarity or distance between those representations, and finally applying a clustering algorithm to form groups. This workflow forms the basis of the standard TSCL pipeline discussed next.

The TSCL pipeline

TSCL follows a general pipeline with three stages: representation building, dissimilarity measurement, and clustering. To visualize these stage, we have the following diagram:

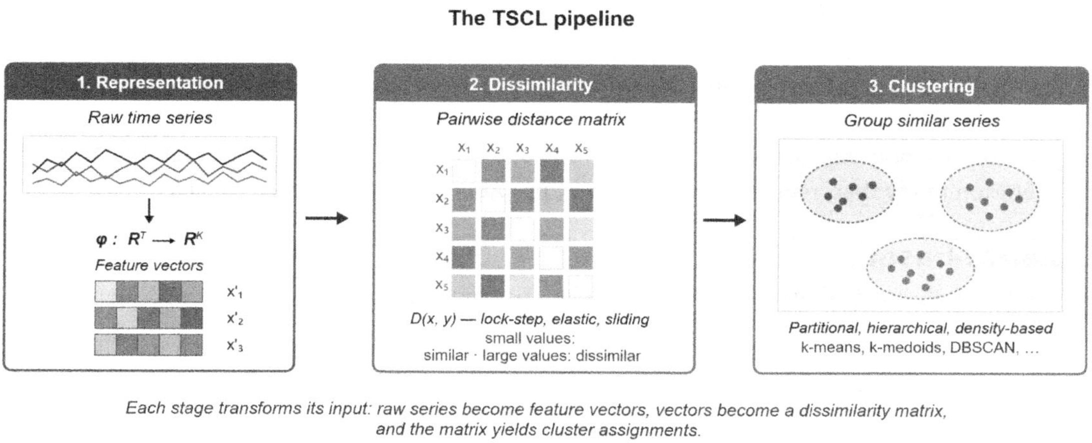

Figure 15.6: Clustering Pipeline: Representation to Dissimilarity to Clustering

The three stages in depth are:

1. **Representation** is a featurization stage. Given a raw time series $\mathbf{x}_i = \{x_{i1}, x_{i2}, \dots, x_{iT}\}$, we transform it using some mapping $\phi\colon \mathbb{R}^T \to \mathbb{R}^K$ (where $K < T$) to get a representation $\mathbf{x}'_i = \{x'_{i1}, \dots, x'_{iK}\}$ of the series in a new feature space. Our aim here is capturing the meaningful parts of our series while ditching the noise. If $\mathbf{x}_i$ is similar to $\mathbf{y}_i$, then $\phi(\mathbf{x}_i)$ should be similar to $\phi(\mathbf{y}_i)$, and vice versa. Representations split into two categories: *numeric* methods that use real-valued arrays or matrices with reduced dimensions (DFT, DWT, neural embeddings), and *symbolic* approaches that discretize data into symbols, enabling text-based techniques like hashing and sequence matching (SFA, SAX).
2. **Dissimilarity** is where we measure distance, comparing how different our transformed series are. It's important that whatever measure we use produces small values for similar representations and large values for dissimilar ones. A few broad categories of dissimilarity measure, which we met in *Chapter 14*:
   - **Lock-step:** One-to-one point correspondence measures, like p-norms and their variants, along with Jaccard (set similarity rather than vector norm).
   - **Elastic:** This is where we compare time points in one-to-many form, typically used when one-to-one mapping isn't informative, e.g. DTW.
   - **Sliding** These slide one entire time series along another to find the position where they match best overall, e.g. **Shape-Based Distance** (**SBD**).
3. **Clustering algorithm:** This is our final stage, where we take our representation distances and cluster the data with one of many clustering algorithms.

Clustering approaches

In reality TSCL approaches are pipelines, built around distance, distribution, subsequence, or representation methods. There is some overlap and confusion in terminology too, as distance-based approaches refer to direct application of dissimilarity measures, while representation-based TSCL refers to representation learning, a deep-learning approach to building representations. It is not an exaggeration to say that, regardless of approach, preprocessing is the key to success.

Distance-based

Distance-centric TSCL treats clustering as a geometric partition under a user-defined metric $D(\mathbf{x}, \mathbf{y})$.

- **Lock-step metrics** (e.g. Euclidean distance) assume exact phase alignment: $D_{\mathrm{ED}}(\mathbf{x}, \mathbf{y}) = \sqrt{\sum_{t=1}^{T}(x_t - y_t)^2}$.
- **Elastic metrics** such as DTW allow flexible alignment to handle phase and rate distortions: $\mathrm{DTW}(\mathbf{x}, \mathbf{y}) = \min_\pi \sum_{(i,j)\in\pi} d\left(x_i, y_j\right)$ These extend to **Longest Common Subsequence** (**LCSS**) and **Move-Split-Merge** (**MSM**), at $O(T^2)$ cost per comparison.
- **Sliding metrics** such as SBD use FFT to achieve $O(T\log T)$ alignment under circular shifts.

With a distance matrix in hand, we apply either **partitional** algorithms (k-means, k-medoids, k-Shape with spectral centroid update) or **hierarchical** linkages (single, complete, Ward) to build a dendrogram. There are advantages to this, such as transparent geometry and metric modularity. However, the pairwise cost grows $O(N^2)$ with dataset size, and this approach is highly sensitive to noise in our sequences. We also have the additional problem of length differences between series. As with classification, distance-based methods show mediocre performance at scale.

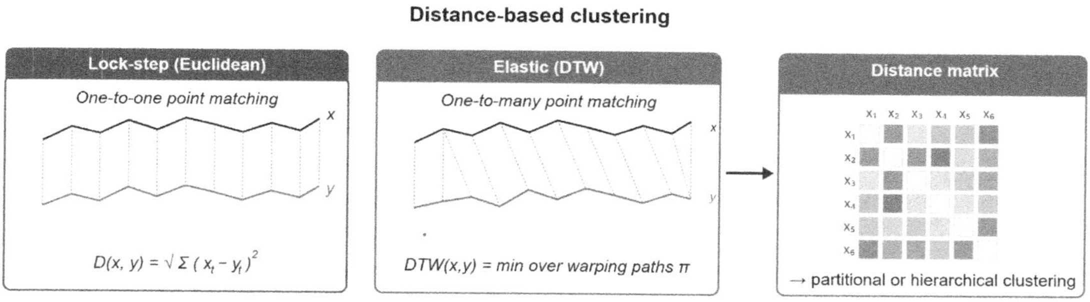

Figure 15.7: Distance-based clustering: lock-step metrics like Euclidean compare points one-to-one, while elastic metrics like DTW allow flexible alignment

Distribution-based

Distribution approaches encode series through **probabilistic or statistical distributions** rather than raw geometry. There are four subcategories here:

- **Model-based** methods fit statistical or probabilistic models like **Gaussian Mixture Models (GMM)**, **Hidden Markov Models (HMM)**, ARIMA or Gaussian-process priors, and then cluster in parameter space via likelihood or selection criteria such as the **Bayesian Information Criterion (BIC)**.
- **Density-based** methods such as DBSCAN, OPTICS, and DPC identify connected dense regions in $\mathbb{R}^d$ after kernel density estimation: $\hat{f}(\mathbf{x}) = \frac{1}{Nh^d}\sum_{i=1}^{N} K\left(\frac{\mathbf{x}-\mathbf{x}_i}{h}\right)$.
- **Feature-based** methods extract global and subsequence statistics through featurization libraries such as `tsfeatures`, `tsfresh`, or `catch22`. These features form a fixed-length vector, to which we apply standard vector clustering.
- **Encoding-based** methods use symbolic transforms (SAX, iSAX, SFA) or learned embeddings, clustering series in the resulting latent manifold.

To some extent, these approaches handle noise and variable length naturally, yet they demand careful model or threshold selection. They can also be fast and provide us with an additional dimension for exploratory analysis.

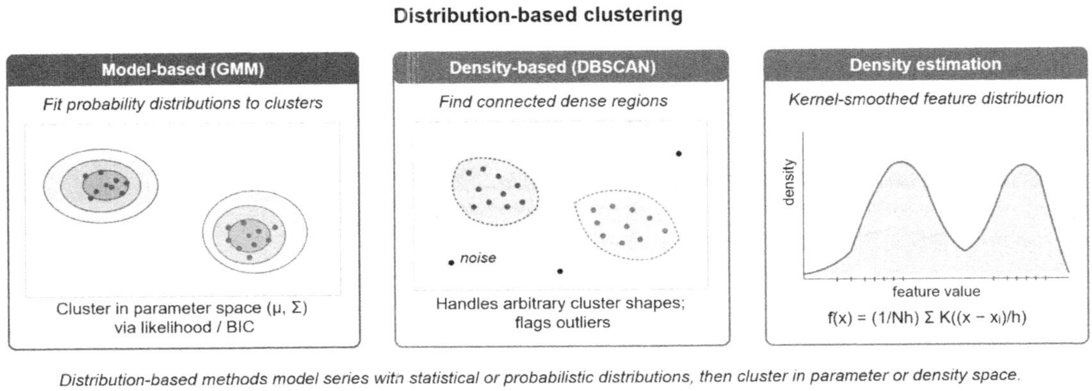

Figure 15.8: Distribution-based clustering: model-based approaches fit probability distributions, density-based approaches find connected dense regions, and both rest on kernel density estimation of the underlying feature distribution

Subsequence-based

The idea TSCL pipelines:subsequence-based" of subsequence-based approaches is to represent each series with localized motifs rather than entire traces.

- **Sliding-window techniques** tile sequences into overlapping windows, cluster those windows (often via DTW or SBD), then aggregate the window clusters to series-level assignments.
- **Shapelet-based techniques** search or learn discriminative subsequences; every series is re-encoded by its distances to k shapelets, yielding a k-dimensional feature vector for downstream clustering.

As with classification, these inherit phase invariance and interpretability, which can make them excellent for grouping series. However, their limiting factor is the combinatorial search for high-quality subsequences. At scale they are expensive and slow; depending on your computer resources you may decide which to use. As a rule of thumb, I find fewer than 1,000 series of length 730 to be manageable.

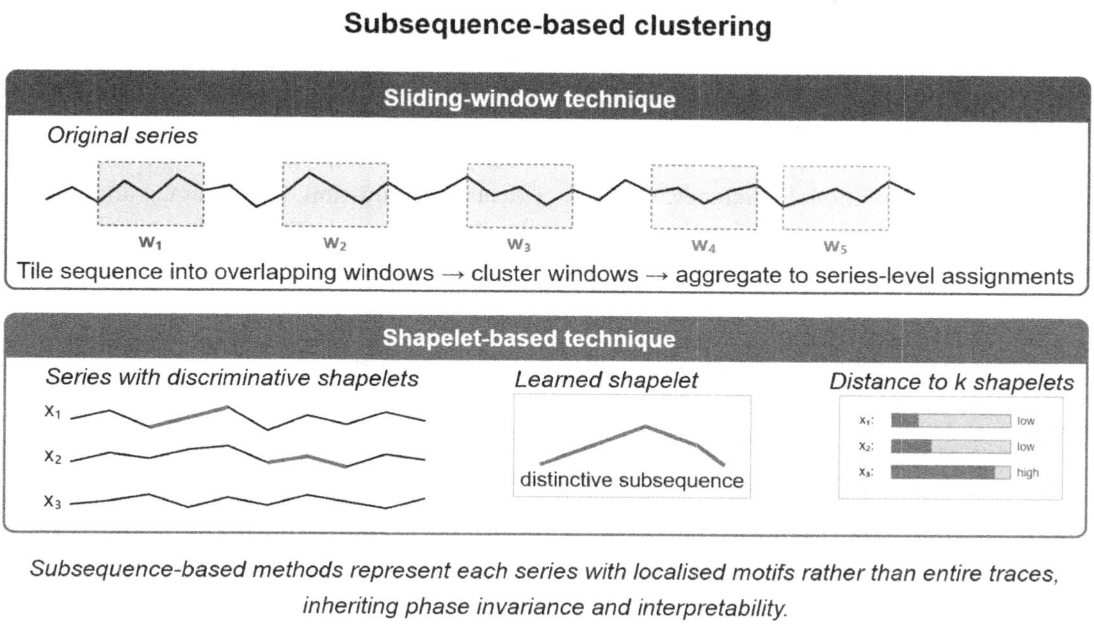

Figure 15.9: Subsequence-based clustering: sliding-window techniques tile a long series into overlapping windows for clustering, while shapelet-based techniques re-encode each series by its distance to a set of learned discriminative subsequence

Representation learning

Deep neural encoders learn a mapping $f_\theta: \mathbb{R}^T \rightarrow \mathbb{R}^d$ that embeds sequences in a manifold $Z = f_\theta(X)$, which is subsequently clustered. In simple terms, these neural networks compress our time series data into lower-dimensional representations (not unlike built features or shapelets), making clustering easier. Two optimization families dominate the literature:

- Comparative-based losses learn by comparing examples. They use techniques like InfoNCE to draw or push series together in a learned space depending on similarity.
- Generative-based losses combine two objectives: reconstructing the original data (like an autoencoder trying to rebuild the input) and a clustering penalty that encourages similar series to group together: $\mathcal{L} = \mathcal{L}_{\text{rec}} + \lambda\,\mathcal{L}_{\text{cluster}}$.

The reconstruction loss ($\mathcal{L}_{\text{rec}}$) ensures the model captures important information, while the clustering regularize ($\mathcal{L}_{\text{cluster}}$) encourages the learned representations to form distinct groups. **Evidence Lower Bound (ELBO)** is a specific type of loss used in variational autoencoders that balances reconstruction accuracy with how well the learned representations follow a desired probability distribution.

Architectures spanning 1D CNNs, LSTM/GRU, Transformer and graph encoders have been adapted for TSCL. Among these, CNN-based architectures generally achieve the best results for univariate time series clustering, offering $O(k \cdot n^2)$ complexity per layer where k is kernel size, compared to RNNs' $O(n \cdot h^2)$ per layer where h is hidden size, or Transformers' $O(n^2 \cdot d)$ per layer scaling (Paparrizos, Yang, and Li 2024). CNNs do well in capturing local temporal patterns through weight sharing while maintaining computational efficiency, which is critical for production deployments, and they tend to perform well on high-dimensional and/or irregular data. That said, GPU cost and hyperparameter sensitivity are not insignificant factors in their utility.

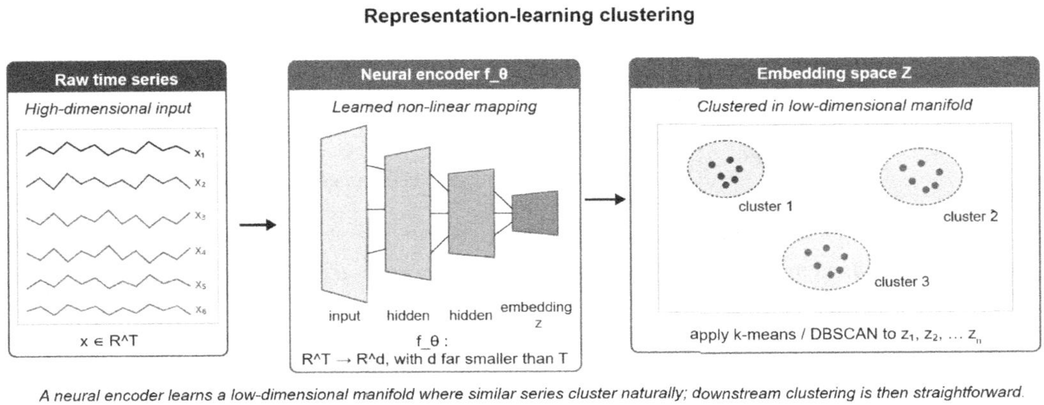

Figure 15.10: Representation-learning clustering: a neural encoder learns a non-linear mapping from raw series to a low-dimensional manifold where similar series cluster naturally and standard algorithms can take over.

Deep-learning approaches to TSCL

There are two general strategies for deep-learning TSCL: two-stage pipelines that separate encoding from clustering and end-to-end frameworks that learn both simultaneously.

Two-stage pipeline

Two-stage pipelines often perform well, achieving good results with a straightforward process. It is the more popular strategy in literature at time of writing, with methods like TS2Vec and TNC being more commonly implemented (Paparrizos, Yang, and Li 2024). General approach is as follows:

Use an unsupervised encoder such as an autoencoder (which learns to compress and reconstruct data), a VAE (adds probabilistic constraints), or a contrastive CNN to transform our series into compact representations $Z = \{z_i\}$. These encoders find 'meaningful' patterns by minimizing reconstruction errors or contrastive losses.

Apply standard clustering algorithms like k-means or DBSCAN to these learned representations Z. Even with a simple k-means clustering, we can achieve excellent results with this approach so long as the encoder has learned good representations that capture multi-scale temporal patterns and structural relationships in the data.

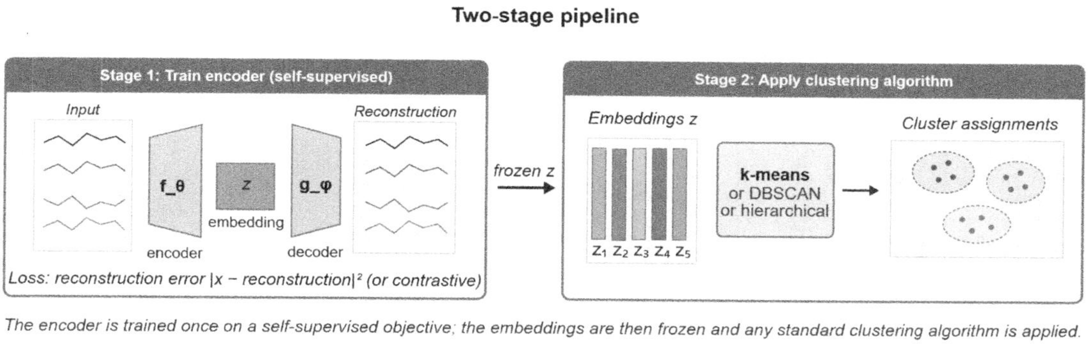

Figure 15.11: Two-stage pipeline: the encoder is trained once on a self-supervised reconstruction or contrastive objective, then its embeddings are frozen and any standard clustering algorithm is applied

End-to-end neural clustering

Frameworks such as DEC, Deep k-Means (CKM), VaDE and ClusterGAN integrate representation and clustering objectives into a single jointly trained system. While these joint approaches can produce tighter cluster boundaries, they introduce complexity through additional hyperparameters and sensitivity to initialization. As an example, the DEC loss uses a Kullback-Leibler divergence term that compares current soft cluster assignments Q (confidence of each point's cluster assignment) with a sharpened target distribution P (our desired cluster assignments):

$$\mathcal{L}_{\text{DEC}} = \text{KL}(P \parallel Q) = \sum_{i,k} p_{ik} \log \frac{p_{ik}}{q_{ik}}$$

Training alternates weight updates, centroid (or mixture) refinement and pseudo-label recalculation, producing embeddings and partitions simultaneously. These one-stage methods often tighten cluster manifolds but introduce additional hyperparameters (KL weight λ, annealing schedules) and dependence on centroid initialisation, necessitating systematic experimentation to ensure convergence. Recent hybrid approaches combining reconstruction and contrastive objectives (like DTCC and TimeCLR) attempt to balance these trade-offs (Paparrizos, Yang, and Li 2024).

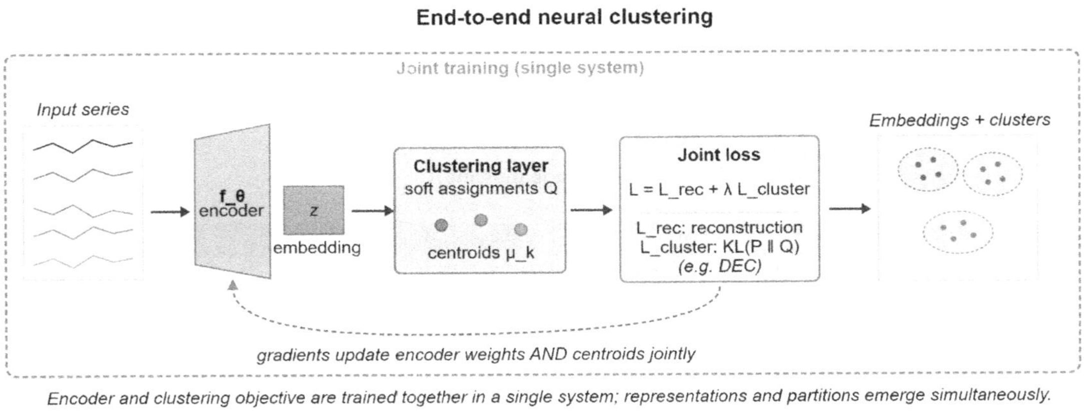

Figure 15.12: End-to-end neural clustering: encoder, embedding, and clustering layer are trained together against a joint loss combining reconstruction and a clustering term such as DEC's KL divergence; gradients update encoder weights and centroids simultaneously

What can we take from this? Distance-based methods suffer from quadratic computational complexity for hierarchical clustering and $O(N^2)$ pairwise distance calculations, while elastic measures like DTW compound this with $O(T^2)$ cost per comparison. Subsequence-based approaches like shapelets provide excellent interpretability and performance at the higher end of the options available, but face combinatorial search complexity that becomes prohibitive at scale. We know of no documented use of shapelets in production featurisation systems at the time of writing. Similarly, deep-learning representation methods like R-Clustering have shown strong performance (performing best on 33 of 112 datasets, with Agglomerative Euclidean clustering as the next best), compared to shapelets for example (Marco-Blanco and Cuevas 2023). But they also require large datasets and substantial GPU resources, along with extensive hyperparameter tuning. There are cases of these being used, but only by companies with substantial compute. This doesn't mean we should discount them entirely, but in many settings they are not practical at scale, nor explainable.

Applying featurization

We are aiming here to create a GFM with N-HITS, which performed best in our transfer learning testing on a smaller version of the M5 dataset (see *Chapter 11*). Our featurization and clustering will be applied to one location (California state, `location 3: CA_3`) of the M5 dataset, which contains 3,038 time series of item sales with pricing data. We focus on a single location to keep computation manageable and the results interpretable; applying more widely just requires location in the static covariates.

The code in this chapter was developed with aeon 1.0.0, neuralforecast 1.7.4, and tsfeatures 0.4.5.

Our primary pipeline extract the features, and clusters them to create labels for our data set the GFM can use:

1. Data preparation
2. Extracting features
3. Basic exploration
4. Clustering
5. Training GFM
6. Evaluation

Data preparation

We start by loading and formatting our data, removing short series, and shifting demand/sales values to avoid infinite values in calculations. You don't have to remove short series, as the features we build can deal with less than 365 periods of varying lengths. Metrics can fail when you have very few values, depending on the calculation. Shifting values by adding 1 undermines zero counts, and metrics relying on them, so if zero counts matter to your use case, it would be best to build separate metrics for them if calculations result in nan and/or inf.

```
df = pd.read_csv(CFG.data_folder / 'm5-forecasting-accuracy' / 'sales_train_
validation.csv')
df['timestamp'] = pd.to_datetime(df['ds'])
df = df.rename(columns={
    'id': 'unique_id',
    'sold': 'y',
    'sell_price': 'price'
})

df['ds'] = pd.to_datetime(df['ds'])
df = df[['unique_id', 'ds', 'y', 'price']].copy()

# Drop any series with fewer than 365 days
df = df.groupby(
    'unique_id'
).filter(lambda x: len(x) >= 365).reset_index(drop=True)
df['y'] = df['y'] + 1  # Shift to avoid zero values
```

After this, we have a table with a `unique_id`, `datestamp` (`ds`), `sales` (`y`), and exogenous (`price`). If you are using your own data, exogenous variable(s) are optional.

```
df.head()
```

The output:

unique_id	ds	y	price
FOODS_1_001_CA_3	2011-01-29	2.0	2.0
FOODS_1_001_CA_3	2011-01-30	3.0	2.0
FOODS_1_001_CA_3	2011-01-31	2.0	2.0
FOODS_1_001_CA_3	2011-02-01	2.0	2.0
FOODS_1_001_CA_3	2011-02-02	2.0	2.0

Table 15.2: First five rows of the M5 FOODS_1 CA_3 dataset after selecting the required columns.

Because we are building and testing models with this data, we need to split it to avoid information leakage. We use a train-test split here, but train-validation-test is a much better choice (see *Chapter 3* for a discussion of evaluation strategies).

```
h = 90  # forecast horizon

cutoff_date = df['ds'].max() - pd.Timedelta(days=h)

train_df = df[df['ds'] <= cutoff_date].reset_index(drop=True)
val_df = df[df['ds'] > cutoff_date].reset_index(drop=True)
```

You'll notice we didn't split by `unique_id`. That's because all the series span at least 365 days and end on the same date, so a single cutoff date applies uniformly. Some series are shown below to help confirm the splitting. We can see this data isn't great in parts, and would certainly be improved at weekly aggregations.

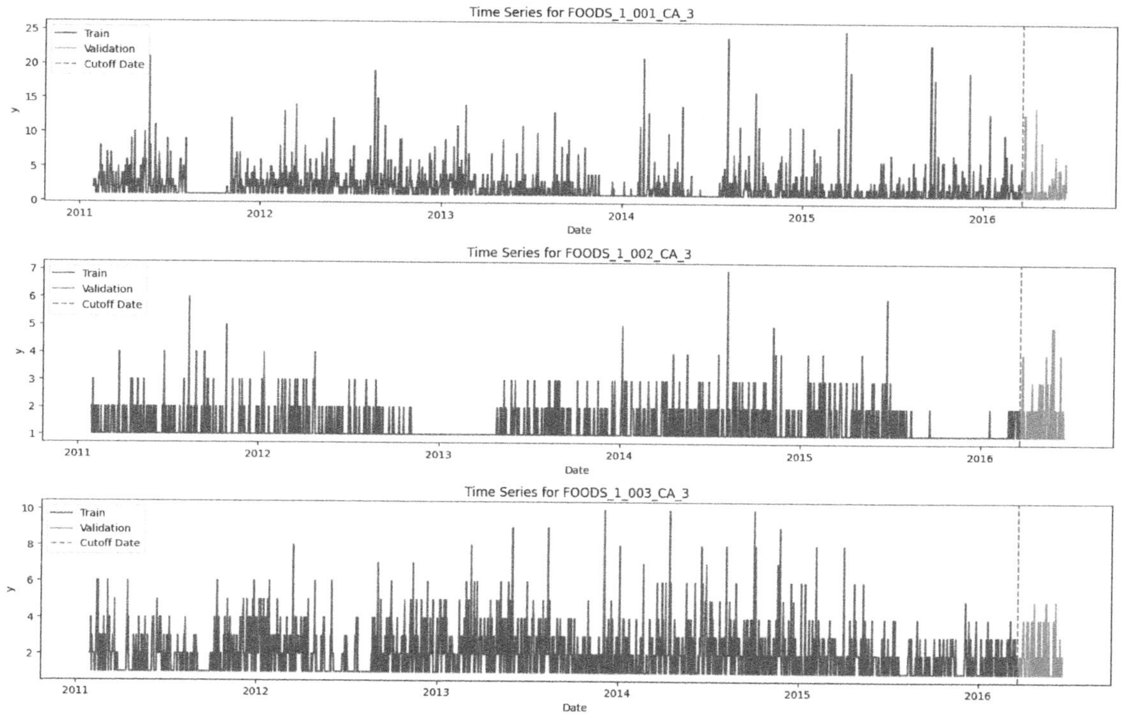

Figure 15.13: Sample M5 CA_3 series with train-test split

Feature extraction

We're now in a position to create our time series features. This can be the easy bit, depending on what features we want. You select a library such as the reduced `hctsa` library `catch22`, or one of the many others: `tsfresh`, `tsfel`, `tslearn`, and so on. Choosing which to use relies on your investigations and experimentation, and on how your data is preprocessed. For example, the flat spots in the series images above and the noise adversely affect the building of some features, which may lead you to aggregate and/or impute data. Regardless, we are going to simply apply Nixtla's tsfeatures, which is a Python implementation of an R package with the same name by Montero-Manso, Talagala, Hyndman and Athanasopoulos.

At this point we are not directly scaling or applying other preprocessing, so feel free to follow along with your own data. If you have sparse or missing data and have not done your wrangling, you may encounter issues like infinite values; you can either deal with these or leave them in. Sometimes it's best to do nothing and remove the problematic features afterwards. `tsfeatures` builds 22 types of features very quickly (Nixtla version), and does a reasonable job of clustering data as seen in earlier M-competitions.

As with most featurization libraries there are different ways to apply `tsfeatures`. In the following code, we import each feature type separately so that we can control the features we build with a list, rather than letting `tsfeatures` build all features. We pass this list to the tsfeatures method, along with the frequencies of the data we want to consider (weekly and yearly), and apply it to `train_df`.

```
from tsfeatures import (
    acf_features, heterogeneity, series_length,
    arch_stat, holt_parameters, sparsity,
    count_entropy, hurst, stability, crossing_points,
    hw_parameters, stl_features,
    entropy, intervals, unitroot_kpss, flat_spots, lumpiness, unitroot_pp,
    frequency, nonlinearity, guerrero, pacf_features
)

default_feats = [
    acf_features, heterogeneity, series_length,
    arch_stat, holt_parameters, sparsity,
    count_entropy, hurst, stability, crossing_points,
    hw_parameters, stl_features,
    entropy, intervals, unitroot_kpss, flat_spots, lumpiness, unitroot_pp,
    frequency, nonlinearity, guerrero, pacf_features
]

feature_df = tsfeatures(
    train_df,
    dict_freqs={'D': 7, 'W': 52},
    features=default_feats
)
```

The output is a pandas DataFrame with a row for each series (`unique_id`; n = 3,038) and the featurised summaries (p = 49 columns).

```
feature_df.head(3)
```

The output of the code is:

unique_id	x_acf1	seas_pacf	nonlinearity	lumpiness	unitroot_pp	guerrero
FOODS_1_001_CA_3	0.074	0.086	0.084	0.040	-1934.51	0.087
FOODS_1_002_CA_3	0.067	0.075	-0.084	0.050	-2043.27	0.038
FOODS_1_003_CA_3	0.073	0.033	0.018	0.054	-1916.87	0.138

Table 15.3: Representative sample of extracted tsfeatures for the first three M5 CA_3 series. Six columns are shown from the full set of 49; each is drawn from a different feature family (autocorrelation, seasonality, nonlinearity, heteroscedasticity, stationarity, and distributional transformation)

We have some features, like `arch_acf`, which have NaN values. Looking at the extent of this problem with `print(feature_df.isna().sum())` we found a lot of time series with NaN values in `hurst`. The option we applied earlier (adding 1 to y) resolved many of these rows. We could have also removed the rows (which would have defeated the point of using featurized clustering) or imputed the values (risky without extensive EDA). Afterwards some columns still had NaN values for a subset of series; this happens when data is not suited to some of the simple statistical model summaries like GARCH. The best thing to do here is remove those columns with `feature_df = feature_df.dropna(axis=1)`. In addition, some columns have no variance (i.e. a single value) so we should remove those with `feature_df = feature_df.loc[:, feature_df.nunique(dropna=False) > 1]`. There is no point in keeping single-value columns as they provide no information for clustering. This takes us from 49 to 37 columns; many of the dropped columns are autocorrelation features that were entirely NaN.

We can get some sense of what the features are doing by creating a correlation heatmap (*Figure 15.14*). This is a complex patchwork of features that have several relationships in the data but no clear dominant patterns (maybe with the exception of the differenced PACF features). This is a good thing: we don't want features to be too well correlated with each other, as we are essentially using them to differentiate time series based on dissimilarity.

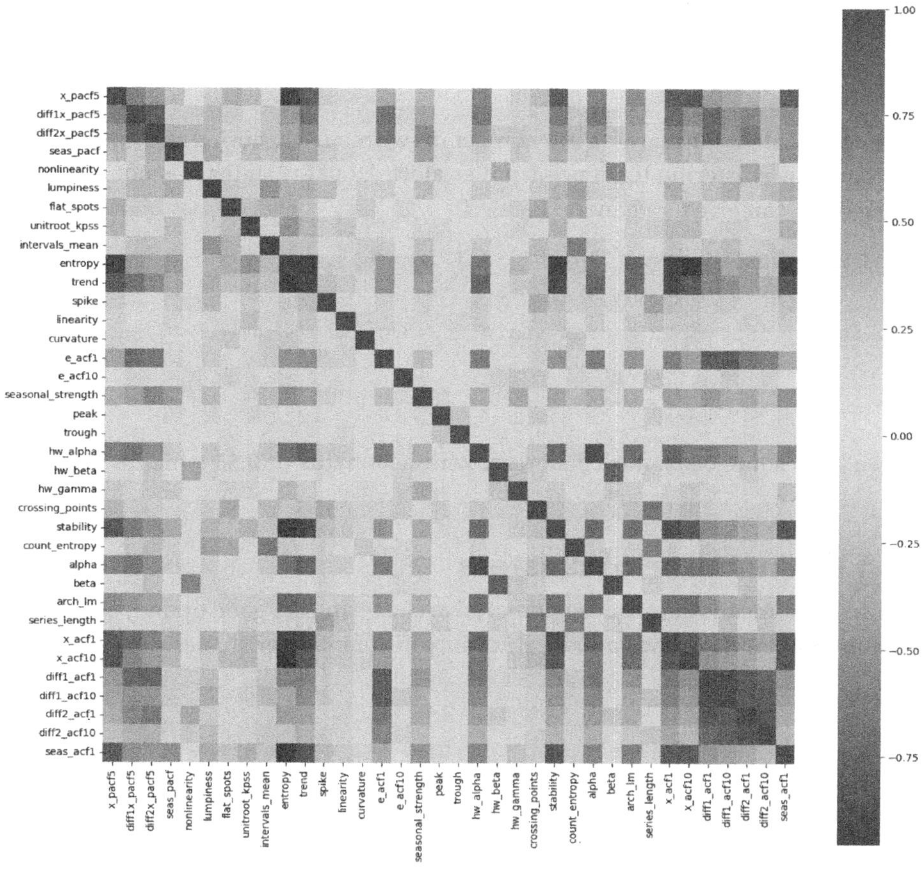

Figure 15.14: Feature correlation heatmap for the 37 retained tsfeatures

We could reduce the number of features with ridge regression to select the best subset, but I tend to find this actually reduces the quality of clustering as it's based on linear relationships which many features do not comply with. A better option can be to create a classification task (after you have created clusters) using non-linear classifiers such as decision trees or SVMs. This isn't great though, as it is a kind of confirmation bias: selecting the best features for preselected clusters. This is more of a problem when you have more features than rows (curse of dimensionality). Luckily we have only 37, so we don't need to bother with feature selection or dimensionality reduction, which can be achieved best with the autoencoding we will discuss later.

Scaling

At this point we simply need to scale our features so that they are on the same scale. Why? Well, as Tivadar Danka mentions in *Mathematics of Machine Learning*, if you have a feature at a higher magnitude than others it is a little like being in a room with loud music while someone is whispering to you from another room. The whisper may be more informative than the loud music, which you'd only know if they were on the same scale.

```
from sklearn.preprocessing import minmax_scale

feature_df_scaled = feature_df.copy()
feature_df_scaled.iloc[:, 2:] = minmax_scale(feature_df_scaled.iloc[:, 2:])
```

Now that we have scaled our features, we can get on with our clustering. Our first task is to check whether the data has clustering tendency (spoiler alert: Hopkins statistic is not always useful with large datasets, but it can give us a general idea).

```
X = feature_df_scaled.drop(['unique_id'], axis=1).values

h2 = hopkins(X, frac=0.1, random_state=562456)
print(f"Hopkins: {h2:.4f}")
Hopkins: 0.854
```

Seems like we have some cause for clustering, with a Hopkins of 0.85 which is well above the 0.5 threshold.

Selecting the number of clusters

We are going to use k-means here, but agglomerative clustering is a good option, as are density-based approaches like DBSCAN if you want more flexibility in your cluster shapes.

```
kmeans_elbow(
    X=X,
    palette=custom_palette,
    k_distortion=(1, 20),
    k_ch=(2, 20),
)
```

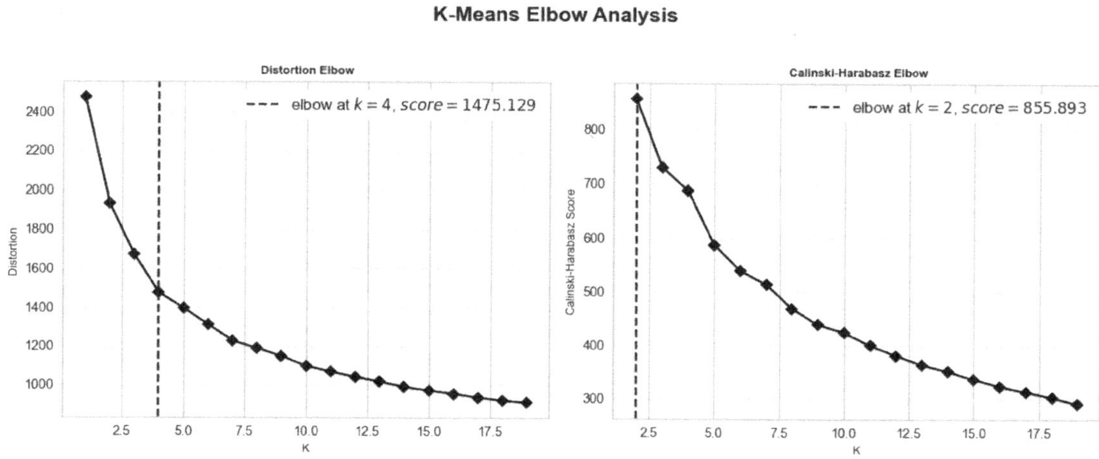

Figure 15.15: Elbow plots for the standardized M5 CA_3 features showing distortion (left) and Calinski-Harabasz index (right)

Calinski-Harabasz (CH) compares the dispersion between clusters to dispersion within clusters. Conversely to distortion, a higher rather than lower CH value means clusters are dense and well-separated.

There are about three patterns to look for with CH and distortion. If the CH values show a clear peak (or a valley for distortion) or an abrupt elbow, we should choose the number of clusters at that point, as it suggests an optimal balance between compactness and separation. If the CH index increases smoothly or remains flat, there is no strong evidence for a particular number of clusters, so domain knowledge and EDA on the clusters should guide the decision. If the CH index fluctuates, we look for local maxima or points where the increase slows down significantly.

Regardless, the CH index here on scaled data suggests k=2, which is ridiculously low. CH is not a good metric for this data. With the unscaled data the CH index suggests 4 or 6, and seems to work better, but the clusters are terrible for modelling (you can test this yourself). For distortion, both graphs show smooth declines from k=2 onwards. This is all uncertain: is it 2, 4 or 8? 20 could be as valid as any other number of clusters. Let's go with 8 and see what happens.

```
kmeans = KMeans(n_clusters=8, init='k-means++', n_init='auto', random_state=42)
y_kmeans = kmeans.fit_predict(X)
```

We'll visualize the results with silhouette scores and t-SNE to get a general idea.

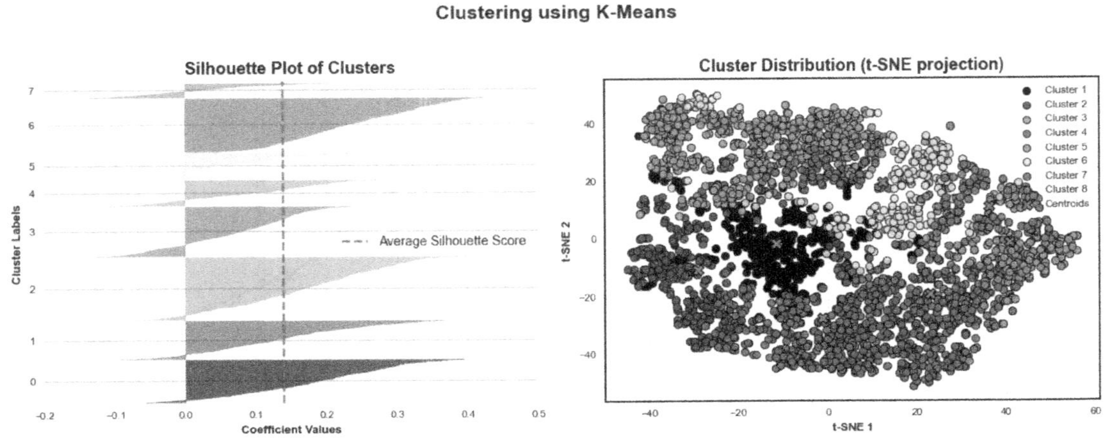

Figure 15.16: Silhouette scores and t-SNE visualisation for k=8 clusters on standardized features

What should be immediately obvious is that the silhouette scores are generally positive for our scaled data, although cluster 7 has about half negative scores. This is a sign that the cluster may not be a good addition, but it could be due to the amount of data we have, or it could be a sign that we should use a different clustering algorithm.

The t-SNE also shows that the clusters are largely distinct, but there is some overlapping of labels. This suggests our approach is not working perfectly, and we should consider:

- A different algorithm, e.g. hierarchical clustering, DBSCAN, Gaussian mixture models, self-organizing maps
- A different number of clusters, e.g. 2, 4, 16, 32
- More data, i.e. more features
- Wrangling the data differently, for example there is missing-at-random (and missing-not-at-random) data that could be imputed, which likely affects performance
- A different dimensionality reduction technique

We can run a more controlled t-SNE and tune the parameters to see if this is more to do with our visualization than the clustering itself. We should remember that t-SNE is a dimensionality reduction technique that here takes 37 features down to 2. Reducing feature space with t-SNE is by no means guaranteed to properly capture relative relationships between data points in 2D space.

```
tsne = TSNE(
    n_components=2,
    perplexity=60,
    n_iter=1200,
    random_state=86743,
    method='barnes_hut',
    init="pca",
```

```
    n_jobs=-1
)

X_tsne = tsne.fit_transform(X)
tsne_df = pd.DataFrame(X_tsne, columns=['tsne1', 'tsne2'])
tsne_df['cluster'] = y_kmeans

plt.figure(figsize=(10, 8))
sns.scatterplot(
    data=tsne_df,
    x='tsne1', y='tsne2',
    hue='cluster', palette=custom_palette,
    s=80, edgecolor='black', linewidth=0.5, alpha=0.8
)
```

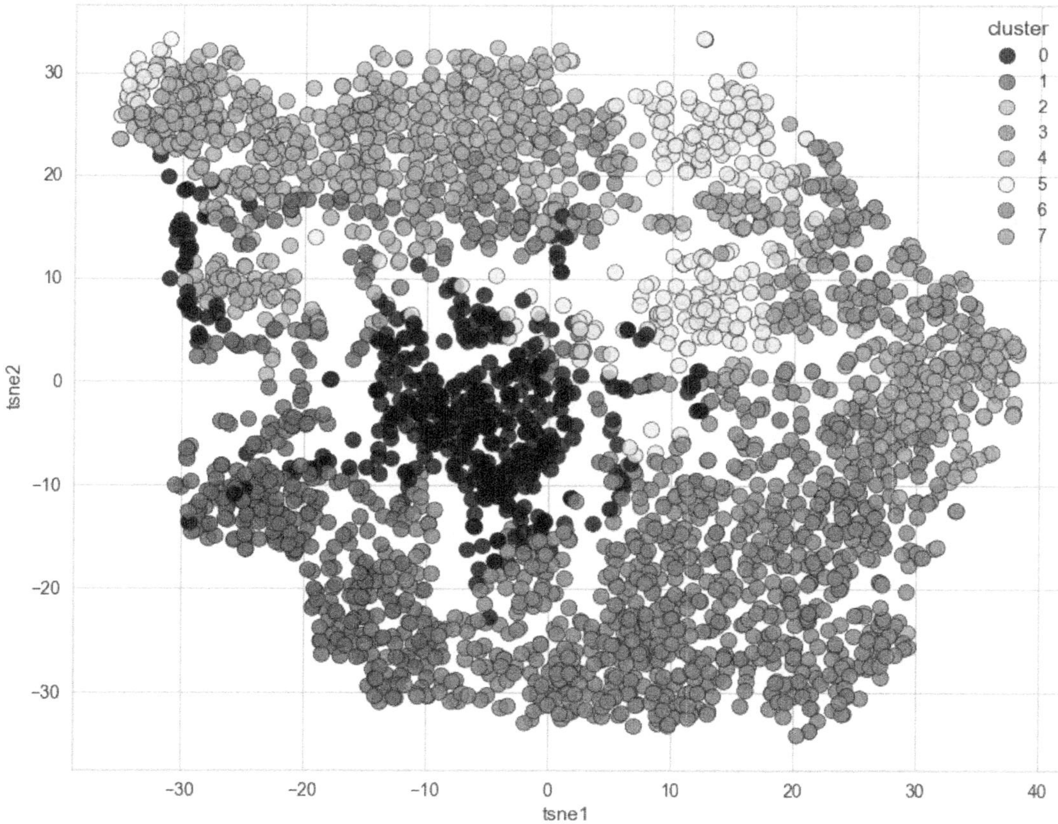

Figure 15.17: Tuned t-SNE visualization of k=8 clusters on standardized features

Overall, despite some pollution in the cluster boundaries, the clusters look reasonably well defined.

Building GFMs with cluster labels

We can get an idea for how well this is working by building a GFM with default parameters that utilizes the cluster labels, and comparing the overall performance with a model that uses no clusters. This gives us a direct measure of how well labels help or hinder the model. First we have to join the cluster labels to our feature dataset.

```
feature_df = pd.concat(
    [
        feature_df.reset_index(drop=True),
        tsne_df.reset_index(drop=True)
    ], axis=1)
```

Then we create a static exogenous feature DataFrame, which is specific to Nixtla libraries. This DataFrame consists of the `unique_id` and the various static identifiers. Here we use clusters, but you could use locations if you had multiple, or business groupings such as luxury brands or services. We one-hot encode the cluster label, as this works better than a single integer vector.

```
static_df = feature_df[['unique_id', 'cluster']].copy()

cluster_dummies = pd.get_dummies(static_df['cluster'], prefix='cluster')
static_df = pd.concat([static_df[['unique_id']], cluster_dummies], axis=1)
```

Now all we need to do is instantiate our models (default hyperparameters) and train them, with and without our `static_df`.

```
max_steps = 3500
early_stop_patience = 50
h = 90

models = [
    NHITS(
        input_size=90,
        h=h,
        max_steps=max_steps,
        scaler_type='standard',
        loss=MAE(),
        hist_exog_list=None,
        futr_exog_list=['price'],
        stat_exog_list=['cluster_0', 'cluster_1', 'cluster_2', 'cluster_3',
                        'cluster_4', 'cluster_5', 'cluster_6', 'cluster_7'],
        early_stop_patience_steps=early_stop_patience,
        learning_rate=0.0001,
        alias='NHITS_with_clusters'
```

```
    ),
    NHITS(
        input_size=90,
        h=h,
        max_steps=max_steps,
        scaler_type='standard',
        loss=MAE(),
        hist_exog_list=None,
        futr_exog_list=['price'],
        stat_exog_list=None,
        early_stop_patience_steps=early_stop_patience,
        learning_rate=0.0001,
        alias='NHITS_no_clusters'
    )
]

nf = NeuralForecast(models=models, freq='D')

nf.fit(
    df=train_df,
    static_df=static_df,
    val_size=90,
    verbose=True
)

predictions = nf.predict(futr_df=val_df, static_df=static_df)
```

If you're interested in the training losses and the full code, please see the notebook in the book repository. We created a lot of error metrics, looking at the overall dataset predictive performance, per-series performance, and performance relative to a naive baseline.

Model	MAE	RMSE	MASE	MAPE	sMAPE	Rel. RMSE
NHITS_with_clusters	1.426	2.012	1.069	0.498	0.207	0.861
NHITS_no_clusters	1.444	2.041	1.064	0.485	0.204	0.888

Table 15.4: Overall error metrics for N-HITS trained with and without cluster identifiers as a static covariate. Relative RMSE is expressed against a seasonal naive baseline; lower is better for all metrics except total_wins, which counts the number of series (out of 12,152) where each model produced the lowest error for a given metric

Per-series wins (number of series where the model achieved the lowest value for each metric):

Model	Rank	MAE wins	RMSE wins	MASE wins	Total wins
NHITS_with_clusters	1	2000	1512	2000	7024
NHITS_no_clusters	2	1038	1526	1038	5128

Table 15.5: Per series wins

After all this we have some slight reductions in MAE and RMSE for both standard and relative metrics, but overall there is only a small difference in how well the clustered model beats the non-clustered model. It is encouraging that both beat the naive model strongly. The individual series performance is negligible for RMSE; I'm inclined to ignore the other metrics due to inherent issues in their calculations relative to this data.

It could be that the clusters could be improved, or that the model is not complex enough to take advantage of the clusters. We applied both of these options in the notebook. First, we applied some hyperparameter tuning using neuralforecast's auto models to the clustered and non-clustered data.

The results show modest improvements with clustering:

Model	MAE	RMSE	MASE	MAPE	sMAPE
AutoNHITS_clustered	1.439	2.029	1.060	0.503	0.202
AutoNHITS_no_clusters	1.428	2.018	1.059	0.502	0.208

Table 15.6: Overall error metrics for the hyperparameter-tuned AutoNHITS variants, with and without cluster identifiers.

Comparing *Table 15.6* against *Table 15.4*, hyperparameter tuning only helped improve the model without cluster labels, and only slightly. As we mentioned before, the cluster labels didn't look great, and we could easily try more clusters. We could also clean up our feature representations by denoising them with an autoencoder.

Denoising with an autoencoder

Autoencoders are self-supervised representation learners. An autoencoder takes an input vector $\mathbf{x} \in \mathbb{R}^d$ and compresses it with an encoder $f_\theta\colon \mathbb{R}^d \to \mathbb{R}^k$ into a latent space $\mathbf{z} = f_\theta(\mathbf{x})$, then reconstructs it with a decoder $g_\phi\colon \mathbb{R}^k \to \mathbb{R}^d$, yielding $\hat{\mathbf{x}} = g_\phi(\mathbf{z})$. The goal of training is to minimize a reconstruction loss, commonly the mean-squared error $\mathcal{L}_{\text{rec}} = \| \mathbf{x} - \hat{\mathbf{x}} \|_2^2$, so the network learns to retain signal while discarding nuisance variation. A common way to think about this is that an art forger is trying to reconstruct a lost piece of artwork (the input), compressing it through their understanding (the encoder and latent space) and then recreating it (the decoder). To succeed they must fool an art historian (the loss function). Unlike in real life, our forger gets multiple attempts until its reconstruction is good enough (loss minimized).

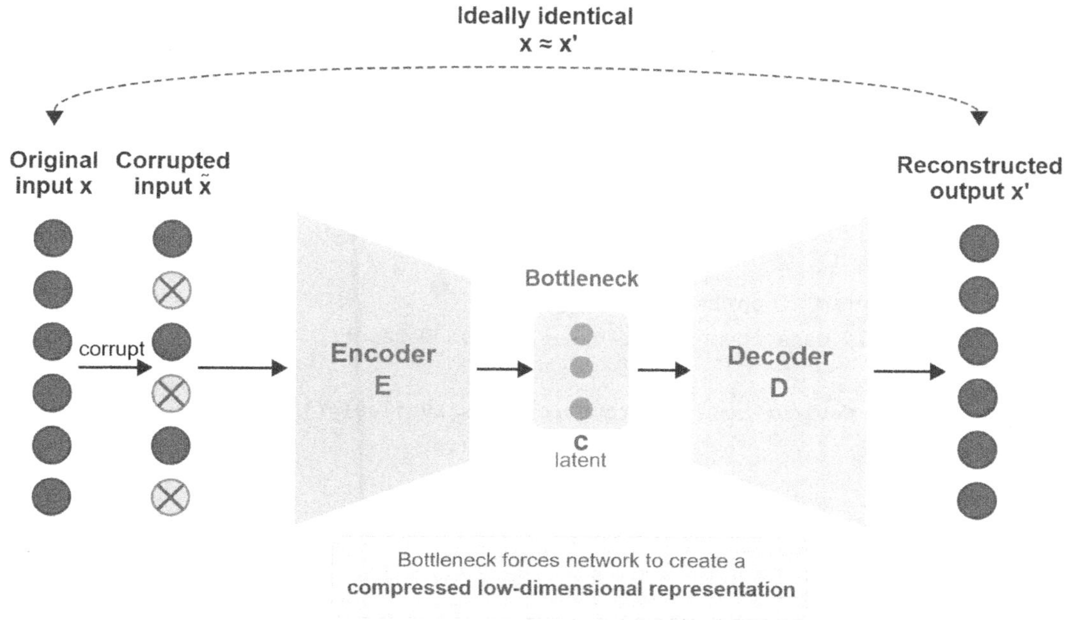

Figure 15.18: Denoising autoencoder architecture diagram

A **denoising autoencoder (DAE)**, as seen in *Figure 15.18*, adds an explicit corruption step in the training loop, typically Gaussian noise on our input, dropout on the hidden units, or both, so we are training our network to reconstruct a clean signal from corrupted inputs. This forces the latent space to store stable structure rather than spurious fluctuations, which can translate into more reliable distance metrics downstream.

In our pipeline we feed the 37-dimensional feature matrix (already min-max normalized) into a modest multilayer autoencoder with layers 128 > 64 > `encoding_dim`, and we set `encoding_dim=input_dim`. Encoder expands to 128 units then compresses back to a 37-dimensional bottleneck at `encoding_dim`, so there's a bottleneck in the expand-then-compress sense but it's not *undercomplete*: code and input share dimensionality. An overcomplete autoencoder like this can learn identity if nothing stops it, which is why we add dropout ($p = 0.1$) on every hidden layer. As our `tsfeatures` are noisy, computed on short and intermittent series, dropout-regularised reconstruction helps keep structure that's common across series and reduces per-series wobble. The `encode()` step gives us the embedding **z** we then cluster with k-means. Hopkins comes out at 0.86 and the elbow is sharper after this step. Using cluster labels based on these encoded features gives a better global RMSE for our N-HITS model. If we'd set $k < d$, the same architecture would double as non-linear dimensionality reduction, giving us a compact code we could reuse for clustering, nearest-neighbor search, or as static side-information in the GFM.

It's worth considering how this compares to alternatives. PCA is a linear projection tied to second-order moments, so it cannot capture the non-linear interactions among features that matter here (for example, when seasonality strength and volatility interact multiplicatively). t-SNE excels at visualization but has no analytical inverse and fails on out-of-sample points, making it unsuitable as a preprocessing step. The autoencoder provides a deterministic mapping $\mathbf{x} \mapsto \mathbf{z}$ and a parametric decoder, both of which scale to millions of series with mini-batch SGD. At inference, the `encode()` method extracts `encoded_features`, a NumPy array ready for elbow analysis. Reducing `encoding_dim` to, say, 7 would simultaneously denoise and compress, enabling experiments on the trade-off between latent dimensionality and GFM accuracy.

We start by preparing the feature data for the autoencoder and scaling it separately.

```
import torch
import torch.nn as nn
import torch.optim as optim
from torch.utils.data import DataLoader, TensorDataset

device = torch.device('cuda' if torch.cuda.is_available() else 'cpu')
print(f"Using device: {device}")

autoencoder_features = feature_df.drop(
    ['unique_id', 'cluster', 'tsne1', 'tsne2'],
    axis=1, errors='ignore'
)

from sklearn.preprocessing import MinMaxScaler
ae_scaler = MinMaxScaler()
ae_scaled_data = ae_scaler.fit_transform(autoencoder_features)
```

Our `TimeSeriesAutoencoder` code initializes weights with Xavier uniform for stable gradients, and uses Adam with weight decay to curb overfitting. The encoder first widens the raw 37-dimensional vector to 128 units before funneling it down to the latent code. This expand-then-compress pattern lets the network discover non-linear feature interactions in a higher-capacity space and then distill only the salient ones into `encoding_dim` units. The decoder mirrors this trajectory, with the trailing Sigmoid keeping reconstructions in [0,1] to match the min-max scaled inputs.

```
class TimeSeriesAutoencoder(nn.Module):
    def __init__(self, input_dim, encoding_dim=7):
        super(TimeSeriesAutoencoder, self).__init__()
        self.input_dim = input_dim
        self.encoding_dim = encoding_dim

        self.encoder = nn.Sequential(
            nn.Linear(input_dim, 128),
            nn.ReLU(),
            nn.Dropout(0.1),
            nn.Linear(128, 64),
            nn.ReLU(),
            nn.Dropout(0.1),
```

```
            nn.Linear(64, encoding_dim)
        )

        self.decoder = nn.Sequential(
            nn.Linear(encoding_dim, 64),
            nn.ReLU(),
            nn.Dropout(0.1),
            nn.Linear(64, 128),
            nn.ReLU(),
            nn.Dropout(0.1),
            nn.Linear(128, input_dim),
            nn.Sigmoid()
        )

        self._init_weights()

    def _init_weights(self):
        for module in self.modules():
            if isinstance(module, nn.Linear):
                nn.init.xavier_uniform_(module.weight)
                if module.bias is not None:
                    nn.init.zeros_(module.bias)

    def forward(self, x):
        encoded = self.encoder(x)
        decoded = self.decoder(encoded)
        return decoded

    def encode(self, x):
        return self.encoder(x)
```

A stricter reading of *denoising autoencoder*, following Vincent et al. (2008), needs an explicit input-corruption step: add Gaussian noise $\epsilon \sim \mathcal{N}(0, \sigma^2)$ to each mini-batch before a forward pass (`batch_data += torch.randn_like(batch_data) * 0.05`). We don't do that here. Our features are already noisy, and the nn.Dropout(0.1) layers inside the encoder and decoder mask 10% of activations with zero on each forward pass — *Bernoulli corruption* of the hidden units which plays a similar role to input noise. If you wanted to try the stricter version it's a one-line change in the training loop. If you add noise to batching you'll want to tune the value.

We wrap the scaled feature matrix in a `TensorDataset`, train for 500 epochs with `batch_size=256`, and record the epoch-level MSE.

```
input_dim = autoencoder_features.shape[1]
encoding_dim = input_dim  # no compressive bottleneck; set lower for
dimensionality reduction
model = TimeSeriesAutoencoder(input_dim, encoding_dim).to(device)

X_tensor = torch.FloatTensor(ae_scaled_data).to(device)
dataset = TensorDataset(X_tensor, X_tensor)
dataloader = DataLoader(dataset, batch_size=256, shuffle=True)

criterion = nn.MSELoss()
optimizer = optim.Adam(model.parameters(), lr=0.0001, weight_decay=1e-5)

epochs = 500
train_losses = []
model.train()

print(f"Training autoencoder: {input_dim} -> {encoding_dim} dimensions")
for epoch in range(epochs):
    epoch_loss = 0.0
    batch_count = 0

    for batch_data, _ in dataloader:
        outputs = model(batch_data)
        loss = criterion(outputs, batch_data)

        optimizer.zero_grad()
        loss.backward()
        optimizer.step()

        epoch_loss += loss.item()
        batch_count += 1

    avg_loss = epoch_loss / batch_count
    train_losses.append(avg_loss)

    if (epoch + 1) % 20 == 0:
        print(f'Epoch [{epoch+1}/{epochs}], Loss: {avg_loss:.6f}')
```

Once the autoencoder is trained we can pull the embedding via `model.encode()` or the full-dimensional reconstruction via `model()`. You can reuse this model rather than retraining it if you have similar data, although you would need to reuse the same `ae_scaler` to transform any new data before feeding it through `model.encode()` so the latent space stays consistent.

```
model.eval()
with torch.no_grad():
    X_tensor = torch.FloatTensor(ae_scaled_data).to(device)
    encoded_features = model.encode(X_tensor).cpu().numpy()

print(f"Encoded features shape: {encoded_features.shape}")
```

Autoencoder hyperparameters

The performance of an autoencoder depends heavily on its architectural and training choices. Parameters such as the encoding dimension, hidden layer widths, dropout rate, and optimizer configuration determine how much information is compressed, how well noise is removed, and how effectively the model generalizes to unseen data.

Table 15.7 summarizes the main hyperparameters used in our implementation, along with their roles and common alternatives that can be explored during tuning.

Component	Default	Purpose	Alternatives
encoding_dim	= 37 (pure denoise)	Controls compression ratio	7–15 for dimensionality reduction
Hidden widths	128 > 64	Balance capacity vs overfitting	{256, 128} or staircase 128 > 64 > 32
Dropout p	0.1	Noise strength	{0.0, 0.2, 0.3} in grid search
Optimizer	Adam, 1e-3	Fast convergence	AdamW or Ranger, schedulers (One-Cycle)
Loss	MSE	Stable for [0,1] inputs	MAE or Huber if outliers dominate?

Table 15.7: Autoencoder hyperparameters used for denoising tsfeatures matrix, with their defaults, the role each plays in the model, and alternatives worth trying in a grid search

A practical recipe is to start with encoding_dim ≈ $d/4$, and monitor validation MSE. If it rises sharply, the bottleneck is too tight. Conversely, if clustering metrics plateau, reduce encoding_dim, deepen the encoder, or raise dropout to promote disentanglement. As usual, our guiding principle is *systematic experimentation*: iterate on capacity and noise until the elbow plot and forecasting metrics converge.

```
print("Analysing optimal clusters for autoencoder features...")
kmeans_elbow(
    X=encoded_features,
    palette=custom_palette,
    k_distortion=(1, 30),
```

```
    k_ch=(2, 30),
)
```

So how does our new model do with the denoised data, more cluster labels, and inheriting the hyperparameters of the tuned model?

Model	MAE	RMSE	MASE	MAPE	sMAPE	Rel. RMSE	Rel. MAE
AutoNHITS_ae_clustered	1.406	1.99	1.05	0.48	0.206	0.830	0.785
AutoNHITS_no_clusters	1.428	2.018	1.059	0.502	0.208	0.854	0.808
NHITS_with_clusters	1.426	2.012	1.069	0.498	0.207	0.861	0.803
NHITS_no_clusters	1.444	2.041	1.064	0.485	0.204	0.888	0.815
NHITS_cluster_specific (hyperopt)	1.472	2.067	1.121	0.537	0.204	0.892	0.831
AutoNHITS_clustered	1.439	2.029	1.060	0.503	0.202	0.906	0.804
NHITS_cluster_specific (default)	1.459	2.059	1.081	0.510	0.202	1.010	0.824

Table 15.8: Summary of every GFM variant tested, ranked by relative RMSE against seasonal naive baseline

The autoencoded clustering approach achieved our best global performance (RMSE = 1.99), with a relative RMSE of 0.83. That's a 2.4% improvement over the default no-cluster baseline (2.04), or 1.0% over the tuned baseline (2.02). Notably, we didn't re-run hyperparameter optimization after switching to the autoencoded cluster labels, and adding or changing static features shifts the input distribution the model sees, which likely shifts the optimal hyperparameters too. A fairer comparison would retune each variant.

Model	MAE	RMSE	MASE	MAPE	sMAPE	Rel. RMSE	Rel. MAE
NHITS cluster_specific (default)	1.459	2.059	1.081	0.510	0.202	1.010	0.824
NHITS cluster_specific (hyperopt)	1.472	2.067	1.121	0.537	0.204	0.892	0.831

Table 15.9: Error metrics for the 20 cluster-specific N-HITS models, with default hyperparameters and with hyperparameters inherited from the AutoNHITS tuning run. Metrics are aggregated across all 12,152 series

Neither of these approaches improves on the global error of our autoencoded-clusters GFM. The default cluster-specific variant in particular is the only model in our experiment to post a relative RMSE worse than the seasonal naive baseline (1.010), which is a useful negative: dropping from a well-regularized global model to twenty smaller ones with no tuning throws away more signal than cluster structures can make up for. That said, the cluster-specific GFMs, both the default and tuned versions, along with N-HITS with no clusters, are actually the best performers at the individual series level; the autoencoded-clusters GFM only wins 1,302 of 12,152 series despite its strong global RMSE. The global error metric is likely being driven by a subset of series where the autoencoded clustering provides a strong signal, and the global model's forecasts are probably fairly conservative, performing well on average but poorly on individual series where the cluster-specific models can specialize.

Given the results of modelling and evaluation, we can see firstly that clustering helps a little with default hyperparameters, but the tuned model with clusters isn't actually better than the tuned model without. The unclustered hyperparameter-tuned model is better than its direct default comparison, which may point to an issue with the validation step (a simple split), a lack of comparisons, missing important hyperparameters like batch size from the tuning, or a failure to increase the upper bound of hyperparameters like `max_steps`. Regardless, increasing the number of clusters and denoising the data gave us some significant improvements at the full model level, which is how we should evaluate this. However, the individual series performance was not fantastic, which does point towards the necessity of using more than one model in a system and selecting between or combining some of them. It's also the case that we should probably have run a new hyperparameter search with the autoencoded clusters, as they are different to the original set.

There are several further improvements we could explore:

- Using multiple models, e.g. a GFM with CatBoost, and ensembling the results together.
- Applying models to clusters, or clusters plus some hierarchical level (to some extent this is what the M5 winners did with their models, without the clustering)
- Allowing training to run for longer.
- Running the autoencoder with different stopping parameters, e.g. a lower learning rate, or changing the autoencoder architecture to improve cluster quality.
- Creating time series features from exogenous variables (e.g. price), if the exogenous variable is predictive of the dependent variable.
- Building more, or more specific, features that depend on domain knowledge and EDA (we also ignored a lot of potentially useful features such as holidays and promotions).
- Cleaning the data properly, as there are outliers, missing values and other issues that could be affecting results.
- Add a zero threshold or zero inflation model to deal with sparsity.

Evaluating clustering results

Evaluating time series clusters presents a problem: without ground truth labels, how do we know if our clusters are meaningful? This is particularly tricky in TSCL, where internal metrics often provide misleading guidance.

Standard internal metrics like the silhouette coefficient, Calinski-Harabasz index, and Davies-Bouldin score attempt to quantify cluster quality through geometric properties. The silhouette coefficient for a point i is defined as:

$$s(i) = \frac{b(i) - a(i)}{\max\big(a(i), b(i)\big)}$$

where $a(i)$ is the mean intra-cluster distance and $b(i)$ is the mean nearest-cluster distance for point i. Values close to 1 indicate well-clustered points, values near 0 indicate points on cluster boundaries, and negative values indicate points likely assigned to the wrong cluster.

These metrics can be about as useful as a chocolate teapot for time series clustering, and there are a few reasons why. High-dimensional feature spaces can make geometric separation misleading, since distances in many dimensions tend to concentrate. Pattern-based similarity doesn't always translate to Euclidean proximity, especially when our features are themselves summaries of temporal structure. And the optimal k suggestions from elbow plots and silhouette analysis often conflict with each other, or simply lack a clear optimum, as we saw with our M5 data where the Calinski-Harabasz index suggested k=2 for scaled features. That is clearly inadequate for capturing the diversity in 3,038 time series, and in fact increasing our clusters to 20 had a more beneficial effect on forecasting performance.

The most reliable evaluation strategy for our purposes is **extrinsic validation**: measuring whether clusters improve performance on the downstream task. For forecasting applications, this means comparing models trained with and without cluster labels:

```
def evaluate_clustering_impact(train_data, clusters, model_class, horizon=90):
    """
    Evaluate clustering by comparing forecasting performance
    with and without cluster information.
    """
    clustered_model = model_class(stat_exog_list=cluster_features)
    clustered_rmse = train_and_evaluate(clustered_model, train_data)

    baseline_model = model_class(stat_exog_list=None)
    baseline_rmse = train_and_evaluate(baseline_model, train_data)

    improvement = (baseline_rmse - clustered_rmse) / baseline_rmse
    return improvement
```

This is essentially the approach we took, with our clustering and modeling being guided by forecasting performance. Our best autoencoded clustering achieved RMSE = 1.99, a 2.4% improvement over the default no-cluster baseline (RMSE = 2.04) or 1.0% over the tuned baseline (`RMSE = 2.02`) neither of which the internal metrics, which suggested k=2, would have predicted.

Adding new static features or switching to encoded cluster labels changes the input distribution a model sees, which is highly likely to shift the optimal hyperparameters likely. A cleaner comparison would re-run the hyperparameter search for each variant rather than reusing a single tuned configuration across them.

Summary

We have learned that time series data may be clustered together using several methods, and that the variety of approaches gives us multiple pathways when we need to group similar temporal patterns. Distance-based methods, including Dynamic Time Warping and Shape-Based Distance, are conceptually simple but suffer from quadratic computational complexity that makes them impractical for large datasets. We can ensemble these together (hierarchical clustering with DTW) for better performance, but this exacerbates the scalability issue.

Subsequence-based methods like shapelets offer excellent interpretability and can capture local patterns that global statistics miss. However, they face combinatorial search complexity that becomes prohibitive at scale, making them suitable primarily for smaller datasets or specialized applications where interpretability is paramount.

Feature-based clustering is the most practical solution for production systems. Using libraries like tsfeatures or catch22, we can transform time series into fixed-length statistical summaries that capture essential characteristics: trend, seasonality, autocorrelation, and distributional properties. This approach scales linearly with dataset size and provides interpretable results, though it requires careful preprocessing and domain knowledge to select meaningful features. In our M5 experiments, feature-based clustering with proper scaling and denoising achieved the best forecasting improvements.

Two-stage pipelines that combine autoencoders with standard clustering algorithms also show considerable promise. Our autoencoder experiments achieved the best overall performance (RMSE = 1.99), and although we used statistical featurization rather than neural networks to create the initial representations, the principle is the same: a learned, denoised representation can outperform the raw feature set. End-to-end deep clustering methods require substantial computational resources, large datasets, and careful hyperparameter tuning, which limits their accessibility for many applications.

Throughout our experiments, preprocessing often mattered as much as the algorithm choice, perhaps more so. Decisions about scaling, detrending, and noise reduction fundamentally shape what patterns our clustering algorithm can discover. Evaluation also remains challenging: internal metrics like silhouette scores provide only rough guidance, with the most reliable validation coming downstream at task performance.

If you take one thing away from this chapter, it should be this: start with feature-based approaches using established libraries, validate through forecasting performance rather than internal metrics, and treat preprocessing as a critical hyperparameter to optimize. The modest improvements we achieved (a 2.4% RMSE reduction over the default baseline, about 1% against the tuned baseline) show that clustering can enhance global forecasting models, though it's not a silver bullet. It is, however, a useful additional option in our toolkit when we need to improve GFM performance at scale.

In the next chapter we move from clustering to embeddings, where we'll see how learned vector representations can capture semantic and temporal structure in ways that complement the featurization approaches we've explored here.

Recommended reading

- Bandara, K., Hewamalage, H., Liu, Y-H., Kang, Y., and Bergmeir, C. 2021. "*Improving the Accuracy of Global Forecasting Models Using Time Series Data Augmentation.*" Pattern Recognition. `https://doi.org/10.1016/j.patcog.2021.108148`.
- Marco-Blanco, J., and Cuevas, R. 2023. "*Time Series Clustering With Random Convolutional Kernels.*" arXiv. `https://doi.org/10.4850/arXiv.2305.1057`.
- Middlehurst, M., Schäfer, P., and Bagnall, A. 2024. "*Bake Off Redux: A Review and Experimental Evaluation of Recent Time Series Classification Algorithms.*" Data Mining and Knowledge Discovery. `https://doi.org/10.1007/s10618-024-01022-1`.
- Paparrizos, J., Yang, F., and Li, H. 2024. "*Bridging the Gap: A Decade Review of Time-Series Clustering Methods.*" arXiv. `https://doi.org/10.4850/arXiv.2412.2082`.
- Vincent, P., Larochelle, H., Bengio, Y., and Manzagol, P-A. 2008. "*Extracting and Composing Robust Features with Denoising Autoencoders.*" Proceedings of the 25th International Conference on Machine Learning - ICML '08. `https://doi.org/10.1145/1390156.1390294`

Get this book's PDF version and more

Scan the QR code (or go to `packtpub.com/unlock`). Search for this book by name, confirm the edition, and then follow the steps on the page.

UNLOCK NOW

Note: Keep your invoice handy. Purchases made directly from Packt don't require an invoice.

16

Embeddings for Time Series

In the previous chapters, we have seen a variety of models and time series tasks. In this chapter, we will guide you through the topic of time series embeddings.

As we have seen time and again in the book, time series data is never just a series of numbers. There is always a context, a meaning, and a story behind the numbers. It is essential to perceive the data in a way that captures the context and digs deeper into the underlying data generating process. Embedding the time series data into a different space is a powerful way to achieve this. As you will see, embeddings are capable of capturing the dynamical system nature of the time series data, thus approximating the data generating process. Embeddings for time series are a very powerful tool for many time series tasks, including forecasting, classification, and anomaly detection. In fact, we have been using time series embeddings in the preceding chapters. In this chapter, we will explore different embedding techniques, build up intuitions of time series embeddings, and understand the importance of embedding in time series tasks. We start with the dynamical system perspective for intuitions, then we will look at the learned embeddings using deep learning models.

In this book:

- A dynamical system perspective
- Learned embedding
- The diversity of time series embeddings

A dynamical systems perspective

We discussed how a swing works in the previous chapters. We have seen how the swing angle changes over time and shows cyclic patterns. As a refresher, we use a cosine wave with amplitude gradually decreasing over time to simulate the swing angle; see the top panel in *Figure 16.1*. There are two naïve but intuitive methods to embed the swing angle time series in a different space. The first method is to create a data point using the current time and its lagged value, i.e., the time series value at particular time steps in the past. Mathematically speaking, for a swing angle $\theta(t)$ at time t, we plot the angle $\theta(t - \Delta t)$ at the previous time step $t - \Delta t$ as the y coordinate and the angle $\theta(t)$ as the x coordinate in the bottom left panel. It is called the delayed embedding. The second method is to apply differentials to the variable. We can plot the change rate of the angle $\Delta\theta(t) = \theta(t) - \theta(t - \Delta t)$ against the angle $\theta(t)$ in the bottom right panel. It is called the **phase portrait**. Strictly speaking, we have assumed that our unit of time is Δt, thus the quantity $\Delta\theta(t)$ is the so-called angular velocity. If you are not familiar with physics, this is similar to the term *velocity* we talk about in our daily life, which indicates how fast a car changes its location, but here angular velocity indicates how fast the swing changes its angle. Comparing the original time series with the two embeddings, we can see that it is easier to observe the inner patterns using the embeddings: Instead of extended values on infinite space, where making correlations for values at left and right extremes is hard, the two embeddings directly show us the patterns in a finite space, and we compactified the information onto this small region. In the figure, on the bottom left panel, we plot the swing angle at t and that at the previous sampling time step. The bottom right panel shows the change rate of the angle vs the angle itself.

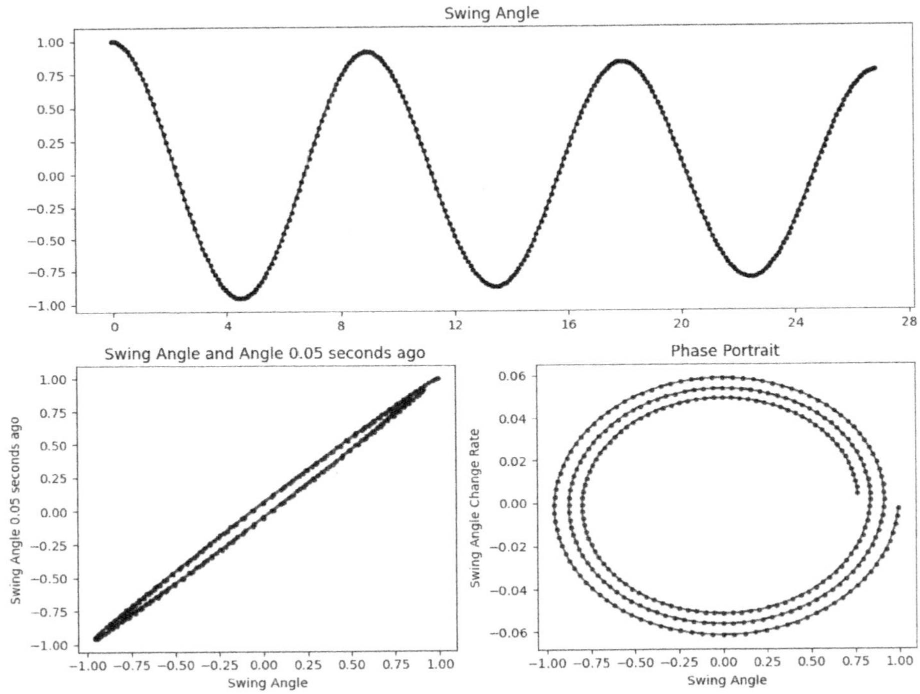

Figure 16.1: Normalize swing angle time series and two different embeddings

In fact, many time series forecasting models utilize such concepts. For example, in statistical forecasting models, the **Autoregressive** (**AR**) model category uses the lagged values of the time series as features. An AR model can have different orders depending on how many historical values are used. An *AR(2)* model uses the current value and one step in the past, which is usually formulated as follows

$$\theta(t_{T+1}) = \phi_T \theta(t_T) + \phi_{T-1} \theta(t_{T-1}) + \epsilon,$$

where T is the current time, and $T + 1$ is one step in the future. The formulation clearly shows that it is essentially the delayed embedding using time-lagged values. The lagged values are simply the y coordinate in the bottom left panel in *Figure 16.1*.

Even in deep learning based forecasting models, we usually formulate the input time series using time delayed embeddings. For example, in the transformer chapter, we formatted our input and target of a time series $\theta(t) = [\theta(t_1), \theta(t_2), \cdots, \theta(t_T + 1)]$ using the following matrix:

$$\mathbf{\Theta} = \begin{bmatrix} \theta(t_1) & \theta(t_2) & \cdots & \theta(t_p) & | & \theta(t_{p+1}) \\ \theta(t_{1+1}) & \theta(t_{1+2}) & \cdots & \theta(t_{1+p}) & | & \theta(t_{1+p+1}) \\ \vdots & \vdots & \ddots & \vdots & | & \vdots \\ \theta(t_{i-p+1}) & \theta(t_{i-p+2}) & \cdots & \theta(t_i) & | & \theta(t_{i+1}) \\ \vdots & \vdots & \ddots & \vdots & | & \vdots \\ \theta(t_{T-p+1}) & \theta(t_{T-p+2}) & \cdots & \theta(t_T) & | & \theta(t_{T+1}) \end{bmatrix}$$

This can be fed into a deep learning forecasting model as input features with context length of p (left of the vertical lines in the matrix) and targets of horizon being 1 (right of the vertical lines in the matrix). It is nothing but a higher dimensional version of what we have shown in *Figure 16.1*. It may seem trivial; however, the idea is stemmed from the so-called Takens' theorem, which states that we can reconstruct the state space of a dynamical system from such delayed embeddings of the time series data.

There is a catch. The analysis of swing angle is easy, because we are not limited by the sampling rate, nor do we need to deal with noise. In real life, we may not have access to high sampling rate data, and the data may be noisy. We show exactly the same plots in *Figure 16.2* but for the Walmart sales data in the M5 dataset. As the data gets noisier and the sampling rate gets lower, the time-delayed embeddings become less straightforward as discussed several times in the previous chapters.

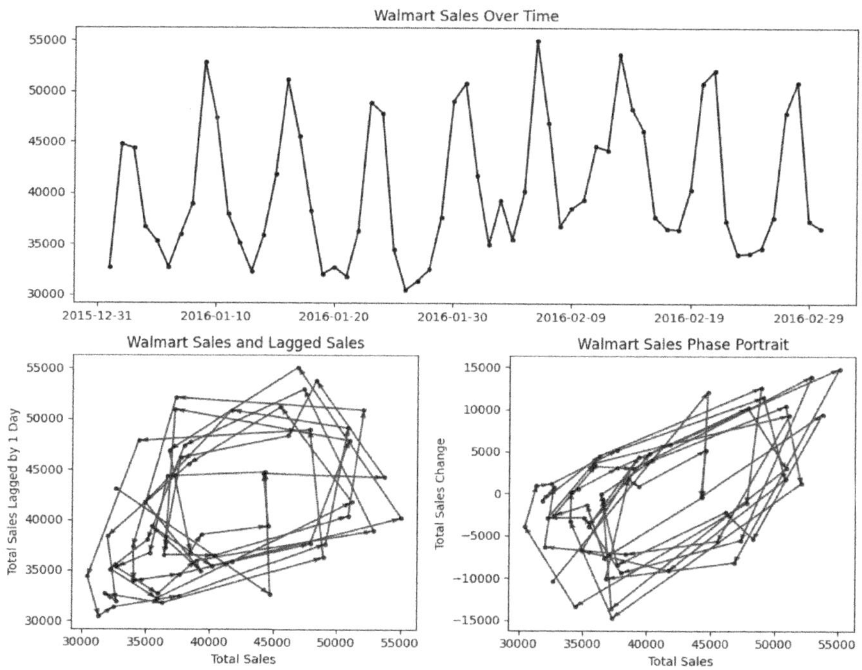

Figure 16.2: Walmart sales in three states (WI, CA, and TX) in the M5 dataset

When compared to Figure 16.1, the embedding pattern is less significant in *Figure 16.2*.

Regardless, the idea of using lagged values or differences to the previous value is useful in many cases. The delayed embeddings are one of the key components in lag-llama, a time series foundation model we mentioned in previous chapters.

Learned embeddings

In most autoencoder architecture-based time series forecasting models, the encoder already learns an embedding of the input time series, specially designed for the specific task and model architecture. In the previous chapters, we have discussed a TimeVAE model that generates synthetic time series data. To dive deeper into how the model works and understand time series embeddings in deep neural networks, we will visualize the learned latent space embedding of input data in the TimeVAE model.

Recall that a TimeVAE model generates the mean and variance of the latent space distribution. We pull this generated mean and reduce the dimensionality to 2D using dimension reduction methods such as PCA for visualization. For easier understanding, we use the naive sine wave dataset as the input to the model.

First, we create a `embedding_extractor` function to extract the latent space embeddings from a trained TimeVAE model.

```
from torch.utils.data import DataLoader
import lightning as L
import torch

def embedding_extractor(
    forecaster: L.LightningModule,
    x: torch.Tensor
) -> tuple[torch.Tensor]:
    """compute the embeddings based on the input

    :param forecaster: the trained forecaster
    :param x: input historical time series,
    """
    forecaster.model.to(x.device)
    z_mean, z_log_var, z = forecaster.model.encoder(
        x.type_as(
            forecaster.model.encoder.z_mean_layer.weight
        )
    )

    return z_mean, z_log_var, z
```

Following *Chapter 12*, our trained model is `vae_model`. For better visualizations, we recreate a batch of the input data, `input_example`, with batch size 400 and time series length 24. A larger batch size is chosen to ensure we have enough data points to visualize. We can extract the latent space embeddings as follows:

```
z_mean_example, _, _ = embedding_extractor(
    vae_model,
    input_example
)
```

`z_mean_example` is our clean version of the embedding. It has dimension (400,1,8). To visualize the input data and the embedding, we reduce the dimension to 2D using PCA:

```
from torchdr import TSNE

n_components = 2

dr_input_result = TSNE(
    perplexity=30,
    n_components=n_components
).fit_transform(
    input_example.squeeze()
)

dr_z_result = TSNE(
    perplexity=30,
    n_components=n_components
).fit_transform(
    z_mean_example.squeeze()
)
```

Refer to color images on the Graphic bundle or the ebook.

We can visualize the 2D embeddings and put them side by side in *Figure 16.3* for comparison.

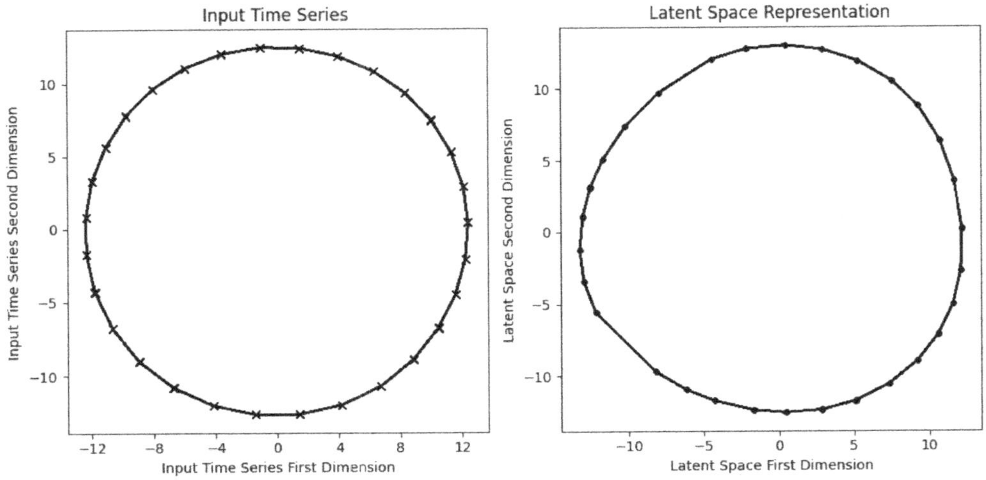

Figure 16.3: Visualization of the input time series and latent space embedding for a TimeVAE model

The *left* panel shows a perfect circle, which is expected as the input is a sine wave. The *right* panel shows the latent space embedding. It maintains the local structure of the input data, but with 1/3 of the original size.

In real-world datasets, the learned embeddings can be more complex. However, the idea is similar. The learned embeddings contain key patterns of the original data that are related to the task at hand. To make embeddings more useful, some models seek more universal embeddings of time series that can be used for different tasks.

Theoretically, embeddings are useful because we assume that real-world data can be represented using a lower-dimensional structure, or the data we collect are often some complicated global structures embedded in a higher-dimensional space. By learning the embeddings, we approximate the actual dynamics of the data-generating process with lower dimensions. This resembles the so-called manifold hypothesis. For example, in language models, embeddings usually refer to embedding in lower-dimensional space.

However, embedding is a trickier concept for time series data. The reality is that it is often useful to embed the data into higher dimensions in dynamical systems. There also exist mathematical theorems that support such ideas. For example, the Whitney embedding theorem, which roughly says that *smooth* shapes on bended surfaces can be perfectly embedded inside a flat but higher dimensional space, or the so-called Euclidean space. It is theoretically okay to have embeddings with higher dimensions if they are useful.

In classical time series analysis, there is the famous **Short Time Fourier Transform** (**STFT**) that transforms the time series into a higher dimensional time-frequency space to find patterns. STFT is a fantastic embedding of many time series data such as audio. Of course, for more sparse and more complicated time series data, STFT may not work well. On the other hand, we can take this idea and use neural networks to create embeddings similar to STFT transformed data. Yue et al. proposed a time series embedding method called TS2Vec that utilizes this idea.

TS2Vec cuts the input time series into time segments, e.g., using a sliding window. Each time segment is then encoded using an encoder. The model can be trained using different tasks including forecasting. We train a TS2Vec model on the swing angle data using the forecasting task using code adapted from the source code of TS2Vec, available at `https://github.com/zhihanyue/ts2vec`. We specified the embedding to be 20-dimensional. The first 1000 time steps of the swing angle data and the corresponding embedding is visualized in *Figure 16.4*. Compared to the original swing angle data, the learned embeddings show more periodic patterns.

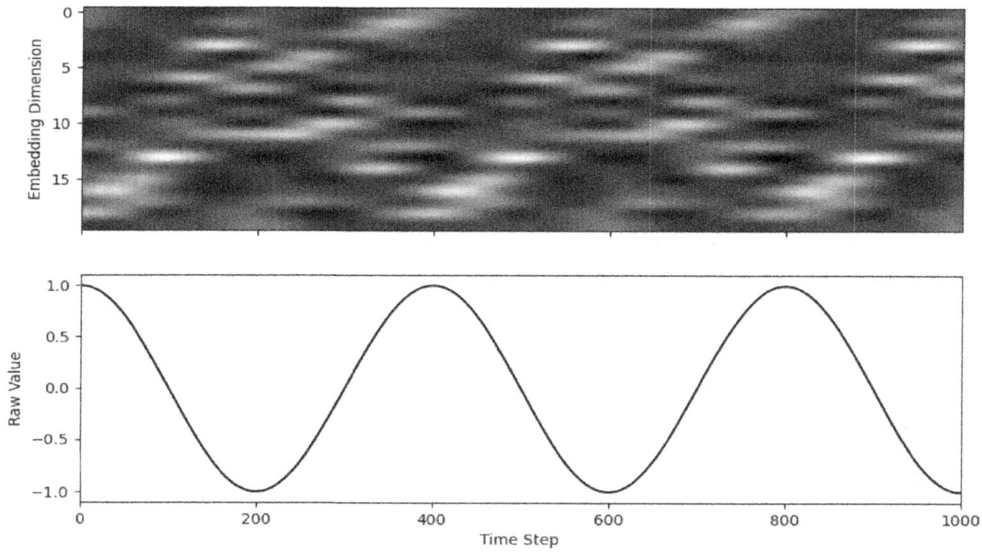

Figure 16.4: Embedding of the swing angle data and the corresponding embedding at each time step

In this figure, the *upper* panel shows the embedding, with the darker color indicating larger values. The vertical axis is the dimension of the embedding; here we have 0 to 19. The *lower* panel shows the original data.

In fact, the visual patterns themselves are useful for anomaly detection. In the event of anomalies, the embedding patterns will be visually disrupted. To get a feeling of the idea, in *Figure 16.5* we plotted out the embeddings for a damped swing, with the swing angle gradually decreasing over time.

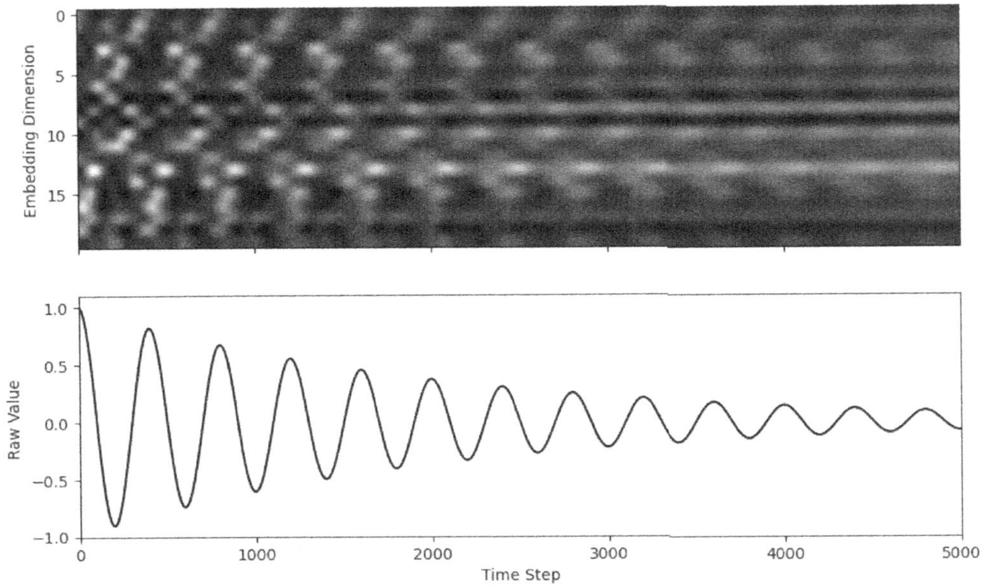

Figure 16.5: Embedding of the swing angle data of a damped swing and the corresponding embedding at each time step

Similar to *Figure 16.4*, the *upper* panel shows the embedding, and the *lower* panel shows the original data. The periodic patterns gradually smear out.

For our naive example, it seems that the embeddings are not quite useful. However, in the event of real-world complicated data, such embeddings can be very useful for downstream tasks such as anomaly detection. We have discussed the key ideas of the TS2Vec model.

The diversity of time series embeddings

Embeddings can take many forms and serve various purposes depending on the specific characteristics of the data and the tasks at hand, especially for time series data. We have discussed two types of embeddings in this chapter: the dynamical system perspective-based time delayed embeddings and the learned embeddings using deep learning models. Apart from these, there are embeddings using language models, for example, discussed in "*Time-LLM: Time series forecasting by reprogramming large language models*" by M. Jin et al.

As much as we want to have a unified embedding of all time series data like language models, it is more challenging for time series data due to the diverse nature of time series data and the enormous types of context, even for time series data that looks similar. In some sense, a fully unified time series embedding is close to a unified theory of everything in physics. However, we have noticed trends that people are building unified time series embedding frameworks, such as the TS2Vec model we discussed above. Such efforts are very valuable, and we expect more progress in this direction in the future.

Summary

We have explored two types of embedding in this chapter: formulating time series data based on the idea of dynamical systems using our intuitions, and learning a time series embedding using a deep learning model. There exist many different ways of embedding time series data. Many times we don't talk about them that much as we have models that function end to end. However, it is crucial to understand the importance of time series embeddings. We didn't include all the details of the implementations; however, you can find the experiments we have carried out in the official repository of this book. PCA, t-SNE and UMAP are popular dimensionality reduction techniques that can be used to *embed* time series data for visualization or clustering. We have not discussed them in this chapter, but they are very useful in practice. As we move on to the rest of the chapters, we will see more about the role of embeddings in a variety of time series tasks.

Recommended reading

- For more discussions on Takens' theorem, see Small, M. 2005. "*Applied nonlinear time series analysis: Applications in physics, physiology and finance*." World Scientific Series on Nonlinear Science Series A. `http://dx.doi.org/10.1142/5722`.

For more advanced discussions on time series embeddings, we suggest you refer to the following papers:

- Chen, X., et al. 2025. "*Rethinking time encoding via Learnable Transformation functions*." arXiv. `http://arxiv.org/abs/2505.00887`.
- Fraikin, A., Bennetot, A., and Allassonnière, S. 2023. "*T-Rep: Representation Learning for Time Series using Time-Embeddings*." arXiv. `https://doi.org/10.48550/arXiv.2310.04486`.
- For applications of time series embedding, time series embedding is widely used in classification tasks; we refer readers to the review paper: Irani, H., Ghahremani, Y., Kermani, A., and Metsis, V. 2025. "*Time series embedding methods for classification tasks: A review*." arXiv. `http://arxiv.org/abs/2501.13392`.
- Jin, M., et al. 2023. "*Time-LLM: Time series forecasting by reprogramming large language models.*" arXiv. `http://arxiv.org/abs/2310.01728`.
- Kazemi, S. M., et al. 2019. "*Time2Vec: Learning a vector representation of time*." arXiv. `http://arxiv.org/abs/1907.05321`.
- Small, M. 2005. "*Applied nonlinear time series analysis: Applications in physics, physiology and finance.*" World Scientific Series on Nonlinear Science Series A. `http://dx.doi.org/10.1142/5722`.
- Yue, Z., et al. 2021. "*TS2Vec: Towards universal representation of time series*." arXiv. `http://arxiv.org/abs/2106.10466`.

Join our community on discord

Join our community's Discord space for discussions with the authors and other readers.

`https://discord.gg/VXpz9YPUTW`

17

Supervised and Unsupervised Anomaly Detection

In data analysis, the goal is not to find perfect patterns, but to identify when patterns break—for it is in the breaks that we find the signal."—Andrew Ng

In previous chapters we built an understanding of temporal patterns and how they can be useful for identifying, segmenting, and modeling data. But what happens when the data we want to model contains observations that are unexpected or unlikely to be repeated? Detecting anomalies and outliers is a critical aspect of time series analysis, particularly when modeling with neural networks where unusual observations can distort learned representations.

Unlike classification, where we assigned data to known categories, or clustering, where we discovered groupings based on similarity, anomaly detection looks to identify exceptions: rare events that do not conform to established temporal patterns. Anomalies are crucial information. We can use them to detect equipment degradation, fraudulent transactions, stock-outs, sales dumps, abnormal cardiological signals, and many more. The challenge in **Time Series Anomaly Detection** (**TSAD**) lies not just in detecting deviations, but doing so reliably when anomalies are rare, varied, context-dependent, and often hidden in noise.

Say we have a manufacturing facility monitoring sensors across production lines. A temperature sensor might show a gradual drift upward, with individually unremarkable values that are in fact an early signal of bearing failure. Each reading may appear normal in isolation, but within the context of time this pattern suggests a potential problem. Similarly, in financial markets a series of trades might each fall within a normal range, yet their sequence, and timing could constitute market manipulation. In retail, sudden demand spikes for specific products might indicate a new emerging trend, necessitating a new model, or simply a stock count correction. How significant any of these anomalies are depends on their relationship to a series's historical patterns and context.

In this chapter, we will cover both unsupervised and supervised anomaly detection for time series, primarily focusing on unsupervised anomaly detection, since in practice labeled anomalies are rarely available in the quantities supervised approaches require.

Unsupervised anomaly detection, like time-series clustering, is difficult, and uncertain; however, many types of anomalies are straightforward to find with simple approaches. The challenge comes when discovering less obvious patterns, which can sometimes be slight changes in our data. A change in heart rate variability may be subtle but signal an issue; it could equally be nothing, something any cardiologist would dismiss. Deciding which is which is not easy.

Supervised anomaly detection, by contrast, is essentially classification. We apply it when we have labeled data, or stored patterns we can use to detect new occurrences. Taking our earlier examples (equipment failures, confirmed fraud, stock-outs, store closures), it is often possible to label these events and then find similar instances in our data, so that we can classify future anomalies. Supervised approaches carry the additional challenge of severe class imbalance, as anomalies typically represent <5% (often <1%) of observations.

We'll also examine how feature engineering approaches from Chapter 14 can be tailored to capture outliers and anomalous behavior, how distance-based methods can be modified to detect outliers, and how we can apply deep learning architectures to learn representations that separate normal from anomalous patterns. We discuss strategies for handling imbalanced data, tuning decision thresholds when false alarms are costly, and evaluating models when accuracy is a meaningless metric. While we could use PyTorch for supervised anomaly detection, we are generally better off reaching for simple calculations and established models first.

What is an outlier or anomaly?

Before we can detect anomalies, we need to be clear about what we are investigating, as the two words are often used interchangeably. Blogs and journal articles on TSAD often conflate *outlier* and *anomaly*, but we'll define them separately to ease our discussion (Blázquez-García et al. 2020).

An outlier is an observation that is extreme relative to the marginal distribution of our data, without reference to temporal order. Classically, this is defined as any observation beyond ± 3.5 standard deviations from the mean (or median) of our data distribution. Specifically, outlier detection asks 'is this value unlikely under the overall distribution $p(y)$?', with the distribution here being static. In *Figure 17.1*, panel (a) shows marginal outlier detection: a collection of observations with no temporal structure, from which we identify a single point that exceeds our static threshold. The outlier is extreme under $p(y)$, which we can also see in the histogram panel relative to the bulk of the data.

Refer to color images on the Graphic bundle or the ebook.

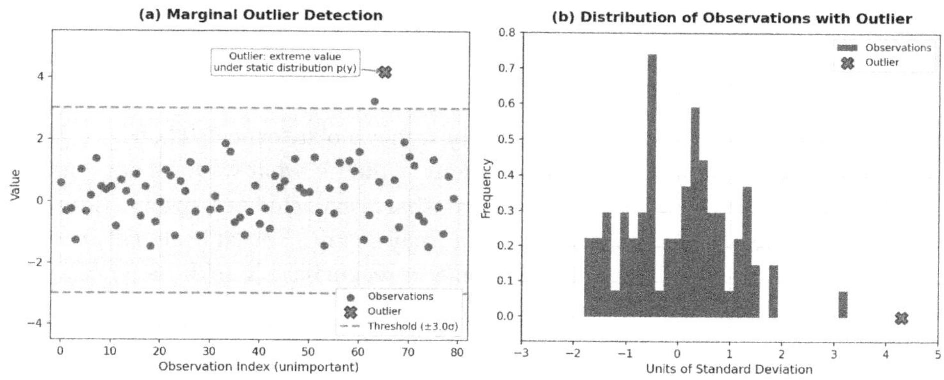

Figure 17.1: A statically defined outlier. Panel (a) shows an observation exceeding a ±3.5σ threshold; the histogram panel shows the same point in the tail of the marginal distribution p(y)

Anomalies in series are observations, or patterns, that are unexpected given temporal and contextual history—a Christmas sales spike for chocolate, for instance, is not an anomaly. Making TSAD is a *conditional* question: *'is this observation unlikely given what came before and what we know about the current context?'* Formally, we ask whether:

$$p(y_t \mid H_{t-1}, x_t)$$

is small under the conditional distribution, where H_{t-1} denotes the historical observations and x_t represents contextual or exogenous variables (day of week, promotional activity, weather, and so on).

By their nature **outliers are often anomalies, but anomalies are not always outliers.** A sensor reading of an oven at 300°C might be an outlier in a system that typically operates at 180–200°C. But a reading of 185°C could be an anomaly if the oven was recently switched off and should be cooling toward ambient temperature. *Figure 17.2* demonstrates this kind of anomaly within a temporal context.

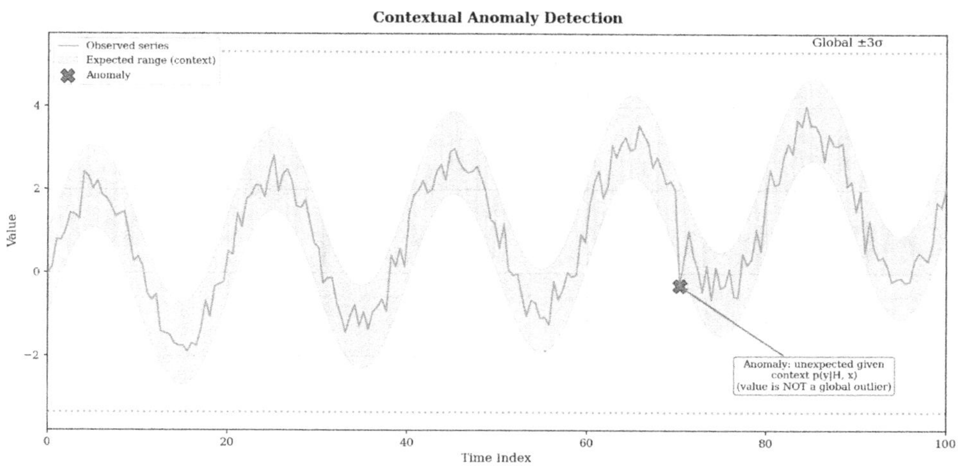

Figure 17.2: A contextual anomaly. The series shows clear seasonal structure; all points fall within the global distribution, but one point near -0.3 sits outside the local temporal context where a value closer to 1 would be expected

We have a time series with clear seasonal structure in which all points fall *within* the global distribution, but one point (≈ -0.3) falls *outside* of the simple temporal context. Values of around 0.3 do appear elsewhere in the series, but at the highlighted point we would expect the value to be closer to 1.

With time-series data we can classify anomalies by how they violate expected behavior, beyond global and contextual. The simplest case is a **point anomaly**, which is what we have demonstrated in the previous graphs. A single observation deviates from what is expected at a specific time index, either globally (*outlier*) or contextually (*anomaly*). The value may be unremarkable globally but inconsistent with local temporal context such as seasonality, regime, or recent history. In the latter case, the anomaly emerges only when conditioning on time-dependent structure, as we saw in *Figure 17.2*.

Let's load some data to work with; feel free to code along with a series of your own. We'll use some pseudo-data inspired by typical patterns of sales for a product with a regularly promotional pricing cycle, which is not an uncommon retail strategy. This gives us a realistic base structure to inject anomalies into.

```
# pip install pandas numpy matplotlib seaborn stumpy pyod catboost
# Load data
file_path = CFG.data_folder / "chapter17" / "pseudo_sales_adj_v1.csv"
data = pd.read_csv(file_path)
data['date'] = pd.to_datetime(data['date'], format='%d/%m/%Y')
data.set_index('date', inplace=True)
data = data.sort_index()
```

The output:

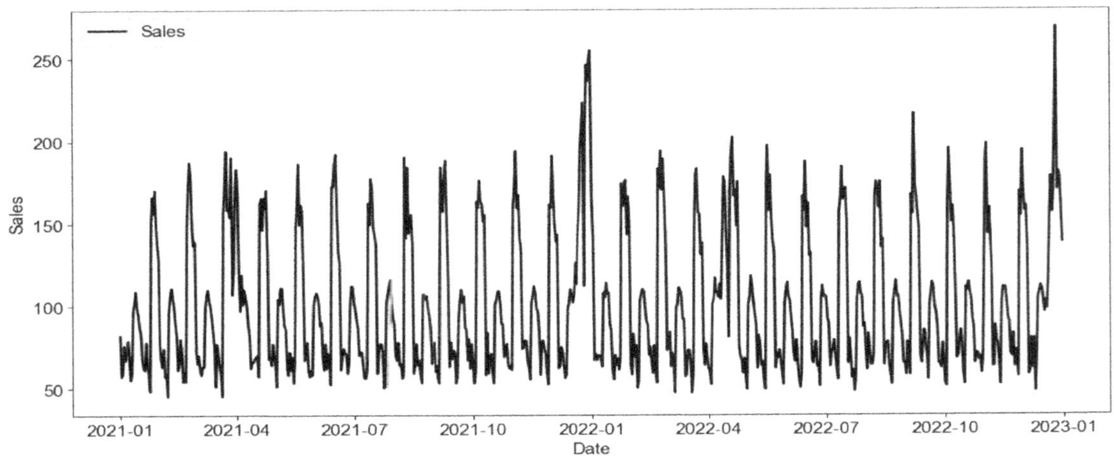

Figure 17.3: Pseudo sales data with no anomalies, showing two Christmas peaks and promotional regime

At this point the data has no anomalies; those two spikes that look like outliers are simply sales around the Christmas holidays. Let's inject a clear point outlier (a large, isolated spike) and a contextual point anomaly (an unremarkable value if in another part of the series but in the wrong placement) so we have clear anomalies to investigate.

```
# Create point outlier and anomaly based on rolling std
point_outlier = pd.Timestamp('2022-10-04')
rolling_std_dev = data['sales'].rolling(window=14).std()
std_dev_rolling = rolling_std_dev.loc[point_outlier - pd.Timedelta(days=1)]
data.loc[point_outlier, 'sales'] = (
    data.loc[point_outlier - pd.Timedelta(days=1), 'sales'] +
    (5 * std_dev_rolling)
)

point_anomaly = pd.Timestamp('2021-08-27') # 2021-07-27
std_dev_rolling = rolling_std_dev.loc[point_anomaly - pd.Timedelta(days=1)]
data.loc[point_anomaly, 'sales'] = (
    data.loc[point_anomaly - pd.Timedelta(days=1), 'sales'] + (1.2 * std_dev_
rolling)
)

# Rerun plot
fig, ax = plt.subplots()
ax.plot(data.index, data['sales'], color=custom_palette[0], label='Sales')
ax.scatter(
    point_outlier,
    data.loc[point_outlier, 'sales'],
    color=custom_palette[2], marker='o', s=100, label='Point Outlier', zorder=5
)
ax.scatter(
    point_anomaly,
    data.loc[point_anomaly, 'sales'],
    color=custom_palette[2], marker='x', s=100, label='Point Anomaly',zorder=5
)

ax.set_xlabel('Date')
ax.set_ylabel('Sales')
ax.legend()
plt.show()
```

The output is:

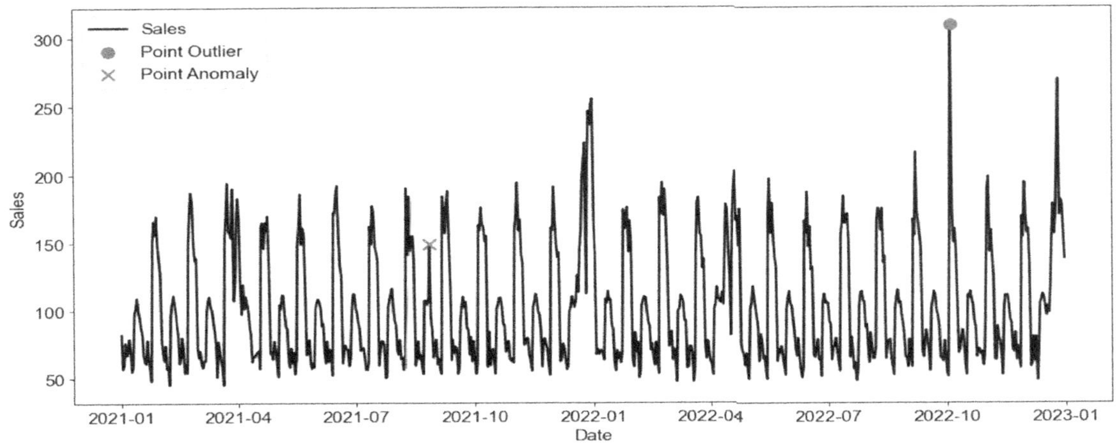

Figure 17.4: Sales data with point outlier and contextual anomaly

To implement this ground-truth labelling in the dataset, we use the following code:

```
data_pointanoms = data.copy()
data_pointanoms['is_anomaly'] = 0
outlier_dates = [point_outlier + pd.Timedelta(days=i) for i in range(1)]
anomaly_dates = [point_anomaly + pd.Timedelta(days=i) for i in range(1)]

data_pointanoms.loc[outlier_dates, 'is_anomaly'] = 1
data_pointanoms.loc[anomaly_dates, 'is_anomaly'] = 1
```

Collective (or sequence-based) anomalies arise when a contiguous subsequence exhibits abnormal joint behavior, even though each constituent point may appear locally plausible. These anomalies are defined by deviations in temporal structure, such as unexpected oscillations, bursts, level shifts, or changes in dynamics, rather than pointwise extremeness. As illustrated in the figure, an anomalous subsequence may lie largely within globally defined marginal bounds yet form an implausible pattern when considered as a sequence.

```
data = pd.read_csv(file_path)
data['date'] = pd.to_datetime(data['date'], format='%d/%m/%Y')
data.set_index('date', inplace=True)
data = data.sort_index()

# Create point outlier and anomaly based on rolling std
point_outlier = pd.Timestamp('2022-06-04')
rolling_std_dev = data['sales'].rolling(window=14).std()

for i in range(3):
```

```
    d = point_outlier + pd.Timedelta(days=i)
    std_dev_rolling = rolling_std_dev.loc[d - pd.Timedelta(days=1)]
    data.loc[d, 'sales'] = (
        data.loc[d - pd.Timedelta(days=1), 'sales'] +
        ((5 - 1.5*i) * std_dev_rolling)
    )

point_anomaly = pd.Timestamp('2021-06-03')
for i in range(3):
    d = point_anomaly + pd.Timedelta(days=i)
    std_dev_rolling = rolling_std_dev.loc[d - pd.Timedelta(days=1)]
    data.loc[d, 'sales'] = (
        data.loc[d - pd.Timedelta(days=1), 'sales'] +
        ((1.1 - 0.2*i) * std_dev_rolling)
    )

# Rerun plot
fig, ax = plt.subplots()
ax.plot(data.index, data['sales'], color=custom_palette[0], label='Sales')

ax.scatter(
    [point_outlier + pd.Timedelta(days=i) for i in range(3)],
    [
        data.loc[point_outlier + pd.Timedelta(days=i), 'sales']
        for i in range(3)
    ],
    color=custom_palette[2], marker='o', s=100, label='Squence outliers',
zorder=5
)
ax.scatter(
    [point_anomaly + pd.Timedelta(days=i) for i in range(3)],
    [data.loc[point_anomaly + pd.Timedelta(days=i), 'sales'] for i in
range(3)],
    color=custom_palette[2], marker='x', s=100,
    label='Sequence anomaly', zorder=5
)

ax.set_xlabel('Date')
ax.set_ylabel('Sales')
ax.legend()
plt.show()
```

On running this code, we get the following output:

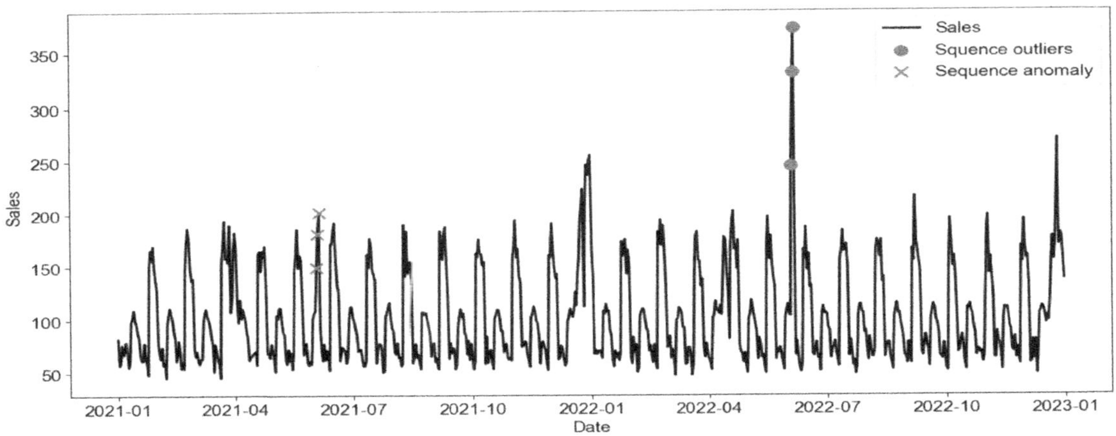

Figure 17.5: Sales with collective anomalies. The circles mark a three-day sequence outlier which is extreme globally; crosses mark a three-day sequence anomaly that sits within normal global bounds but is contextually anomalous

Contextual anomalies, whether point or sequential, do not align with the expected temporal structure of the positional index they are situated in. This type of anomaly often emerges after accounting for long-term dynamics rather than short-term fluctuations, along with structural components such as trend, seasonality, and exogenous variables. Generally, 'context' can be defined by latent representations: rolling statistics, learned autocorrelation structure, regime membership, or latent embeddings of recent history. Fundamentally most approaches in this chapter can be reduced down to variations in how we define temporal context.

Evaluating TSAD systems

Regardless of how a TSAD algorithm works internally, evaluation follows a common framework. The algorithm produces an anomaly score for each time point (or subsequence), we apply a threshold to convert those scores into binary predictions, and we compare the predictions against ground truth labels. *Figure 17.6* illustrates this:

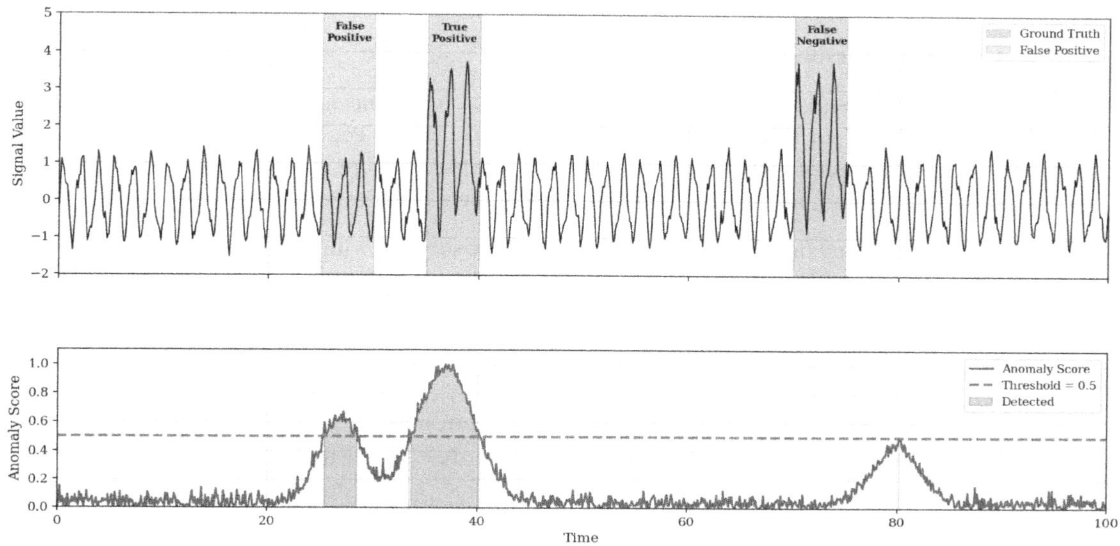

Figure 17.6: Evaluating an anomaly detector

In the figure, the upper panel shows a series with two ground-truth anomaly regions shaded. The lower panel shows a detector's anomaly score over time, with a decision threshold as a dashed line. One region is correctly flagged (TP), one spurious region is flagged where no anomaly exists (FP), and a second true anomaly is missed because the score only rises after the window has closed (FN)

In this example, we observe three types of outcomes:

- **True Positive:** Our detector correctly identifies the first anomaly, with the anomaly score exceeding the threshold during the ground-truth window.
- **False Positive:** The detector flags a region around t=27 where no ground-truth anomaly exists, perhaps responding to noise or a pattern that resembles but isn't actually anomalous.
- **False Negative:** Failure to identify a second anomaly. The anomaly score rises after the ground-truth window, too late to be useful.

There is a problem with this framework that reveals why standard accuracy is meaningless for TSAD. With anomalies comprising less than 1% of observations, a detector that predicts 'normal' for everything achieves >99% accuracy while being **completely useless**. We need metrics that account for the cost asymmetry between false positives (wasted investigation) and false negatives (missed failures), and that handle the temporal nature of anomaly windows. *A comparative summary* will explore appropriate evaluation metrics in detail.

Academic issues

In 2022 Wu and Keogh published *Current Time Series Anomaly Detection Benchmarks are Flawed and are Creating the Illusion of Progress*. Liu and Paparrizos (Liu and Paparrizos, n.d.) have extended this critique. Several points raised are worth understanding for implementation, before investing time developing a TSAD system:

- Many benchmark datasets are trivially solvable with a single threshold rule or a few lines of code. Neural-network papers claiming large F1 improvements were often beating poorly tuned statistical baselines.
- Many contain subjective or incorrect labels, penalising algorithms for finding real anomalies the dataset creators missed.
- Point adjustment, a common evaluation method, artificially inflates F1 scores, making neural networks appear far more effective than they are.

It would seem that simple methods such as rolling thresholds, statistical profiling, isolation forest, and matrix profile often are better choices as they fail loudly and understandably, which is valuable in production

.For readers who want to test methods on harder ground, the UCR Time Series Anomaly Archive provides genuinely challenging benchmarks that address several of the flaws Wu and Keogh identified. For a recent survey on TSAD and evaluation, it's worth reading Paparrizos et al. (2025).

Rule-based approaches

What do we mean by rules-based? We simply mean creating a rule about your data that you can calculate without training a model. For example, let's go back to our data with point anomalies and add some zeros to simulate data drop-offs. This could be out of stock or simply missing data; either way we may need to resolve it.

You can see these below where we define a stockout as anything below `5` (in practice, we'd simply filter for `sales < 1`). Using a simple floor rule `like .filter(['sales'] < 1)` is very effective and fast. You can go the other way with a ceiling value like `.filter(['sales'] > 270)` to catch unusual spikes. Of the thirty one anomalies we have created, the rules here detect twenty nine

```
stockouts_data = data_pointanoms.copy()

# Define stock-out dates
single_day_stockouts = [
    '2021-03-15',
    '2021-07-04',
    '2022-01-12',
    '2022-11-20'
]
```

```
stockout_sequences = [
    pd.date_range(start='2021-05-10', end='2021-05-17'),  # 1 week outage
    pd.date_range(start='2022-08-01', end='2022-08-14'),  # 2 week outage
    pd.date_range(start='2022-12-24', end='2022-12-26')   # Holiday logistics
                                                             failure
]

all_stockout_dates = pd.DatetimeIndex(single_day_stockouts).append(
    [period for period in stockout_sequences]
)

stockouts_data.loc[stockouts_data.index.isin(all_stockout_dates), 'sales'] =
0.0
stockouts_data.loc[stockouts_data['sales'] == 0, 'is_anomaly'] = 1

# Rules-based thresholds
floor_threshold = 5
ceiling_threshold = 270

total_anomalies = stockouts_data['is_anomaly'].sum()
n_above = (stockouts_data['sales'] > ceiling_threshold).sum()
n_below = (stockouts_data['sales'] < floor_threshold).sum()

# Point anomaly dates (from data_pointanoms)
point_outlier_date = pd.Timestamp('2022-10-04')
point_anomaly_date = pd.Timestamp('2021-08-27')

fig, ax = plt.subplots()
ax.plot(stockouts_data.index, stockouts_data['sales'],
           color=custom_palette[0], label='Sales')
ax.axhline(y=floor_threshold, color=custom_palette[2],
           linestyle='--', linewidth=2,
           label=f'Stock-out rule (x < {floor_threshold})')
ax.axhline(y=ceiling_threshold, color=custom_palette[2],
           linestyle='--', linewidth=2,
           label=f'Stock-dump rule (x > {ceiling_threshold})')
ax.scatter(
    point_outlier_date,
    stockouts_data.loc[point_outlier_date, 'sales'],
    color=custom_palette[2], marker='o', s=100, label='Point Outlier', zorder=5
)
```

```
ax.scatter(
    point_anomaly_date,
    stockouts_data.loc[point_anomaly_date, 'sales'],
    color=custom_palette[2], marker='x', s=100, label='Point Anomaly', zorder=5
)
ax.set_xlabel('Date')
ax.set_ylabel('Sales')
ax.set_title(f'Rules based—Detected above: {n_above}, below: {n_below}, out of {total_anomalies}')
ax.legend()
plt.show()
```

The output of this code is:

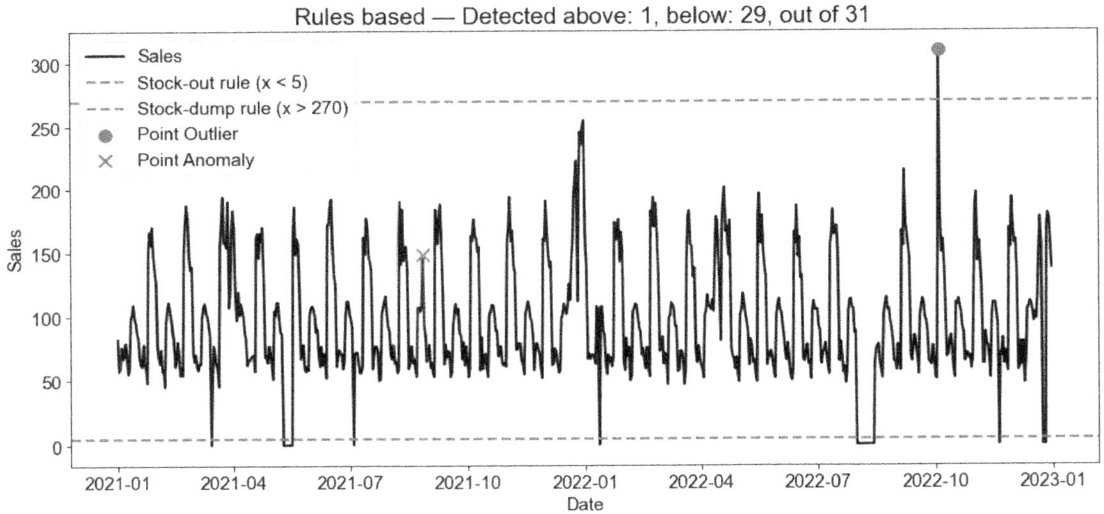

Figure 17.7: Rules-based anomaly detection with floor and ceiling

This is similar to a statistical control process, like an X-chart (or Shewhart control chart). The difference is that rules-based approaches rely on domain knowledge (e.g., 0 sales may indicate a stockout), while X-charts typically use statistical calculations such as $\bar{x} \pm 3\sigma$ ($\mu \pm 3$ std) from a baseline (mean here). Because we are using the whole data set to define μ and σ, this amounts to a simple static outlier detector that catches only 4 values, only one of which is a true outlier, and it fails to pick up on the 0 values.

```
mean = stockouts_data['sales'].mean()
std = stockouts_data['sales'].std()

upper_limit = mean + 3 * std
lower_limit = mean - 3 * std
```

```
total_anomalies = int(stockouts_data['is_anomaly'].sum())
n_above = (stockouts_data['sales'] > upper_limit).sum()
n_below = (stockouts_data['sales'] < lower_limit).sum()
detected_above = stockouts_data[stockouts_data['sales'] > upper_limit]
detected_below = stockouts_data[stockouts_data['sales'] < lower_limit]

fig, ax = plt.subplots()
ax.plot(stockouts_data.index, stockouts_data['sales'], color='black',
linewidth=2, label='Sales')

if not detected_above.empty:
    ax.scatter(detected_above.index, detected_above['sales'],
               color=custom_palette[2], marker='o', s=100,
               edgecolor='black', linewidth=1, zorder=5,
               label=f'Detected Above ({len(detected_above)})')

if not detected_below.empty:
    ax.scatter(detected_below.index, detected_below['sales'],
               color=custom_palette[3], marker='x', s=100,
               linewidth=2, zorder=5,
               label=f'Detected Below ({len(detected_below)})')
ax.scatter(
    point_anomaly_date,
    stockouts_data.loc[point_anomaly_date, 'sales'],
    color=custom_palette[2], marker='x', s=100, label='Point Anomaly', zorder=5
)

ax.axhline(y=mean, color=custom_palette[5], linestyle=':', linewidth=4,
           label=f'Mean (μ = {mean:.2f})')
ax.axhline(y=upper_limit, color=custom_palette[2],
           linestyle='--', linewidth=2,
           label=f'Upper (μ + 3σ = {upper_limit:.0f})')
ax.axhline(y=lower_limit, color=custom_palette[2],
           linestyle='--', linewidth=2,
           label=f'Lower (μ - 3σ = {lower_limit:.0f})')
ax.set_xlabel('Date')
ax.set_ylabel('Sales')
ax.set_title(f'X-Chart (μ ± 3σ)—Detected above: {n_above}, below: {n_below},
out of {total_anomalies}')
ax.legend()
plt.show()
```

The output is:

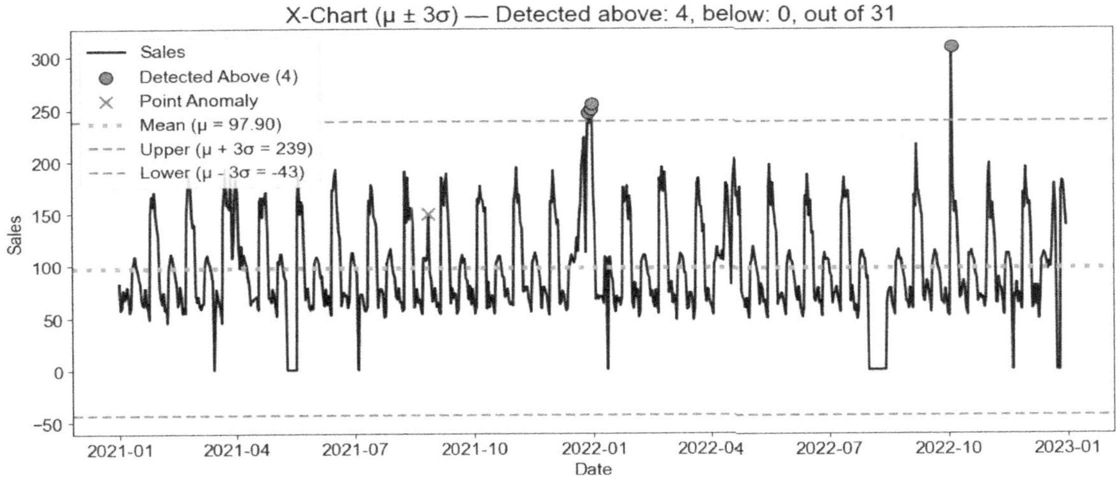

Figure 17.8: X-Chart static threshold detection

We can adjust the σ value for the lower or upper boundary with domain knowledge, to produce results similar to our rules-based approach. While the statistical approach can be useful when we don't have enough knowledge of a process, we should probably consider hybridising it with rules-based approaches.

To partially account for seasonal spikes or drops, we can simply add conditional rules, i.e. ignore values (direct values or statistical) above X on certain dates (bank holidays, Christmas, etc.), or have different boundaries based on promotion depth. As long as we have event calendars and good domain understanding, this can work. It's often faster and more effective than modeling, even at scale.

There are obvious limitations for both approaches, the most severe being:

- Direct rules may not scale well across multiple series with different structures
- As the number of rules increases they may become more complex, and even start to conflict
- A value of 155 may look perfectly normal within our rule or statistical bounds, but if the day it fell on had a large promotion, contextually we'd expect sales around 180
- Time series are dynamic and can change significantly relative to information we can control

Adaptive thresholds

Static thresholds assume our data's structure does not change (much) over time, which is fine in some settings. But what if we have trends, seasonalities, and other exogenous variables influencing our data (as we do in our example)? In addition, the process might gradually drift or have slowly evolving variance. Instead of manually updating our thresholds, we can apply rolling windows.

It is similar to X-chart processing: μ is still computed with some defined σ bounds, but we add a window of n observations which moves across a series.

Now is a good time to fire up a notebook and test this on your own data. In this example, a 21-day rolling window is used, with a lower bound of `1.5 std` and an upper bound of `3 std`. A shift is also added to avoid look-ahead bias.

```
# Adaptive thresholds: Rolling μ ± 3σ
window = 21  # 3 weeks

# Calculate rolling statistics (shift to avoid lookahead bias)
stockouts_data['rolling_mu'] = stockouts_data['sales'].rolling(window).mean().
shift(1)
stockouts_data['rolling_sigma'] = stockouts_data['sales'].rolling(window).
std().shift(1)

# Dynamic thresholds
stockouts_data['upper_limit'] = stockouts_data['rolling_mu'] + 3 * stockouts_
data['rolling_sigma']
stockouts_data['lower_limit'] = stockouts_data['rolling_mu'] - 1.5 * stockouts_
data['rolling_sigma']

# Detect anomalies
stockouts_data['detected'] = (
    (stockouts_data['sales'] > stockouts_data['upper_limit']) |
    (stockouts_data['sales'] < stockouts_data['lower_limit'])
)

# Check which detections match actual anomalies
true_positives = (stockouts_data['detected'] &
                  (stockouts_data['is_anomaly'] == 1)).sum()
false_positives = (stockouts_data['detected'] &
                   (stockouts_data['is_anomaly'] == 0)).sum()
false_negatives = (~stockouts_data['detected'] &
                   (stockouts_data['is_anomaly'] == 1)).sum()

precision = true_positives / (true_positives + false_positives)
recall = true_positives / (true_positives + false_negatives)

# Which dates were detected vs missed?
detected_dates = stockouts_data[stockouts_data['detected']].index
known_anomaly_dates = stockouts_data[stockouts_data['is_anomaly'] == 1].index

correctly_detected = detected_dates.intersection(known_anomaly_dates)
missed = known_anomaly_dates.difference(detected_dates)
false_alarms = detected_dates.difference(known_anomaly_dates)
```

The common approaches are:

- Simple moving average, giving equal weight to all observations in a window
- Exponentially Weighted Moving Average (EWMA), where recent observations are weighted more heavily

We have just used simple MAs here. An EWMA can do slightly better if you want to try it out.

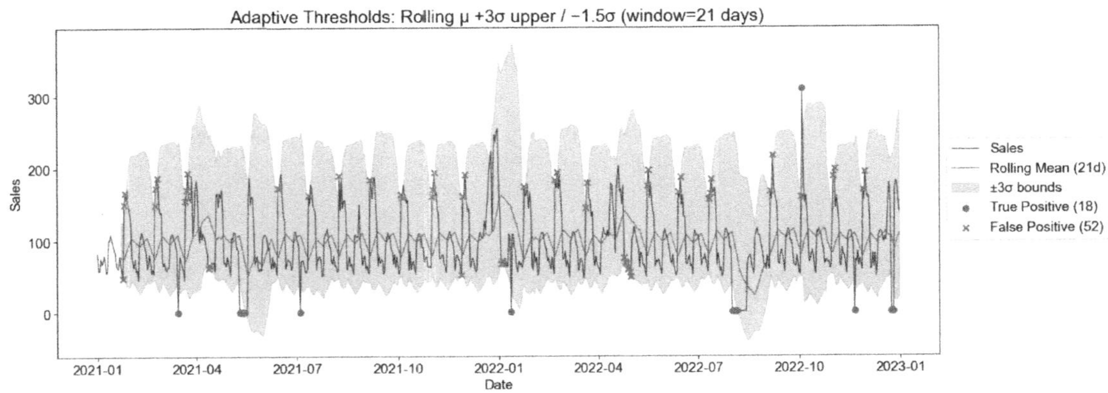

Figure 17.9: Adaptive threshold rolling mean with bounds

Despite our discussion earlier about single-point anomalies like 0 values being just one of the patterns we want to detect, from here on we'll print out precision and recall throughout this section. As you'll see, they give a good indication of performance.

You can see in our chart that this simple approach performs poorly, as bounds are shifted away from seasonal peaks, resulting in 52 false positives (mostly seasonal peaks when the sale price is lowest). Increasing window size (i.e., **MA = 28**) smooths the threshold, reducing false positives to 2 and increasing true detection from 18 to 19. Of course, we might as well say 2 genuine detections, as most anomalies are simply easy-to-filter 0 values.

Approaches like this work well for stable processes with gradual drift, which makes them popular in manufacturing and infrastructure monitoring. But our 'adaptive' threshold is operating on all y values varying by MA only, which is reasonable for sensors but not so much for consumer demand. When we have strong exogenous effects like events or price, we need more information in our system. So why not hybridize our approach with some rules and conditions based on the main exogenous variable (price):

- Non-promotional pricing
- Shallow discount pricing
- Deep discount pricing

We apply the thresholds based on these conditions with the standard $\sigma = 3$, and drop the rolling windows here as they're clearly not suitable.

If we just applied this we would detect 25 of the 30 outliers, which isn't awful but still misses easy-to-detect data. We could add an additional rule when we are confident that values shouldn't fall below some proportion of an average: $x_t \geq \alpha\mu$, meaning any value (x_t) that is below or equal to a proportion (α) of some average (μ). We can be more specific by adding a conditional r ($x_t \geq \alpha\mu_r$), i.e. a mean based on *regular* values (lower), to capture our lower bound.

The following code snippet creates these conditions and captures the accuracy of outliers detected relative to known anomalies. We don't use the lower bound rule based on sigma here to make the thresholds easier to see, but typically you'd want that too.

```
alpha = 0.4 # 40% of mean (regular sales condition)

def get_regime(row):
    if row['price'] == 5.0:
        return 'regular'
    elif row['price'] == 3.5:
        return 'promo_low'
    else: # price == 2.5
        return 'promo_high'

stockouts_data['regime'] = stockouts_data.apply(get_regime, axis=1)

# σ upper bound per regime
regime_stats = stockouts_data.groupby('regime')['sales'].agg(['mean', 'std'])
regime_stats['upper'] = regime_stats['mean'] + 3 * regime_stats['std']
stockouts_data['upper'] = stockouts_data['regime'].map(regime_stats['upper'])

# Regular-anchored proportional floor
mu_regular = stockouts_data.loc[stockouts_data['regime'] == 'regular',
'sales'].mean()
stockouts_data['prop_lower'] = alpha * mu_regular

# Bounds using proportional floor and std based upper bound based on regime
stockouts_data['detected'] = (
    (stockouts_data['sales'] < stockouts_data['prop_lower']) |
    (stockouts_data['sales'] > stockouts_data['upper'])
)


# Confusion vs ground-truth is_anomaly (expects 0/1)
is_true = stockouts_data['is_anomaly'].fillna(0).astype(int) == 1
is_det  = stockouts_data['detected'].astype(bool)

tp = (is_det & is_true).sum()
```

```
fp = (is_det & ~is_true).sum()
fn = (~is_det & is_true).sum()

precision = tp / (tp + fp) if (tp + fp) else 0.0
recall    = tp / (tp + fn) if (tp + fn) else 0.0

# values for inspection
detected_dates = stockouts_data.index[is_det]
known_anomaly_dates = stockouts_data.index[is_true]
correctly_detected = detected_dates.intersection(known_anomaly_dates)
missed = known_anomaly_dates.difference(detected_dates)
false_alarms = detected_dates.difference(known_anomaly_dates)
```

As you can see we do fairly well, getting all the outliers (**recall** = **1**), but we have 11 false positives (Christmas demand). Flagging and removing or imputing important data is a no-no if you want to keep your job!

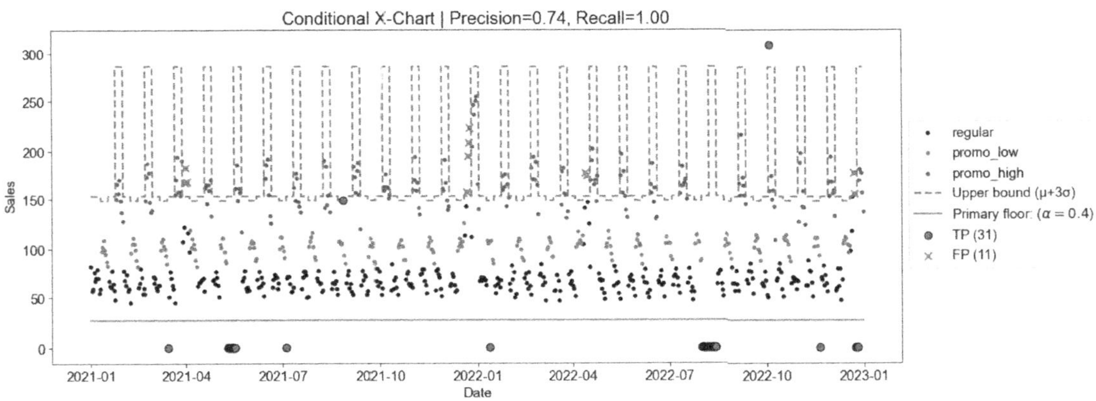

Figure 17.10: Conditional X-chart detection results

Obviously simple rules stacked with statistical processes can be extremely useful. But when data has strong trends, *y* drift, multiple seasonalities, and/or exogenous drivers like price, maintaining these approaches can be difficult. An alternative is to profile our data by building a model of expected behavior and examining residuals.

Statistical profiling

The conditional X-chart we built is partway to a model. By specifying thresholds relative to exogenous variables, we are saying *normal* observations are context dependent. Obviously stacking conditions has a ceiling; having price, promotions, day-of-week effects, and seasonality all interacting, the number of conditions multiplies and the boundaries between them become arbitrary. At some point you're no longer writing rules, but approximating a model, badly.

A cleaner solution is to explicitly build a model. Statistical profiling formalizes this: fit a model (your choice) of expected behavior, and treat its residuals as the detection signal. An anomaly is not a value that exceeds some global threshold; it is a value that is unexpectedly large or small given everything the model knows. Formally, at each time step (t) we compute:

$$e_t = y_t - \hat{y}_t$$

where $\hat{y}_t$ is a model's predicted value given available context: seasonality, trend, price, promotional depth. The residual e_t is the portion a model cannot explain. Under normal conditions, residuals behave like noise, roughly centred at zero with stable variance. When something anomalous happens, we see spikes in residual information. In theory we are applying anomaly detection on a potentially *cleaner* signal, compared to a raw series. As we mentioned before, an anomaly is an unlikely observation given what we know; $p(y_t \mid H_{t-1}, x_t)$ is small. Model profiling can make this more explicit, with a model handling conditioning, leaving us with the remainder to inspect.

Contamination problem

Building a profile model on data that contains anomalies can be circular, as the model absorbs them into its parameters, impacting predictions. As a result, residuals for anomalous observations are reduced.

That's the contamination problem in the abstract. Let's be concrete about how it manifests in real pipelines, because the failure mode is subtle and the consequences can be severe.

The most common pattern we see is labelling systems that use rules or profiling models in production. Rules are written at system launch and work well for six months, then the underlying process drifts (a supplier changes baseline stock levels, a pricing strategy shifts) and the rules no longer function accurately. They fire on 'normal' observations and miss real ones. Someone updates the rules, but the labeled data in the training set still reflects the old definitions. Later, supervised models train on corrupted labels and learn the wrong decision boundary. Any additional labels are stored but the previous anomalies continue to propagate into the distribution used by the profiling method going forward.

Another typical failure is when a profiling model is retrained monthly on, say, the previous 90 days of data. Three months ago there was a week-long stock disruption which was not labeled. The profiling model now incorporates a week of (near-)zero sales as being within normal operating range. As a result, the next disruption produces modest residuals.

Detection and labelling infrastructure are tightly coupled, and both require maintenance. Neither is a one-time engineering task.

Building profiles

Our example data has clear structure, with strong pricing effects at three levels, day-of-week seasonality, and annual patterns. We will use `statsforecast` with AutoARIMA as our profile model; if you're coding along, feel free to experiment with other algorithms. Our AutoARIMA setup is a SARIMAX with exogenous regressors capturing feature context (i.e., price, holidays) along with autocorrelation structure and trend.

We fit on all data—including anomalies—which as previously stated can cause the model to incorporate anomalies as part of the model, causing its residuals for anomalous periods to shrink toward zero, masking the anomaly.

```
## Contaminated profile
from statsforecast import StatsForecast
from statsforecast.models import AutoARIMA

# Prepare statsforecast long format
sf_data = stockouts_data[['day_of_week', 'week_num', 'promotion', 'price',
'sales', 'is_holiday', 'is_anomaly']].copy()
sf_data = sf_data.reset_index().rename(columns={'date': 'ds', 'sales': 'y'})
sf_data['unique_id'] = 'product_1'

# Exogenous columns
exog_cols = ['day_of_week', 'week_num', 'promotion', 'price', 'is_holiday']

# Training df: unique_id + ds + y + exogenous columns together
train_cols = ['unique_id', 'ds', 'y'] + exog_cols

# Future exogenous—one row for the throwaway h=1 forecast
last_row = sf_data.iloc[-1]
last_date = sf_data['ds'].max()
X_future = pd.DataFrame({
    'unique_id': ['product_1'],
    'ds': [last_date + pd.Timedelta(days=1)],
    'day_of_week': [(last_date + pd.Timedelta(days=1)).dayofweek],
    'week_num': [last_row['week_num']],
    'promotion': [last_row['promotion']],
    'price': [last_row['price']],
    'is_holiday': [False]
})

sf_contaminated = StatsForecast(
    models=[AutoARIMA()],
    freq='D',
    n_jobs=1
)

# Exogenous go IN df, X_df is future-only
sf_contaminated.forecast(
    df=sf_data[train_cols],
```

```
    X_df=X_future,
    h=1,
    fitted=True
)

fitted_c = sf_contaminated.forecast_fitted_values()

# Reset index if fitted_c has unique_id/ds as index
fitted_c = fitted_c.reset_index()

sf_data = sf_data.merge(
    fitted_c[['unique_id', 'ds', 'AutoARIMA']],
    on=['unique_id', 'ds'],
    how='left'
)

stockouts_data['y_hat_contaminated'] = sf_data['AutoARIMA'].values
stockouts_data['resid_contaminated'] = stockouts_data['sales'] - stockouts_
data['y_hat_contaminated']
# Z-score detection on contaminated residuals
mu_c  = stockouts_data['resid_contaminated'].mean()
sig_c = stockouts_data['resid_contaminated'].std()

stockouts_data['z_contaminated'] = (
    (stockouts_data['resid_contaminated'] - mu_c) / sig_c
)

threshold = 3.0
stockouts_data['detected_contaminated'] = stockouts_data['z_contaminated'].
abs() > threshold

# Evaluate against ground truth
is_true = stockouts_data['is_anomaly'].fillna(0).astype(int) == 1
is_det  = stockouts_data['detected_contaminated'].astype(bool)

tp_c = (is_det &  is_true).sum()
fp_c = (is_det & ~is_true).sum()
fn_c = (~is_det & is_true).sum()

precision_c = tp_c / (tp_c + fp_c) if (tp_c + fp_c) else 0.0
recall_c    = tp_c / (tp_c + fn_c) if (tp_c + fn_c) else 0.0
```

```
print(f"Contaminated-Precision: {precision_c:.2f}  Recall: {recall_c:.2f}")
print(f"  TP: {tp_c}  FP: {fp_c}  FN: {fn_c}")

# Output:
Contaminated-Precision: 0.69  Recall: 0.35
  TP: 11  FP: 5  FN: 20
```

The following plot shows the contaminated profile model with Z-score detection:

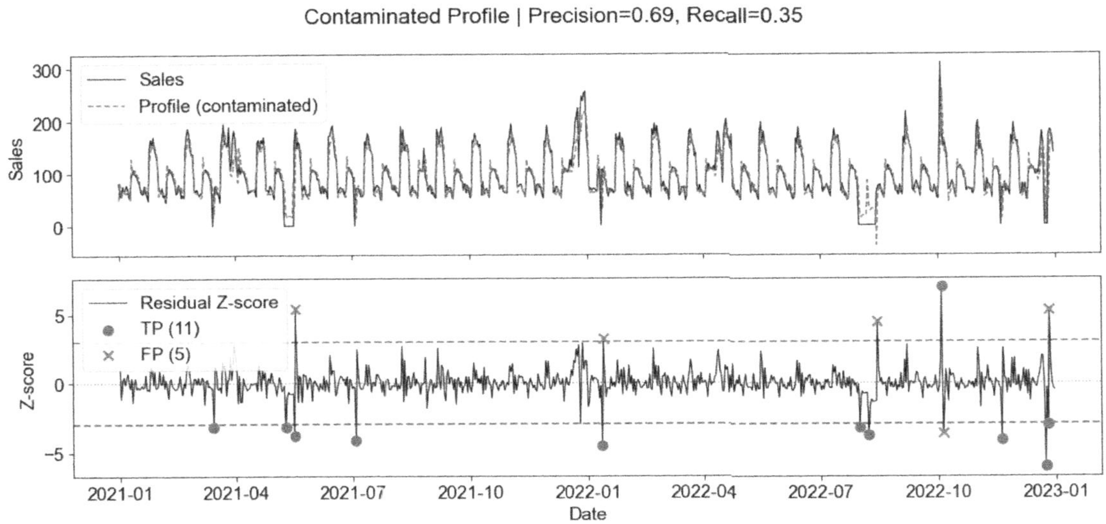

Figure 17.11: Contaminated profile model with Z-score detection

This does not do well compared to our rules + statistical process approach. It does get most zeros and the global point outlier, but incorporates several false positives, giving us mediocre precision and poor recall.

Residuals as the detection signal

With fitted values in hand, computing the anomaly score is straightforward. We standardize residuals using statistics from the clean training period—the model's normal operating envelope—and flag observations where the Z-score exceeds a threshold.

$$z_t = \frac{e_t - \mu_e}{\sigma_e}$$

We are going to use these residuals later so we will update the naming conventions.

```
# rename cols
stockouts_data['y_hat'] = stockouts_data.pop('y_hat_contaminated')
stockouts_data['residual'] = stockouts_data.pop('resid_contaminated')
stockouts_data['residual_z'] = stockouts_data.pop('z_contaminated')
```

```
stockouts_data['detected_profile'] = stockouts_data.pop('detected_
contaminated')
```

The residual panel is noisy and does not immediately show us what was invisible in our raw series; we can see the model is not particularly overfit either. But once price regime and seasonal structure are accounted for, we can see some spikes around the stockouts. The Christmas peaks that plagued our conditional X-chart largely disappear, as the model expected high sales on those dates.

The One-Class gaussian

Z-score profiling may feel familiar; if it does, it's because it is a one-class Gaussian method, which is the simplest member of the one-class classifier family. We fit a model of normality from 'normal' data and flag observations where the likelihood of the residual under that model falls below a threshold. In the univariate case:

$$f_{\mu_e,\sigma_e}(e_t) = \frac{1}{\sigma_e\sqrt{2\pi}}\exp\left(-\frac{1}{2}\left(\frac{e_t-\mu_e}{\sigma_e}\right)^2\right)$$

Flagging at $|z| > 3$ is equivalent to flagging observations whose likelihood falls below some threshold τ. Adjusting τ trades precision against recall—the same tension as tuning σ in the X-chart, but on a signal that could be more informative.

Z-scoring residuals, one-class Gaussian, and the X-chart applied to model residuals are essentially the same operation written in three different vocabularies. Understanding this allows us to consider the nature of our generalizations. If residuals are non-Gaussian, i.e. heavy-tailed during deep promotions, we can swap distributional assumptions. If we have multivariate residuals (volume, transaction size, returns simultaneously), we can generalize to a Multivariate Normal with Mahalanobis distance:

$$d_M(\mathbf{e}_t) = \sqrt{(\mathbf{e}_t - \boldsymbol{\mu})^\top \boldsymbol{\Sigma}^{-1}(\mathbf{e}_t - \boldsymbol{\mu})}$$

Because $\boldsymbol{\Sigma}$ encodes the covariance of normal residuals, our detector should pick up on which joint deviations are expected or not, which we may consider an improvement over thresholding each residual independently.

Isolation forest

One-class Gaussian processes assume that *residuals follow a Gaussian distribution*. When this holds true, it works fairly well. When it doesn't, we get skewed residuals, multimodal normal behavior, and non-linear interactions between features. In many cases you may check your residuals and find you need a more flexible approach. **Isolation Forest** is often a practical solution.

Standard decision trees split data to maximize class separation by exhaustively searching features and thresholds that best separate known categories, guided by information gain or Gini impurity. Isolation Forests do the opposite: they split data by randomly choosing a random feature and random value for splitting (cutting) within the feature's observed range. Then they measure how many splits (cuts) it takes to isolate each observation into its own partition.

Anomalies are *rare and structurally different from 'normal' observations*, so they tend to occupy sparse regions of feature space, *away* from a denser cluster of 'typical' points. An isolated point in a sparse region needs very few random cuts to separate it from everything else. 'Normal' observations cluster together in dense regions; each random cut peels off a small group of similar neighbours, requiring many cuts to separate any one observation from the cluster. Anomaly scores for an isolation forest are the inverse of the average isolation depth across many trees.

$$s(x,n) = 2^{-\frac{E[h(x)]}{c(n)}}$$

where $h(x)$ is the path length (number of splits to isolate x) and $c(n)$ is the average path length of an unsuccessful search in a Binary Search Tree with n observations, used as a normalisation factor. Points consistently isolated in few cuts produce $s \approx 1$ (highly anomalous). Points requiring many cuts score $s \approx 0$ (deeply normal). Scores around 0.5 indicate uncertainty.

	Standard Decision Tree	**Isolation Tree**
Goal	Separate classes	Isolate individual points
Split criterion	Gini/Information gain	Random feature, random threshold
Requires labels?	Yes	No
What depth tells you	Feature importance	How easy a point is to isolate
Short path means	Simple rule found	Point is anomalous

Table 17.1: Standard decision vs isolation tree

In its base form the Isolation Forest is unsupervised. But it can be used in a **semi-supervised** mode by training exclusively on clean data and scoring a full series. When labels are available, the anomaly score becomes a valuable feature for a **supervised classifier**, which we will demonstrate shortly. This provides a strong incentive for tracking and saving anomalies rather than running detection from scratch.

We'll use the pyod (Python Outlier Detection) library which provides several anomaly detection algorithms behind a consistent API, making it straightforward to test methods and compare results without rewriting your pipeline.

```
# pip install pyod
from pyod.models.iforest import IForest
```

Isolation forest – unsupervised

When you have no labels at all, what we refer to as a cold-start (e.g. a new product, a freshly deployed sensor), you have no clean subset to train on. In that case you may pass a full series directly to your IF model and accept that the contamination parameter is doing some heavy lifting, in that it's guessing how often anomalies occur (i.e. 0.05 = 5% of your observations).

Important parameters for IForest HP tuning are: '`n_estimators`', '`max_samples`', '`max_features`', '`contamination`', although without labels, tuning is something of a guess.

```
from pyod.models.iforest import IForest

X_if_v0 = stockouts_data[['day_of_week', 'week_num', 'promotion', 'price',
'sales', 'is_holiday']]

iso_full = IForest(
    n_estimators=500,
    max_samples=0.5,
    max_features=0.5,
    contamination=0.08,
    random_state=456
)
iso_full.fit(X_if_v0)

stockouts_data['if_unsup_score'] = iso_full.decision_function(X_if_v0)
stockouts_data['if_unsup_label'] = iso_full.predict(X_if_v0)
stockouts_data['detected_if_unsup'] = stockouts_data['if_unsup_label'] == 1

is_det_iu = stockouts_data['detected_if_unsup'].astype(bool)
tp_iu = (is_det_iu &  is_true).sum()
fp_iu = (is_det_iu & ~is_true).sum()
fn_iu = (~is_det_iu & is_true).sum()

prec_iu = tp_iu / (tp_iu + fp_iu) if (tp_iu + fp_iu) else 0.0
rec_iu  = tp_iu / (tp_iu + fn_iu) if (tp_iu + fn_iu) else 0.0
print(f"IF (fully unsupervised)—Precision: {prec_iu:.2f}  Recall: {rec_
iu:.2f}")

# Output:
IF (fully unsupervised)—Precision: 0.51  Recall: 0.97
```

The plot:

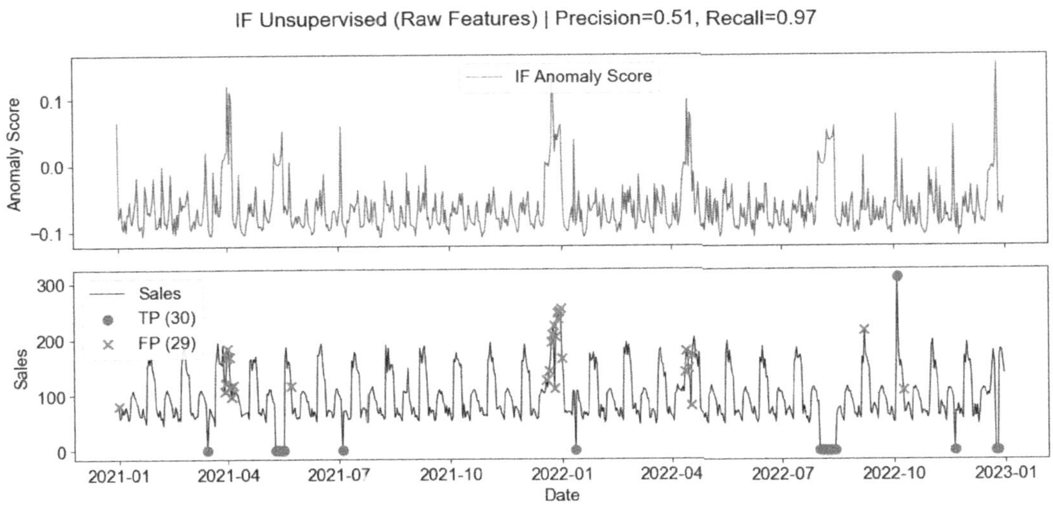

Figure 17.12: Isolation Forest unsupervised detection

Recall of 0.97 tells us there is signal in the data. The false positives cluster around promotional transitions and vacation peaks where the feature space is genuinely sparse but not anomalous. The model has no concept of expected promotional uplift because it has no profile model. Can the profiling residuals give IF a cleaner input signal?

Standard isolation forest on residuals

We apply the same method to the residual feature space built during profiling. If the profile model has done its job, the residuals should be a cleaner signal—anomalies without the promotional noise.

```
X_if_v1 = stockouts_data[['residual', 'residual_z', 'price']].fillna(0)

# Best params from tuning above
iso_resid = IForest(
    n_estimators=500,
    max_samples=0.8,
    max_features=0.5,
    contamination=0.08,
    random_state=456
)
iso_resid.fit(X_if_v1)

stockouts_data['if_score'] = iso_resid.decision_function(X_if_v1)
stockouts_data['if_label'] = iso_resid.predict(X_if_v1)
stockouts_data['detected_if'] = stockouts_data['if_label'] == 1
```

```
is_det_if = stockouts_data['detected_if'].astype(bool)
tp_if = (is_det_if &  is_true).sum()
fp_if = (is_det_if & ~is_true).sum()
fn_if = (~is_det_if & is_true).sum()

prec_if = tp_if / (tp_if + fp_if) if (tp_if + fp_if) else 0.0
rec_if  = tp_if / (tp_if + fn_if) if (tp_if + fn_if) else 0.0
print(f"IF on residuals—Precision: {prec_if:.2f}  Recall: {rec_if:.2f}")

# Output:
IF on residuals—Precision: 0.25  Recall: 0.48
```

The resultant plot is:

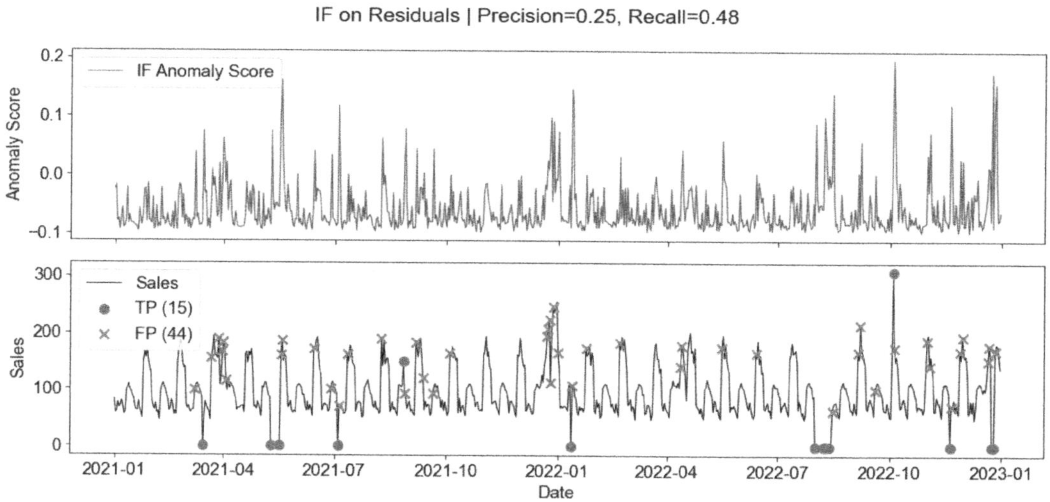

Figure 17.13: Isolation Forest on residuals

Recall collapses to **0.48**. Why? Our AutoARIMA profile model accounted for a lot of variance already and may have incorporated some contextual outliers into our model, shrinking residual signal that the IF needs to detect anomalies. Our unsupervised IF, on raw y, sidestepped this entirely. Standard IF's axis-aligned splits may also struggle when anomalies are defined by combinations of residual magnitude and price regime.

Extended isolation forest

Standard IF splits/cuts are always perpendicular to a single feature axis, which can cause issues in detecting anomalies that are unusual across combined feature dimensions rather than singular. As a result, anomalies on the edge of individual feature distributions can be artificially high, as an artifact of rectilinear partitioning geometry.

Extended Isolation Forest (**EIF**) resolves this by replacing perpendicular cuts with random hyperplane splits. Allowing decision boundaries to be at any angle through feature space, this produces more uniform coverage, handling multidimensional anomalies more reliably. pyod provides EIF via its **Isolation-based Nearest Neighbor Ensemble** (**INNE**) module, which applies a related isolation principle without the axis-aligned constraint.

```
from pyod.models.inne import INNE

X_if_v0 = stockouts_data[['day_of_week', 'week_num', 'promotion', 'price',
'sales', 'is_holiday']].copy()

inne_model = INNE(
    n_estimators=500,
    max_samples=0.5,
    contamination=0.08,
    random_state=456
)
inne_model.fit(X_if_v0)

stockouts_data['eif_score'] = inne_model.decision_function(X_if_v0)
stockouts_data['eif_label'] = inne_model.predict(X_if_v0)
stockouts_data['detected_eif'] = stockouts_data['eif_label'] == 1

is_det_eif = stockouts_data['detected_eif'].astype(bool)
tp_eif = (is_det_eif &  is_true).sum()
fp_eif = (is_det_eif & ~is_true).sum()
fn_eif = (~is_det_eif & is_true).sum()

prec_eif = tp_eif / (tp_eif + fp_eif) if (tp_eif + fp_eif) else 0.0
rec_eif  = tp_eif / (tp_eif + fn_eif) if (tp_eif + fn_eif) else 0.0
print(f"INNE (Extended IF)—Precision: {prec_eif:.2f}  Recall: {rec_eif:.2f}")

# Output:
INNE (Extended IF)—Precision: 0.05  Recall: 0.10
```

The difference between standard and extended isolation methods matters most in higher-dimensional feature spaces. In our three-feature example, the gap will be modest. Where rotational splits earn their keep is when you feed in a full window of lagged residuals, a catch22 feature vector, or any space where anomalies are defined by subtle multivariate combinations rather than single-feature extremes. If you find standard Isolation Forest producing high anomaly scores near the boundaries of your training distribution regardless of where your known anomalies sit, INNE or a dedicated EIF implementation is the natural upgrade.

INNE's rotational flexibility is unnecessary here. Our feature space is six-dimensional. In higher-dimensional spaces (like lagged residual windows), the axis-alignment constraint of standard IF becomes a real limitation and INNE earns its keep (Bouman, Bukhsh, and Heskes, n.d.). On six features, it does not.

Local outlier factor

Statistical profiling and isolation forest both operate globally: they model what 'normal' looks like across the full dataset and score deviations from that global standard. But not all anomalies are globally extreme. A point may be surrounded by neighbors that are similarly deviant but would not flag under a global method because the neighborhood itself has shifted.

Local Outlier Factor (LOF) attempts to address this by estimating local density via k nearest neighbours and comparing each point's density to its neighbours' densities. Formally, the LOF score for observation x is:

$$\mathrm{LOF}_k(x) = \frac{1}{|N_k(x)|} \sum_{o \in N_k(x)} \frac{\mathrm{lrd}_k(o)}{\mathrm{lrd}_k(x)}$$

Where $N_k(x)$ is a set of k nearest neighbours and $\mathrm{lrd}_k(x)$ is a local reachability density, which is roughly inverse to the average reachability distance to those neighbours. A LOF score near 1 means a point's density matches its neighbours (*normal*). Scores significantly above 1 indicate a point is in a sparser region than its surroundings (*locally anomalous*).

In a residual feature space, LOF detects observations whose residual pattern is unusual relative to its immediate neighborhood rather than the full history. This is useful when data has multiple normal regimes with different densities, a situation where global methods may over-flag the sparse group and under-flag anomalies within dense groups.

```
from pyod.models.lof import LOF

lof_model = LOF(
    n_neighbors=50,
    contamination=0.05,
    algorithm='auto'
)
lof_model.fit(X_if_v0)

stockouts_data['lof_score'] = lof_model.decision_function(X_if_v0)
stockouts_data['lof_label'] = lof_model.predict(X_if_v0)
stockouts_data['detected_lof'] = stockouts_data['lof_label'] == 1

is_det_lof = stockouts_data['detected_lof'].astype(bool)
tp_lof = (is_det_lof &  is_true).sum()
fp_lof = (is_det_lof & ~is_true).sum()
fn_lof = (~is_det_lof & is_true).sum()
```

```
prec_lof = tp_lof / (tp_lof + fp_lof) if (tp_lof + fp_lof) else 0.0
rec_lof  = tp_lof / (tp_lof + fn_lof) if (tp_lof + fn_lof) else 0.0
print(f"LOF-Precision: {prec_lof:.2f}  Recall: {rec_lof:.2f}")

# Output:
LOF-Precision: 0.83  Recall: 0.97
```

The `n_neighbors` parameter is the main parameter you'd want to tune. If it's too small (e.g. 20) LOF becomes overly sensitive to noise; too large and it reverts to global behavior. Iterating through values to work out a good fit can be useful here; if you have reliable labeled anomalies in your data set already you can tune `n_neighbors` with data splits. We found 50 to work extremely well with this data set.

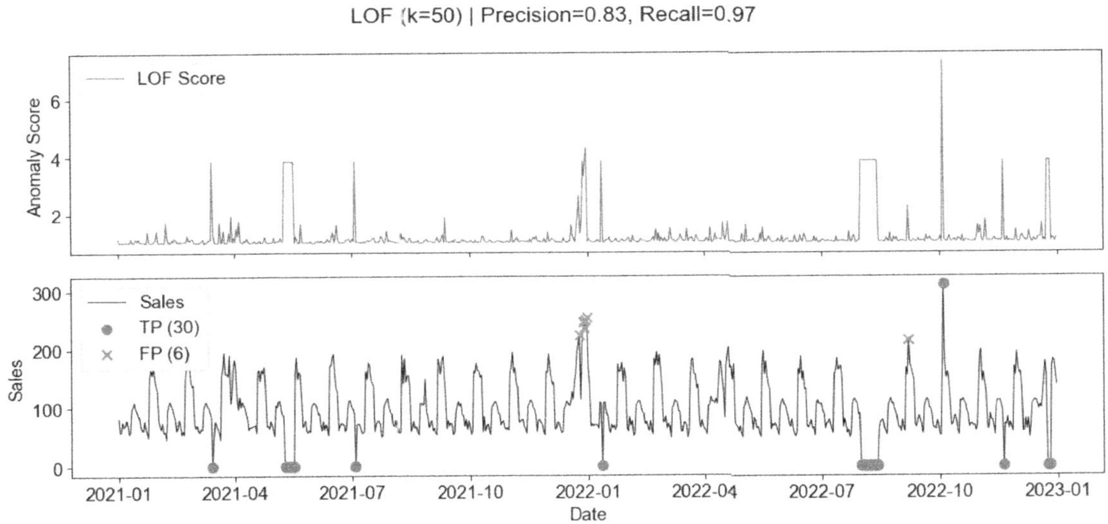

Figure 17.14: Local Outlier Factor detection

With most of the unsupervised methods run we can compare performance with precision, recall and F1 scores in a single table.

Method	Features	Precision	Recall	F1
Z-score Profile (contaminated)	resid_contaminated	0.74	1.00	0.85
IF—Raw Features (unsupervised)	sales + context (no model)	0.51	0.97	0.67
IF—Residuals	residual, residual_z, price	0.25	0.48	0.33
INNE (Extended IF)—Residuals	residual, residual_z, price	0.05	0.10	0.07
LOF—Residuals	residual, residual_z, price	0.83	0.97	0.90

Tabe 17.2: Comparing performances

LOF on raw features with a wider neighborhood achieves the best unsupervised F1 (0.90), with both high precision and recall. The Z-score profiler on contaminated residuals also performs well (0.85), though its perfect recall reflects the manually set threshold. The fully unsupervised IF trades precision for recall, if you're not squeamish about imputing perfectly fine data it's not a bad way to go. Residual-based methods underperformed, highlighting the issue with contaminated profiling. The unsupervised scores remain valuable not as final decisions but as features for a supervised model that can calibrate them against confirmed labels.

Matrix profile

All the methods discussed so far share a common architecture:.

- Build a model or distribution of normal behavior.
- Score new observations against it.
- Flag deviations.

Under this, our *model* can be rules-based, statistical process-based, density-based, or tree-based. Fundamentally, all these approaches try to find observations that are unusual relative to normality.

Matrix profiling is a bit different, in that it looks at how similar each subsequence of a series is to every other subsequence. A subsequence with no similar counterpart in a series is a *discord*; a pattern that does not repeat within a series. Discords are anomalies under this framing. Conversely, subsequences that appear repeatedly are *motifs*, the *normal* patterns of a series.

Matrix Profiling can work well for collective anomalies spanning multiple time steps that are anomalous as a sequence, not at any individual point. For our univariate example series this means catching shapes that point-based methods miss, i.e. the two three-day spikes we added (contextual and global). In these cases no single observation, especially for the contextual sequence outliers, is extreme. The anomaly is defined by shape, not the magnitude.

Subsequences vs Shapelets

In Chapter 14 on TS classification we encountered *shapelets*. These are subsequences that have been *selected* for their discriminative power because their presence or absence best separates classes.

A Matrix Profile operates on *subsequences*: every contiguous window of m points extracted by sliding across the series. All subsequences are scored equally; there is no selection or learning.

For a time series T of length n and a chosen subsequence window length m, we compute a Matrix Profile P—a vector of length $n - m + 1$ where each entry $P[i]$ stores a z-normalised Euclidean distance between subsequence $T_{i:i+m}$ and its nearest neighbor in T, excluding trivial self-matches.

$$P[i] = \min_{j:|i-j|\geq m} d_z\left(T_{i:i+m}, T_{j:j+m}\right)$$

High values of $P[i]$ mean subsequence i has no close match anywhere in our series—it is isolated, anomalous. A companion Matrix Profile Index $I[i] = \text{argmin}_j d_z(\cdot)$ stores which subsequence is the nearest neighbor, pointing from each window to its best match. Motifs are pairs with low profile values (close matches); discords are indices with high profile values (no close match).

We can use stumpy's implementation of the STOMP algorithm for computing Matrix Profiles in $O(n^2)$ time with GPU acceleration available. For series with between 700-1000 points, like our example of 2 years of retail data, the computation is near-instantaneous. It's fast with much longer series too.

Application to residuals

Unlike other unsupervised methods which sometimes benefited from operating on residuals, stripping out *explainable structure* to isolate unexplained signals, Matrix Profiling does not have to work on residuals. We find it works best when building subsequence comparisons on raw signal, sales here. Seasonal patterns that repeat weekly will match other weeks to produce low profile values. Weeks with novel shapes, such as stockouts, demand shocks, or supply disruptions, should lack a match, surfacing as discords.

> You may want to try some transformations to clean up your data before running Matrix Profiling; we don't do that here.

We apply `stumpy.stump` directly to the sales series of our dataset. We set the window size (m) to 4 weeks, and the threshold to the 95th percentile. We also set `normalize=False` in stump; we did this as it causes stumpy to use the **Approximate All-Pairs Motif** (AAMP) algorithm, which uses raw Euclidean distance instead of z-scores. It's recommended when anomalies are about amplitude rather than shape-based. This can help if you are using residuals from profile modeling for your Matrix Profile. In this example we found it worked better to remove normalisation; with `normalisation = True` we saw precision of 0.19 and recall of 0.23.

```
import stumpy

# Matrix Profile on raw sales—no profile model needed
sales = stockouts_data['sales'].values.astype(float)
m = 28  # four weekly cycles

mp = stumpy.stump(sales, m=m, normalize=False)  # AAMP: raw Euclidean distance, amplitude-based
profile_values = mp[:, 0].astype(float)

# Align back to full index
mp_index = stockouts_data.index[:len(profile_values)]
stockouts_data['mp_score'] = np.nan
```

```
stockouts_data.loc[mp_index, 'mp_score'] = profile_values

mp_threshold = np.nanpercentile(profile_values, 95)
stockouts_data['detected_mp'] = stockouts_data['mp_score'] > mp_threshold

is_det_mp = stockouts_data['detected_mp'].fillna(False).astype(bool)
tp_mp = (is_det_mp &  is_true).sum()
fp_mp = (is_det_mp & ~is_true).sum()
fn_mp = (~is_det_mp & is_true).sum()

prec_mp = tp_mp / (tp_mp + fp_mp) if (tp_mp + fp_mp) else 0.0
rec_mp  = tp_mp / (tp_mp + fn_mp) if (tp_mp + fn_mp) else 0.0
print(f"Matrix Profile (univariate)—Precision: {prec_mp:.2f}  Recall: {rec_
mp:.2f}")

# Output
Matrix Profile (univariate)—Precision: 0.28  Recall: 0.32
```

The plot:

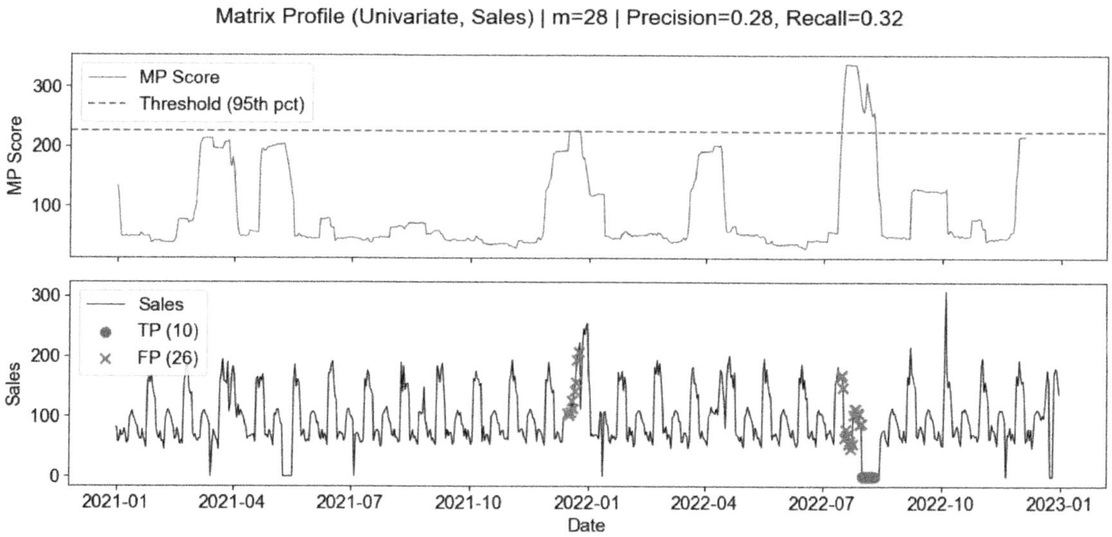

Figure 17.15: Univariate Matrix Profile discord detection

Precision (0.28) and recall (0.32) are not fantastic, better than when normalised though. And while our univariate Matrix Profile has found 36 discords, only 10 are true anomalies. In part this is a structural limitation. Our sales series has strong exogenous features (price changes, promotions, and holidays) which legitimately drive shifts and shape changes that are *rare but not anomalous*. Christmas holidays being flagged is a key issue. The stockouts could just be picked out by a separate rule. The Matrix Profile **correctly** identifies two (of three) Christmas periods as discords, with no direct close matches but they aren't anomalies. Univariate Matrix Profile has no concept of *why* a pattern is unusual. It only knows *that* a subsequence shape is rare.

When Does normalize=False Help?

Non-normalized distance works best when amplitude shift *is* the anomaly, and a series has a consistent baseline. Monitoring data (CPU usage, temperature sensors) fits this pattern well. A sudden spike from 10% to 90% is anomalous because of its magnitude.

Multidimensional matrix profile

If we can give the algorithm both sales *and* price shapes simultaneously, then a promotional uplift—where both sales rise and price drops—will match other promotional uplift windows. The sales-during-stockout pattern, where sales collapse but price remains unchanged, has no such match and should surface as a genuine discord.

`stumpy.mstump` (Yeh, Kavantzas, and Keogh (2017)) extends the algorithm to multiple dimensions simultaneously. Rather than computing pairwise distances on a single series, it computes multidimensional subsequence distances across d aligned time series. The include parameter forces specific dimensions (sales here) to participate in the distance calculation, and the discords=True flag reverses the internal distance matrix so that the profile favors large distances—surfacing discords rather than motifs.

We get an output `mps` that is a $d \times (n - m + 1)$ array, with row k containing the k-dimensional matrix profile, where distance is averaged across the best k dimensions. For discord detection we use the row corresponding to the number of dimensions we want to consider jointly.

Z-Normalizing step functions

Matrix Profiling relies on z-normalized Euclidean distance, which compares shapes rather than absolute magnitudes. But what happens with a step function like price, which might stay exactly at £5.00 for weeks? A perfectly flat subsequence has zero variance, so z-normalisation would divide by zero; stumpy handles this by collapsing flat windows to a straight line of zeros.

```
# Build multi-dimensional input, with each row = one dimension
# Sales is still target; but we add price for exogenous context
T_multi = np.array([
    stockouts_data['sales'].values.astype(float),
    stockouts_data['price'].values.astype(float),
```

```
])

m = 28

# Appling include=[0] forces sales (row 0) to always be in the subspace
# discords=True reverses distances so profile favours large values
mps, indices = stumpy.mstump(
    T_multi, m=m, include=[0], discords=True, normalize=True)

# Using a 2-dimensional profile (row index 1, i.e. k=2: both sales and price)
# This profile considers both dimensions jointly
profile_multi = mps[1]  # k=2 dimensional profile

mp_index = stockouts_data.index[:len(profile_multi)]
stockouts_data['mp_multi_score'] = np.nan
stockouts_data.loc[mp_index, 'mp_multi_score'] = profile_multi

mp_multi_threshold = np.nanpercentile(profile_multi, 95)
stockouts_data['detected_mp_multi'] = (
    stockouts_data['mp_multi_score'] > mp_multi_threshold)

is_det_mpm = stockouts_data['detected_mp_multi'].fillna(False).astype(bool)
tp_mpm = (is_det_mpm &  is_true).sum()
fp_mpm = (is_det_mpm & ~is_true).sum()
fn_mpm = (~is_det_mpm & is_true).sum()

prec_mpm = tp_mpm / (tp_mpm + fp_mpm) if (tp_mpm + fp_mpm) else 0.0
rec_mpm  = tp_mpm / (tp_mpm + fn_mpm) if (tp_mpm + fn_mpm) else 0.0
print(f"Matrix Profile (multi, sales+price)—Precision: {prec_mpm:.2f}  Recall:
{rec_mpm:.2f}")

# Output:
Matrix Profile (multi, sales+price)—Precision: 0.33  Recall: 0.39
```

The output is the following plot:

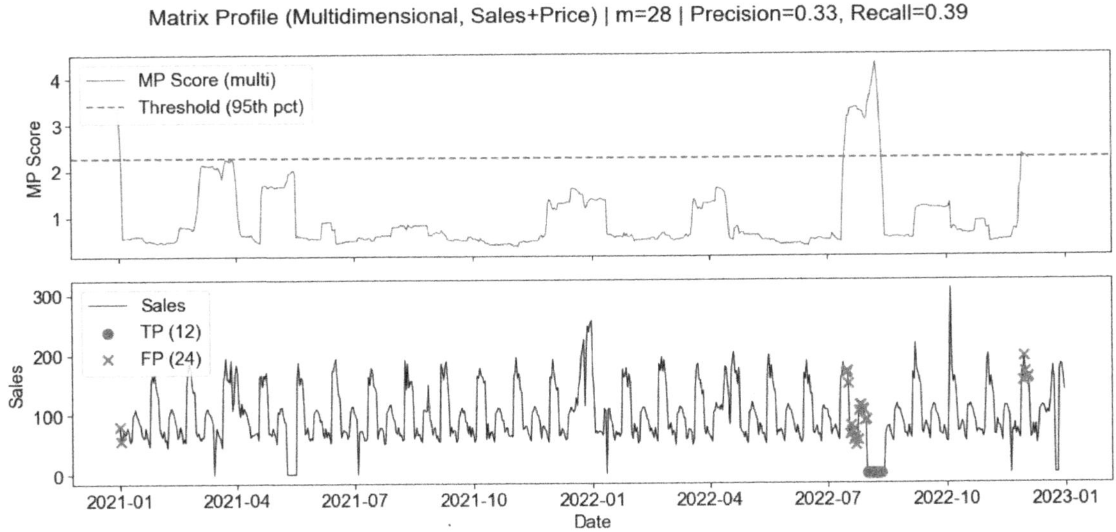

Figure 17.16: Multidimensional Matrix Profile (sales + price)

By including price as a second dimension for profiling, we have raised precision (0.33) and recall (0.39). But we are still far away from LOF and IF. One further option stumpy provides is annotation vectors (AVs), which allow us to add per-subsequence weights in [0,1] that down-weight anomaly scores for known-explainable patterns (e.g., holidays), via element-wise multiplication with a matrix profile. We tried suppressing vacation windows this way, but it only nudged precision to 0.35. Still worth experimenting with for your own use-case.

mstump mechanics

mstump returns mps, a $d \times (n - m + 1)$ array where row k is the k-dimensional matrix profile. Row 0 is the best single-dimension profile, row 1 considers the best two dimensions, and so on. The include parameter constrains which dimensions must participate—include=[0] forces sales to always be included, so the algorithm selects the most informative additional dimensions rather than potentially ignoring your target variable. With discords=True, the internal distance sorting is reversed so that the profile values favor large distances (discords) rather than small ones (motifs).

Choosing the window length

For matrix profiling, window length m is your primary hyperparameter. You can think of it as similar to a moving average: smaller windows are more sensitive to localized data movements, larger windows smooth toward broader trends.

- Small windows (e.g., $m = 3$ for daily data) make your profile overly sensitive, resulting in trivial matches; we see inflated profile values everywhere, making it harder to find genuine discords.
- Large windows (e.g., $m = 90$) cause a matrix profile to capture anomalies at higher levels (i.e. quarterly), but make us blind to weekly resolution.

For our retail data, $m = 28$ days is a principled choice. It captures four weekly cycles, covering the typical price and demand patterns. The motifs—recurring normal monthly rhythms—will have short profile distances, while genuine anomalies (like a multi-day stockout collapsing the expected 28-day shape) will register clearly as discords.

```
# Compare profile scores across window lengths
sales = stockouts_data['sales'].values.astype(float)
fig, axes = plt.subplots(3, 1, sharex=True,
                         figsize=(CFG.img_dim1, CFG.img_dim2 * 1.4))

for ax, m_val in zip(axes, [3, 7, 14]):
    mp_val = stumpy.stump(sales, m=m_val, normalize=False)
    pv = mp_val[:, 0].astype(float)
    idx = stockouts_data.index[:len(pv)]

    ax.plot(idx, pv, color=custom_palette[1],
            linewidth=0.8, label=f'm={m_val}')
    ax.axhline(np.percentile(pv, 95),
               color=custom_palette[2], linestyle='--',
               linewidth=1, alpha=0.7)
    ax.set_ylabel('MP Score')
    ax.legend(loc='upper left')

axes[-1].set_xlabel('Date')
fig.suptitle('Matrix Profile Score—Sensitivity to Window Length m')
plt.tight_layout()
plt.show()
```

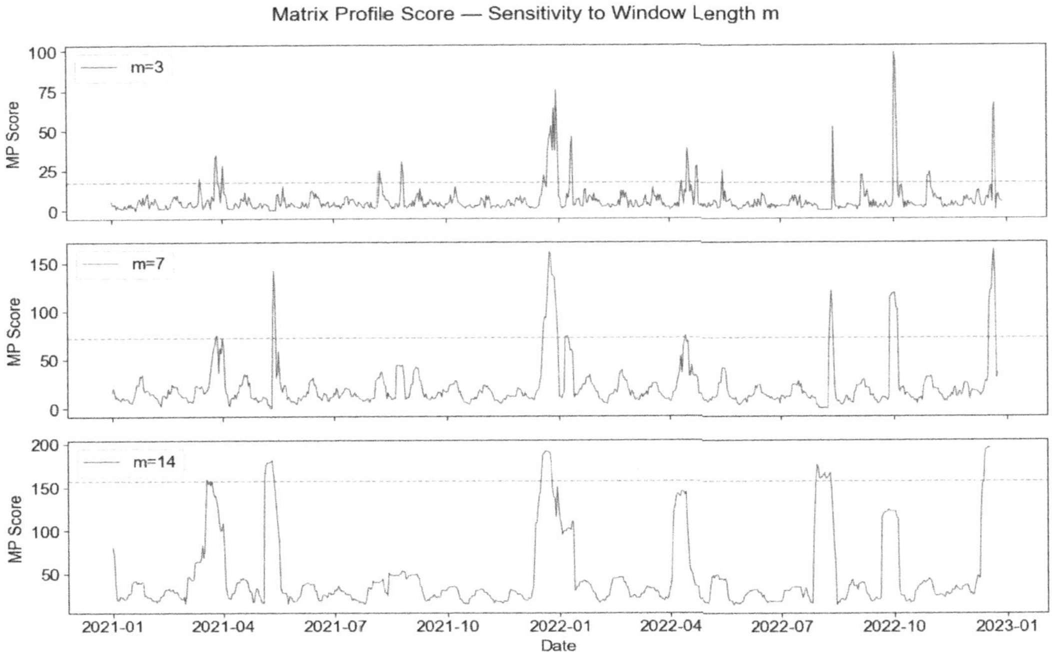

Figure 17.17: Matrix Profile score sensitivity to window length

You may have noticed that we could search across multiple window sizes simultaneously to find discords that persist regardless of scale. This is exactly what the Pan Matrix Profile (`stumpy.stump`) does. A simple version of this is to compute profiles at several window lengths and compare results.

```
sales = stockouts_data['sales'].values.astype(float)
window_sizes = [7, 14, 21, 28]

# Compute profiles at each scale and build consensus
raw_profiles = np.full((len(window_sizes), len(sales)), np.nan)

for i, m_val in enumerate(window_sizes):
    mp_val = stumpy.stump(sales, m=m_val, normalize=False)
    pv = mp_val[:, 0].astype(float)
    pv[np.isinf(pv)] = np.nan
    # Min-max normalise each profile to make scales comparable
    pv_norm = (pv - np.nanmin(pv)) / (np.nanmax(pv) - np.nanmin(pv))
    raw_profiles[i, :len(pv_norm)] = pv_norm

consensus = np.nanmax(raw_profiles, axis=0)
idx = stockouts_data.index[:len(consensus)]
```

```
fig, ax = plt.subplots(figsize=(CFG.img_dim1, CFG.img_dim2))
ax.plot(idx, consensus, color=custom_palette[1], linewidth=0.8,
        label='Consensus (max across scales)')
ax.axhline(np.nanpercentile(consensus, 95), color=custom_palette[2],
           linestyle='--', linewidth=1.5, label='95th percentile')
ax.set_xlabel('Date')
ax.set_ylabel('Normalised MP Score')
ax.set_title('Multi-Scale Matrix Profile Consensus')
ax.legend(loc='upper left')
plt.tight_layout()
plt.show()
```

The resultant plot is as follows:

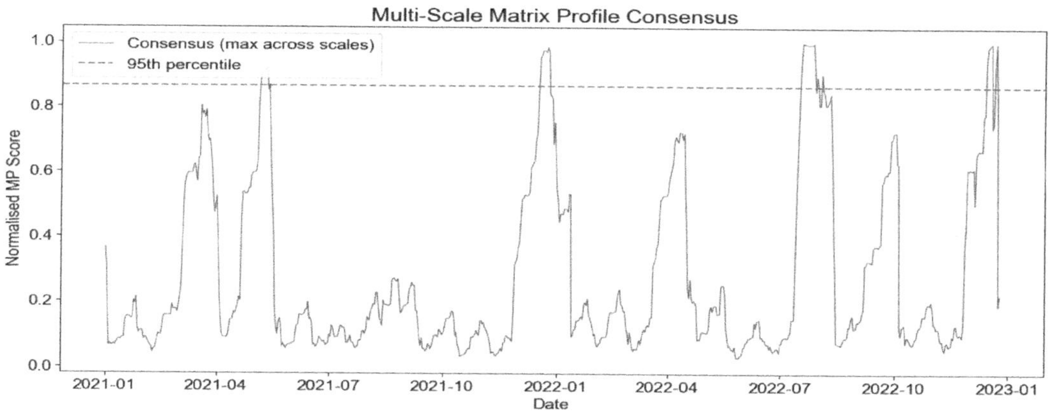

Figure 17.18: Multi-scale Matrix Profile consensus across window lengths

Plotting the profile score across multiple window lengths is a quick diagnostic. If high-scoring regions are consistent across m values, you have found a robust discord. If they shift significantly with m, the anomaly is sensitive to scale and you should be cautious about over-interpreting it.

Matrix profile vs. Point-Based methods

Matrix Profile and point-based methods (Z-score, Isolation Forest, LOF) are complementary rather than competing. Point-based methods detect observations that are individually extreme or locally isolated. Matrix Profile detects patterns, subsequences, that are globally unusual. The methods will disagree precisely on the most interesting cases: a smooth ramp-down to zero over a week will produce modest point-based scores (no individual day is extreme) but a high matrix profile score (no other week in the history has this shape). A single spike will produce a high Z-score but may not produce a high matrix profile score if similar spikes have occurred before.

In a production system, running both and surfacing cases where they agree gives high-confidence anomalies. Cases where they disagree are worth human review and are often where the most interesting domain-specific anomalies live.

Alternative approaches

Matrix Profiling's performance here is underwhelming, but there are additional options in stumpy that we can try:

- AB-joins against a clean reference.
- Have evaluation respect window-based profiling, rather than point-based.

Let's take a look at both of these.

AB-Join: comparing against a reference period

A self-join compares every subsequence in a series against every other subsequence in the same series. If an anomaly is rare enough, it will stand out as a discord, but so will anything else that happens only once, like unique vacation patterns. An alternative is an AB-join, under which we compare subsequences in the full series (T_A) against a known-clean reference (T_B). Any subsequence in T_A that has no close match in T_B is flagged.

This is appealing but requires access to and knowledge of clean data. If we know which periods are anomaly-free, we have already solved half our problem. For completeness, we have added the code below.

```
# Clean reference data
clean_mask = (stockouts_data.index >= "2021-01-13") & (stockouts_data.index <=
"2022-07-31")
sales_clean = stockouts_data.loc[clean_mask, "sales"].values.astype(float)

mp_ab = stumpy.stump(T_A=sales, m=m, T_B=sales_clean, ignore_trivial=False)
profile_ab = mp_ab[:, 0].astype(float)

mp_nn_index = stockouts_data.index[:len(profile_ab)]  # <-- alignment index

stockouts_data["mp_ab_score"] = np.nan
stockouts_data.loc[mp_nn_index, "mp_ab_score"] = profile_ab

mp_ab_threshold = np.nanpercentile(profile_ab, 95)
stockouts_data['detected_mp_ab'] = stockouts_data['mp_ab_score'] > mp_ab_
threshold

is_det_ab = stockouts_data['detected_mp_ab'].fillna(False).astype(bool)
tp_ab = (is_det_ab &  is_true).sum()
fp_ab = (is_det_ab & ~is_true).sum()
fn_ab = (~is_det_ab & is_true).sum()

prec_ab = tp_ab / (tp_ab + fp_ab) if (tp_ab + fp_ab) else 0.0
```

```
rec_ab  = tp_ab / (tp_ab + fn_ab) if (tp_ab + fn_ab) else 0.0
print(f"MP AB-join (sales vs clean ref)—Precision: {prec_ab:.2f}  Recall: {rec_
ab:.2f}")

# Output:
MP AB-join (sales vs clean ref)—Precision: 0.33  Recall: 0.39
```

An AB-join should improve precision by reducing false positives from rare-but-normal patterns that exist in the reference period. However, it introduces a strong assumption about what *clean* means and performs poorly when the reference period is contaminated. We achieved low precision (0.33) and poor recall (0.39).

Window-Overlap evaluation

There is a subtler issue with how we have been evaluating Matrix Profile throughout this section. Point-based methods (Z-score, IF, LOF) flag individual timestamps, and our ground truth labels individual timestamps, so precision and recall are straightforward. Matrix Profile flags *windows*: when $P[i]$ exceeds a threshold, the anomalous subsequence spans $[i, i + m)$. But our evaluation only marks index i as detected, ignoring the remaining $m - 1$ days in that window.

If a three-day stockout sits at positions 15-17 within a 28-day window starting at position 2, evaluation marks position 2 as detected and positions 15-17 as ground truth, giving zero overlap. This systematically deflates both precision and recall for Matrix Profile relative to point-based methods.

The correction is to expand each detection to its full window before computing overlap:

```
def window_overlap_eval(detected_starts, is_true, m, n):
    """
    Expand window-start detections to full [i, i+m) coverage,
    then compute point-level precision and recall against ground truth.
    """
    detected_expanded = np.zeros(n, dtype=bool)
    for i in np.where(detected_starts)[0]:
        detected_expanded[i:min(i + m, n)] = True

    tp = (detected_expanded &  is_true.values).sum()
    fp = (detected_expanded & ~is_true.values).sum()
    fn = (~detected_expanded & is_true.values).sum()

    prec = tp / (tp + fp) if (tp + fp) else 0.0
    rec  = tp / (tp + fn) if (tp + fn) else 0.0
    return prec, rec

n = len(stockouts_data)
```

```
# Re-evaluate univariate MP with window overlap
prec_mp_w, rec_mp_w = window_overlap_eval(
    is_det_mp.values, is_true, m, n
)
print(f"MP univariate (window-overlap)—Precision: {prec_mp_w:.2f}  Recall: {rec_mp_w:.2f}")

# Re-evaluate multidimensional MP with window overlap
prec_mpm_w, rec_mpm_w = window_overlap_eval(
    is_det_mpm.values, is_true, m, n
)
print(f"MP multi (window-overlap)—Precision: {prec_mpm_w:.2f}  Recall: {rec_mpm_w:.2f}")

# Output:
MP univariate (window-overlap)—Precision: 0.17  Recall: 0.48
MP multi (window-overlap)—Precision: 0.15  Recall: 0.55
```

Window-overlap evaluation will typically improve recall substantially (more ground-truth days covered by expanded windows) while precision may decrease (expanded windows cover more non-anomalous days). This is exactly what we find with both the univariate and multivariate recalls increasing to 0.48 and 0.55 respectively, while precision drops to 0.17 and 0.15. No single evaluation method is correct (see (Paparrizos et al. 2022) for discussion).

Fair Comparison Caveat

Window-overlap numbers should not be directly compared with IF, LOF, or Z-score precision/recall. Those methods flag individual points; Matrix Profile flags windows. Overlap evaluation corrects for the alignment problem within the MP family but introduces a different bias (expanded windows inflate coverage). For a fair cross-method comparison, either apply the same window expansion to all methods or use a range-based metric like point-adjusted F1, which is common in TSAD benchmarks.

Where matrix profile fits

Matrix Profile underperforms here as a consequence of our data structure. This retail series is driven by exogenous features (price, promotions, holidays) that create legitimately diverse sales shapes. Matrix Profiles don't measure *why* a shape is unusual; they only highlight *that* a shape is rare. While our example data is not a good fit, for series with strong self-contained signals they are excellent: sensor telemetry, heartbeat monitoring, seismological data, and power systems. In those contexts, discords genuinely represent anomalous behavior rather than explained variation.

For exogenous-driven time series, Matrix Profile serves best as a complementary signal rather than a standalone detector.

A comparative summary

Rather than compare methods on a single plot, which would be misleading given that each targets different anomaly types and operates at different points in the analysis pipeline, the table below summarises key characteristics of each approach covered in this chapter. For an empirical comparison across a range of datasets, both the UCR TSAD Archive and the TimeEval benchmark provide good resources (Paparrizos et al. 2022).

Method	Supervision	Assumption	Best for	Main weakness
Rules / floor-ceiling	None	Domain knowledge	Known failure modes (stockouts, outages)	Does not scale; brittle to drift
X-chart (static)	None	Gaussian marginal	Stable processes	Fails with trend, seasonality
Adaptive threshold	None	Locally Gaussian	Gradual drift, manufacturing	Strong exogenous effects
Conditional X-chart	None	Gaussian within regime	Discrete external drivers (price)	Requires regime knowledge
Statistical profile (Z-score)	Semi (clean data)	Gaussian residuals	Series with known structure	Non-Gaussian residuals
LOF	Semi (clean data)	Local density	Multiple normal regimes	Sensitive to n_neighbors
Isolation Forest	Semi / Unsupervised	Anomalies are sparse	General purpose, few assumptions	Axis-aligned splits
INNE / Extended IF	Semi / Unsupervised	Anomalies are sparse	High-dim or combined-feature anomalies	Slower than standard IF
Matrix Profile	None	Temporal self-similarity	Collective / subsequence anomalies	Window length selection
Supervised (CatBoost)	Full (labels)	Learnable residual patterns	When confirmed labels exist	Requires quality labels

Table 17.3: A comparative summary

Supervised anomaly detection

Supervised anomaly detection is simply classification, as any good academic will tell you. You have observations, labels, on which you train a model that distinguishes anomalous from non-anomalous observations. *Chapters 14 and 15* covered some aspects of classification methodology, so we won't repeat that here. What is worth discussing is what makes the anomaly detection variant of classification distinct from the general case, and what a practical pipeline looks like.

Two things distinguish supervised anomaly detection from general classification problems: severe class imbalance and label quality. Both are structural rather than incidental, and both directly affect how you build and evaluate supervised detectors.

The label quality problem

Labels in anomaly detection are rarely clean. They come from one of three sources, each with its own contamination issue.

- **Operational incident logs** are accurate for large, impactful events (a week-long stockout, a confirmed sensor failure) but can miss subtle anomalies that don't rise to the threshold of a ticket. The positive class is high-precision but recall can be low.
- **Rules-based labelling pipelines** are the most common source, and the most dangerous. Rules are written at a point in time, reflecting a data distribution at that moment. As the underlying process drifts, rules generate stale labels. Flagged observations are no longer anomalous (false positive labels) and new anomaly patterns are missed (false negatives).
- **Prior unsupervised detector output** is used when no other labels exist. If your profiling model drifted before retraining, or was itself trained on contaminated data, the labels it generated carry that error forward into your supervised model. This can result in compounded contamination, which makes you wonder if foundational anomaly detection is ever possible. Regardless, a bad unsupervised model produces bad labels; a supervised model trained on bad labels produces confidently wrong predictions.

The practical response is to audit before training. But I have three million time series, you say. We should then inspect a random sample of our positive-class (anomaly) labels and verify them against raw data. If more than ~20% of your positive labels are wrong, a supervised model will likely underperform a well-tuned unsupervised approach. You're better off fixing labels first.

Class imbalance

With anomalies comprising well under 5% of observations, often under 1%, a classifier that predicts "normal" everywhere achieves very high accuracy while being completely useless. This is why accuracy is a meaningless metric for anomaly classification, and why you should never use it as your optimization target. Some practical remedies are:

- `class_weight='balanced'` (or `auto_class_weights='Balanced'` in CatBoost). This reweights the loss function by inverse class frequency, making a missed anomaly proportionally more costly than a missed normal observation. It is almost always worth doing and costs nothing

- **Threshold adjustment**: most classifiers produce a probability score, and a default threshold of 0.5 implicitly assumes balanced classes. Lowering it (say, to 0.2) increases recall at the cost of precision. Use a precision-recall curve on a validation set to find the threshold that reflects your actual cost tradeoff.
- **Oversampling the minority class: DO NOT DO THIS.** Once widely used (and researched!) but now considered poor practice. SMOTE creates synthetic points by interpolating in feature space, producing examples that are statistically plausible but temporally incoherent. Better alternatives are to augment using domain knowledge (jitter known anomaly windows, vary their duration) or simply collect more real labels.

Gradient boosting on residual features

The most effective supervised approach in our experience, and the one that connects most naturally to the profiling work earlier in this chapter, is to build a feature matrix from profile model residuals and train a gradient boosting classifier on it. A profile model removes known structure. What remains in the residuals is the unexplained variation. A supervised classifier trained on residual features learns to distinguish residual patterns associated with confirmed anomalies from patterns that are just normal noise, things like seasonal peaks that happened to breach a threshold, measurement jitter, or genuine promotions that the model underestimated.

Below we have made a few specific design choices.

- We use `TimeSeriesSplit` rather than random cross-validation. This is because our rolling features (`resid_roll_std, sales_roll_mean`) are computed from past observations and shifted by one step to avoid look-ahead. Random shuffling would be fine if we didn't include these kinds of features; as it stands they would leak information about the temporal neighborhood of test points into training.
- We use CatBoost with `auto_class_weights='Balanced'`, which automatically reweights the loss function inversely proportional to class frequency. With anomalies representing roughly 4% of observations, an unweighted classifier would achieve 96% accuracy by predicting *normal* for every row. Balanced weighting forces the model to treat a missed anomaly as roughly 24x more costly than a false alarm during training, which prevents it from ignoring the minority class (outliers). The threshold adjustment step after training lets us fine-tune the actual precision-recall tradeoff.

CatBoost was selected for our classifier, but you could easily use another algorithm i.e. XGBoost, LightGBM, or a regularized logistic regression. What is important here is having good labels, a strong temporal structure and optionally a well-explained model leaving us with good residual signal.

```
from catboost import CatBoostClassifier
from sklearn.model_selection import TimeSeriesSplit
from sklearn.metrics import classification_report, precision_recall_curve
import pandas as pd
import numpy as np
```

```python
def build_supervised_features(df, window=14):
    """
    Feature matrix for supervised residual profiling.
    All rolling features are shifted by 1 to avoid lookahead bias.
    """
    feats = pd.DataFrame(index=df.index)

    # Core residual signal
    feats['residual']          = df['residual']
    feats['residual_z']        = df['residual_z']
    feats['residual_abs']      = df['residual'].abs()

    # Local residual statistics
    for w in [7, 14, 28]:
        feats[f'resid_roll_mean_{w}'] = (
            df['residual'].rolling(w, min_periods=3).mean().shift(1)
        )
        feats[f'resid_roll_std_{w}'] = (
            df['residual'].rolling(w, min_periods=3).std().shift(1)
        )
        feats[f'resid_roll_max_{w}'] = (
            df['residual'].abs().rolling(w, min_periods=3).max().shift(1)
        )

    # Scores from unsupervised detectors as features
    for col in ['if_unsup_score', 'if_score', 'eif_score', 'lof_score']:
        if col in df.columns:
            feats[col] = df[col]

    # Context
    feats['price']             = df['price']
    feats['promotion']         = df['promotion']
    feats['sales_roll_mean']   = (
        df['sales'].rolling(14, min_periods=3).mean().shift(1)
    )
    feats['day_of_week']       = df.index.dayofweek
    feats['month']             = df.index.month

    return feats.fillna(0)
```

```
feature_df = build_supervised_features(stockouts_data)
labels     = stockouts_data['is_anomaly'].fillna(0).astype(int)

# Time-series split
tscv = TimeSeriesSplit(n_splits=4)

all_preds  = np.zeros(len(labels))
all_probs  = np.zeros(len(labels))
test_mask  = np.zeros(len(labels), dtype=bool)

for train_idx, test_idx in tscv.split(feature_df):
    X_tr = feature_df.iloc[train_idx].values
    X_te = feature_df.iloc[test_idx].values
    y_tr = labels.iloc[train_idx].values
    y_te = labels.iloc[test_idx].values

    clf = CatBoostClassifier(
        iterations=300,
        learning_rate=0.05,
        depth=4,
        auto_class_weights='Balanced',
        eval_metric='F1',
        random_seed=42,
        verbose=0
    )
    clf.fit(X_tr, y_tr, eval_set=(X_te, y_te))

    all_preds[test_idx]  = clf.predict(X_te)
    all_probs[test_idx]  = clf.predict_proba(X_te)[:, 1]
    test_mask[test_idx]  = True

y_test_all  = labels.values[test_mask]
y_pred_all  = all_preds[test_mask].astype(int)
y_prob_all  = all_probs[test_mask]

print(classification_report(y_test_all, y_pred_all,
                            target_names=['Normal', 'Anomaly']))

# Output:
              precision    recall  f1-score   support
```

```
      Normal       1.00      0.99      1.00       562
     Anomaly       0.81      1.00      0.90        22

    accuracy                           0.99       584
   macro avg       0.91      1.00      0.95       584
weighted avg       0.99      0.99      0.99       584
```

The four blocks in `build_supervised_features` each give our classifier a different view on the same observation. The core residual signal (`residual`, `residual_z`, `residual_abs`) is the raw 'what the profile couldn't explain'; this is our primary anomaly evidence. The local residual statistics (rolling mean, std, and max at 7, 14, and 28 days, all shifted by one) provide context, because a residual of +20 means very different things depending on whether the surrounding fortnight has been quiet or noisy. The unsupervised detector scores (`if_unsup_score`, `if_score`, `eif_score`, `lof_score`) are second opinions from earlier in the chapter; our classifier learns when to trust them and when to override them. Finally the context features (`price`, `promotion`, `sales_roll_mean`, `day_of_week`, `month`) let our model learn that a large negative residual in late December under a deep promotion is structurally different from the same residual on a quiet Tuesday in February. Every rolling feature is shifted by one step so that no feature at time t contains y_t itself.

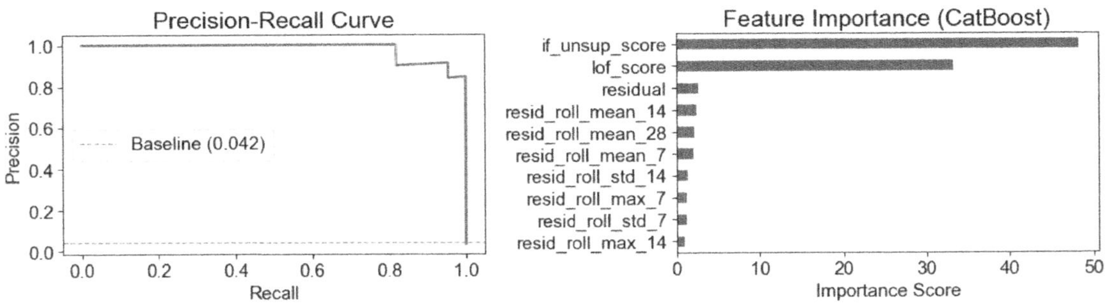

Figure 17.19: CatBoost precision-recall curve and feature importance

The feature importance panel tells us which signals the model is actually using. If `if_unsup_score` or `if_score` dominate, the supervised model is largely deferring to unsupervised detectors and the supervised wrapper adds little beyond threshold calibration. Some influence from `resid_roll_std` suggests the classifier has learned temporal patterns from residuals that no single unsupervised method captures on its own. We could instead, build more features, and labels with the raw data, which would avoid the need for ensembled output.

Threshold adjustment

CatBoost's default threshold of 0.5 assumes equal misclassification costs. In anomaly detection this is rarely appropriate, as the cost of missing real stockouts often exceeds false alarms. The precision-recall curve in the figure above shows the full tradeoff surface; here we select the threshold that maximizes F1 as a reasonable default.

```
from sklearn.metrics import f1_score

thresholds = np.arange(0.1, 0.9, 0.01)
f1_scores = [
    f1_score(y_test_all, (y_prob_all >= t).astype(int), zero_division=0)
    for t in thresholds
]

best_thresh = thresholds[np.argmax(f1_scores)]
best_f1     = max(f1_scores)

y_pred_tuned = (y_prob_all >= best_thresh).astype(int)
tp_sup = ((y_pred_tuned == 1) & (y_test_all == 1)).sum()
fp_sup = ((y_pred_tuned == 1) & (y_test_all == 0)).sum()
fn_sup = ((y_pred_tuned == 0) & (y_test_all == 1)).sum()

prec_sup = tp_sup / (tp_sup + fp_sup) if (tp_sup + fp_sup) else 0.0
rec_sup  = tp_sup / (tp_sup + fn_sup) if (tp_sup + fn_sup) else 0.0

print(f"Best threshold: {best_thresh:.2f}")
print(f"Supervised (tuned): Precision: {prec_sup:.2f}  Recall: {rec_sup:.2f}
F1: {best_f1:.2f}")

# Output:
Best threshold: 0.55
Supervised (tuned): Precision: 0.91  Recall: 0.95  F1: 0.93
```

If false negatives are expensive (missed equipment failures, missed fraud) you want a lower threshold to maximize recall. If false positives are expensive (wasted analyst time, unnecessary operational interventions) you want a higher threshold to maximize precision. There is no universally correct answer; the threshold is a business decision, not a model decision.

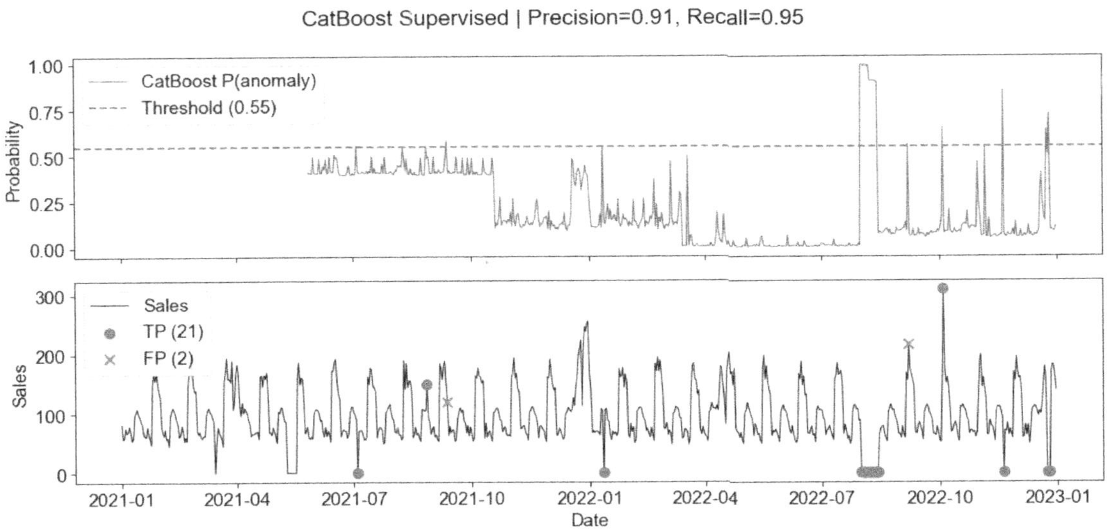

Figure 17.20: CatBoost supervised detections with tuned threshold

Method	Type	Precision	Recall	F1
Z-score Profile	Unsupervised	0.74	1.00	0.85
IF: Raw Features	Unsupervised	0.51	0.97	0.67
IF: Residuals	Unsupervised	0.25	0.48	0.33
INNE: Residuals	Unsupervised	0.05	0.10	0.07
LOF: Residuals	Unsupervised	0.83	0.97	0.90
CatBoost (supervised, tuned threshold)	Supervised	0.91	0.95	0.93

Table 17.4: Comparing all approaches

CatBoost wins clearly here, but this result is an artifact of the task. Our anomalies were generated programmatically and we have certainty in them. We have a strong temporal structure and relationships in our data, even where they don't repeat in calendar position, noise is low. With clean labels, repeated anomaly patterns, and a profile model that already explains most of the variance, a supervised classifier on residual features has very little left to do. It is an easy task, that shows us how critical structure and labelling are.

We should think of this as a calibration layer that sits on top of unsupervised detection. In production, labels come from unsupervised detectors and human-in-the-loop review (the contamination problem returns); novel anomalies by definition should not repeat in your training data. We will return to this self-supervised detection in Chapter 18.

Some practical considerations:

- **Hyperparameter sensitivity**: contamination has the largest effect in isolation-based methods; it sets the decision threshold on anomaly scores, not the model structure. Halving `max_samples` and `max_features` to 0.5 while raising contamination to 0.08 lifted the unsupervised IF's recall from 0.61 to 0.97. For supervised models, threshold adjustment on the probability output typically matters more than tree hyperparameters themselves (Soenen, Wolputte, and Perini 2021).
- **The contamination-imputation loop**: Start with a generous contamination rate (high recall, accept false positives), review flagged observations, then impute confirmed anomalies with interpolated values. Some values were already wrong; replacing them should improve signal rather than fabricate it. Normal data is abundant enough to anchor any reasonable interpolation. Retrain a profile on the imputed series and re-score. Each iteration sharpens the residuals and tightens detection boundaries. Human confirmation at each step is essential; uncertain flags stay in the data untouched. Depending on your case, you could pair this with forecast accuracy, being sure to leave validation sets clean; otherwise, you're marking your own homework.

Forecast-First anomaly detection

Throughout this chapter we have built detection methods individually (rules, adaptive thresholds, statistical profiles, Isolation Forests, LOF, Matrix Profile) and compared them side by side. Most of these methods share the same underlying architecture, though; they are stages in a single composable pipeline, not competing alternatives. How you build, maintain, and extend a detection system in practice follows directly from this.

This pattern is sometimes called *forecast-first* or *model-first* anomaly detection. Libraries like Darts (Unit8) and `statsforecast` (Nixtla) package it as a first-class abstraction. Once you see the structure, the design decisions at each stage become much clearer.

The pipeline

Anomaly detection pipelines have five stages, illustrated in *Figure 17.21*. Each stage is independent and changeable, meaning you can change a profile model or scorer without changing another part of the same pipeline.

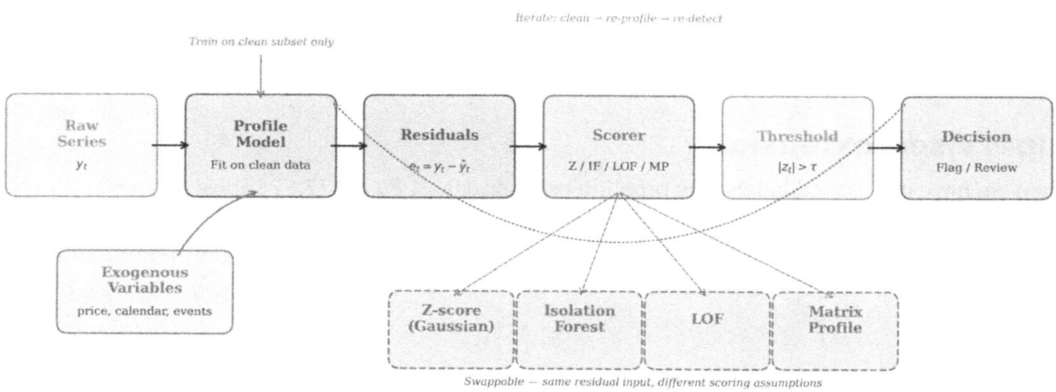

Figure 17.21: Forecast-first anomaly detection pipeline

The pipeline in action, looks like this:

1. **Fit a profile model of** expected behavior on clean data. This is the conditioning step, encoding what $\hat{y}_t$ should look like given seasonality, trend, exogenous regressors, and autocorrelation structure. We used AutoARIMA with price as an exogenous variable, but MSTL, regression models, even a well-specified EWMA all work here. The critical requirement is that you train on data you trust to be anomaly-free
2. **Extract residuals**: compute $e_t = y_t - \hat{y}_t$ across the full series. The residual strips out everything the model can explain, leaving unexplained variation. Under normal conditions this looks like centered noise; under anomalous conditions it spikes.
3. **Score the residuals**. This is where the methods we covered become interchangeable: Z-score assumes Gaussian residuals, Isolation Forest scores by isolation depth, LOF scores by local density, Matrix Profile scores by subsequence dissimilarity, and a supervised classifier scores using learned residual patterns from confirmed labels. All take the same input and produce the same output (an anomaly score per observation). Swapping the scorer is a one-line change once the residual infrastructure exists.
4. **Threshold those scores into binary decisions**. This is where $|z_t| > \tau$, the 95th percentile of matrix profile values, or the contamination parameter in pyod all live. The threshold encodes your cost tradeoff between false positives and false negatives.
5. **Decide and act**: flag, review, clean, or escalate. In production this feeds into alerting systems, dashboards, or automated data cleaning pipelines. The output often feeds back into stage 1; once you have confirmed anomalies, you exclude them from the next round of model training.

The iteration loop

The pipeline is not run once. In production, the feedback loop is the most important part of the architecture. The cycle is: profile, detect, confirm, clean, re-profile, re-detect. Each iteration refines both the profile model and the labeled anomaly set. This also helps guard us against model drift, as data may have legitimate changepoints, i.e., a household reduces its power consumption by installing solar panels, requiring an update to your underlying model.

Your pipeline is also where contamination problems may occur. If you skip the confirmation step, anomalies may become treated as ground truth, thereby excluded from the next training set, and your system's definition of normal will be gradually corrupted. Human review, or at a minimum a stable holdout of confirmed anomalies, is necessary to keep the loop calibrated.

Supervised extension

When we have confirmed labels, the pipeline extends easily. *Figure 17.22* shows a supervised variant: unsupervised scores become features alongside residuals and contextual variables, and a gradient boosting classifier learns which residual patterns correspond to confirmed anomalies versus noise.

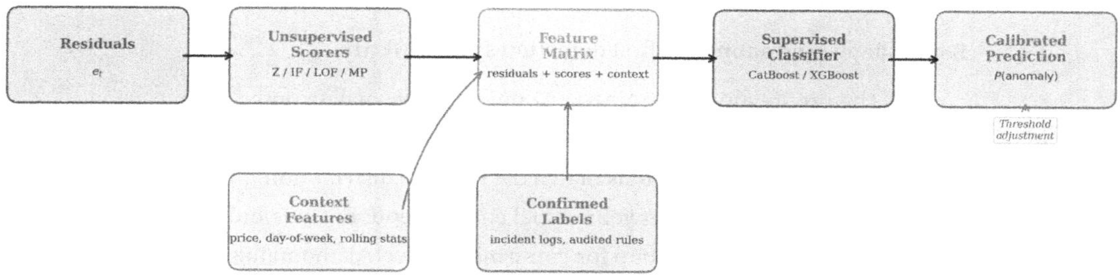

Figure 17.22: Supervised extension: unsupervised scores as features

This is exactly what `build_supervised_features` and our CatBoost classifier implement. Feature importance tells you whether a supervised model is adding value beyond unsupervised scorers. If `if_score` dominates, the classifier is largely deferring to isolation forest. If `resid_roll_std` or contextual features dominate, the classifier has learned temporal patterns that no single unsupervised method captures on its own.

Composability in practice

The practical value of thinking in pipeline terms is that it separates concerns cleanly. When detection performance degrades, and it will because the underlying process drifts, you can diagnose which stage is failing. Residuals that are large and structured (visible seasonality remaining) mean the profile model is underfit; fix stage 1. Clean residuals but too many false positives means the scorer's assumptions don't match the residual distribution; try a different scorer or adjust its parameters. Sensible scores but wrong binary output means the threshold is mis-calibrated. And if everything looks correct on historical data but fails on new data, the process has drifted and you need to retrain the profile.

This modularity also means you can run multiple scorers in parallel on the same residual signal and surface cases where they agree (high confidence) versus disagree (worth human review). In our comparison section, Z-score, Isolation Forest, and INNE all operated on the same residuals. In a production system you would ensemble their outputs or flag disagreements rather than choosing a single method.

Library ecosystem

Several libraries package this pipeline as a high-level API. Darts provides `ForecastingAnomalyModel`, which wraps any of its forecasting models with pluggable scorers (Wasserstein distance, KMeans on subsequence windows, norm-based scoring) and returns anomaly scores directly. Nixtla's statsforecast tutorial demonstrates the same pattern using MSTL with prediction intervals, where observations outside the interval are flagged. Both implement the forecast-first architecture described here.

Understanding the components matters more than any specific API. If you have followed this chapter, you already know how to build every stage from scratch, which means you can adapt it to your tooling, your data, and your operational constraints rather than being locked into a single library's abstraction.

Before deploying a forecast-first detection system, verify:

- The profile model is trained on confirmed clean data only
- Residuals on the training period look like centred noise (no remaining structure)
- The scorer's assumptions match the residual distribution (plot the residuals first)
- The threshold reflects your actual cost tradeoff, not a default value
- There is a feedback loop for confirming detected anomalies and retraining

Summary

In this chapter we covered the fundamentals of time series anomaly detection, progressing from simple rules-based approaches through statistical profiling to supervised classification. We distinguished between outliers, which are statistically extreme values under a marginal distribution, and anomalies, which are observations that are unexpected given temporal and contextual history. We also examined point, collective, and contextual anomaly types.

We built detection systems of increasing sophistication: static thresholds and X-charts, adaptive rolling windows, conditional regime-based bounds, and statistical profiling using AutoARIMA residuals. We saw how the contamination problem (training on data that contains the anomalies you are trying to detect) can silently degrade detection performance, and why fitting profiles on confirmed clean data is essential.

We applied unsupervised methods including Isolation Forest, INNE, Local Outlier Factor, and Matrix Profile to both raw features and model residuals, finding that method effectiveness depends heavily on data structure. LOF performed best in our retail example, while Matrix Profile underperformed as it is better suited to data with strong temporal self-similarity (e.g. electrocardiogram output). Finally, we combined unsupervised scores with residual features in a supervised CatBoost classifier, demonstrating how the forecast-first pipeline (profile, extract residuals, score, threshold, decide) provides a composable and maintainable architecture for production anomaly detection systems.

Recommended reading

- Blázquez-García, A., Conde, A., Mori, U., and Lozano, J. A. 2020. "*A Review on Outlier/Anomaly Detection in Time Series Data.*" arXiv. `http://arxiv.org/abs/2002.04236`.
- Bouman, R., Bukhsh, Z., and Heskes, T. n.d. "*Unsupervised Anomaly Detection Algorithms on Real-World Data: How Many Do We Need?*".
- Liu, Q., and Paparrizos, J. n.d. "*The Elephant in the Room: Towards A Reliable Time-Series Anomaly Detection Benchmark.*".
- Paparrizos, J., Boniol, P., Liu, Q., and Palpanas, T. 2025. "*Advances in Time-Series Anomaly Detection: Algorithms, Benchmarks, and Evaluation Measures.*" Proceedings of the 31st ACM SIGKDD Conference on Knowledge Discovery and Data Mining V.2. `https://doi.org/10.1145/3711896.3736565`.

- Paparrizos, J., Kang, Y., Boniol, P., Tsay, R. S., Palpanas, T., and Franklin, M. J. 2022. "*TSB-UAD: An End-to-End Benchmark Suite for Univariate Time-Series Anomaly Detection*." Proceedings of the VLDB Endowment. `https://doi.org/10.14778/3529337.3529354`.
- Soenen, J., Van Wolputte, E., and Perini, L. 2021. "*The Effect of Hyperparameter Tuning on the Comparative Evaluation of Unsupervised Anomaly Detection Methods.*".
- Wu, R., and Keogh, E. 2022. "*Current Time Series Anomaly Detection Benchmarks are Flawed.*" arXiv. `https://arxiv.org/abs/2009.13807`.
- Yeh, C-C. M., Kavantzas, N., and Keogh, E. 2017. "*Matrix Profile VI: Meaningful Multidimensional Motif Discovery.*" 2017 IEEE International Conference on Data Mining (ICDM). `https://doi.org/10.1109/ICDM.2017.66`

18

Self-Supervised Learning for Time Series

In the previous chapters, we have explored various models and techniques for time series data. From a high level perspective, we have seen two main types of learning paradigms: supervised learning and unsupervised learning. For the majority of the time series forecasting models we have discussed, they rely on splitting the time series into input and target sequences, and then training the model to predict the target sequence given the input sequence. This is a form of supervised learning, where we have a clear distinction between the input and the target.

Meanwhile, in *Chapters 11, 15, and 17*, we have also discussed unsupervised learning for time series, where we do not specify a clear target or label, and instead we are trying to detect the differences or anomalies in the time series data. This fits into the unsupervised learning paradigm, where we are trying to learn from the data without explicit labels.

Apart from the two main learning paradigms, we also briefly touched upon pre-trained models, aka foundation models. These models are pre-trained on a large amount and diverse data, and from a downstream task perspective, they are used as zero-shot learners. This is an extreme form of transfer learning, which we have discussed in *Chapters 11*. However, given that *more is different*, it is worth mentioning the pre-trained foundation models separately.

In fact, there exist two other learning paradigms, which are semi-supervised learning and self-supervised learning. Semi-supervised learning utilizes both labeled and unlabeled data. There are different approaches to achieve it, such as creating new labels for the unlabeled data by using the information of labeled data to create pseudo-labels for the unlabeled dataset.

The other paradigm is self-supervised learning, our focus in this chapter. In self-supervised learning, we do not need labels at all. Instead, we create a supervised learning problem from the data itself, without relying on external labels. If we rethink about time series forecasting problems we have mentioned, we sort of resemble the idea that time series data itself is providing the supervision. We don't have explicit target for time series forecasting problem, the structure of the data allows us to create tasks that the model can learn from: we create our target by slicing the time series data into input and target based on the task and our model requirements. The difference between the supervised learning methods we discussed, and self-supervised learning is subtle. In the forecasting methods we discussed before, the way we slice the data into input and target is clearly defined and required. However, self-supervised learning is usually not directly resolving the tasks, instead, it creates intermediate tasks to obtain useful representations of our data. By this standard, the forecasting methods we discussed are not self-supervised learning. However, the way we are forecasting is already shedding light on the concept of self-supervised learning as we are creating our input and target from the raw dataset. Self-supervised learning is similar but focuses on generating the representations of data instead of directly using the model for predictions. Self-supervised learning is particularly useful for time series data, as we often have a lot of unlabeled time series data in a variety of data generation processes, and we want to leverage that data to learn useful representations or to pre-train models.

In this chapter, we learn:

- Why do we need representations of time series?
- Self-supervised learning
- Contrastive predictive coding

Why do we need representations of time series?

From a practical perspective, representations of time series data are useful for downstream tasks such as forecasting, classification, and anomaly detection. By incorporating the representations in downstream tasks, the result from self-supervised learning is often useful for improving the performance of the downstream tasks. For example, in the PatchTST model (Y. Nie, et al., "*A time series is worth 64 words: Long-term forecasting with transformers*"), self-supervised learning is used to pre-train the model to get the representations of the time series data, and then the pre-trained model is used to perform forecasting tasks.

From a theoretical perspective, learning representations of time series data is often useful to uncover the underlying structure of the data generation process. In the work by Jiaxi Hu et al., "*Towards Physics-guided Time Series Embedding*," the authors proposed the postulation of "*Embedding Duality*," and we quote the authors: hidden state representation in deep time series models is equivalent to the underlying dynamical system structure of the data in phase space. As we have discussed before, a time series forecasting model that captures the data generating process generates good forecasts. If a model that generates time series representations that is equivalent to the dynamical system structure of the data in the phase space, we are approximating the data generating process. As we have seen in previous chapters, especially chapters 8 and 16, uncovering the underlying structure of the data generation process is often crucial as it implies that our model is parsimonious.

Self-supervised learning

On the surface, it may seem similar to unsupervised learning, but what is special is that we are creating a supervised learning problem from the data itself and generating representations for downstream tasks, which opens more doors compared to traditional unsupervised learning methods. In a review work, *Self-Supervised Learning for Time Series Analysis: Taxonomy, Progress, and Prospects*, Kexin Zhang and colleagues categorize self-supervised learning into three main categories: generative methods, contrastive methods, and adversarial methods.

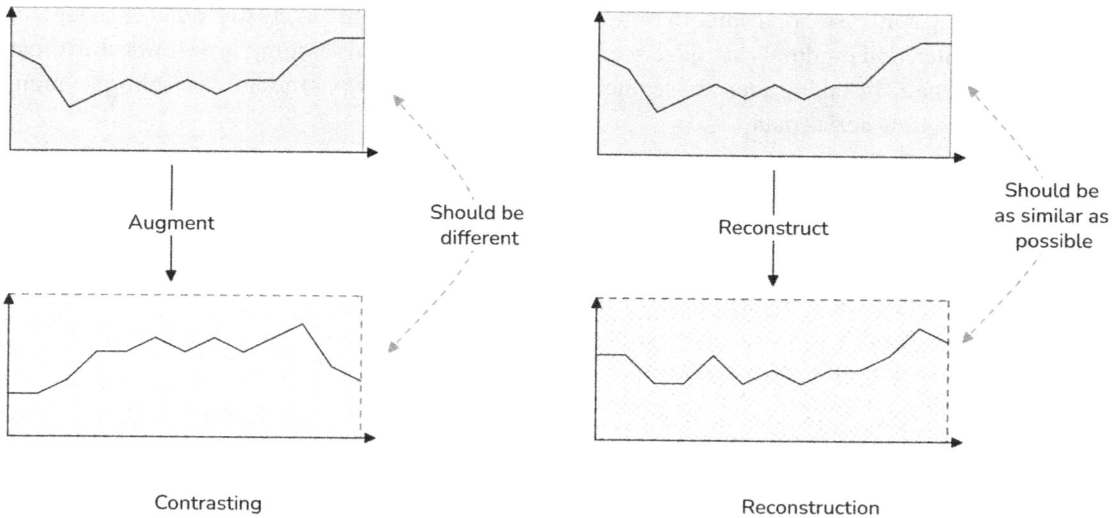

Figure 18.1: Examples of contrastive method and reconstruction method

The *left panel* shows an augmentation-based contrastive method, where we simply reversed and flipped the time series, and constructed a new time series that should be different from the original one, creating contrast. The *right panel* shows the idea of reconstruction, where we apply a model to reconstruct the time series. Note that these two examples are by no means exhaustive.

In *Chapter 16*, we have discussed autoencoder models to obtain embeddings of time series data; the method is in fact a generative method of self-supervised learning. Generative methods rely on reconstructing the input data in the process. Since reconstruction of the input is a very low-level and local task, the model learns to capture the local structure of the data. While this can be useful in many scenarios, missing high-level and global structure can be a problem for tasks such as time series classification.

Apart from generative method by reconstruction, another idea is to compare different segments or views of time series data. The key is to compare the representations of known time series segments or views that we know are similar or different. For example, we can create the so-called **positive** and **negative** pairs of time series and use a loss function to bring the representations of positive pairs closer and push the representations of negative pairs apart. See *Figure 18.1* for an example. There are also different ways of categorizing contrastive methods; here we follow Kexin Zhang and colleagues' categorization in their work *Self-Supervised Learning for Time Series Analysis: Taxonomy, Progress, and Prospects*. We only discuss three of the five categories mentioned in the paper:

- **Augmentation contrast**: Imagine we have an image of a cat, if we rotate the image by 90 degrees, we can say that the original image and the rotated image are different views of the same underlying cat. Similarly, for time series data, we can apply different augmentations to create different views of the same underlying time series. For example, to create positive pairs, we can add a small random noise to the original time series, then the original time series and the new augmented one should be considered as similar. In fact, time series data augmentation is a very wide area of research. It is not only useful for contrastive learning but also a great method to enhance the training data for supervised learning.

- **Sampling contrast:** For a smooth time series dataset, it is highly likely that adjacent segments are similar, and randomly sampled segments are different. This assumption allows us to create positive pairs by taking adjacent segments, and negative pairs by randomly sampling segments from the time series data.
- **Prediction contrast:** Instead of contrasting with different time series in their raw space, we can also contrast in the latent space. **Contrastive Predictive Coding** (**CPC**) encodes the input time series into a latent representation, and then uses an autoregressive model to evolve the latent representation forward in time. The representation of the future time steps is then assumed to contain similar information to the encoded representation of the current time step, and thus they are considered as positive pairs, while randomly sampled time steps are considered as negative pairs. In this chapter, we will dive deeper into this idea.

To summarize the discussions so far, *Table 18.1* compares generative and contrastive self-supervised learning methods for time series data:

	Generative	**Contrastive**
Objective	Casting constraints on the generation output to recreate or predict raw time series data	Comparing samples, pulling representations of similar pairs closer and pushing dissimilar pairs apart
Architecture Examples	Usually an encoder-decoder architecture: encoder for compression and decoder for recreation	Encoder and contrastive loss
Challenges	Complicated models to reconstruct details in raw time series, losing high-level patterns.	Proper selection of positive and negative pairs.

Table 18.1: Comparison of the generative and contrastive methods

Both generative methods and contrastive methods mentioned above have fixed objectives; the objectives only depend on the input. Adversarial methods, also known as GANs, are different because their objectives are dynamically dependent on the models themselves. To clarify this, let's first look at how GANs work. GANs leverage two components to play a game: one component, the so-called **generator**, generates new time series data based on some specific seed, while the second component, the so-called **discriminator**, decides whether the data generated by the generator is similar to the raw time series data. The two components will play the game to achieve the result that the generator generates data points that the discriminator cannot tell apart from the raw time series data. The simplest adversarial methods depend on the minimax game, which is to train the generator to fool the discriminator while minimizing the discrimination error of the discriminator. In other words, the generator and discriminator are opponents of each other. The key to adversarial methods is the dynamic goal that is constantly evolving because the generator and discriminator are evolving based on how good their opponent is.

Contrastive predictive coding

Following the brief introduction of CPC in the previous section, apart from building up the contrast using predictions, as shown in *Figure 18.2*, another key component of CPC is the contrastive loss function—often implemented as the InfoNCE loss—which will be discussed in depth shortly.

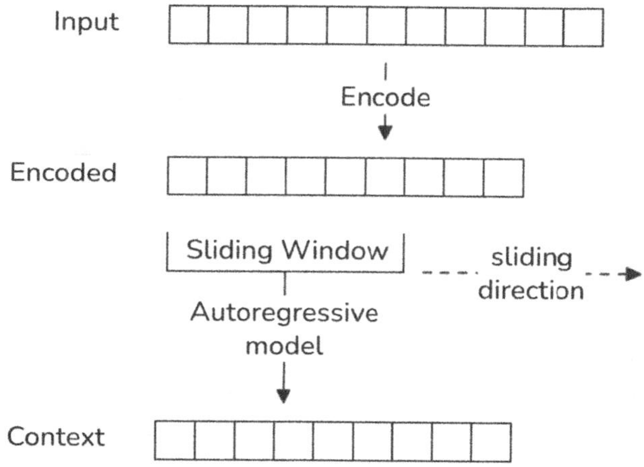

Figure 18.2: The structure of CPC

A CPC model encodes the input time series, then predicts next time steps on the encoded representation. Since the context is built using an autoregressive method, each element actually contains historical information up to the most recent time step contained in the sliding window.

Imagine we are trying to find mature bananas among a bunch of mixed raw and mature bananas in a blurry photo. One strategy is to look at the color of bananas. We try to tune the colors of the photo so that the yellow bananas are more distinguishable from the green bananas, just by color. This is similar to the idea of InfoNCE loss we are using here, where we are trying to maximize the mutual information between the encoded representation and the predictions of encoded representation based on the so-called context, i.e., evolved representation using autoregressive models as shown in *Figure 18.2*. Mathematically speaking, given a similarity function $s(q, k)$, where q is the query and k is the key, we can write down the InfoNCE loss as:

$$\mathcal{L} = -\mathbb{E}\left[\log \frac{\exp\big(s(z_0, z^+)\big)}{\exp\big(s(z_0, z^+)\big) + \sum_j \exp\left(s\left(z_0, z_j^-\right)\right)}\right],$$

where z_0 is the encoded representation at a particular time step, z^+ is the positive sample, i.e., prediction based on the context, and z_j^- are the negative representations we sampled. Let's assume we have one negative sample to build up our intuitions. In the extreme case where there is 0 similarity between the positive sample and the original representation, and the similarity between the negative samples and the original representation is also 0, the loss will be $-\log(1/2)$. Similarly, we can go through all four extreme cases, as shown in Table 18.2. It is clearly shown that the loss is minimized when the similarity between the positive sample and the original representation is high, and the similarity between the negative sample and the original representation is low.

similarity between z_0 and z^+	similarity between z_0 and z^-	Loss
0	1	$-\log(1/(e+1))$
0	0	$-\log(1/2)$
1	1	$-\log(1/2)$
1	0	$-\log(e/(e+1))$

Table 18.2: Demonstration of InfoNCE loss with one positive and one negative sample

In this table, the rows are sorted by the loss from largest to smallest.

With this last piece of the puzzle in place, we can now implement our simple version of CPC for time series data. We will implement a simple version of CPC with a downstream classification task to validate that CPC provides useful representations for downstream tasks.

Data preparation

Let's start with preparing the data. We will use the `Lightning DataModule` to prepare all the datasets and dataloaders. To make the code more readable and maintainable, we will first define some configuration models using `pydantic`:

```
from pydantic import BaseModel, Field


class DataConfig(BaseModel):
    """Dataset location and dataset identity settings.

    :param dataset_name: name of the dataset to use.
        This should correspond to a subdirectory in
        `data_dir` that contains the dataset files.
    :param data_dir: directory where the dataset
        subdirectories are located.
    """

    dataset_name: str = "ECG200"
    data_dir: str = "data"
```

We will prepare the dataset in a specific folder, so we have to tell the data module where to find it by specifying the `dataset_name` and `data_dir`.

Let's start with the *dataset*. The details about the tensor shapes and the expected input format are described in the docstring of the dataset class.

```
import torch
from torch.utils.data import Dataset
import numpy as np
from loguru import logger


class CPCDataset(Dataset):
    """Simple dataset wrapper for CPC tensors.

    Expected input object format:
    - `samples`: array/tensor with shape `[N, T]`
      or `[N, C, T]`.
    - `labels`: array/tensor with shape `[N]`.

    :param dataset_loaded: Dictionary containing
        `samples` and `labels` arrays/tensors.
        Samples are converted to `float32`
        and labels to `int64`.
    """

    def __init__(self, dataset_loaded: dict):
        super().__init__()

        samples = dataset_loaded["samples"]
        labels = dataset_loaded["labels"]

        if isinstance(samples, np.ndarray):
            samples = torch.from_numpy(samples)
        if isinstance(labels, np.ndarray):
            labels = torch.from_numpy(labels)

        if samples.ndim == 2:
            logger.debug(
                "Adding channel dimension to "
                f"samples with shape {samples.shape}"
            )
            samples = samples.unsqueeze(1)
        elif samples.ndim == 3:
            if list(samples.shape).index(min(samples.shape)) != 1:
```

```
                samples = samples.permute(0, 2, 1)

        self.samples = samples.float()
        self.labels = labels.long()

    def __len__(self):
        """Return number of samples `N`."""
        return self.samples.shape[0]

    def __getitem__(self, index):
        """Return one sample-label pair.

        :param index: Sample index.
        :returns: Tuple `(x, y)` where `x`
            has shape `[C, T]` and `y`
            is a scalar class label tensor.
        """
        return self.samples[index], self.labels[index]
```

With the dataset, we can set up the `data loader` and data module for Lightning. As before, we will first define the configuration models for data loader and data module.

```
class DataloaderConfig(BaseModel):
    """DataLoader settings for train and validation loaders.

    :param batch_size: number of samples per batch.
    :param drop_last: whether to drop the last incomplete batch.
    :param num_workers: number of subprocesses to use for data loading.
    :param val_split: fraction of the dataset to use for validation.
    """

    batch_size: int = 64
    drop_last: bool = False
    num_workers: int = 0
    val_split: float = 0.1


class DataModuleConfig(BaseModel):
    """Top-level configuration object for CPC training.

    :param data: dataset configuration as in `DataConfig`.
    :param dataloader: dataloader configuration as in `DataloaderConfig`.
```

```
    """

    data: DataConfig = Field(default_factory=DataConfig)
    dataloader: DataloaderConfig = Field(default_factory=DataloaderConfig)
```

As we will have to split the training data into train and validation sets, we will implement a simple stratified split function to ensure that the class distribution is preserved in both sets. The function takes in the labels, the fraction of samples to use for validation, and a random seed for reproducibility, and returns the train and validation indices.

```
def _stratified_split(
    labels: np.ndarray, val_fraction: float,
    random_state: int = 42
) -> tuple[np.ndarray, np.ndarray]:
    """Stratified train/val index split without sklearn.

    :param labels: Array of class labels.
    :param val_fraction: Fraction of samples to use for validation.
    :param random_state: Random seed for reproducibility.
    :returns: Tuple of train and validation indices.
    """
    rng = np.random.default_rng(random_state)
    train_idx, val_idx = [], []
    for cls in np.unique(labels):
        cls_indices = np.where(labels == cls)[0]
        rng.shuffle(cls_indices)
        n_val = max(1, round(len(cls_indices) * val_fraction))
        val_idx.extend(cls_indices[:n_val].tolist())
        train_idx.extend(cls_indices[n_val:].tolist())
    return np.array(train_idx), np.array(val_idx)
```

The data module is set up as follows. Our data module will load the training and test data from disk, split the training data into train and validation sets, and create dataloaders for each set.

```
import lightning as L
from torch.utils.data import DataLoader, Subset
from pathlib import Path


class CPCDataModule(L.LightningDataModule):
    """
    PyTorch Lightning DataModule for CPC training.
```

```
    :param config: Configuration object containing
        dataset and dataloader settings.
    """
    def __init__(self, config: DataModuleConfig):
        super().__init__()
        self.config = config
        self.train_loader = None
        self.val_loader = None
        self.test_loader = None
        self.input_channels = None

    def setup(self, stage=None):
        if self.train_loader is not None:
            return

        dataset_path = Path(
            self.config.data.data_dir) / self.config.data.dataset_name

        train_data = torch.load(dataset_path / "train.pt")
        test_data = torch.load(dataset_path / "test.pt")

        full_train_ds = CPCDataset(train_data)
        test_ds = CPCDataset(test_data)

        self.input_channels = int(full_train_ds.samples.shape[1])

        labels = full_train_ds.labels.numpy()

        train_idx, val_idx = _stratified_split(
            labels,
            val_fraction=self.config.dataloader.val_split,
            random_state=42,
        )
        train_ds = Subset(full_train_ds, train_idx)
        val_ds = Subset(full_train_ds, val_idx)

        dl_kwargs = dict(
            batch_size=self.config.dataloader.batch_size,
            num_workers=self.config.dataloader.num_workers,
        )
        self.train_loader = DataLoader(
```

```
            train_ds,
            shuffle=True,
            drop_last=self.config.dataloader.drop_last,
            **dl_kwargs,
        )
        self.val_loader = DataLoader(
            val_ds, shuffle=False, drop_last=False, **dl_kwargs
        )
        self.test_loader = DataLoader(
            test_ds, shuffle=False, drop_last=False, **dl_kwargs
        )

    def train_dataloader(self):
        return self.train_loader

    def val_dataloader(self):
        return self.val_loader

    def test_dataloader(self):
        return self.test_loader
```

Model implementation

The model implementation is also done in a similar order: we define the configs, then the model, and then the `LightningModule`. But before we dive into the model implementation, let's first talk about the loss function. The `InfoNCE` loss is implemented as a separate class, as it is quite self-contained and a bit more complex than a typical loss function.

```
import torch
import torch.nn as nn
import torch.nn.functional as F


class CPCCriterion(nn.Module):
    """InfoNCE loss used by Contrastive Predictive Coding.

    For each prediction horizon ``k``, the criterion
    predicts future encoded vectors from context,
    then contrasts one positive against sampled negatives.

    :param dim_context: Context feature dimension ``C_ctx``.
    :param dim_encoded: Encoded target feature dimension ``C_enc``.
    :param n_predictions: Number of future steps to predict.
```

```
    :param n_negatives: Number of negative samples per query.
    :param temperature: Logit temperature scaling factor.
    """

    def __init__(
        self,
        dim_context: int,
        dim_encoded: int,
        n_predictions: int = 12,
        n_negatives: int = 128,
        temperature: float = 0.07,
    ):
        super().__init__()
        self.n_predictions = n_predictions
        self.n_negatives = n_negatives
        self.temperature = temperature
        self.predictors = nn.ModuleList(
            [
                nn.Linear(dim_context, dim_encoded, bias=False)
                for _ in range(n_predictions)
            ]
        )

    def _sample_negatives(
        self, encoded_pool: torch.Tensor, n_queries: int
    ) -> torch.Tensor:
        """Sample negative encoded vectors.

        :param encoded_pool: Flattened encoded with shape ``[B*T, C_enc]``.
        :param n_queries: Number of query vectors to match.
        :returns: Negative samples with shape
            ``[n_queries, n_negatives, C_enc]``.
        """
        idx = torch.randint(
            low=0,
            high=encoded_pool.size(0),
            size=(n_queries, self.n_negatives),
            device=encoded_pool.device,
        )
        return encoded_pool[idx]
```

```
    def forward(
        self, context: torch.Tensor, encoded: torch.Tensor
    ) -> torch.Tensor:
        """Compute averaged InfoNCE loss across valid prediction horizons.

        :param context: Context tensor with shape ``[B, T, C_ctx]``.
        :param encoded: Encoded tensor with shape ``[B, T, C_enc]``.
        :returns: Scalar loss tensor.
        :raises RuntimeError: If no prediction horizon is valid, usually
            because ``T`` is too short relative to ``n_predictions``.
        """
        _, seq_len, _ = encoded.shape
        encoded_pool = F.normalize(encoded.flatten(0, 1), dim=-1)

        total_loss: torch.Tensor | None = None
        used_steps = 0

        for k in range(1, self.n_predictions + 1):
            if seq_len - k <= 0:
                break

            c_t = context[:, : seq_len - k, :]
            z_pos = encoded[:, k:, :]

            z_hat = self.predictors[k - 1](c_t)
            z_hat = F.normalize(z_hat.flatten(0, 1), dim=-1)
            z_pos = F.normalize(z_pos.flatten(0, 1), dim=-1)

            n_queries = z_hat.size(0)
            pos_logits = torch.sum(z_hat * z_pos, dim=-1, keepdim=True)

            z_neg = self._sample_negatives(encoded_pool, n_queries)
            neg_logits = torch.einsum("nc,nkc->nk", z_hat, z_neg)

            logits = (
                torch.cat([pos_logits, neg_logits], dim=1)
                / self.temperature
            )
            targets = torch.zeros(
                n_queries, dtype=torch.long, device=logits.device
            )
```

```
            step_loss = F.cross_entropy(logits, targets)
            total_loss = (
                step_loss
                if total_loss is None
                else total_loss + step_loss
            )
            used_steps += 1

        return total_loss / used_steps
```

The `configs` for the model are defined as follows. We have touched upon the torch models to be used before, so we will skip the detailed explanations of the parameters as the `docstrings` already provides the necessary information.

```
class ConvLayerConfig(BaseModel):
    """Single 1D convolution specification.

    Parameters define one convolution block as
    `(in_channels, out_channels, kernel_size, stride, padding)`.
    """

    kernel_size: int
    stride: int
    padding: int


class EncoderConfig(BaseModel):
    """Encoder architecture and normalization settings."""

    hidden_size: int = 256
    conv_layers: list[ConvLayerConfig] = Field(
        default_factory=lambda: [
            ConvLayerConfig(kernel_size=10, stride=5, padding=3),
            ConvLayerConfig(kernel_size=8, stride=4, padding=2),
            ConvLayerConfig(kernel_size=4, stride=2, padding=1),
            ConvLayerConfig(kernel_size=4, stride=2, padding=1),
            ConvLayerConfig(kernel_size=4, stride=2, padding=1),
        ]
    )


class ARConfig(BaseModel):
```

```
    """Autoregressive model selection and parameters.

    We use a GRU as the autoregressive model.
    """

    hidden_size: int = 256
    n_levels: int = 1


class LossConfig(BaseModel):
    """CPC/InfoNCE objective configuration.

    :param n_predictions: number of future time steps to predict.
    :param negative_sampling_ext: number of negative samples.
    :param temperature: temperature parameter for InfoNCE loss.
    """

    n_predictions: int = 12
    negative_sampling_ext: int = 128
    temperature: float = 0.07
```

We need two key components to build a CPC model: an *encoder* to encode the raw time series data into a latent representation, and an *autoregressive model* to evolve the latent representation forward in time to build up the context. The encoder is implemented as a stack of 1D convolutional layers. Our model code is adapted from the CPC implementation in "*Unsupervised Representation Learning for Time Series: A Review*" by Meng et al.

```
from functools import cached_property
from typing import Literal


class CPCEncoder(nn.Module):
    """Convolutional encoder for time-series CPC.

    Input is expected as ``[B, C_in, T]``
    and output as ``[B, H, T_enc]``,
    where ``H`` is ``size_hidden`` and ``T_enc``
    depends on convolution specs.

    :param input_channels: Number of input channels ``C_in``.
    :param size_hidden: Number of output channels for each convolution
        block.
```

```
    :param conv_specs: List of convolution settings as
        ``(kernel_size, stride, padding)``. If ``None``, a default
        5-layer downsampling stack is used.
    """

    def __init__(
        self,
        input_channels: int = 1,
        size_hidden: int = 256,
        conv_specs: list[tuple[int, int, int]] = [],
    ):
        super().__init__()

        blocks: list[nn.Module] = []
        in_channels = input_channels
        for kernel_size, stride, padding in conv_specs:
            blocks.extend(
                [
                    nn.Conv1d(
                        in_channels,
                        size_hidden,
                        kernel_size,
                        stride=stride,
                        padding=padding,
                    ),
                    nn.BatchNorm1d(size_hidden),
                    nn.ReLU(),
                ]
            )
            in_channels = size_hidden

        self.network = nn.Sequential(*blocks)
        self.output_dim = size_hidden

    @cached_property
    def output_dimension(self) -> int:
        """Return the encoder feature dimension ``H``."""
        return self.output_dim

    def forward(self, x: torch.Tensor) -> torch.Tensor:
        """Encode raw sequence features.
```

```
        :param x: Input tensor with shape ``[B, C_in, T]``.
        :returns: Encoded tensor with shape ``[B, H, T_enc]``.
        """
        return self.network(x)
```

The autoregressive model is implemented using GRU. As the authors (A. van den Oord, Y. Li, and O. Vinyals) stated in the original CPC paper, *any type of encoder and autoregressive model can be used in the proposed framework*. We choose GRU here because GRU is fast and stable.

```
class CPCAR(nn.Module):
    """Recurrent autoregressive context model for CPC.

    Operates on encoded sequences in ``[B, T, C]`` format and returns
    contextualized representations in ``[B, T, C_ctx]``.
    """

    def __init__(
        self,
        dim_encoded: int,
        dim_output: int,
        n_levels_gru: int = 1,
    ):
        """Initialize recurrent AR module.

        :param dim_encoded: Input feature size ``C``
            from the encoder.
        :param dim_output: Hidden/context size ``C_ctx``
            of the recurrent module.
        :param n_levels_gru: Number of recurrent layers.
        """
        super().__init__()
        self.rnn = nn.GRU(
            dim_encoded, dim_output,
            num_layers=n_levels_gru,
            batch_first=True
        )

    @cached_property
    def output_dimension(self) -> int:
        """Return context feature dimension ``C_ctx``."""
        return self.rnn.hidden_size
```

```
    def forward(self, x: torch.Tensor) -> torch.Tensor:
        """Run recurrent AR on encoded sequence.

        :param x: Encoded tensor with shape ``[B, T, C]``.
        :returns: Context tensor with shape ``[B, T, C_ctx]``.
        """
        x = torch.flip(x, [1])
        self.rnn.flatten_parameters()
        x, _ = self.rnn(x)
        x = torch.flip(x, [1])

        return x
```

Training preparation

With all the elements ready, we can now compose the encoder and autoregressive model into a complete CPC model. Note that here we are matching the so-called context and encoded features in the latent space, as discussed before.

```
class CPCModel(nn.Module):
    """Compose encoder and autoregressive context model."""

    def __init__(self, encoder: nn.Module, ar: nn.Module):
        """Initialize CPC model.

        :param encoder: Module mapping ``[B, C_in, T]`` to
            ``[B, C_enc, T_enc]``.
        :param ar: Module mapping ``[B, T_enc, C_enc]`` to context features
            ``[B, T_enc, C_ctx]``.
        """
        super().__init__()
        self.encoder = encoder
        self.ar = ar

    def forward(
        self, batch_data: torch.Tensor
    ) -> tuple[torch.Tensor, torch.Tensor]:
        """Run encoder then AR network.

        :param batch_data: Input batch with shape ``[B, C_in, T]``.
        :returns: Tuple ``(context, encoded)`` where ``context`` has shape
            ``[B, T_enc, C_ctx]`` and ``encoded`` has shape
```

```
            ``[B, T_enc, C_enc]``.
        """

        encoded = self.encoder(batch_data).permute(0, 2, 1)
        context = self.ar(encoded)
        return context, encoded
```

To make the training easier, we now create the `LightningModule` for the CPC model. But first, let's define the `config` for the training. These are not as important as the model and data configs, but they are the tiny details that make the training process smoother and more maintainable.

```
import yaml


class OptimizerConfig(BaseModel):
    """Optimizer hyperparameters. We use Adam method.
    """

    learning_rate: float = 2e-4
    weight_decay: float = 1e-6


class TrainerConfig(BaseModel):
    """Trainer runtime options, including epoch count and output path."""

    epochs: int = 100
    output_dir: str = "lightning_logs/ts_cpc"


class RuntimeConfig(BaseModel):
    """Execution environment settings such as seed and device."""

    seed: int = 0
    accelerator: Literal["auto", "cpu", "gpu", "mps"] = "auto"
    devices: int = 1


class TrainConfig(BaseModel):
    """Top-level configuration object for CPC training."""

    data: DataConfig = Field(default_factory=DataConfig)
    dataloader: DataloaderConfig = Field(
        default_factory=DataloaderConfig
```

```
    )
    encoder: EncoderConfig = Field(
        default_factory=EncoderConfig
    )
    ar: ARConfig = Field(default_factory=ARConfig)
    loss: LossConfig = Field(default_factory=LossConfig)
    optimizer: OptimizerConfig = Field(
        default_factory=OptimizerConfig
    )
    trainer: TrainerConfig = Field(
        default_factory=TrainerConfig
    )
    runtime: RuntimeConfig = Field(
        default_factory=RuntimeConfig
    )

    @classmethod
    def from_yaml(cls, path: Path) -> "TrainConfig":
        """Load a training configuration from YAML.

        :param path: Path to a YAML file containing
            any subset of ``TrainConfig`` fields.
        :returns: Parsed configuration with defaults
            applied to omitted fields.
        """
        with open(path) as f:
            data = yaml.safe_load(f)
        return cls(**(data or {}))
```

The `LightningModule` is the routine wrapper around the model and criterion.

```
class CPCLightningModule(L.LightningModule):
    """Lightning wrapper around CPC model and criterion.

    This module handles model construction, train/validation
    steps, and optimizer configuration.

    :param config: Hierarchical training configuration.
    :param input_channels: Number of input channels
        in each sample ``x`` where ``x``
        has shape ``[B, C_in, T]``.
    """
```

```
    def __init__(self, config: TrainConfig, input_channels: int):
        super().__init__()
        self.config = config
        self.save_hyperparameters(config.model_dump())

        encoder = CPCEncoder(
            input_channels=input_channels,
            size_hidden=config.encoder.hidden_size,
            conv_specs=[
                (l.kernel_size, l.stride, l.padding)
                for l in config.encoder.conv_layers
            ],
        )

        ar = CPCAR(
            dim_encoded=config.encoder.hidden_size,
            dim_output=config.ar.hidden_size,
            n_levels_gru=config.ar.n_levels,
        )

        self.model = CPCModel(encoder, ar)
        self.criterion = CPCCriterion(
            dim_context=ar.output_dimension,
            dim_encoded=encoder.output_dimension,
            n_predictions=config.loss.n_predictions,
            n_negatives=config.loss.negative_sampling_ext,
            temperature=config.loss.temperature,
        )

    def _shared_step(self, batch, split: str):
        """Run one shared forward/loss step for train or validation.

        :param batch: Tuple ``(x, y)`` where ``x``
            has shape ``[B, C_in, T]`` and
            ``y`` is the label tensor.
        :param split: Metric prefix, typically ``"train"`` or ``"val"``.
        """
        batch_x, _ = batch
        batch_x = batch_x.float()
        context, encoded = self.model(batch_x)
```

```
        loss = self.criterion(context, encoded)
        self.log(
            f"{split}_loss",
            loss,
            on_step=False,
            on_epoch=True,
            prog_bar=True,
            batch_size=batch_x.size(0),
        )
        return loss

    def training_step(self, batch, batch_idx):
        """Run one training step.

        :param batch: Mini-batch tuple ``(x, y)``.
        :param batch_idx: Batch index within the epoch.
        :returns: Scalar training loss.
        """
        return self._shared_step(batch, "train")

    def validation_step(self, batch, batch_idx):
        """Run one validation step.

        :param batch: Mini-batch tuple ``(x, y)``.
        :param batch_idx: Batch index within the validation epoch.
        """
        self._shared_step(batch, "val")

    def predict_step(
        self, data, batch_idx: int = 0
    ) -> tuple[torch.Tensor, torch.Tensor]:
        """Run one prediction step.

        :param data: Mini-batch tuple ``(x, y)``.
        :param batch_idx: Batch index within the prediction epoch.
        :returns: Tuple of context and encoded tensors.
        """
        x = data.float()
        context, encoded = self.model(x)
        return context, encoded
```

```
    def configure_optimizers(self):
        """Create optimizer for model and criterion parameters.

        :returns: Adam optimizer configured from ``config.optimizer``.
        """
        return torch.optim.Adam(
            list(
                self.model.parameters()
            ) + list(
                self.criterion.parameters()
            ),
            lr=self.config.optimizer.learning_rate,
            weight_decay=self.config.optimizer.weight_decay,
        )
```

Training and evaluation

We have prepared a very handy `TrainConfig` class that can easily load `configs` from a YAML file. This allows us to keep all the training settings in one place and makes it easier to reproduce the training process. Suppose we put our config file in `configs/config.binaryheartbeat.yaml`; it can be loaded like this:

```
config = TrainConfig.from_yaml(
    Path("configs/config.binaryheartbeat.yaml")
)
```

In this example, we are using the `BinaryHeartBeat` dataset, which is a dataset of heartbeat signals with two classes: **normal** (110 samples) and **abnormal** (299 samples). We choose this dataset because it is a relatively small dataset that can be trained quickly, and the time series is quite long, which allows us to have more valid prediction horizons in CPC. The dataset is available at: `https://www.timeseriesclassification.com/description.php?Dataset=BinaryHeartBeat`. We have already downloaded the dataset, converted the train, and test samples to torch tensors, and put it in the data folder, so we can directly load it using the data module we implemented before. The `config` itself is:

```
data:
  dataset_name: BinaryHeartbeat
  data_dir: data

dataloader:
  batch_size: 32
  drop_last: false
  num_workers: 0
  val_split: 0.1
```

```
encoder:
  hidden_size: 64
  conv_layers:
    - {kernel_size: 10, stride: 5, padding: 3}
    - {kernel_size: 8,  stride: 4, padding: 2}
    - {kernel_size: 4,  stride: 2, padding: 1}
    - {kernel_size: 4,  stride: 2, padding: 1}

ar:
  hidden_size: 200
  n_levels: 1

loss:
  n_predictions: 8
  negative_sampling_ext: 128
  temperature: 0.07

optimizer:
  learning_rate: 0.0002
  weight_decay: 0.000001

trainer:
  epochs: 300
  output_dir: lightning_logs/binaryheartbeat_cpc

runtime:
  seed: 137
  accelerator: auto
  devices: 1
```

In this `config`, we use four convolutional layers in the encoder. The `config` can be used to fit the model. We first prepare the training by creating the output folder, just in case, and then we create the data module and the `LightningModule`.

```
import os

L.seed_everything(config.runtime.seed, workers=True)
os.makedirs(config.trainer.output_dir, exist_ok=True)

data_module = CPCDataModule(
    DataModuleConfig(
        data=config.data,
```

```
            dataloader=config.dataloader,
        )
    )
    data_module.setup(stage="fit")
    lightning_module = CPCLightningModule(
        config=config, input_channels=data_module.input_channels
    )
```

Finally, we create the trainer and call the `.fit` method to start the training process. We ignore all the fancy callbacks and logging settings for simplicity, but they can be easily added to the trainer if needed.

```
trainer = L.Trainer(
    max_epochs=config.trainer.epochs,
    accelerator=config.runtime.accelerator,
    devices=config.runtime.devices,
    default_root_dir=config.trainer.output_dir,
    log_every_n_steps=5,
)
trainer.fit(lightning_module, datamodule=data_module)
```

Though the implementation is tedious, our model is in fact quite small, with only `119 K` parameters, using the yaml `config` shown previously.

Once the model is trained, we can create a simple MLP classifier to predict the normal/abnormal labels. We will compare the performance of the classifier trained on the raw time series data and the one trained on the encoded representations from the CPC model. We expect the classifier trained on the encoded representations to give us better performance. The MLP classifier is implemented as follows. We will not go into the details of the implementation as it is quite straightforward, but the core of the classifier is the following:

```
nn.Sequential(
    nn.Linear(input_dim, hidden_dim),
    nn.ReLU(),
    nn.Dropout(dropout),
    nn.Linear(hidden_dim, hidden_dim // 2),
    nn.ReLU(),
    nn.Dropout(dropout),
    nn.Linear(hidden_dim // 2, n_classes),
)
```

and a cross-entropy loss `nn.CrossEntropyLoss`.

Input Data	Accuracy	F1 macro	F1 weighted
CPC representations	0.7300	0.6601	0.7306
raw input	0.7292	0.5456	0.6792

We observe that the classifier trained on the CPC representations is better than the one trained on the raw input, especially in terms of F1 macro score and F1 weighted score, as expected. Note that F1 scores improved quite a bit compared to accuracy, this is because CPC representations helped the model performance within classes, as our dataset is skewed, the F1 score is more sensitive to such improvements. One questions we can ask is that we spent a lot of computing power to calculate the representations, it is for sure producing better performance. However, it is trickier than that. CPC relies on the autoregressive model to perform the forecasting task, as the "predictive" in the name suggests. We choose this dataset because the time series is quite long and gives us a lot of space to perform the forecasting task. If the time series is too short, we will not have enough valid prediction horizons to build up the contrast, and the performance of CPC might not be improved. To see this in action, we also trained the same model on the *ECG5000* dataset (`length = 140`, available at `https://www.timeseriesclassification.com/description.php?Dataset=ECG5000`) and the *ECG200* dataset (`length = 96`, available at `https://www.timeseriesclassification.com/description.php?Dataset=ECG200`), which are much shorter than the *BinaryHeartBeat* dataset (`length = 18530`). The performance of CPC on those two datasets are barely better than the raw input, as shown in *Table 18.3*.

Dataset	Input Data	Accuracy	F1 macro	F1 weighted
ECG5000	CPC representations	0.9305	0.6628	0.9264
ECG5000	raw input	0.9229	0.6570	0.9249
ECG200	CPC representations	0.8689	0.8466	0.8679
ECG200	raw input	0.8778	0.8557	0.8765

Table 18.3: Performance comparison of MLP classifiers trained on CPC representations and raw input for ECG5000 dataset and ECG200 dataset

Summary

In this chapter, we have discussed the paradigms of deep learning, and explained the different methods of self-supervised learning for time series. We discussed in detail the idea Contrastive Predictive Coding (CPC) method for self-supervised learning on time series data, and implemented our own version of it. We also demoed the role of CPC representations for downstream time series classification tasks.

Throughout this book, we have journeyed from the fundamental building blocks of PyTorch and all sorts of models, covering different architectures, and techniques. We discussed classical models, deep learning models such as transformers, and generative approaches such as diffusion models, and this final chapter on self-supervised learning. But what is the principle of a good model? It is cliche but a good model is one that can approach the data generating process and capture the underlying structure of the data. For time series data, it is especially so as time series data is often generated by dynamical systems and maps part of the physical world. Understanding the data generating process is the key to building good models for time series data.

The field of deep learning for time series is moving fast. The methods in this book are steppingstones, not destinations. We hope the combination of theory, implementation, and hands-on experimentation has given you both the tools and the intuition to keep exploring — and perhaps to contribute something new of your own. Thank you for taking this journey with us. Mastering time series data is notoriously challenging, but you now have the foundation to tackle it head-on. We wish you the very best in your future projects, and we look forward to seeing what you build. Happy forecasting.

Recommended reading

There are many other self-supervised learning methods that we can explore, such as SimCLR based models. To get a comprehensive overview of the self-supervised learning methods for time series data, we recommend reading the following two papers:

- Zhang, K., et al. 2023. "*Self-Supervised Learning for Time Series Analysis: Taxonomy, Progress, and Prospects.*" arXiv.org. `https://arxiv.org/abs/2306.10125`
- Meng, Q., Qian, H., Liu, Y., Xu, Y., Shen, Z., and Cui, L. 2023. "*Unsupervised Representation Learning for Time Series: A Review.*" arXiv.org. `https://arxiv.org/abs/2308.01578.`

Some others we recommend are:

- Hu, J., et al. 2024. "Toward Physics-guided Time Series Embedding." arXiv. `https://arxiv.org/abs/2410.06651.`
- Nie, Y., et al. 2022. "*A time series is worth 64 words: Long-term forecasting with transformers.*" arXiv. `http://arxiv.org/abs/2211.14730`
- To dive deeper into the CPC models: Oord, A. van den, Li, Y., and Vinyals, O. 2018. "Representation Learning with Contrastive Predictive Coding." arXiv.org. https://arxiv.org/abs/1807.03748
- Chapter, but we didn't discuss the details. The following paper is a good starting point to learn about the topic: Wei, L., and E. Keogh. 2006. "*Semi-supervised time series classification.*" ACM SIGKDD. `http://dx.doi.org/10.1145/1150402.1150498.`

Join our community on discord

Join our community's Discord space for discussions with the authors and other readers.

`https://discord.gg/VXpz9YPUTW`

19

Unlock Your Exclusive Benefits

Your copy of this book includes the following exclusive benefits:

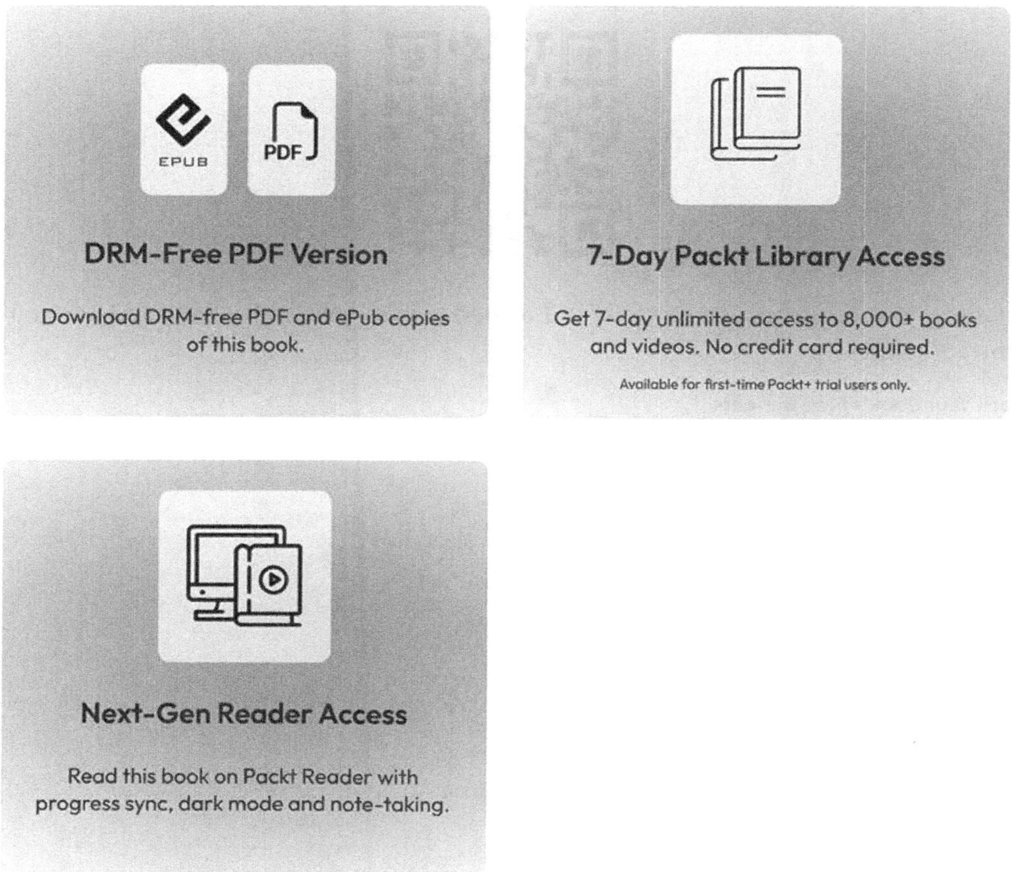

Follow the guide below to unlock them. The process takes only a few minutes and needs to be completed once.

Unlock this Book's Free Benefits in 3 Easy Steps

Step 1

Keep your purchase invoice ready for *Step 3*. If you have a physical copy, scan it using your phone and save it as a PDF, JPG, or PNG.

For more help on finding your invoice, visit `https://www.packtpub.com/en-us/unlock?step=1`.

> **Note:** If you bought this book directly from Packt, no invoice is required. After *Step 2*, you can access your exclusive content right away.

Step 2

Scan the QR code or go to `packtpub.com/unlock`.

On the page that opens (similar to *Figure 19.1* on desktop), search for this book by name and select the correct edition.

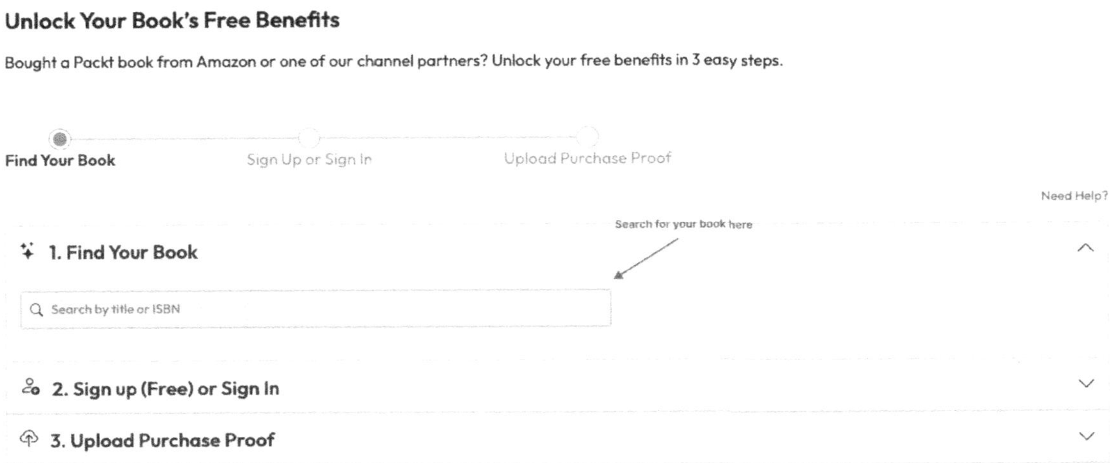

Figure 19.1: Packt unlock landing page on desktop

Step 3

After selecting your book, sign in to your Packt account or create one for free. Then upload your invoice (PDF, PNG, or JPG, up to 10 MB). Follow the on-screen instructions to finish the process.

Need Help

If you get stuck and need help, visit `https://www.packtpub.com/unlock-benefits/help` for a detailed FAQ on how to find your invoices and more. This QR code will take you to the help page.

Note: If you are still facing issues, reach out to `customercare@packt.com`.

packtpub.com

Subscribe to our online digital library for full access to over 7,000 books and videos, as well as industry leading tools to help you plan your personal development and advance your career. For more information, please visit our website.

Why subscribe?

- Spend less time learning and more time coding with practical eBooks and Videos from over 4,000 industry professionals
- Improve your learning with Skill Plans built especially for you
- Get a free eBook or video every month
- Fully searchable for easy access to vital information
- Copy and paste, print, and bookmark content

At www.packtpub.com, you can also read a collection of free technical articles, sign up for a range of free newsletters, and receive exclusive discounts and offers on Packt books and eBooks.

Other Books You May Enjoy

If you enjoyed this book, you may be interested in these other books by Packt:

Machine Learning for Trading, Third Edition

Stefan Jansen

ISBN: 978-1-80324-697-0

- Transform raw data into predictive alpha factors, validated with leak-proof cross-validation
- Master advanced models, from Gradient Boosting Machines to Transformers, Graph Neural Networks, and Reinforcement Learning agents
- Harness Generative AI, Retrieval Augmented Generation, and Causal Inference to make models interpretable, auditable, and compliant with regulatory standards
- Build production-ready trading infrastructure using MLOps, feature stores, and model monitoring to transition research into live capital deployment safely

Machine Learning for Time Series with Python, Second Edition

Ben Auffarth

ISBN: 978-1-83763-133-9

- Visualize time series data with ease
- Characterize seasonal and correlation patterns through autocorrelation and statistical techniques
- Get to grips with classical time series models such as ARMA, ARIMA, and more
- Understand modern time series methods including the latest deep learning and gradient boosting methods
- Choose the right method to solve time-series problems
- Become familiar with libraries such as Prophet, sktime, statsmodels, XGBoost, and TensorFlow
- Understand both the advantages and disadvantages of common models
- Evaluate high-performance forecasting solutions

Packt is searching for authors like you

If you're interested in becoming an author for Packt, please visit `authors.packt.com` and apply today. We have worked with thousands of developers and tech professionals, just like you, to help them share their insight with the global tech community. You can make a general application, apply for a specific hot topic that we are recruiting an author for, or submit your own idea.

Share your thoughts

Now you've finished *Time Series with PyTorch*, we'd love to hear your thoughts! Scan the QR code below to go straight to the Amazon review page for this book and share your feedback or leave a review on the site that you purchased it from.

https://packt.link/r/1805128183

Your review is important to us and the tech community and will help us make sure we're delivering excellent quality content.

Index

E

F

N

O

T

Made in the USA
Monee, IL
16 September 2026

63035040R00332